ACS SYMPOSIUM SERIES **484**

Food Safety Assessment

John W. Finley, EDITOR
Nabisco Brands, Inc.

Susan F. Robinson, EDITOR
American Chemical Society

David J. Armstrong, EDITOR
U.S. Food and Drug Administration

Developed from a symposium sponsored
by the Division of Agricultural and Food Chemistry
at the 200th National Meeting
of the American Chemical Society,
Washington, D.C.,
August 26–31, 1990

American Chemical Society, Washington, DC 1992

Library of Congress Cataloging-in-Publication Data

Food safety assessment / John W. Finley, editor, Susan F. Robinson, editor, David J. Armstrong, editor.

p. cm.—(ACS symposium series, ISSN 0097–6156; 484)

"Developed from a symposium sponsored by the Division of Agricultural and Food Chemistry at the 200th National Meeting of the American Chemical Society, Washington, D.C., August 26–31, 1990."

Includes bibliographical references and index.

ISBN 0–8412–2198–7

1. Food adulteration and inspection.

I. Finley, John W., 1942– . II. Robinson, Susan F., 1946– . III. Armstrong, David J., 1942– . IV. American Chemical Society. Division of Agricultural and Food Chemistry. V. American Chemical Society. Meeting (200th: 1990: Washington, D.C.) VI. Series.

TX531.F658 1992
363.19'26—dc20

91–45099
CIP

The paper used in this publication meets the minimum requirements of American National Standard for Information Sciences—Permanence of Paper for Printed Library Materials, ANSI Z39.48–1984. ♾

PRINTED IN THE UNITED STATES OF AMERICA

ACS Symposium Series

M. Joan Comstock, *Series Editor*

Foreword

THE ACS SYMPOSIUM SERIES was founded in 1974 to provide a medium for publishing symposia quickly in book form. The format of the Series parallels that of the continuing ADVANCES IN CHEMISTRY SERIES except that, in order to save time, the papers are not typeset, but are reproduced as they are submitted by the authors in camera-ready form. Papers are reviewed under the supervision of the editors with the assistance of the Advisory Board and are selected to maintain the integrity of the symposia. Both reviews and reports of research are acceptable, because symposia may embrace both types of presentation. However, verbatim reproductions of previously published papers are not accepted.

Contents

Evaluation of Specific Foods

Indexes

Preface

EVALUATION OF THE SAFETY, or more correctly, the assessment of risk of new products or ingredients can be a tedious and expensive process. The authors in this volume present an overview of the considerations that must be addressed in the process of determining the risk of a new food material. This book presents the reader with a contemporary discussion of the principles and issues that are involved in the safety evaluation of food, ingredients, and new processes for manufacture and distribution of food products.

The authors and editors want to provide the reader with some contemporary guidelines for food safety evaluation. First, one must determine the anticipated use of the new food or ingredient, determine the potential human exposure, and study the structure–activity relationship. Then, after determining the chemistry of the compound in vitro and in vivo, testing can be initiated. Finally, one can do clinical testing before public exposure. With each stage of testing, a set of clearly defined questions should be asked, and appropriate testing should be designed to answer these questions, minimizing animal use, time, and costs.

This book brings together many of the world's experts in the field of safety evaluation. The questions and insights they offer should be most useful to anyone concerned about food safety evaluation.

Disclaimer

This book was co-edited by David Armstrong in his private capacity. No official support or endorsement by the U.S. Food and Drug Administration is intended or should be inferred.

JOHN W. FINLEY
Nabisco Brands, Inc.
East Hanover, NJ 07936

SUSAN F. ROBINSON
American Chemical Society
Washington, DC 20036

DAVID J. ARMSTRONG
U.S. Food and Drug Administration
Washington, DC 20204

August 16, 1991

Perspectives: Past and Present

Chapter 1

Food Safety Assessment

Introduction

John W. Finley[1] and Susan F. Robinson[2]

[1]Nabisco Brands, Inc., 200 Deforest Avenue, East Hanover, NJ 07936
[2]American Chemical Society, 1155 16th Street, NW, Washington, DC 20036

The absolute safety of a food or an ingredient can never be guaranteed. However, with appropriate precautions during development, through manufacture into products, during processing and final preparation, and in distribution, the risk from any food can be kept to an absolute minimum. It is impossible to test 100% of all foods for every possible variable or contaminant. Nutritional abuse and overcompensation by consumers are also beyond the control of the food manufacturer.

As our society has moved from an agrarian society to a urban society, there have been equally significant changes in the eating habits and the nature of our diets. In an agrarian society, significant portions of the food supply were produced and consumed on the family farm. Frequently these products were consumed fresh or nearly fresh or were preserved by home processing techniques. These home processing techniques for preservation clearly lacked the sophistication of modern processing plants and were therefore prone to human error. Such errors frequently resulted in spoilage and, occasionally, illness or death from microbial contamination. Urbanization has resulted in consumer demand for a wider variety of foods than were ever imagined 100 years ago, products that must also be nutritious and safe. This variety of wholesome food is not only delivered thousands of miles from its origin, but it is delivered at a lower percentage of annual income than at any point in history.

Food technology employs many processing tools to make this food delivery system a successful reality. A number of incidental and deliberate ingredients are used to facilitate the delivery of wholesome and organoleptically pleasing food. Pesticides are applied to certain crops to protect against infestation of the crop and to ensure economically viable production as well as wholesomeness. Antioxidants and antimicrobials are frequently used in susceptible foods to ensure safety during processing, distribution, and storage. Antioxidants, flavors, and colors are added to foods to help improve the masticatory and organoleptic quality of the final food product. All of these materials, as well as novel or unique means of processing foods must be thoroughly tested to ensure that any risk to the consumer is minimized. In reality, these additives have received much more rigorous testing than most common foods that we take for granted. We have now reached a point in our food distribution system where

0097–6156/92/0484–0002$06.00/0

anywhere in the western world one can have a year-round supply of tropical fruit, fresh vegetables, and fresh fish. A tour through the supermarket clearly illustrates our opportunity to take home a magnificent variety of prepared foods that can be prepared in minutes in a microwave oven to provide a gourmet meal. As technology advances and knowledge grows, new and unique food ingredients and processes are being developed and their impact on food safety and quality must be evaluated.

The first group of chapters in this book provide an overview of food safety issues. This section contains an interesting historical perspective on food safety evaluation from ancient times through modern safety testing by Sanford Miller. In his chapter, Dr. Miller not only discusses what has been done to assess safety historically, but also provides an insightful view on why the testing was done and how it evolved to our current system for assessing safety. Public concerns related to our current food supply are discussed by Fred Shank. Dr. Shank points out that our food is safe, but we need to do a much more effective job of informing the public about risk. If we can provide better assurances to the public through improved in-plant testing programs at critical points [for example, the Hazard Analysis Critical Control Point (HACCP) system] and improve public communication as to the safety of our food supply, the consumer, the processor,and government agencies will all benefit. Dr. Shank points out that the Food and Drug Administration (FDA) will work with researchers in a cooperative spirit to accomplish this goal.

If the current food supply is safe, how do we assess the risk of new foods and ingredients? Whenever we attempt to evaluate the risk or safety of a new food or ingredient (either a direct additive or an incidental additive such as a pesticide), the type of potential hazard that can be raised must be determined. For example, with a new food ingredient, we must decide what potential chemical or biological hazard is associated with the proposed ingredient. If such a hazard exists, at what concentration does the substance become toxic? Next, we must determine the nutritional impact of the ingredient in the diet. Does the substance add nutrients or does it negatively impact the absorption or utilization of a nutrient? Does the proposed ingredient provide any benefit to the diet (a difficult aspect to determine)? This perception of risk/benefit is often nebulous and difficult to measure, but, in his chapter, Michael Pariza offers an objective discussion of the current status of the important issues in this area.

Evaluating the issues around pesticides and pesticide residues in foods and the potential problems surrounding the incorporation of agential-engineered plant pesticides are additional topics covered by the next group of authors. Biotechnology has provided us with a wide variety of new foods and ingredients ranging from wheats, with improved yield and disease resistance, to high-solids tomatoes, to plants which can produce drugs. The food safety evaluation community must identify reasonable and proper means to evaluate these products to ensure continued safety in the food supply. As the technology improves and products become ready for incorporation into the food chain, we must be able to ensure safety during processing with both current and new products.

The use of computers in the assessment of risk is covered by the next chapters. In this section, various approaches are discussed to predict and quantitate the nature of risk of a specific food or ingredient. The quantization of exposure to a material is critical in determining the extent of the risk. Two chapters describe techniques to assess the exposure of any food, ingredient, or component in the food supply. Drs. Lepparulo-Loftus and Abrams use various databases to estimate how much of any food or component in a food is consumed by the population. This estimation is the first critical step in assessing the risk associated with a new ingredient. The questions faced in this area are (1) for what purpose will the ingredient be used, (2) what will be the concentration of the ingredient in a given food, and (3) how much of that food will be consumed by the consumer.

Computer modeling provides an excellent means of estimating the toxicity of a chemical compound before any biological testing. Studying quantitative structure-activity relationships (QSAR) relates structural information for a chemical component with known biological activities and provides a rapid assessment of the likelihood of a compound being mutagenic or carcinogenic. Armed with this critical information, a researcher can design the correct testing protocol to determine if a compound is indeed a problem or even consider alternative ingredient production approaches that are less likely to present problems. Michael Smithing describes an intuitive program to predict a variety of forms of toxicity in a broad range of species from bacteria through mammals. Like QSAR, this program provides the researcher with a rapid means of screening compounds for potential hazards. The technique looks at segments of molecules and compares them to compounds of known toxicity to establish a likelihood of toxicity for a particular compound. Computer modeling provides a unique opportunity in risk assessment of the safety of food ingredients. It provides accurate models to predict potential negative effects and an excellent means to predict the potential exposure of the population to a particular component. Having made these assessments, the next step is to design a testing program to assess the safety of a new ingredient.

Several authors discuss guidelines for safety evaluation testing. Every food, ingredient, or process is unique and requires different approaches to testing. The researcher must discuss plans for safety evaluation studies with organizations such as the FDA to ensure that the best approaches are being taken to answer critical questions about the safety of the ingredient. Asking the sometimes difficult questions about a compound and what steps are necessary to evaluate its risk are most important and should be done as early as possible in the program. George Pauli points out that the burden of demonstrating safety rests with the company that seeks authorization to use the material or process. Authorization is then granted on a generic basis for the use of the compound in a particular application. Dr. Pauli provides an excellent guide as to which questions should be asked and then answered by the researcher to demonstrate the safety of an ingredient.

Otho Easterday and his coauthors provide a priority ranking system for guidelines for the evaluation of flavor ingredients. This ranking system is accepted by national and international organizations. Examples of applications of this approach are included in the chapter.

Alan Rulis discusses the concept of the threshold of recognition, or how to evaluate ingredients that offer minimum risk. This important concept deals with materials that find their way into the food supply in ultrasmall quantities and, based on statistical evaluation, do not present hazards at the levels at which they are found. All chemicals have a threshold of toxicity. When contaminants are substantially below that level, testing is usually not required. The level of technology in analytical chemistry has reached a point where we must now apply caution in overinterpretation of results. For example, when the Delany Clause (forbidding the addition of any compound that is known to induce cancer in man or animals) was added to the Food, Drug, and Cosmetic Act in 1958, analytical techniques were not as sensitive and many compounds could not be detected at or below their threshold of biological activity. Now, we have clearly surpassed that level of detection, and laws that have zero tolerance for toxic (or, in the case of the Delany amendment, carcinogenic) compounds need to be reconsidered. A real challenge for the scientific community is to effectively communicate this need for change to lawmakers and the general public. After decades of discovering carcinogens in many unexpected places, now researchers must convince the public that these compounds may not always be dangerous.

The addition of food coloring has been a controversial topic for decades. Regardless of the nature of the ingredient, the food, or the process being evaluated, each case is unique and the "Redbook" provides a guideline to evaluation of the material. This section of the book helps researchers develop a clear testing program to answer critical questions about the safety of a product when consumed.

When the decision has been made as to which biological tests must be run to assess the safety of a material, the specifics of these tests must be planned. At this point, the researcher needs to consider the possible outcome of the planned test and, more importantly, how to plan the tests to ensure that the right questions have been answered.

The next group of authors address the critical biological tests that are usually included in the assessment of a new ingredient. D. G. Hattan discusses how to plan acute and chronic tests to maximize the information obtained and minimize the use of animals. This chapter also addresses the difficult issues in designing meaningful testing to assess macro-ingredients such as fat substitutions or bulking agents. These ingredients cannot be tested with the 100 or 1000 safety factors usually applied to trace ingredients such as flavors or colors. Dr. Hattan discusses the emerging approaches for evaluation of such materials.

Next, John Kirshman deals with current trends in animal testing. Important concerns about the adequacy of present testing and approaches for evaluating complex mixtures such as foods are discussed. New trends in food safety testing offer the researcher innovative and unique approaches to testing which may provide answers to some of the difficult questions we are now facing. Researchers are also concerned with the accuracy of animal models and whether we can identify better test models that will more accurately model humans.

The ultimate test model is the human being. Walter Glinsmann presents an extensive discussion of the considerations surrounding human testing Many aspects must be carefully considered before embarking on a clinical testing program. In addition to the safety issues, ethical, legal, and regulatory consequences must be considered. Clinical testing can be useful (1) in establishing the appropriateness of animal models to assess human toxicological endpoints, (2) in defining human tolerance and appropriate endpoints, (3) in estimating safety more precisely when a large safety factor cannot be calculated from a no-effect or no-adverse-effect level determined in animals, (4) when inherited or disease-related conditions may adversely impacted by the product and relevant animal models are not available to assess these impacts, and (5) when complex adverse health effects (e.g. neurobehavioral, food allergy, or food sensitivity) are associated with the product.

Drs. Lachance and Milner discuss nutritional effects on food safety evaluation. Dr. Milner emphasizes the impact of diet and nutritional status on prevention of carcinogenesis. Dr. Lachance points out that nutritional status is critical to the evaluation of new ingredients. He also discusses the importance of over-nutrition or food abuse as a safety consideration. As with many other authors in this book, he emphasizes the need for improved consumer education. All nutrition, health care, and food safety professionals must help with consumer education, particularly as related to abuse of foods.

With the major emphasis and attention focused on the chemical safety of new ingredients or foods, the microbiological aspect of food safety is often overlooked. Several authors provide insight into some current microbiological issues related to both current and new food products. Some of the issues to be considered are minimally processed foods, newly identified organisms which grow at or near refrigeration temperatures, products of fermentation and biotechnology, preservative-free foods, and environmentally acceptable packaging. These seemingly contradictory concerns must all be considered in the development and safety assessment of new foods, particularly those in which bacterial growth is likely. Myron Solberg explores several alternative opportunities for microbiological control in foods. Drs. Wood and Pohland discuss the problem of mycotoxins and contaminants in foods and feeds. Up to 25% of the world's food crops may be contaminated with mycotoxins. A rational approach for dealing with this problem is discussed.

The last group of chapers provides the reader with a mixture of specific applications related to the safety evaluation of foods. C. M. Bergholz discusses efforts to evaluate the safety of the macro-ingredient Olestra. She presents an interesting approach to testing an ingredient for which normal safety factors are not practical. A fat substitute such as Olestra could be used at up to 20% of the diet and would be impossible to test using normal safety factors. This section also contains useful prototype studies in the evaluation of food colors, nitrates, nitrites, and N-nitroso compounds.

Drs. Addis and Hassel discuss safety issues that make assessment of antioxidants difficult and sometimes confusing. First, although antioxidants can be shown to be safe even at very high doses, some biological activity has been observed. Second, direct health benefits have been reported, such as mediation of atherosclerosis. Third, antioxidants inhibit the formation of lipid oxidation products, products that can have very serious health implications.

This book will provide some contemporary guidelines for food safety evaluation: One must determine the anticipated use of the new food or ingredient, determine the potential human exposure, and study the structure-activity relationship. After determining the chemistry of the compound in vitro and in vivo, testing can then be initiated.

Finally, clinical testing is done before any public exposure. With each of these stages of testing, a set of clearly defined questions should be asked and appropriate testing designed to answer these questions. Working closely with the appropriate governmental organizations at every step in the process will help ensure that the appropriate questions are asked and answered.

RECEIVED November 22, 1991

Chapter 2

History of Food Safety Assessment

From Ancient Egypt to Ancient Washington

Sanford A. Miller

Graduate School of Biomedical Sciences, The University of Texas Health Science Center at San Antonio, 7703 Floyd Curl Drive, San Antonio, TX 78284–7819

It is the burden of each generation to believe that the world, and more particularly, insight, began at the time of their birth. It is difficult for young people to recognize that their knowledge is a transition point in a long history that began at the beginning of time. Fundamentally, the problems faced by modern man differ only in degree from those of our ancestors. In 600 B.C., Lao-Tzu, the germinal Taoist wrote, "He who has extensive knowledge is not a wise man." Knowing does not give one insight nor does it provide directions as to what to do with that knowing. The capability of modern science to detect increasingly smaller numbers of molecules, for example, does not of itself provide any better understanding of the biological meaning of those small numbers of molecules or, in turn, their significance to human health. This, lack of insight, in turn, leads to confused or inappropriate public policy. The study of history, however, can provide the beginning of understanding by permitting the comparison of the importance and significance of events. This is true for food safety as it is for world politics. Unfortunately, the problem often is in utilizing these insights rather than in their development.

A Short History of Food Safety

To begin this process of developing insight in the history of food safety evaluation, let us start at the beginning.

> *Now the serpent was cunning, more cunning than any creature that God, the Eternal, had made; he said to the woman, "And so God has said you are not to eat fruit from any tree in the park." The woman said to the serpent, "We can eat fruit from the trees in the park, but as far as the tree in the center of the park, God has said, 'You must not eat from it, you must not touch it lest you die.' "No," said the serpent to the woman, "You shall not die; God knows on the day you eat from it, your eyes will be open and you will be like gods, knowing good from evil." So when the woman saw that the tree was good to eat and delightful to see, desirable to look upon, she took some of the fruit and ate it; she also gave*

0097–6156/92/0484–0008$06.00/0

> *some to her husband and he ate it. Then the eyes of both were open and they realized that they were naked; so they stitched some fig leaves together and made themselves girdles.* (Genesis 3:1-17).

As a result of their violation of the regulations, Adam and Eve were cast out of the Garden of Eden, carrying with them the burdens of woe and sorrow for all of mankind.

This brief vignette describes practically all we know about food safety and regulation. The first regulatory authority identified a hazardous food and issued regulations concerning its use. The first huckster raised doubts in the credibility of the regulator and convinced the first public that there was more to be gained by consumption of this banned product than by following the regulations. The first violation was therefore met with what can only be considered the first regulatory action. Little has changed with the exception that over the millennia, for a variety of reasons, much of the primary emphasis has shifted from health to economics and back again, although at all times these issues were intertwined.

There are many references to food safety in the Bible. The dietary laws of Moses are considered by some to be a direct reflection of tribal judgement based upon human experience. It is argued that the Book of Leviticus prohibited the use of pork or the meat of any scavenger or deceased animal possibly as a result of observations that consumption of these products frequently leads to human disease (*1*). Perhaps the best known Biblical reference to food safety was the phrase "death in the pot" which became the rallying cry against food adulteration at the end of the 18th Century, according to Peter Hutt (*2*). This is a reference to the Prophet Elisha, who having come to the land of Gilgal had his servant put a pot on the fire and boil some herbs and other greens for the Guild of Prophets. One of the servants found a wild vine and included its fruit in the pot. The pottage was then poured out for the meal. As they were eating, they noticed the strange fruit and cried out, "Oh, man of God, there is death in the pot." As a result, they would not eat the pottage until Elisha, apparently an early toxicologist, was able to detoxify the food in the pot.

Because of the importance of the food trade to the Roman Empire, Roman civil law included broad edicts against any kind of commercial fraud or contamination of food. Nevertheless, because of the lack of sensitive detection methods, a wide variety of techniques were in common use to adulterate foods in ancient Rome. Cato described a method to determine if wine had been watered. He also discussed processes for making wine "mild and sweet." He provided instructions for preparing salted and pickled meat, cheese or fish, and a detailed recipe for salting ham (*3*).

China had a similar concern in assuring the safety of food. Chinese medicine has a long history of preventive medicine in which food and diet played a central role (*4*). The Confucian Analects (Lun Yu) of 500 B.C. contained stringent warnings against the consumption of spoiled or contaminated food, "sour rice, discolored fish or flesh, insufficiently stored or cooked." In 2 A.D. Chang Chung-Ching published a manual for Safety Regulations for Food (Chin Kuei Yao Lueh) (*5*) that incorporated many of the earlier Confucian prohibitions.

As with any widespread activity basic to a culture, attempts at fraud and adulteration of food were widespread during ancient times. Pliny the Elder writing in the 1st Century A.D., documented numerous examples of food adulteration. He deplored the greed of merchants who "spoil everything with frauds and adulterations." He went on to say, "So many poisons are employed to force wine to suit our taste - and we are surprised that it is not wholesome."(*6*). Pliny was an advocate of "simple food" - a return to the "old" ways - who had concluded that the greatest aid to health was moderation in food and all things. It is sometimes difficult to distinguish between Pliny's concern for fraudulent or unsafe modification of traditional food and his philosophical opposition to any change including those which lead to improved products in terms of quality and quantity (*2*).

Similar contradictions could be described for every period of human existence. Peter Hutt traced the history of food law through all periods of history from the Bible, through the fall of the Roman Empire, the Dark Ages, the 17th and 18th Century England, and into modern times (*2*). For example, he discusses the fact that the early English statutes were directed primarily at assuring that a given quantity of a food would be sold at a given price. However, as Mr. Hutt points out, it soon became apparent that the price of food cannot be controlled without adequate regulation of the quality. As a result, in 1266, Parliament enacted the Assize of Bread which prohibited the sale of any staple food product that is "not wholesome for man's body." No modern statute, he said, has found a better or more inclusive language to convey the legislative directive to prohibit unsafe food.

The history of food regulation in the United States, has followed a similar course. To a significant extent the demand for a national food safety act began in the middle of the 19th Century. However, it wasn't until Upton Sinclair graphically portrayed his view of where technology had brought the meat industry in his novel, The *Jungle*, that the outcry became widespread and overwhelming. In 1906, after many previous failures, Congress enacted the first national legislation in the United States regulating the safety, economic integrity and labelling of food. While the 1906 act was sufficient for addressing some of the more flagrant abuses and problems of the time, it was soon made obsolete by the growth of science and technology. By 1930, a wide variety of pesticides and other agricultural chemicals were in common use and the use of additives was more prevalent in food processing. An increasingly urbanized and industrialized society was becoming dependent on an increasingly sophisticated food industry to ensure an abundant and economical food supply (*6*).

In 1938, these concerns were addressed by Congress in the enactment of the Federal Food Drug and Cosmetic Act which, though amended many times, remains the basic statute governing food regulation in the United States. In many ways, the primary goal of this Act, in contrast to the 1906 Act, was safety, but with regard for the availability of useful chemicals.

During the 1940's and 1950's, science, and subsequently technology, developed rapidly resulting in a proliferation of food chemicals and new processes. Advances also occurred in toxicology and more systematic approaches were adopted for evaluating the safety of food substances. It soon became evident to FDA and the food industry, as well as to Congress, that more formal pre-market reviews of these materials were necessary to ensure safety and maintain public confidence. As a result, in 1958, Congress enacted the Food Additives Amendment, and in 1960, the Color Additives Amendment which established as law the basic proposition that new food chemicals should be tested by their proponents and reviewed and approved by FDA prior to marketing (*6*).

These amendments contained in their legislative history a definition of safety which said that proponents of new chemicals must prove to "a reasonable certainty that no harm would result from consumption of the chemical under its intended conditions of use." This "reasonable certainty of no harm" safety standard was intended to be very protective, but it was not absolute. Congress recognized that, as a matter of science, safety could not be proven to be an absolute certainty. It thus chose a standard that relied on the informed judgement of experts and could provide considerable safety while not blocking innovation.

Nevertheless, the key feature of these pre-market approval statutes was the now famous Delaney Clause, which incorporated a less judgmental standard, in that it provided that "no food or color additive may be deemed safe that has been found to induce cancer when ingested by man or animals." Since 1958, much of the history of FDA food safety regulatory activities have involved efforts to interpret and rationalize the mandate of the Delaney Clause (*7, 8*).

Science and Law

In briefly reviewing the history of the regulation of the food supply and the evaluation of its safety, it seems clear that advances in science and technology were important stimuli for modification of law. The increasing productivity of agriculture and a food processing industry based upon successful exploitation of expanding scientific insights were major determinants of the growth of urban multiplexes. These developments, in turn, increased the dependence of the consumer on a complex, highly structured and regulated food supply. With each step in the direction of increased urbanization, consumers became even further separated from the sources of their food supply, decreasing their ability to depend on their own knowledge and experience to assure food safety. This turn put increasing pressure on science and technology to assure that safety. Peter Hutt has argued that virtually all changes in the evaluation process for food safety are related to changes of science, rather than of law (*9*). He further argues that laws regulating food safety have remained basically the same. What changes is the way in which the laws are interpreted and enforced. Nevertheless, there are examples of regulation and law forcing science to develop technologies to enable enforcement. It is clear that the enactment of the Delaney Clause and the subsequent anti-cancer laws and regulations were important determining factors in the explosion in research on mechanisms associated with carcinogenesis and the identification of carcinogens (*10*). In any case, the result of this dynamic interplay between science and law has been the development of a complex, often archaic set of regulations based upon varying legislation and precedent elaborated with the goal of making certain that the consumer receives exactly what is promised and does so in such a way that health is not impaired.

Art To Science — Food Safety Evaluation

Most of the history of food safety evaluation was in the domain of human experience and judgement. Other than the senses, early scientists had few tools to determine if substances in foods or foods themselves were hazardous for humans. Nevertheless, the ancients were remarkable in the number of correlations between chemicals and health that they did make. For example, the earliest record of Egyptian medicine, the *Ebers Papyrus*, dated around 1500 B.C., contains information that actually extends back many centuries before (*11*). Of the more than 800 recipes and formulations given, many contained recognized poisons or herbs, such as hemlock, opium, heavy metals, and so on.

The development of observational epidemiology and toxicology continued with Hippocrates and his students (*12*). Much of their work reflected concern with the purity of air, water, and food, and contained remarkable descriptions of the effect of environment on public health, an issue that was addressed by a number of writers in ancient Greece and Rome. As a result, the public water supply and sewage systems in Rome demonstrated sound appreciation that pure water and pure food was essential to good health.

Hippocrates is credited as the founder of modern medical science because he related all health and disease to natural, rather than supernatural causes. About the same time, the "Chinese Hippocrates," Pien Chio published a treatise making the same associations (*13*). Moreover, both Hippocrates and Pien Chio recognized that there were useful techniques to mitigate the effect of naturally occurring poisons and adulterants by controlling absorption of toxic materials. Nevertheless, until a number of other factors were put into place, food safety evaluation remained basically the retrospective observation of events rather than a predictive, preventive process. More importantly, the nature of the observational process restricted evaluation to acute toxic phenomenon. It is difficult for people to intuitively relate a phenomenon observed several years after an event to that event. Thus, the recognition of chronic toxic affects also had to await the development of more sophisticated insights.

The development of the art of evaluating the safety of foods required converting the process from observation to science. It is a long and complex history that proceeded along four principle lines: (a) the development of the principle of the dose response, (b) the development of the principle of test to target species prediction (animal studies), (c) the development of analytical chemistry and its application to foods, and (d) the development of microbiology (the demonstration that microorganisms are a major contributor to food hazards). Two additional sub-areas have to be added to this list. First, the development of any rational generic hazard assessment requires a natural progression of food toxicology from a phenomenalogic, observational science to the elucidation of mechanism. Second, the development of statistical processes based on dose effect permitting the estimation of relative human risk.

Dose–Response/Risk Assessment

In the middle of the 16th Century, the German chemist, Paracelsus (Philippus Theophrastus Bombastus Von Hoenheim) wrote, "All substances are poisons, there is none which is not a poison. The right dose differentiates a poison and a remedy." This aphorism articulated, for the first time, the concept of the dose response curve, the beginning of modern toxicology and food safety evaluation. It is interesting to note that the Paracelsus also predicted quantitative analysis. Although there is much fantasy and mystical speculation in Paracelsus' books, a detailed evaluation of his work suggests that he recognized that a chemistry could be organized to quantitate the presence of various substances in foods and other materials (*14*). Nevertheless, Paracelsus did not directly advance the science of risk assessment or regulatory decision-making. He correctly pointed out that there is a line that divides a safe and an unsafe dose, but he offered no criteria or insight for determining how or where to draw the line. Nevertheless, Paracelsus was pivotal in the development of modern science, standing between the metaphysics and magic of classic antiquity and the emerging science of the 17th and 18th Century. In addition, to his contributions in articulating the principle of dose response and the possibility of quantitative analysis, Paracelsus also argued the need for experimentation in establishing responses to chemicals. This, in itself, was remarkable in that the scholasticism of his era actively discouraged experimentation since it might compel reevaluation of authority, i.e., Aristotle, Hippocrates, and Galen.

It took five further centuries, for the principles established by Paracelsus to take their next step. In 1927, Trevan described the dose-response curve based upon the characteristic sigmoid response of biological systems (*15*). As a result, he also introduced the concept of the LD-50. In 1935, Bliss established the usefulness of expressing dosage in log rather than linear units and, subsequently demonstrated utility of the probit, making simpler the calculations of toxic doses (*16*). Because of the ease with which these approaches could be used, it is not surprising that the initial attempts to establish tolerances for food additives were based upon quantal dose responses and therefore largely reflected the concentration on acute toxicity rather than longer term chronic responses. The need to consider long-term responses lead to the recognition that the classic dose-response and its derived values, e.g. LD-50, was not sufficient for this purpose. This, in turn, stimulated research for a more appropriate way to express dose-response for chronic effects.

It was not until the late 1920's when the increasing use of insecticide sprays led to concerns over chronic ingestion of lead and arsenic from residues left on fruits and vegetables that the need for longer term, chronic toxicity testing became evident. These early studies were often considerably less than a lifetime. Because of a high rate of disease and the lack of information concerning the nutritional requirements of test animals, basic work in toxicology, particularly in the emerging fields of nutritional toxicology, was essential for the design and performance of chronic studies. By 1949, the FDA monograph, Procedures for the Appraisal of the Safety of Chemicals in Foods, Drugs, and Cosmetics, effectively mandated the performance of chronic toxicity tests for substances to be added to food (*17*). Questions of interpretation became more difficult as measurements became more sensitive. Toxicologists were increasingly able to detect biochemical and physiological changes whose biological and toxicological significance was unclear. The "no effect level" or NOEL became the basic determinant of regulatory decisions.

Concern mounted over carcinogenicity and reproductive toxicology. In 1949, the FDA monograph made no mention of carcinogenicity studies while the 1955 monograph, included a separate section on carcinogenicity that contained the following statement: "Positive results in these animal tests can be taken as creating a suspicion that the chemical under study may be carcinogenic for man but do not prove it to be so"(*18*). The 1955 monograph also recognized the difficulty in detecting weak carcinogens. As a result, it suggested the advisability of testing in two species of genetically appropriate animals and the importance of histological evaluation of tumor bearing animals. By 1959, following passage of the 1958 Food Additive Amendments, the monographs contained a separate section on carcinogen screening, and added sections on dietary factors, the proper number of animals, the evaluation of malignancy, and many other scientific issues to be considered in such testing (*19*).

By 1970, it had become clear that the traditional approach of establishing "no effect levels" based upon chronic toxicity tests was becoming so complicated and difficult to interpret, that the method was, for many substances, almost useless. For carcinogens particularly, the problem of interpretation became particularly acute. In an effort to resolve the question and to make the regulatory response to the toxicological evaluation more rational, the development of a mathematical analytical method seemed reasonable. The result, was the statistical risk assessment method which was first applied by FDA in the development of its Sensitivity of Method regulations in 1973 (*20*).

Thus, we have proceeded from the observational epidemiology of Hippocrates through the exposition of the dose-response relationship by Paracelsus through the period of tolerances developed on the basis of acute toxic responses. The process continued to evolve with the development of chronic toxicity testing and the recognition that extrapolation methods based upon the threshold concepts would not work well with carcinogens having very low or no thresholds of response thus requiring the use of statistical extrapolation methods to determine human hazard.

Test To Target Species — Animal Models

From this discussion of chronic toxicity testing, it is clear that the establishment of modern food safety evaluation required, in addition to dose-response theory, the development of appropriate animal models, i.e. the principle of test to target species prediction. Animal experimentation was commonly practiced by ancient Greek and Roman scientists. One could argue that many of the erroneous theories of human anatomy and physiology were the result of attempting to inappropriately carry observations in animals directly to humans. Animal experimentation remained essentially *ad hoc*, actively discouraged in the Middle Ages, becoming acceptable in the 18th and 19th Centuries. As a result of the growing acceptability of animal studies, scientists became aware of the differences in anatomy and physiology between animals and humans. One consequence of this awareness was a growing distrust of animal testing as a model for human biology. Alexander Pope's aphorism, "the proper study of mankind is man," is a reflection of this view. As a result, virtually all risk evaluation relied on human observation or actual experimentation on human subjects. Although some testing was conducted in animals, it was regarded as unreliable and insufficient for any final human safety determinations. In the United States, Dr. Harvey Wiley's early experiments on food additives were conducted on human volunteers - the famed poison squad. Between 1902 and 1904, feeding experiments were conducted on 12 young men using several preservatives then found in the American food supply. It soon became clear to Dr. Wiley that studies in animals, not humans, would have to provide the scientific basis for food safety evaluations. Before the House Committee on Interstate and Foreign Commerce in 1906, Dr. Wiley noted that determining the physiological effects of chemicals in human subjects was not as easy or as straight forward as with animals. Dr. Wiley, questioned by Congressman James D. Mann about the strength of his conclusions concerning the safety of borax, said: "My conclusion is that the cells must have been injured, but I have no demonstration of it because I could not kill the young men and examine their kidneys"(*21*). In 1955, Arnold J. Lehman and Geoffrey Woodard quoted A. L. Tatume on the necessity for animal studies: "People are rather unpredictable and don't always die when they are supposed to and don't always recover when they should. All in all, we must depend heavily on laboratory experimentation for sound and controllable basic principles"(*18*).

Of course systematic epidemiological investigations, including fortuitous observations from environmental exposures, provided, in some cases, a sound basis for the assessment of risks. The difficulty in accumulating the data made such procedures of limited value. Moreover, they were clearly deficient as a feasible procedure for the pre-market detection of food-borne chronic risks. Thus, the need for animal experimentation was, and is, abundantly clear. Nevertheless, many of the same problems plaguing early researchers are faced by researchers today, i.e., the philosophic and pragmatic difficulties of translating studies in animals to man (*22*). Whatever else is true, man is simply not a big rat, and as a result, erroneous conclusions can be drawn if the differences among species is not recognized and explicitly considered in the interpretation of test data and the development of food safety decisions.

Analytical Chemistry

The ability to detect and quantify the presence of toxic and adulterating substances in food is an essential component of modern food safety evaluations. Although Pliny, Galen and their contemporaries could identify gross adulteration of foods, they were unable to detect substances at moderate or low levels, nor were they able to quantitate their findings. The work of Paracelsus at the end of the 15th and the beginning of the 16th Century marked the beginning of quantitative analysis (*14*). It was not until the 18th Century, however, that chemistry began to emerge in its modern form (*23*). The German chemist, Andreas Marggraf, developed the wet method of analytical chemistry permitting more complex, precise, and sensitive analysis. With the work of

Lavoisier in the latter half of the 18th Century followed by the explosion of organic chemistry in France and Germany during the 19th Century, chemistry reached the stage of development in which chemical theory, if well practiced, could be applied to the special problems of assuring the safety of the food supply. In one sense, the chemistry of food safety may be said to begin with the famous English chemist Robert Boyle, in the second half of the 17th Century. Boyle developed new tests for food adulteration, many based on the newly discovered principles of specific gravity. As a result, the field of chemistry developed beyond the point of opinion based upon uncertain tests such as burning, sight, taste, and smell, and instead relied upon objective and reproducible criteria. When publicized, these new chemical tests lead, in turn, to increased public concern with the safety of the food supply. Pamphlets and newspaper articles warned about the destructive ingredients to health found in food and drink. It is not necessary to reiterate the large number of often ingenious techniques used by adulterers to modify food (*24*). Pepper, for example, had always been adulterated by mustard husks, pea flour, and juniper berries, and so on. Tea and spices were counterfeited on such a grand scale that special laws were passed to try and deal with this particular problem. While most of the materials used for adulteration were innocuous, except to the pocketbook, some were quite dangerous. Canned green beans, for example, were often colored with copper sulfate, as was china tea. It was not until the German-born, English chemist, Frederick Accum, published a "A Treatise on Adulteration of Food and Culinary Poisons" that the public became fully aware of what was already known in many legal and governmental circles (*26*). Accum described in detail and at length the numerous kinds of adulteration practiced on food and the various methods available to detect them. More important, he argued that many of these adulterations were highly deleterious to the public health. Although Accum himself was forced to flee England as a result of the controversy over the publication of his book, the public had been alerted and efforts to utilize Accum's techniques to ensure the quality of the food supply were initiated.

The continued development of methods of analysis for the contamination of the food supply was, in part, related to the increasing sophistication of agricultural and food technology, and in part, to the growing conviction on the part of the public and in the U.S. Congress that the food supply somehow was becoming less safe. In the 1940's, with the development of electronic photometry, colorimetric analysis permitted levels of detection and quantification in parts per thousand. By 1970, the availability of separation techniques, such as electrophoresis and chromatography and devices such as spectrophotometers reduced these levels to parts per million. In the 1980's with mass spectroscopy, electron spin resonance and so on, detection and quantification in parts per billion became routine. Today levels two to three orders of magnitude lower are available for many substances. In terms of detection alone, virtually any small number of molecules can be detected and identified with modern technology. The growth in sophistication of analytical chemistry has reached the point that the problem is no longer in detection, but rather in determining the biological significance of such small numbers of molecules.

Microbiology

The final historical link in the development of food safety evaluations is microbiological hazards. It is unfortunate that, today, discussions of food safety appear to focus mainly on chemical safety rather than the much more important and immediate problems of microbiological hazard. It has long been recognized by culinary historians that microorganisms were utilized extensively by early man, in particular for the production of beer and bread. What is not so well recognized is that, from classical times, physicians were aware that disease could be spread by contagion and by air-borne infection, which they generally identified with bad smells. Without understanding the cause of disease, physicians and scientists well recognized the association of contaminated food and outbreaks of many diseases. The Confucian Analects speak of prohibited foods such as "sour rice, spoiled fish and flesh, bad smelling food" and so on (*13*). Nevertheless, the standard of hygiene in private homes, markets, and other places where food was prepared was extremely low. The growth of urban centers and the need for central markets to supply the exploding population of these developing towns and cities resulted in increasing hazard from food-borne disease. When food safety was a function of individual households, only that household was at risk. With the development of urban markets, whole cities were endangered. As Tannahill described, "it the warehouses of the great food merchants, the waste materials of slaughter houses, the refuse flung aside by stall holders in the markets and householders, were all breeding grounds for pestilence and a haven for the omnivorous black rat, carriers of plague carrying fleas, which, by the 13th Century, infested most of the new towns of Europe"(*25*).

Some towns such as Gottingen in Central Germany enforced street cleaning regulations from as early as 1330. Unfortunately, the refuse collected under these circumstances, was often swept into the rivers near the towns (*25*). It seems likely that the fish in these rivers were seriously contaminated and must have been potent carriers of diseases such as dysentery, paratyphoid, and cholera. Salmonellosis and staphylococci infection also must have been wide spread. Botulinum was also epidemic, particularly in Southern France and the Pyrenees.

It seems clear that the association between illness and specific, contaminated foods is probably much older than the identification of chemical food adulteration. Nevertheless, the modern history of bacterial food-borne disease is much younger and probably can be dated to 1857 when Louis Pasteur published his germinal paper on lactic acid fermentation (*27*). By 1861, Pasteur had refuted the concept of spontaneous generation. In the process of examining these contemporary issues, Pasteur founded the science of microbiology. Among other things, he showed how to sterilize a liquid and how to keep it sterile and demonstrated the importance of developing pure cultures for research. By 1876, Robert Koch, using Pasteur's principles, published the first proof that a specific microorganism could cause a specific disease in an animal model. In 1881, he published a technique for isolating pure cultures that lead to the isolation and characterization of the causal organisms of all the major bacterial diseases known at the time. By 1884, he had published what has become known as Koch's Postulates. However, these concepts, developed by Pasteur and Koch, were not applied generally to the problems of food-borne disease until several decades later. The discovery by Nicholas Appert that heat processing and vacuum packaging could preserve food for extended periods of time lead to the development of the modern food processing industry. From Appert's publication at the end of the 18th Century until the end of the 19th Century, "Appertizing" was an art based upon experience and talent rather than a science. As a result, many containers became contaminated and exploded or were contaminated with toxin producing pathogenic organisms. In 1897, Samuel C. Prescott and William Lyman Underwood, published a paper in MIT's *Technology Quarterly*, which, for the first time, demonstrated that food-borne disease was a result of microbial contamination and that heat processing resulted in the destruction of these organisms (*28*). This lead to the establishment of the time-temperature relationships necessary to ensure predictable and consistent commercial sterility of heat-processed food. Without the work of Underwood and Prescott in applying the principles developed first by Pasteur and later by Koch, the modern food processing industry would never have developed when it did.

As indicated, the recognition that food spoilage could cause disease, lead to elucidation of the scientific principles of sanitation and food sterilization. It was these concepts that lead to the development in many countries of sanitary commissions whose function was to establish and ensure the application of hygienic standards for food establishments. In the United States, such regulations were developed by the Department of Agriculture at the end of the 19th Century. After 1906, such regulations were incorporated under the Pure Food and Drug Act and Meat Inspection Act. To a significant extent, the development and implementation of these regulations were limited by the ability of microbiology to rapidly identify the causes of food-borne disease. Most diarrhetic diseases in the United States were reported as of "unknown etiology" because of the inability of microbiologists to rapidly identify many pathogenic organisms. One consequence of this deficiency was a wide-spread false sense of security in the food industry exacerbated by an inappropriate belief that the widespread use of refrigeration was an impregnable line of defense. At that time, little knowledge of psychrophilic organisms was available. The result was over confidence and a tendency to believe that microbiologic food-borne disease was essentially controlled, at least in the industrialized world.

The revolution in molecular biology in the 1950's and 1960's caused a major disruption of this sense of security, and more importantly, resulted in a major rethinking of our attitudes towards microbiological hazards. Two especially significant papers were published which lead to a major change in the way in which microorganisms could be identified. This, in turn, lead to a re-evaluation of the role of microorganisms in food-borne disease. The first of these appeared in 1974 and reported the use of a restriction enzyme to create a functional genetic element bearing genes from two species of bacteria (*29*). This was followed in 1975 by a paper which reported the fusion of mouse myeloma cells to produce the first functional hybridoma (*30*). This was the start of hybridoma technology, the development of monoclonal antibodies and genetic probes. These two papers, without question, lead to a revolution in food microbiology. For the first time, it was possible to construct antibodies to specific strains of microorganisms and rapidly detect their presence in food. One result of these investigations was the recognition that the incidence of food-borne disease in developed, industrialized countries, was much larger than anyone had previously considered. In the U.S. for example, recent estimates indicate that as many as 20 to 80 million cases of foodborne diarrhetic disease occur each year (*31*). In the developing world, it probably represents the most important source of morbidity and mortality, exceeding that of small pox, HIV and other better publicized pathogens. It may be true that food-borne disease is the major factor impeding progress in the developing world (*32*).

Modern Times

By the beginning of the 1980's, virtually all of the components of modern food-safety evaluation were in place. Nevertheless, rather than producing increased confidence on the part of the regulator in making decisions, and increased confidence on the part of the consumer in the safety of their food supply, the result, unfortunately, has been, on the part of the public, increasing concern, that their food supply is unsafe, and on the part of food safety evaluators that the solution of each set of problems simply results in the appearance of other problems more difficult to resolve. In 1934, T. S. Elliott, wrote in his play, *The Rock*, "Where is the wisdom we have lost in knowledge, where is the knowledge we have lost in information"(*33*). With each step in increasing the depth of our understanding of the chemistry and biology of food safety, we have increased the information load to the point that we have difficulties in interpreting our observations and making understandable the results of our interpretations. For example, Table 1 attempts to provide a sense of the dimensions of contemporary food safety evaluations. Ranging from acute to chronic exposure and from quantal to quality of life parameters, it deals only with chemical safety and probably represents an investment of a decade in time and many millions of dollars in cost.

Even with this extensive evaluation, we are still left with several problems that are difficult to resolve. Food safety technology has been devoted primarily to determining the hazards associated with substances added to food such as food additives, contaminants, pesticide residues and so on. As a result, current protocols deal largely, but, not exclusively, with quantal phenomenon. Only recently have we begun to consider the impact of nonquantal phenomenon such as behavior, emotionality, physical performance, and so on. Much more significantly, we are not confident in our ability to evaluate the safety of foods as differentiated from food additives. Food additives are generally used at a small fraction of the diet. As a result, protocols permit exposure at levels 10 to 100 times the predicted exposure. Such exaggerated doses cannot be used for foods. This issue has become of greater significance with the development of modern gene manipulation techniques and the possible development of new species of food plants and animals. One possible approach to the evaluation of food as differentiated from food additives is shown in Table 2. What distinguishes this set of requirements from those shown in Table 1 is: (a) the increasing dependence on chemistry, (b) the need for human investigations prior to general marketing of the material. It also puts greater emphasis on knowledge of metabolism and prediction rather than the phenomenologic foundation of traditional toxicology.

Table 1. Types of Toxicological Tests

1. Acute tests (single exposure or dose)
   A. Determination of median lethal dose (LD_{50})
   B. Acute physiological changes (blood pressure, pupil dilation, etc.)
2. Subacute tests (continuous exposure or daily doses)
   A. Three-month duration
   B. Two or more species (one nonrodent)
   C. Three dose level (minimum)
   D. Administration by intended or likely route
   E. Health evaluation including body weights, complete physical examination, blood chemistry, hematology, urinalysis, and performance tests
   F. Complete autopsy and histopathology on all animals
3. Chronic tests (continuous exposure or daily doses)
   A. Two-year duration (minimum)
   B. Two species selected for sensitivity from previous tests
   C. Two dose levels (minimum)
   D. Administered by likely route of exposure
   E. Health evaluation including body weights, complete physical examination, blood chemistry, hematology, urinalysis, and performance tests
   F. Complete autopsy and histopathology on all animals
4. Special tests
   A. Carcinogenicity
   B. Mutagenicity
   C. Teratogenicity
   D. Reproduction (all aspects other than teratogenicity)
   E. Potentiation
   F. Skin and eye effects
   G. Behavioral effects

Table 2. Strategies for Safety Testing of Food

Chemical Analysis
(a) Known compounds
(b) Pattern recognition
In Vitro Modeling
(a) Nonmammalian systems (e.g., mutagen testing)
(b) Sequential mammalian tissue models including modification of metabolism of standard substances
Computer simulations
(a) Activity–structure relationships
(b) Kinetic modeling
Traditional Safety Testing
(a) Impact on standard test substances
(b) Impact on stressed system
Human Studies
(a) Comparative molecular, pharacokinetics and pharmacodynamic models
(b) Impact on standard test substances
(c) Impact on stressed systems

Conclusions

The earliest evaluators of food safety depended almost entirely on human observation and professional experience and judgement to reach their conclusions. Today, although the questions we ask and the processes we use may appear to be more sophisticated, they are fundamentally the same as those society has asked and used from the start. The major difference however may be in the area of judgement. With increasing uncertainty and the resultant decreased credibility in the scientific community, the role of nonscientific issues has become of greater significance in the development of food safety regulatory policy. This is not to say that food today is unsafe or that scientists cannot or should not use judgement in formulating their recommendations to policy makers. Rather it is to say that with increasing recognition that absolute judgements or absolute safety are unattainable, more questions are being asked about the credibility of scientists making these judgements. The result is that politics, economics, and social values are playing an increasingly dominant role in the development of food safety issues. As shown in Table 3, the formal requirements for FDA rule making as described in the Administrative Procedures Act are only a small fraction of the statuatory issues FDA must consider in its rule making. Moreover, the FDA must also consider an enormous number of nonstatuatory concerns as well as the impact of the views of an extroardinary number of constituencies ranging from, congress to foreign governments to the American people. Finally, when all is done, the possible intervention of the courts must be considered. It is not surprising that science appears to play a secondary role in some regulatory decisions. This is a social and political matter of concern, for, in the end, it is only science that can bring an objective standard to the judgement process.

Even in this respect we have come full cycle. In the earliest days of food safety evaluations, theology, organized religion, the political administration of the rising cities were more important in determining what was safe and what was permitted in the market place than the opinions of scientists. What has been added to the equation is the intrusion of the adversarial process, a strategy that is devoted more to victory than truth. The effect of this strategy on food safety is illustrated by analogy to contemporary political campaigns where the mud slinging pun and phrase is of greater value than issues, facts or insight. The addiction of politicians, public interest groups, the media and industry groups to the use of these techniques for the attainment of their narrow goals can only result in beseiged and weakened regulatory agencies. Under these conditions, the public can only expect the closing of the regulatory gate to innovation and the rigid application of literally interpreted law and regulation, a situation in which all will suffer.

It is clear, that from a historical point of view, we have come to an important crossroads in the development of food safety evaluations. We know more than we ever have about the biological process of life. It is time that we begin applying that knowledge to the issues of mechanisms of food safety. We must now begin to remove the distinctions among chemistry, toxicology, nutrition and microbiology - the four

Table 3. Factors Possibly Modifying FDA Regualtory Decisions

I *Administrative Procedure Act (APA)*		*II* *Non-APA* *Statutes*	*III* *Non-Statutory* *Agency Factors* *Confounding Factors*	*IV* *Non-Statutory* *Non-Agency*
Informal *Rulemaking*	*Formal* *Rulemakin* *Adjudication*			
(1) Proposal (2) Public comment (3) Final order (4) Court review (Standard: arbitrary or capricious	(1) Proposal (2) Public Comment (3) Final order (4) Formal trial-type hearing (5) Final order (6) Cour review (Standard: substantial evidence in the record	(1) Individual Health and safety Statute (a) Safety Standard (b) Control Mechanism (c) Required additional procedures (2) Executive Order 12291 (3) Regulatory Flexibility Act (4) Paperwork Reduction Act (5) National Environmental Policy Act (6) Freedom of Information Act (7) Federal Confidentiality Act (8) Federal Advisory Committee Act (9) Privacy Act (10) Conflict of interest Laws (11) Relation to other Federal Health and Safety laws (EPA, OSHA, FDA, CPSC, etx.) (12) Pending Regulatory Reform Legislation	(1)General Background/history (2) Current Regualtory philosophy (3) Constituency (4) Prior Agency Relations with individuals/public (5) Nature of proposal (a) Broad/narrow (b) Limited/major change (c) Scientific support/controversy (6) Budget/resource limitations (7) Agency personnel (8) Other priorities (9) Prior precedent (10) Additional non-statutory procedures (a) Advance notice (b) Public meetings (c) Informal discussion (d) Legislative style hearings (e) Regulatory negotiation (f) Advisory Committee (g) Guidelines/principles (h) Scientific Board of Inquiry (i) NAS/FASEB	(1) Congessional Oversight (a) Legislative Committees (b)Oversight Committees (c) Appropriations Committes (d) GAO (e) OTA (2) Executive Branch Oversight (a) Department (b) OMB (c) White House (3) Media attention: (4) Independent action by: (a) Public interest groups (b) Unions (c) Industry trade associations (d) Competitors (e) Professional Societies (f) Individual Scientists (g) Other Federal Health and Safety agencies (h) Other Federal Research Agencies (i) State aagencies (j) Foreign Governments (5) Public emotion (6) Public impact on (a) Industry (b) consumers (c) workers
Judicial Interpretation a. Proposal must includedata and rationale b. Final order must respond to all comments c. Formal trial-type hearingmust be conducted like any civil litigation d. Any person has standing to participate and to appeal to the courts				

Source: Peter Hutt, Robert Scheuplein

major components of the scientific evaluations of the safety of foods. We must begin to think of an integrated, mechanistic food safety evaluative process, that will permit generic and ultimately predictive outcomes for food safety evaluations. The problem, if we do not develop more objective procedures and increase the credibility of the scientific community, is that the role of nonscientific issues and the use of adversarial tactics will become even more important than they are today. The final result can only be decisions based upon the goals of lawyers, politicians, and economists rather than an objective evaluation of the data and needs for the public welfare. In the end, it is the public that will suffer.

Bibliography

1. Simmons, F.J. *Eat Not This Flesh*; V. Wisc. Press: Madison Wisc, 1961.
2. Hutt, P.B. & P.B. Hutt, II, *Food, Drug Cosmetic Law J.* **1984**, *39*, 3
3. Hutt, P.B. *Ann Rev. Nut.* **1984**, *4*, 1
4. Lu Gwei-Djen & J. Needham **Isis**. *1951*, *42*, 13.
5. Needham, J. J. *Hist. Med. & Allied Sci.* **1962**, *17*, 429.
6. Miller, S.A. & M.R. Taylor In *International Food Regulation Handbook*; Middlekauf R.G. and P. Shubik, Eds.; Marcel Dekker, Inc.: New York, 1989, pg. 7.
7. Merrill, R. *Yale J. Reg.* **1985**, *5*, 1.
8. Miller, J.A. *Food Drug Cosmetic Law J.* **1990**, *45*, 17.
9. Hutt, P.B. *J. Assoc. Off. Anal. Chem.* **1985**, *68*, 147.
10. Miller, S.A. & K. Skinner *Fund. Appl. Toxicol.* **1984**, *4*, S426.
11. Doull, J. & M.C. Bruce In *Toxicology: The Basic Science of Poisons*; Klaassen, C.D., M.O. Amdur, & J. Doull, Eds.;, MacMillan: New York, 1986, p. 3.
12. Singer, C.J. *Greek Biology & Greek Medicine*; Clarendon Press: Oxford, 1922.
13. Needham, J. In *Clerks & Craftsman in China and the West*; J. Needham, Univ. Press: Cambridge, 1970, p. 263.
14. Thorndike, L. *A History of Magic & Experimental Science*; Columbia Univ. Press: New York, 1941, Vol. V, pp. 617-651.
15. Trevan, J.W. *Proc. Royal Soc. London (Biol)* **1927**, *101*, 483,
16. Bliss, C.L. *Ann. Appl. Biol.* **1935**, *22*, 134
17. FDA., Proc. For the Appraisal of tne Appraisal of Safety in Foods, Drugs & Cosmetics, AFDO, (1949)
18. Lehman, A.J. *Food, Drug, Cosmetic Law J.* **1955**, *10*, 679.
19. Div. Of Pharm., FDA, HEW., "Appraisal of the Safety of Chemicals in Foods, Drugs & Cosmetics., AFDO (1959)
20. Fed. Reg., 38:19226, July 19, 1973
21. U.S. Congress, House Committee on Interstate & Foreign Commerce, Hearings, the Pure Food Bills, Feb. 13, 1906, pp. 282-283.
22. Oser, B.L. *J. Toxicol. Environ. Hlth.* **1981**, *8*, 521.
23. Leicester, H.M. *The Historical Background of Chemistry*; Wiley: New York, 1956.
24. Tannahill, R. *Food in History*; Stein & Day: New York, 1973, pp. 344-346.

25. *Ibid.*, pp. 203-204
26. Accum, F., A Treatise on Adulterations of Foods & Culinary Poisons, London, 1820.
27. Brock, T., Ed.; *Milestones in Microbiology*; Prentice-Hall: New Jersey, 1961.
28. Prescott, S.C. & W.L. Underwood *Tech. Quart.* **1897**, *10*, 183
29. Chung, A.C.Y. & S.N. Cohen *Proc. Nat'l Aca. Sci.* **1974**, *71*, 1030.
30. Kohler G. & C. Milstein. *Nature* **1975**, *256*, 495.
31. Archer, D.L. & J.E. Kvenberg *J. Fd. Protection* **1985**, *48*, 887.
32. World Health Organization. "The Role of Food Safety in Health & Development," Report of a Joint FAO/WHO Expert Committee on Food Safety, WHO Tech. Report Ser. No. 705, Geneva , 1975.
33. Eliot, T.S. *The Rock*; Faber & Faber, Ltd.: London , 1934.

RECEIVED August 15, 1991

Chapter 3

What Is Safe Food?

Fred R. Shank and Karen L. Carson

Center for Food Safety and Applied Nutrition, U.S. Food and Drug Administration, 200 C Street, SW, Washington, DC 20204

The American food supply is safe. The proper perspective on food safety, however, must encompass both the latest scientific knowledge and public perceptions. We know that the greatest long-range health risks stem from the food choices we make, yet the focus of the media and Congress remains on minute traces of pesticides or other contaminants that present negligible risk. Resolving this dichotomy requires several approaches: possible amendment of the Delaney Clause, enhanced knowledge about the composition of foods and the effect of dietary choices on health, and increased reliance on Hazard Analysis Critical Control Point systems as a tool in enhancing the safety of food. Most importantly, we must develop effective means of communicating to consumers the benefits and associated risks inherent in the food supply.

In contemplating the common goal of human health shared by medical professionals and food scientists, Dr. Samuel O. Thier, President of the National Academy of Sciences' (NAS) Institute of Medicine observes that "safe and nutritious food is going to become progressively more important in the protection of health and the improvement of aging, and we have to be able to coordinate our activities" (*1*). Although he was speaking about coordinating efforts between medical professionals and food scientists and technologists, efforts to achieve safe food have no bounds in the scientific world, industry, or the community at large. As Dr. Thier so eloquently points out, in the years ahead, food is going to play a vastly more important role in protecting health and extending that health into old age.

Evidence is accumulating that dietary choices carry potential for modulating the detrimental effect of either native or adventitious food constituents. To take full advantage of this potential, scientists must be in a position to allow inquiring minds free reign to investigate. Rapidly evolving technology in chemistry, biochemistry, toxicology, and other sciences has provided the tools to translate scientific innovation into nutritionally improved and safer foods. The exercise of those tools will continue to be thwarted, however, until there is a meeting of the minds, that is, an understanding, between political, social, and scientific sectors as to what safe food is.

The reality is that there is no answer to the question, "What is safe food?" But it is also a reality that food is safer today than in the past and that scientific efforts will continue to make it safer in the future.

"Safe food" is different things to different people. Consumers, presumably, have very strictly defined expectations of "safe food"; they expect it to be risk-free—period. They associate increased risk with increased use of added substances such as pesticides and food additives. On the other hand, the definition endorsed by scientists, public health officials, and international organizations is more closely linked to the reality of food composition. These groups expect safe food to provide maximum nutrition and quality while posing a minimal threat to public health. They don't expect food to be risk-free, but they do expect any risks that are present to be minimal.

These two sets of expectations are sufficiently far apart to create problems for all segments of the food industry, broadly defined as consumers, manufacturers, regulators, and other public health officials. To reach a common ground, that is, a point where the majority of the scientific *and* nonscientific population share a common and realistic view of what safe food is, will require a number of changes in the way we currently conduct business, ranging from legislative changes to improved communication. Consumers have not, in the past, been well educated about food.

The concept of "safe food" changes with rapidly evolving chemical, toxicological, and biomedical sciences. As a result, what was considered "safe" yesterday may not be satisfactory today. The lead content of the food supply provides a case in point.

Goals Change

Knowledge about the adverse effects of lead has been considerably refined over the past two decades, primarily because of technological advances in toxicology and expanded knowledge about the toxicology of chemical compounds in the food supply. Concern, particularly about the health and development of young children, has been generated by new evidence indicating that dietary lead levels thought safe several years ago are now being shown to be toxic for infants and children. In a 1988 report to Congress, the Agency for Toxic Substances and Disease Registry, a unit of the federal Centers for Disease Control, concluded that there is "little or no margin of safety" between levels of lead we now find in the blood of large segments of the population and levels associated with toxic risk (*2*).

Improved science has provided the basis for reevaluating old and establishing new threshold levels. In 1979 the Food and Drug Administration (FDA) was using a recommended tolerable total lead intake from all sources of not more than 100 μg/day for infants up to 6 months old and a level of not more than 150 μg/day for children from 6 months to 2 years of age (*3*). On the basis of toxicity data obtained in the interim, FDA is now using a considerably lower range of 6 to 18 μg/day as a provisional tolerable range for lead intake from food for a 10 kg child (*4*).

Initiatives to reduce the level of lead in foods, such as the move to eliminate lead-soldered seams in food cans begun in the 1970s, as well as efforts to eliminate leachable lead from ceramicware glazes, have resulted in a steady decline in dietary lead intake. Food and water still contribute undesirable quantities of lead to the diet, however. Data from FDA's Total Diet Study indicate a reduction in mean dietary lead intake for adult males from 95 µg/day in 1978 to 9 µg/day in the period 1986-88 (Gunderson, E., FDA, personal communication, 1990).

Reducing the contribution of dietary lead from sources such as lead-soldered cans, ceramicware, wine from bottles with lead capsules, and dietary supplements such as calcium is a step in the right direction, but other actions will also be needed. Use of leaded gasoline declined markedly in response to concern about its effect on the environment, but other environmental issues continue to pose challenges. It is necessary to reduce the total lead burden introduced through other manufactured products such as paints, glazes, and pipings, as well as natural sources such as lead minerals leached into groundwater.

Thus, lead is a chemical that demonstrates that "safe" is not a static concept, but a dynamic reflection of research and innovation in the scientific world. In the future, we can expect the so-called "chemical safety" of food to continue to evolve.

Consumer Perceptions

Consumer perceptions about hazards in the food supply are not always synchronized with true food safety issues. Pesticide residues were ranked as the number one serious hazard by consumers in a 1990 survey conducted by the Food Marketing Institute (*5*). Pesticides were followed by antibiotics and hormones in poultry and livestock, irradiated foods, nitrites in food, additives and preservatives, and artificial coloring. This ranking, which has remained the same since 1987, helps explain the tremendous political and social attention paid to some of these perceived hazards. This ranking does not reflect the concerns of scientists and other public health officials, the majority of whom place microbiological hazards and natural toxicants at the top of the list. Nevertheless, pesticides and additives are issues that scientists must consider as part of the overall picture of food safety.

Complicating the pesticide issue is Congressional eagerness to respond to consumer concerns by stressing, and thus giving credence to, those concerns by increasing Congressional scrutiny of agency actions and demands for agency resources. In 1989, FDA analyzed 18,113 samples in its pesticide residue surveillance program, 10,719 or 59% of which were imported (*6*); 99% of the domestic samples and 97% of the import samples were not violative. The majority of the violations that did occur were residues of approved pesticides in commodities for which the pesticides were not registered. Although the number of samples analyzed in 1989 increased over 1988, the violation rates for both years were similar (*7*).

What do all these data mean? In essence they mean that pesticide residue levels in the food supply are generally well below Environmental Protection Agency (EPA) tolerances. Likewise, they indicate that, generally, pesticides are being properly used. Moreover, the relatively constant violation rate in a larger number of samples indicates these conclusions are not happenstance.

Nevertheless, there is continuing consumer concern about and Congressional reaction to pesticide residues in the food supply. Essential to breaking this seemingly never-ending circle, and the increased burdens it imposes on public health agencies, is the accumulation of scientific information, i.e., data, on which to base decisions and actions. Data are necessary to perform more accurate risk assessments and to provide consumers with a clear understanding of where actual risks lie.

Better risk assessments will be essential to food safety programs in the future. Whether the substance of concern is a pesticide residue or another contaminant such as lead, when data are insufficient, risk assessments are often based on worst-case assumptions. In the case of pesticide residues, for example, it may be assumed that a particular pesticide is used on all commodities for which it is approved and that the residue is present at its tolerance level on each of those commodities. This is a gross overestimation of pesticide use and provides an exaggerated estimate of risk. "However, the EPA routinely uses these conservative assumptions to account for gaps in information about actual exposure and uncertainties about health effects" according to the NAS 1987 publication *Regulating Pesticides in Food: The Delaney Paradox* (*8*). Similar overestimations will probably be reported in an upcoming NAS report from the Committee on Pesticides in Diets of Infants and Children, because specific dietary intake data are not always available. The report is expected to be issued soon.

Worst-case scenarios will continue to play a major role in assessing risks in the food supply until data are available to provide a more accurate view of the incidence and quantities of these substances in the food supply.

A change in food safety legislation is essential to our further understanding of "what safe food is."

Legislative Changes

Current legislation governing food precludes the addition of any substance that is found to induce cancer in man or animal. This is a zero-tolerance and is known as the Delaney Clause. When the Delaney Clause was added to the Food, Drug, and Cosmetic Act in 1958, zero may have been a reasonable goal for analytical chemistry and toxicology. Today it is not. With virtually every breakthrough in methodology and analytical technology, zero is pushed lower. Because the zero-risk standard is unattainable, it is probably reasonable to assume that the standard is not being applied as vigorously as it might be.

Moreover, the zero-risk requirements of the Delaney Clause have unfairly led consumers to believe that a risk-free food supply is a real possibility. The disservice to consumers that discussions about a pesticide-free food supply cause, by nurturing those beliefs, cannot be overemphasized. The ethical and more realistic approach is to talk about pesticide residues being present in the food supply within prescribed limits, that is, below tolerances set by EPA. As surveillance and monitoring programs are expanded to encompass a larger cross section of the raw and processed food supply, a clearer perspective is gained on the relative risks associated with pesticide residues and a clearer understanding that there is no such thing as "zero risk."

In considering the question "What is safe food?" risk concepts should not be applied only to pesticides and heavy metals like lead. They apply equally to food additives, to migrating packaging constituents, to potentially hazardous compounds induced by food processing and packaging, and to naturally present toxic food components. All of these have the potential for introducing an element of risk into food or altering the risk already inherent in the food. Often the source of the risk can be controlled or eliminated during the food production chain if the vulnerable points are identified and control mechanisms are established.

Mechanisms for Asssuring The Safety of Food—HACCP

Hazard Analysis Critical Control Point (HACCP) is a powerful tool in economically producing safe food of high quality. If a safe product of high quality is the goal, then the basic system must be designed toward that end. This is what HACCP is about.

HACCP is essentially a critical and comprehensive analysis of individual food production systems, from the field to the store shelf. "The HACCP system consists of (a) determining hazards and assessing their severity and risks; (b) identifying critical control points; (c) developing criteria for control and applying preventive/control measures; (d) monitoring critical control points; and (e) taking immediate action to correct the situation whenever the criteria are not met" (*9*).

While most contemporary discussions of HACCP dwell on microbial contamination, including the recently published voluntary seafood program of FDA and the National Marine Fishery Service (*10*), the HACCP concept is appropriate for other types of potential contamination: chemicals, insect infestation, and filth.

From FDA's viewpoint, the HACCP system has great potential as an alternative to traditional establishment inspection because it does not rely on endpoint inspection, but on application of preventive measures throughout the production/distribution system. Moreover, it ensures, and may improve, the quality of a product while strengthening the manufacturer's ability to continuously produce safe products. In the seafood program currently being developed, it will be the government's role to review system parameters and operating procedures, to provide selective auditing of the system's records, including verification by laboratory analysis, and to provide for appropriate enforcement. Thus, a partnership of sorts is created between industry and government with industry shouldering responsibility for the production of safe food and government ensuring that safety.

Naturally Occurring Toxicants

Any discussion of what constitutes "safe food" cannot ignore natural toxicants, whether inherent or induced. It is important to recognize that the macrocomponents of the diet, particularly the fat and protein content, as well as the numerous anticarcinogens and carcinogenic inhibitors present in our food, appear to interact with other dietary components to modulate carcinogenic risk. An estimated 35% of all cancer deaths have been attributed to diet, exclusive of food additives and alcohol (*11*).

Although the diet is the source of natural carcinogens, mutagens, and other toxic substances capable of exerting an adverse affect on human health (*12*), this statement should not be made or received in a vacuum. It must be emphasized that dietary choices, as well as natural predilection, play an important role in determining whether, or to what degree, toxic substances will have an adverse affect on health. Acknowledgment that the food supply does contain naturally present toxicants provides the opportunity to expand knowledge about those substances and define their interactions with other constituents of the diet, as well as chemical changes in foods as the result of processing and packaging techniques. The body of information about natural toxicants and dietary choices that may modulate the development of chronic diseases, while growing, is in need of expansion.

Innovation and technological advances in toxicology, the biomedical sciences, pharmacology, and the chemical sciences provide the tools to enhance understanding of the relationships between natural compounds and detrimental or positive effects on health, as well as how these compounds are formed. This knowledge is fundamental if FDA is to ask the "right" questions in the search for answers about inherent food safety and dietary interactions.

Positive Use of Natural Components—Designer Foods

Another aspect of food safety that is attracting more attention pertains to those components of food which have the potential to modulate carcinogenesis, as well as other disease conditions. Major research programs, designed with this goal in mind, are under way to examine the relationships between components of foods and disease conditions. The National Cancer Institute's program on "designer foods" is a good example. This is a $50 million program to test the anticarcinogenic properties of phytochemicals. Substances being tested include concentrations of the active components of garlic, flax seed, citrus fruit, and licorice root.

A transition is occurring in the perception of "nutrition." A few years ago, nutrition was essentially the interrelationship of nutrient intake with growth, health maintenance, and prevention of deficiency diseases. Today, the concept of nutrition has evolved to recognize the relationships of food components, often not nutritive, as causative agents of chronic diseases, such as cardiovascular disease and cancer, and we are on the edge of discovering more about how and what food components prevent chronic diseases and are even potential treatments of those conditions (*1*). This is a radical change in perspective.

As the body of research on naturally occurring food components reveals the potential for their use in combatting the onset of certain disease conditions, the potential for abusing the use of those substances through over-supplementation increases. Some naturally occurring food components with anticarcinogenic activity at one concentration themselves become toxic at greater concentrations, and often the safe zone between toxic and beneficial is very small; both selenium (*13*) and vitamin A (*14*) are notable examples. Both have shown anticarcinogenic activity at low intake levels, but are toxic at higher intakes (*13*, *15*). More and more frequently dietary guidelines advise higher consumption of specific foods, such as broccoli and other cruciferous vegetables, for the benefits of their naturally present constituents in modulating the onset of disease conditions.

Risk Communication

Changing from a zero-risk legislative standard and providing information about pesticide residues, natural toxicants, and designer foods present some real challenges in consumer communication.

Risk communication makes the link between scientific decisions and the consumer. The interrelated issues of risk, public policy, and risk communication are of paramount importance to educating the public about their food supply.

Successful risk communication hinges on an educated public. Consumers must understand the many facets of the food safety issue in order for actions—both by consumers and by government—to be reasonable. This in no way should be construed as saying that FDA is shirking its responsibilities to consumers or any other group. The development and use of techniques to reduce to the lowest levels possible, or eliminate, those potential hazards in the food supply that can be eliminated or reduced is a realistic goal. FDA will continue to strenuously enforce the law and ensure the safety of the food supply, as it has in the past, in pursuit of that goal. The exercise of that responsibility would be eased, however, by a more enlightened consumer population.

Consumers must be approached from *their* perspective. Most important is translating scientific data and information about food safety into terms people can understand, so they can assimilate this information into the personal information bank they use to make selections and decisions geared to their individual health needs or desires. If scientists cannot translate the science behind the decisions they make and the actions taken or not taken, then those sciences—food chemistry, food engineering, food technology, or whatever science—are being practiced largely for the scientists' own benefit.

According to Dr. Thier, this is not a situation peculiar to food science. He fears that "... food scientists and technologists are doing the same thing in the nutrition area that we have done in the health area. We have become so excited about our biology that we have forgotten that biology poorly translated into changes in behavior is biology that is wasted to a great extent" (*1*).

Thus, no matter how many elegantly engineered new food products or chemically generated new ingredients are created with the consumer's health in mind, if they don't understand why scientists are not targeting their efforts to rapidly reduce levels of consumer-perceived risks, such as pesticides and food additives, from the food supply, all other efforts are for nought.

Conclusion

The question remains, "What is safe food?" Today's answer isn't necessarily tomorrow's answer, nor should it be if we are to strive for scientific and technological progress. The definition of "safe food" must reflect the technological advances in the multitude of sciences that underlie the production of safe food. Ways of ensuring the safety of the food supply and of delivering information to put relative risks into perspective must be a continual search. Legislative changes that bolster the concept of "safe food" rather than zero-risk food must be championed. As research on natural toxicants and natural substances that modulate the detrimental effects of those toxicants increases over the next few years, the concept of "safe food" will change.

There is clearly an important role for chemists and chemical engineers to play in ensuring the continued safety of the food supply. Active participation in the public arena, offering an experienced voice on the many issues concerning food safety, is a highly important role. By background and training, chemists and chemical engineers are uniquely qualified to analyze and interpret, in an unbiased manner, the issues concerning food safety. Those with the unique communication talents needed to convey risk information to consumers are essential to a realistic perception of the food supply. Development of biodegradable packaging materials, recycling techniques, and new food chemistry safety tests for use on-line in manufacturing plants are important elements of the "safe food" picture in the future.

To sum up, the issues surrounding the concept of safe food are complex, and they are becoming even more so as environmental concerns increase. The food supply is safer now than ever before. Chemists and chemical engineers play a vital role; however, food safety requires the cooperation of all disciplines and all segments of the food science community—industry, government, academia, and other professional groups—in scientific endeavors and in effective communication. Only in this way can the technological advances be made that will allow us to identify and solve the issues of today and the future.

Literature Cited

1. Thier, S. O.; *Food Technol.* **1990**, *44*(8), 26-34.
2. *The Nature and Extent of Lead Poisoning in Children in the United States; A Report to Congress,* Agency for Toxic Substances and Disease Registry, U.S. Department of Health and Human Services/Public Health Service, Atlanta, GA, **1988**.
3. Food and Drug Administration, "Lead in Food: Advanced Notice of Proposed Rulemaking: Request for Data," *Fed. Regist.* August 31, **1979**, *44*, 51233.
4. Food and Drug Administration, "Lead from Ceramic Pitchers: Proposed Rule," *Fed. Regist.* June 1, **1989**, *54*, 23485.
5. *Trends: Consumer Attitudes & the Supermarket 1990*; Food Marketing Institute: Washington, DC, **1990**; p 58.
6. *Food and Drug Administration Pesticide Program: Residues in Foods - 1989,* Food and Drug Administration, Washington, DC, **1990**.
7. *Food and Drug Administration Pesticide Program: Residues in Foods - 1988,* Food and Drug Administration, Washington, DC, **1989**.
8. *Regulating Pesticides in Food: The Delaney Paradox*, National Academy of Sciences, Washington, DC, **1987**; p 32.
9. Bryan, F. L.; *Food Technol.* **1990**, *44*(7), 70-7.
10. Food and Drug Administration and Department of Commerce, "Seafood Inspection: Advance Notice of Proposed Rulemaking: Request for Public Comment," *Fed. Regist.* June 27, **1990**, *55*, 26334.

11. Doll, R.; Peto, R. *J. Natl. Cancer Inst.* **1981**, *66*, 1193-308.
12. Scheuplein, R. J. In *Progress in Predictive Toxicology*; Clayson, D. B.; Munro, I. C.; Shubik, P.; Swenberg, J. A., Eds.; Elsevier Science Publishers B.V.: Amsterdam, **1990**; p 351.
13. Combs, G. F.; Combs, S. B. In *The Role of Selenium in Nutrition*; Academic Press, Inc.; New York, **1986**; pp 443-53.
14. *The Surgeon General's Report on Nutrition and Health*, U.S. Department of Health and Human Services, Washington, DC, **1988**; p 209.
15. Hayes, K. C. In *Impact of Toxicology on Food Processing*; Ayres, J. C.; Kirschman, J. C., Eds.; AVI Publishing Co.: Westport, CT, **1981**; p 254.

RECEIVED August 15, 1991

Risk Assessment

Chapter 4

Risk–Benefit Perception

Michael W. Pariza

Food Research Institute, Department of Food Microbiology and Toxicology, University of Wisconsin, Madison, WI 53706

The results of laboratory studies on how experts and non-experts view risks are summarized. The concepts of "risk assessment" and "risk perception" are defined and discussed. Various aspects of food safety are considered from the standpoint of risk assessment versus risk perception.

Detecting and responding to potentially deleterious environmental conditions is an essential attribute of living systems. In lower organisms such response takes the form of instinctual reactions which may be appropriate in certain situations but not others. For example, when deprived of food rodents will instinctively run, presumably in search of new food sources. While such activity may be well suited to natural settings, it is hardly appropriate in situations where the animal is confined to a cage but given access to a running wheel. Under such conditions, in the absence of food, rats will not conserve energy by remaining inactive but rather will run themselves to death, literally (*1*).

By contrast, humans have the capacity to tailor their response to a perceived risk, and even to alter the environment rather than simply respond to it. The key, of course, is understanding the nature of the potential risk so that the response is appropriate.

Risk Perception versus Risk Assessment

Experts and laymen tend to view risks differently (*2*). Experts give great weight to technical considerations and statistical probabilities, a process called "risk assessment". Laymen, by contrast, rely on intuitive judgments, referred to as "risk perception". Laboratory research has shown that non-expert judgement is shaped by a lack of understanding of probabilistic processes, the inability to recognize biased media coverage, misleading personal experiences, and the anxieties of living in a complex world (*2*). The result is faulty judgement which can work in both directions, sometimes overestimating, sometimes underestimating, the true magnitude of the risk in question. It's worth noting here that experts, when dealing with matters outside the areas of their expertise, are just as prone to biased judgement as the general public.

0097–6156/92/0484–0036$06.00/0

Much has been written about why non-experts respond the way they do (*2*). For example, one of the worst cases is the perception by many of the risks of nuclear power, described as unknown, dread, uncontrollable, inequitable, catastrophic, and likely to adversely affect future generations. Indeed, it's hard to imagine more fearsome descriptors. On the other hand, automobile accidents are viewed more favorably. It may seem to a casual observer that many of the same descriptors used for nuclear power risks are applicable to automobile accident risks, but experts in the psychology of risk perception say that familiarity and choice (or lack thereof) are important factors. Hence the familiar risk one chooses to take (driving or riding in an automobile) seems less threatening than the unfamiliar risk one seemingly has no control over (a nuclear reactor malfunction).

Another important consideration is the issue of risk being used as a surrogate for other concerns, particularly its use in trying to achieve certain social or political objectives. In such circumstances, the entire discussion of risk becomes irrelevant and should be recognized as such.

Food Safety: The Expert View

The expert view of food safety is reflected in the ranking developed by the United States Food and Drug Administration (FDA) (Table 1) (*3*). According to the FDA, the most important food safety considerations are microbial contamination and nutritional imbalance. Environmental contaminants, naturally-occurring toxicants, pesticide residues, and food additives are far less significant. These conclusions— these risk assessments— are based on the usual things that go into expert judgments, that is, technical considerations relating to the scientific nature of the risk in question (for example, what is known about pathogenic food-borne microorganisms and their toxins; under what conditions do specific pathogens grow and produce toxins), and statistical probabilities (for example, the probability that a food will be contaminated with a dangerous microorganism or microbial toxin; the probability that someone eating the food will become ill; the probability that serious illness or death will result). For numbers 1 and 2 in Table 1, the risk assessment exercise is straightforward. There is little doubt that people are made ill through encounters with microbial pathogens (*4*)

Table 1. U.S. Food and Drug Administration Ranking of Food Safety Issues

1.	Contamination by Microbial Pathogens
2.	Nutritional Imbalance
3.	Environmental Contaminants
4.	Naturally-occurring Toxicants
5.	Pesticide Residues
6.	Food Additives

SOURCE: Reference 3.

NOTE: Of these issues, numbers 1 and 2 are by far the most important.

or by adopting unhealthy dietary habits (*5,6*). However, assessments of the risks for items 3-6 are based mostly on extrapolations from high-dose animal experiments to human experience, where harm cannot be objectively demonstrated. Hence the risks are theoretical in nature.

Food Safety: The Public Perception

Consumer surveys show that the public views this area differently. For example, according to a poll conducted by the *Seattle Times* (*7*), 57% of the respondents were "very concerned" about pesticides in food, and another 27% "somewhat concerned". For food additives, 48% were "very concerned" and 39% "somewhat concerned". Sixty percent were "very concerned" about seafood pollutants, 26% "somewhat concerned".

Concerns about microbial contamination in general were not probed, but in response to a specific question about salmonella in poultry, 59% indicated that they were "very concerned", 23% "somewhat concerned". Nutritional considerations were not mentioned.

Exaggerated concern about the health risks of food additives and pesticides is, of course, not limited to the U.S. populace. Sato (*8*) reports that the majority of Japanese consumers think these substances are major causes of cancer, more important than cigarette smoking.

Persons with scientific expertise in food safety may find these consumer poll results distressing. Why does the public respond like this? The answer, of course, lies in what risk perception specialists have been telling us (*2*). When one is not familiar with the technical intricacies of an issue—in other words, when one is not "expert"—then intuitive judgments become dominant. Intuitive judgments, in turn, are shaped by a variety of factors many of which have little or nothing to do with logic or deductive reasoning.

Having said this, I'll now discuss the FDA's ranking of food safety issues (Table 1) in terms of where I think the public is coming from, and how things got this way. I freely admit that the views are speculations, personal opinions that are as yet untested scientifically.

Risks versus Benefits

Microbial contamination is the top issue on the FDA's list, and as discussed previously this based on the measurable incidence of food borne illness in the U.S. In general, this is not a new issue for the public. Most people have heard about food poisoning dangers, some have even had personal experience with it. In general they recognize the importance of using common sense when preparing food, and probably reward the appearance of cleanliness when selecting a restaurant. Hence food borne illness is a familiar issue with at least an element of choice associated with it. Certain aspects, like salmonella in poultry, may appear to be new, unfamiliar, and caused by negligence, but in general, when asked about food safety concerns, dangerous microorganisms are unlikely to spring to mind. Like getting into an automobile, microbial contamination does not seem so dangerous.

What about nutritional imbalance? This is number two on the FDA's list, because there are real problems associated with adhering to poor dietary advice. Obesity, at least in part diet-related, is also an important area of health concern (*5*). My guess is that the public does not recognize this as a food safety problem at all, and in a sense this conclusion is correct. In the U.S., nutritional imbalance is a matter of choice. People choose to eat inadequate diets, and sometimes even think there are benefits to be gained from such diets. Since it's a matter of choice based in part on the perception of benefit, nutritional imbalance is probably even less likely to cause concern than the automobile ride.

On to number 3 (Table 1), which for the purposes of this discussion will be linked with numbers 5 and 6. From the public's perspective, we're now getting on thin ice. Contamination, pesticides, food additives— these are strange unknown quantities that involve very little choice, and they certainly get plenty of bad press to boot. To make matters worse, the benefit side of the equation is murky. Is contamination really an unavoidable byproduct of civilization, they ask. Do we really need pesticides and food additives? There are plenty of voices which for a variety of reason are more than willing to provide simple, reassuring answers. Yes, we can completely avoid contamination, they say. No, we don't need pesticides or food additives— these are just part of a chemical-agricultural industry conspiracy. Indeed, the Natural Resources Defense Council has gone so far as to criticize the University of California system for suggesting that food prices will go up if pesticides are banned (*9*).

Misplaced emphasis is liable to directly affect number 4 (Table 1), naturally-occurring toxicants. Right now the public doesn't seem too concerned about this topic, perhaps because it's seen as being tangled up with natural, said by activists and advertisers alike to be good. But even if they are not so good, naturally-occurring toxicants are unavoidable products of nature, not malevolent byproducts of human activity.

The problem is that this may change as a result of pressures to reduce or eliminate pesticides, coupled with resistance to biotechnology. Traditional plant breeding techniques can be used to enhance insect resistance by increasing the levels of naturally-occurring toxicants (*10*). For example, a variety of insect resistant celery was introduced into commerce which exhibited sharply elevated levels of psoralins (psoralins are both mutagenic in bacteria and carcinogenic in laboratory animals) (*11*).

Biotechnology can be used to introduce host specific insecticides, such as the toxin from *Bacillus thuringiensis*, which affects only certain species of insects (*10*). But again, many members of the public may conclude that traditional breeding is safer than biotechnology simply because of familiarity.

Conclusions

There is obviously a vast gulf between expert and non-expert opinion on the subject of food safety. We can study it, we can try to understand it. Doing something that will change it is the real challenge.

Literature Cited

1. Potter, V. R. *Cancer Res.* **1975**, *35*, 2297-2306.
2. Slovic, P. *Science* **1987**, *236*, 280-5.
3. Schmidt, A.M. Address given at the symposium, "Food Safety-A Century of Progress," celebrating the hundredth anniversary of the Food and Drugs Act at London, Oct. 20-21, 1975; *Food Cosmetics Toxicol.* **1978**, *16* (Suppl.2), 15.
4. Cliver, D. O., Ed. *Foodborne Diseases;* Academic Press: New York, **1990**.
5. Pariza, M. W.; Simopoulos, A. P., Eds. *Calories and Energy Expenditure in Carcinogenesis; Am. J. Clin. Nutrit. Supplement* **1986**,*45*,151-372.
6. *Diet and Health*, Report No. 111, Council for Agricultural Science and Technology: Ames, IA, **1987**.
7. *Seattle Times.* Poll conducted by Elway Research, **1989**.
8. Sato, S. In *Mutagens and Carcinogens in the Diet*; Pariza, M.W., Aeschbacher, H.-U., Felton, J. S., Sato, S., Eds; Wiley-Liss: New York, 1990, pp. 295-306.
9. *Food Chemical News* **1990**, *32*(#24), 30-1.
10. International Food Biotechnology Council. *Regulatory Toxicol. Pharmacol.* **1990**, *12*, s1-s196.
11. Ames, B.N.; Gold, L. S. *Proc. Natl. Acad. Sci. U.S.A.* **1990**, *87*, 7777-81.

RECEIVED August 15, 1991

Chapter 5

Toxicological Evaluation of Genetically Engineered Plant Pesticides

Potential Data Requirements of the U.S. Environmental Protection Agency

J. Thomas McClintock, Roy D. Sjoblad, and Reto Engler

Office of Pesticide Programs, Health Effects Division, U.S. Environmental Protection Agency, Washington, DC 20460

Many of the genetically engineered plant pesticides developed to date are rapidly approaching the stage in development where the next phase will include large-scale field testing, evaluation under FIFRA, and clearance for direct/indirect human consumption under FFDCA. Mammalian toxicological data requirements for safety evaluation should be based on the nature of the specific products arising from the newly introduced genetic material and the expected routes of significant human exposure. For example, minimum data requirements might be required on a food plant genetically engineered to synthesize only a single pesticidal protein known to be non-toxic to humans. A more complex scheme of studies might be required for transgenic plants in which major metabolic pathways have been affected to produce non-proteinaceous pesticide products.

Since the early 1980's, the introduction and expression of chimeric genes in plant cells has been possible, especially through the use of *Agrobacterium*-mediated transformation. Such technology has been used to genetically engineer plants to express pesticidal substances. The most important or recognized examples involve transgenic plants engineered to confer insect resistance (*Bacillus thuringiensis* delta-endotoxin) and tolerance to viral infections (tomato and tobacco mosaic virus). Such transgenic plants have been evaluated under greenhouse conditions and in small-scale field tests which have shown promising results.

In a cooperative review process established in 1986 with the U.S. Department of Agriculture's Animal and Plant Health Inspection Service (USDA/APHIS), EPA's Office of Pesticide Programs (OPP) has reviewed 46 small-scale transgenic plant field test notifications involving 9 different crops and 14 distinct pesticidal genes mostly from bacteria and plant viruses. Common to all of these small-scale field tests have been (i) the total acreage (generally less than 2 acres), (ii) the provisions of containment, whereby the test site is adequately confined and monitored to prevent or minimize dissemination of transgenic pollen or plant parts, and (iii) crop destruction so that following research purposes the food or feed crop does not enter commerce. It should be emphasized that to date no adverse effects have been noted.

Research efforts on transgenic plants have focused on food or feed crops which have been engineered to incorporate or produce pesticidal substances not occurring naturally in the plant. OPP has been coordinating the issues concerning transgenic plants as they might be regulated under the Federal Insecticide, Fungicide, and Rodenticide Act (FIFRA) which requires the registration of all pesticides prior to their sale, distribution and use. In addition, OPP administers certain portions of the Federal Food, Drug, and Cosmetic Act (FFDCA) that pertain to pesticidal residues in food and feed including establishment of tolerances or an exemption from the requirement for a tolerance for pesticide substances or residues in foods. Many of the genetically engineered plant pesticides developed to date are rapidly approaching the stage in development where the next phase will include large-scale field testing. The Agency's policy on regulatory oversight of transgenic plant pesticides is currently under development, and, the progression to commercial use of transgenic plant pesticides will likely involve regulatory responsibilities for OPP consistent with FIFRA oversight of pesticides.

For the purposes of scientific discussion and to ensure clarity, several key terms should be defined. Under FIFRA, a pesticide is legally defined as "...any substance or mixture of substances intended for preventing, destroying, repelling, or mitigating any pest, or intended for use as a plant regulator, defoliant, or desiccant..." OPP has adopted a working definition of "transgenic plant pesticides" to include those plants genetically altered via the introduction of genetic material for the purpose of imparting or increasing the production of a pesticide. The "pesticide active ingredient" is the pesticidal substance(s) produced from, or modified as a result of the direct introduction of genetic material. The "pesticidal product" includes the active ingredient and any substance(s) directly produced from, or modified as a result of the introduction of genetic material. The appropriate focus for human toxicity evaluation is on the "pesticidal product" including the "active ingredient." Other information, as described below, is needed to effectively evaluate potential risks associated with human exposure.

For product assessment, OPP has divided the pesticidal active ingredient into two categories: proteinaceous pesticides and nonproteinaceous pesticides. This approach is based on the fact that plant proteins, whether characterized or not, are significant components of human diets and are susceptible to acid and enzymatic digestion to amino acids prior to assimilation. Presuming that the new proteinaceous products are adequately characterized, minimum human health concerns would exist unless (i) the proteins have been implicated in mammalian toxicity; (ii) exposure to the protein, although never implicated in mammalian toxicity through the different routes of exposure, has not been documented, or (iii) "novel" proteins are created via modification of the primary structure of the natural protein pesticide. Nonproteinaceous plant pesticides (none of which have yet been submitted for review) may be evaluated separately or in a manner analogous to that for conventional chemical or biochemical pesticides.

The purpose of this chapter is to set forth potential data and information appropriate to the evaluation of human health risks associated with the wide-spread use and distribution of transgenic plants modified to produce new pesticidal products. The fundamental information necessary to evaluate such products would comprise a thorough description of the source and nature of the inserted genes or gene segments, and a description of the novel proteins encoded for by the genetic material. Presuming that the encoded proteins have been characterized adequately, this information would allow for a reasonable prediction of toxicology issues and for the type of data essential to the evaluation of potential risks. If the function of the inserted genetic material is to alter the level of an endogenous pesticidal component, then a characterization and description of these substances would be essential for risk evaluation.

Product Characterization

Product characterization embraces four basic areas: (i) identification of the donor organism(s) and the gene sequence(s) to be inserted into the recipient plant; (ii) identification and description of the vector or delivery system used to move the gene into the recipient plant; (iii) identification of the recipient organism, including information on the insertion of the gene sequence; and (iv) data and information on the level of expression of the inserted gene sequence. This information is critical for assessing potential risks to humans and domestic animals when exposed to pesticide-containing plants. Potential data/information that OPP feels is necessary for a risk evaluation are listed below. It should be noted that since the establishment of the cooperative review process with USDA/APHIS in 1986, the utility of this information/data has evolved from the first notification and has been included in many, if not all of the submissions reviewed by OPP.

Source of Pesticidal Genetic Material

1. Identity of the donor organism(s) using the most sensitive and specific methods available.
2. Identity of the pesticidal genetic material including a description of any modification to the regulatory or control region of the gene(s).

Pesticide Products

1. Identity and characterization of the protein/peptides encoded by the inserted genetic material.
2. Identity and characterization of the non-proteinaceous active pesticidal ingredients resulting directly from the introduction of the genetic material.

Vector System

1. A description of the vectors, the identity of the organisms used for the cloning of vectors, and a description of the methodologies used for the assembly of all vectors. The description should include size (kilobase), appropriate restriction endonuclease sites, location and function of all relevant gene segments, all modifications [e.g. restriction site alteration(s), deletions of transposition function, disarming of Ti plasmid] and the final delivery system.

2. A description of the gene segment(s) transferred to the plant.

Recipient Plant

1. Identity and taxonomy of recipient plant, to cultivar and line, if cultivated, or variety, if not cultivated.
2. A description of significant characteristics or traits, including i) any previous genetic (transgenic) alterations; ii) life cycle; iii) mode of reproduction and dissemination; and iv) geographical distribution.
3. A description of the methods used to delivery the gene sequence(s) to the plant and confirmation of the insertion of this genetic material into the recipient plant.

Gene Expression in the Plant

1. A description of whether the inserted genes are expressed constitutively or if the genes are inducible; localization and expression in plant parts; and an estimation of the number of gene copies.
2. Gene expression during the plant's life cycle.

Product Analysis and Residue Chemistry.

1. The proposed mode of action of the pesticidal product.
2. Concentration of pesticidal product in the plant and plant parts and analytical method(s) used for quantification.
3. If known, the potential for translocation of the pesticide products in the plant.

Physical and Chemical Properties. In the event that the genetic manipulation is for the purpose of producing *de novo* non-proteinaceous products, data or information on physical or chemical properties of the pesticidal material may be relevant for risk assessment.

Toxicology

The information obtained from the product characterization data can be used to establish the level of mammalian toxicology data necessary to determine the potential risks associated with human and domestic animal exposure to transgenic plant pesticide products. Key factors determining the extent of data requirements would include the nature of the pesticidal product (i.e. proteinaceous or nonproteinaceous) and whether or not the use pattern will result in dietary and/or nondietary exposure. To illustrate this point, three major categories of transgenic plants have been delineated. The first category (Category I) includes plants genetically engineered to contain genes or gene segments that produce a new proteinaceous substance(s) as the active ingredient. For example, plants which have been engineered to contain the delta-endotoxin gene from *B. thuringiensis* would be considered in this category since the proteinaceous toxin is the active ingredient. The second category (Category II) includes those plants engineered such that the function of the inserted genetic material alters the level of an endogenous pesticidal ingredient. In this instance, the pesticidal ingredient may or may not be proteinaceous. Examples would include plants engineered to contain a secondary metabolite from another plant, an intrinsic compound with increased synthesis, or a totally synthetic gene of non-biological origin producing a pesticidal compound. The final category (Category III) includes those plants engineered such that the plant produces *de novo* a nonproteinaceous pesticide.

The oral route is expected to be the predominant route of exposure to food or feed crops engineered to express pesticidal properties. For all food or feed plants producing proteinaceous pesticidal ingredients (i.e. Categories I and/or II) mammalian toxicity could be assessed by acute oral studies (Table1). The requirement for dermal irritation/toxicity might be met by reporting of any observed dermal toxicity or irritation effects, or after adverse reactions from skin contact during the manufacturing process or during applicator exposure. Reporting of dermal toxicity and/or irritation effects also may be sufficient for non-food use, but may depend on the extent of exposure from handling.

For plants that are genetically engineered to produce nonproteinaceous pesticidal components (i.e. Categories II and/or III) and are intended to be used as a food or feed source, the oral route again would be expected as the primary route of exposure (Table 1). The potential toxicity of the pesticidal products could be assessed by oral toxicity studies (acute, subchronic, chronic or other feeding studies). If plants in Categories II and III are engineered to produce volatile pesticide components, pulmonary exposure might be significant, irrespective of whether there is a food use. Dermal exposure to nonproteinaceous products in Category II and III plants for food/feed or nonfood/nonfeed use may be limited to the reporting of dermal toxicity and/or irritation effects.

The limited routes of significant exposure to transgenic plant pesticide products should simplify the appropriate toxicology testing. Based on the information outlined above, the mammalian toxicity studies for proteinaceous pesticide products, in general, would be less than that anticipated for nonproteinaceous pesticides. Again, this approach is based on the fact that plant proteins are susceptible to acid and enzymatic digestion to amino acids prior to assimilation. However, the basic framework for analyzing potential toxicological issues and/or risks relies to a great extent on the product characterization data and information. Table 2 provides a summary of the data and information that would be relevant to the evaluation of transgenic plants expressing pesticidal properties.

Additional Issues and Concerns

In 1988, the International Food Biotechnology Council (IFBC) was organized to develop criteria and procedures to evaluate the safety of foods produced through genetic modification. The scope of the IFBC included transgenic plants as a major category of food products derived from genetic engineering. The final IFBC report identified specific concerns which included the use of antibiotic resistance markers in transgenic plants and the potential for gene transfer (*1*). IFBC emphasized that the safety of the expression product of a new gene or gene segments should be the focus of concern when evaluating transgenic plants.

Table 1. Potential Mammalian Toxicology Requirements

Plant Category	*Use*	*Data Requirements*
Category I and II (Proteinaceous Products Only)		
	Food	Acute Oral Studies; Reporting of Observed Dermal Toxicology or Irritation Effects
	Non-Food	Reporting of Observed Dermal Toxicology or Irritation Effects
Category II and III (Nonproteinaceous Products)		
	Food	Oral Studies (Acute, Subchronic, Chronic Feeding or Other Studies); Reporting of Observed Dermal Toxicology/Irritation Effects; Pulmonary Studies[a]
	Non-Food	Reporting of Observed Dermal Toxicity or Irritation Effects; Pulmonary Studies[a]

[a] Only if the pesticidal product components in the plant are volatile.

Table 2. Summary of Data and Information Necessary for the Evaluation of Transgenic Plants Expressing Pesticidal Properties

	Proteinaceous		*Nonproteinaceous*	
Discipline	*Food*	*Nonfood*[a]	*Food*	*Nonfood*[a]
Product Characterization	X	X	X	X
Human Exposure				
Oral	X	O	X	O
Pulmonary	O	O	X[b]	X[b]
Dermal	O	O	O	O

X - Data and information is necessary
O - Data and information not needed
[a]Members of this category may include certain engineered food/feed plants whose pesticidal products and intentionally added inerts have been demonstrated not to occur in edible portions of the plant or the plant is not used for food or feed.
[b]Only if the pesticidal product components in the plant are volatile.

As defined above, the pesticide product expressed by the transgenic plant includes not only the active ingredient but any substance(s) directly produced from the introduction of genetic material. In all submissions reviewed by OPP to date, such inserted sequences have included the gene encoding the pesticidal active ingredient and an antibiotic resistance marker gene. Consequently, the use of antibiotic markers and their products has raised some concerns regarding human health.

The major concern has focused on the probability of antibiotic resistance gene transfer from the transgenic plant to animals and/or microorganisms and the potential of creating resistance in humans to medically important antibiotics upon chronic exposure to low levels of antibiotics. The probability of such a transfer would be a function of the mobility of the portion of the genome in which the inserted gene resides. If gene transfer occurred, whereby plant DNA was incorporated into bacterial cells, the novel DNA sequence would not be recognized by the recipient host unless specific transcriptional signals were in place. To date, the *Agrobacterium*-mediated plant constructs have demonstrated permanent and stable incorporation of the marker gene into the genome of plants and would not be transferred by any known biological mechanism (*2*). In any event, current research has focused on the use of alternative marker genes which include: bioluminescence (*lux*), color reaction (*lac* ZY), the use of gene probes to specific gene sequences, and herbicide resistance.

Summary

This chapter sets forth potential data and information appropriate to the evaluation of human health risks associated with the wide-spread use and distribution of transgenic plants modified to produce new pesticidal products. The fundamental information necessary to evaluate such products should comprise a detailed description of the inserted genetic material, the nature of the specific products arising from the newly introduced genetic material, and whether or not the use pattern results in dietary or nondietary exposure. This information should allow for a reasonable prediction of toxicology issues and the type of data and information relevant for a human health risk assessment.

Since the establishment of the cooperative review process with USDA/APHIS, most of the product characterization data and information has been included in the submissions reviewed by OPP. At this time specific requirements for mammalian toxicity studies have not been established; however, the information outlined above should provide the useful scientific framework for establishing appropriate data requirements. In addition, OPP's regulatory policy with respect to transgenic pesticidal plants is currently being developed. The process of formalizing product identity/characterization and toxicology data requirements will involve public participation and comments; peer review from outside scientists; and responses to public and peer review comments.

Literature Cited

1. International Food Biotechnology Council Report. In *Biotechnologies and Food: Assuring the Safety of Food Produced by Genetic Modification*. 1990 International Food Biotechnology Council: Washington, DC, 1990, 186-189.
2. Schell, J.. *Science* **1987**, *237*, 1176-1182.

RECEIVED November 5, 1991

Chapter 6

Evaluating Pesticide Residues and Food Safety

Henry B. Chin

National Food Processors Association, 6363 Clark Avenue, Dublin, CA 94568

The United States without a doubt has the world's most abundant and safe food supply. The plenty which graces American dinner tables is the envy most of the world. This largess is made possible by fertile soil and modern agricultural practices which have included the use of pesticides. Unfortunately, this largess is not without its price. Real or perceived questions of food safety as a result of the possible presence of pesticide residues foods have been posed with increasing vociferousness in recent years. In the ranking of the general public's concerns, adverse health effects as the result of pesticide residues in the diet ranks near the top. Yet, most scientific experts (*1*) rank food safety and pesticide residues, low on the list of concerns. More pressing and significant are health problems relate to food consumption patterns, i.e., amount and types of foods consumed, and food borne pathogenic microorganisms (*2*).

It is not the purpose of this chapter to provide a detailed analysis of the mechanics of risk assessment as they related to food safety and pesticide residues. Others are more expert at the evaluation of potency and the rationales involved in various mathematical models. Rather the purpose of this paper is to discuss food safety and pesticide residues in general terms by examining the components of the risk assessment process and attempting to relate this analysis to some more general observations related to food safety.

In the course of this analysis we will examine the individual technical elements, i.e., the analytical data on residues, food consumption information and the toxicological data which have played a role in the food safety debate.

The evaluation of pesticide residues in food safety is based upon the estimation of potency of the residue and the estimation of the exposure. The exposure estimate is in turn derived from estimates of the level of the residue in various foods and the likely consumption of food containing those residues.

Exposure estimates for people in occupational settings are fairly straight forward. Workers can be fitted with various kinds of monitors to determine their actual exposure or exposures can be fairly accurately estimated using environmental monitors. The relative success which researchers have had in predicting occupational

0097–6156/92/0484–0048$06.00/0

diseases can be at least partially attributable to the ability to measure and/or predict exposure fairly accurately.

In contrast dietary exposures are extremely difficult to estimate with accuracy. It is obviously impractical to monitor the daily diet of a sufficiently large population group to get details on portion sizes and foods eaten on an ongoing basis. It is even less feasible to analyze portions of the meal to obtain actual residue data on that diet.

Level of Pesticide Residues in the Diet

It is frequently said that analytical chemists can find decreasingly smaller amounts of chemicals, suggesting that safety concerns have arisen because of increased knowledge of the composition of foods. But realistically, analytical chemistry has had a very tough time keeping up with the food safety debate. Table I shows the calculated levels of significance of some chemicals along with estimated analytical detection limits. Clearly, analytical science is hard pressed to keep up with the potency extrapolations from animal studies. It seems that in many situations the level of significance is below our ability to detect that chemical. Thus, in many situations analytical chemistry cannot, by itself, provide unqualified assurance about the lack of risk. Just because something isn't detected, is not sufficient, for the needs of the risk assessor, to draw the conclusion that a risk does not exist. The reasonableness of this situation will be discussed later, but this causes many assumptions and compromises in the use of data which tend to exaggerate the risk.

Table I. Comparison of No Significant Risk Levels and Method Detection Limits (MDL) (mcg./L in Drinking Water)

Chemical	*No Significant Risk Level*	*MDL*
Benzidine (and its salts)	0.0015	0.08
N-Nitrosodi-n-butylamine	0.05	10
N-Nitrosodiethylamine	0.01	10
N-Nitrosodimethylamine	0.015	10
2,3,7,8-Tetrachloro-dibenzo-para-dioxin (TCDD)	0.0000025	0.00044

Source: Yi Y. Wang. "Evaluation of Analytical Methods in Water for the Chemicals Listed Under the California Safe Drinking Water and Toxic Enforcement Act of 1986 (Proposition 65)," California State Department of Health Services.

The data from most surveys of pesticide residues in foods show that residue levels are not distributed in a statistically uniform manner about a mean value. Typically 70 to 100 % of the samples which are analyzed contain no detectable residues. Thus, the appearance of the curve which describes the concentration and frequency of occurrence of residue levels in foods is greatly affected by the analytical method. For the purposes of risk assessment, rather than assume that the residue level in samples with no detectable residues is zero, a value equivalent to one-half of the limit of quantification (LOQ) is often assigned to samples containing no detectable residues. Figures 1 and 2 compare a hypothetical situation where 90% of the samples analyzed contained no detectable residues of a certain pesticide. In Figure 1 the LOQ is 0.1 ppm, which is not unusual for data collected for surveillance purposes. Since most tolerances are greater than 0.1 ppm, it would not be unusual for analysts to use 0.1 ppm as a cut-off value for the reporting of data. In Figure 2 the LOQ is 0.01 ppm, which more closely represents the current state of the analytical art. The comparison assumes that the portion of the samples containing no detectable residues is the same. The "true" amount of residue in these samples is zero, a value which cannot be measured analytically. As will be discussed later, these zero values can in many instances be independently assigned on the basis of pesticide application reports.

An alternate way of depicting the influence of the limit of quantification (LOQ) on the calculated mean residue level is shown in Figure 3 where the calculated mean residue level is plotted versus LOQ for the hypothetical situation where 90% of the samples had no detectable levels of a residue. In this example the mean residue level of the 10% of the samples with detectable residues is 1 ppm. In this example, the calculated mean residue level decreases by approximately 40% as the LOQ decreases by 90%.

These examples illustrate that the LOQ must be the lowest feasible in order to produce risk estimates which are not artificially elevated.

Surveillance data, with its typically high detection or quantification limits, is an excellent means of determining compliance with agricultural regulations but that the data, if used for doing risk assessment, could be a source of grossly overestimated concerns about the safety of pesticide residues in the diet. Analytical data for risk assessment purposes should properly be derived only from surveys which were designed for that specific purpose, with analytical methods selected to provide the appropriate level of sensitivity. This would obviously be an expensive undertaking, since for the most part screening methods would not be applicable.

If the amount of residues in foods cannot be entirely accurately estimated from analytical data alone, then what other options are available? The use of tolerance or theoretical maximum residue levels would only compound the errors which were discussed above. A possible option is to examine pesticide usage patterns and use these patterns to correct the data for the deficiencies in the analytical method. In this manner a proportion of the samples, in proportion to the percent of the crop which had not been treated with the pesticide of interest, with no detectable residues would be assigned a zero value. The importance of this adjustment cannot be over emphasized. Examination of pesticide application records will show that it is rare that 100% of a crop is treated with a given pesticide. Several dozen pesticides are often registered for use on a given commodity. Several will provide control for the same pests. It is becoming uncommon for growers to apply chemicals indiscriminately. For an example, individual insecticides and fungicides which are used on tomatoes grown for canning are typically used on less than 1% to about 25% of the entire crop. Since a chemical is used only to a limited extent, it makes sense that a portion of the samples with no detectable residue should be assigned a residue level of zero, as opposed to a level determined by the sensitivity of the analytical procedure.

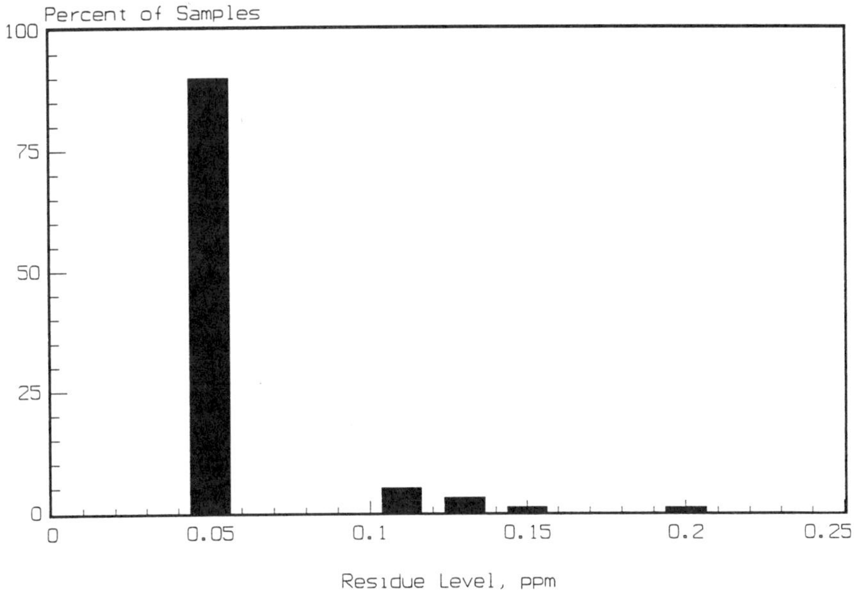

Figure 1. Hypothetical distribution of residue levels wherein the LOQ is 0.1 ppm. The average of the positive samples is 0.13 ppm, and the calculated mean level is 0.058 ppm.

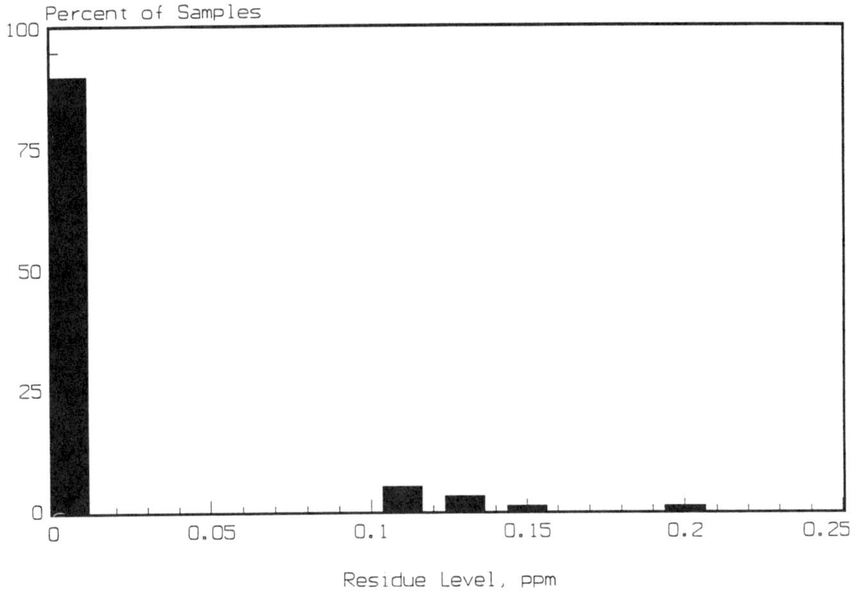

Figure 2. Same distribution as shown in Figure 1, except that the LOQ is now 0.01 ppm. The calculated mean residue level is now 0.018 ppm.

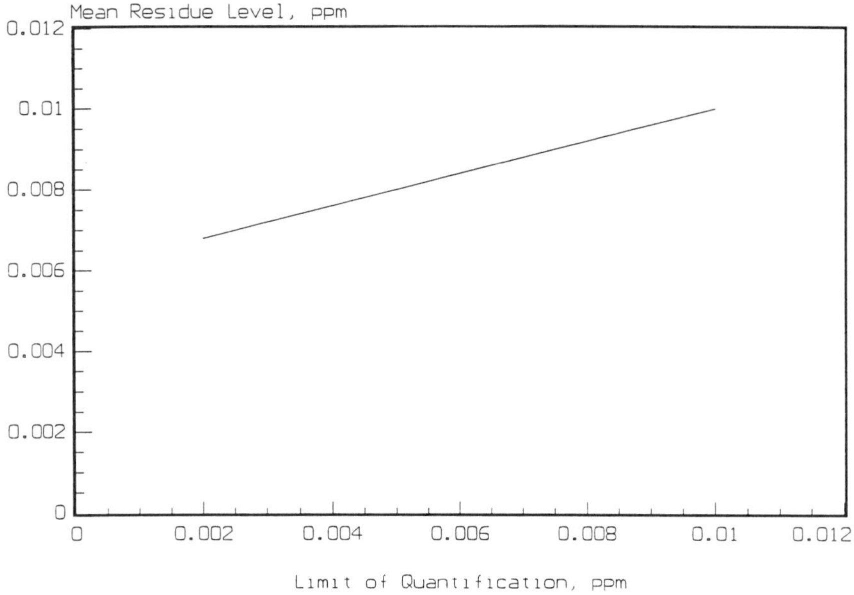

Figure 3. A plot showing the influence of LOQ on the calculated mean residue level. No residue was detected in 90% of the samples. Average of the positive samples was 1 ppm. The mean residue level decreases by 40% as the LOQ decreases by 90%.

When the limitations of the LOQ of the analytical method can compromise the statistical soundness of the database, the use of the percent of crop treated data represents an independent attempt to ensure the statistical integrity of the data. The significance of the statistical integrity can be illustrated by the following hypothetical example.

The following hypothetical situation involving a residue which is not detected in a statistically based survey in 90% of the samples serves to illustrate some of the points made above. The LOQ of the method is assumed to be 0.1 ppm and the average residue in the positive samples is assumed to be 0.13 ppm. We will assume that the carcinogenic potency, using EPA's nomenclature of Q^*, is 0.1 $(mg/kg/d)^{-1}$, and that the average daily consumption is 1.5 gm/kg/d. The average residue calculated after assigning a value of 0.05 ppm to all samples with no detectable residues would be 0.058 ppm. This would calculate to a risk of 8.7×10^{-6} (8.7 in one million). If the average residue level is recalculated using an adjustment for the percent of crop treated, the average level becomes 0.018 ppm. The resulting risk then becomes 2.7×10^{-6}. If we further assume that with improved analytical methodology or with a redesigned analytical program the LOQ could be improved to 0.01 ppm and that the proportion of samples with no detected residues remains the same, the calculated mean residue level becomes 0.013 ppm and the calculated risk drops to 2×10^{-6}.

In a second example, if all samples tested had no detected residues, the average residue using 1/2 the LOQ would be 0.05 ppm with a calculated risk of 7.5×10^{-6}. Adjusting that for 90% of the crop not treated, results in a residue of 0.005 ppm and a ten fold reduction in the calculated risk to 0.75×10^{-6}.

A further refinement on the use of percent crop treated information is to examine the application schedule. Pesticides which were used only pre-plant, mainly herbicides and dormant sprays, are less likely to be present on a raw agricultural commodity as compared to chemicals which were applied post-emergent. However, this refinement has not been used and there are obvious difficulties associated with how to incorporate these temporal factors into the food safety evaluation process.

Food Consumption Information

The second part of the exposure estimate is the estimation of the amount of food consumption. The amount of air inhaled can be estimated with reasonable accuracy. On a daily basis the amount of inhaled air could be proportioned among the work place, the home, and the general environment. In contrast the amount of food consumed by an individual varies day by day as do the kinds of foods being consumed.

Estimates of dietary exposures are based in part on food consumption surveys. Typically, in these surveys participants are asked to describe what they have eaten during a three or five day period. It should be borne in mind that most food consumption surveys were not designed to assess the intake of individual contaminants which may be dispersed in non-uniform, minuscule amounts in the diet, but rather to assess the nutritional status of a population. In fact these surveys are reliable indicators of nutritional intake, since while individual diets vary widely on a daily basis in terms of menu items, they vary very little in terms of total caloric or nutrient intake. Conversely, in assessing the intake of a food additive or pesticide residue there is a somewhat greater dependence upon the reliability of the food consumption pattern, i.e., the intake of an additive is dependent upon the appearance of a food containing that additive or substance in the diet. For health risks associated with chronic exposures, like cancer, the exposure model used for risk assessment assumes that an individual consumes some portion of every food form every day for 70 years. Clearly individual diets, with the exception of a few dietary staples, do not contain the same menu items on a daily basis. It is also obvious that given the low incidence of detectable pesticide residues in most food surveys that the residues cannot be distributed in a homogeneous manner throughout the diet. These food consumption surveys can fairly accurately predict the intake of a nutrient like vitamin C, since a given food group containing vitamin C will likely be consumed on a daily basis, irregardless, of whether it is apples or oranges, and the probability of that apple or orange containing vitamin C is high. Exposure to pesticide residues, however, is dependent upon the appearance of the specific item in the diet. Pesticides which are used on oranges are not necessarily used on apples and are not necessarily present on all oranges.

An example of the problems which can be encountered in estimating food consumption, particularly in the use of 95th percentile consumption estimates, is provided by an examination of caloric intake. On the basis of seven day food diaries, Morgan et al.(*3*) showed that individual daily caloric consumption for children ranged from 400 to 5100 kcal while the seven days average ranged from 1100 to 4500 kcal per day. The median caloric intake did not vary appreciably between individual daily consumption and averages based upon seven days food consumption (2000 vs. 2100). These estimated amounts of caloric intake are consistent with recommendations from the National Academy of Sciences. If food energy values are substituted for pesticide residues, caloric intake can also be estimated from the Tolerance Assessment System (now called the Dietary Risk Evaluation System) used by EPA. The mean caloric intake determined in this manner is consistent with that determined by Morgan et al. The 95th percentile consumption is approximated by the mean plus two standard deviations. Unfortunately the standard deviation of average daily consumption is quite large. For some food items the standard deviation is nearly comparable to the mean value (*4*). Thus the 95th percentile consumption estimate will be two to nearly three times the mean value. The data seems to indicate that caloric intake for those children would be in excess of 4000 calories per day. This would appear to be an abusive situation wherein the risk from obesity would far out weigh any food safety concerns. Some authorities have advocated the use of the 95th percentile consumption as a means to not underestimate the risk. While the rationale has a certain degree of logic statistically, it is arguable whether such a pattern of food intake is sufficiently plausible (for life time exposures) to merit much consideration in the evaluation of food safety.

Potency Estimation

The basis of most potency estimates are animal studies where the animals are exposed to fairly high levels of the chemical as compared to levels which could be found in the diet and the subsequent extrapolation of these results to predict effects in humans. Very elegant statistical models have been developed to aid in this extrapolation. We can measure residue levels. We can do surveys to determine consumption, but the potency estimates are neither directly measured nor can they be directly validated from simple observations. It has been demonstrated that such animal studies can present reasonable estimates of potency and consequently risk resulting from certain occupational exposures where exposure levels can be high, but the evidence that such methods provide realistic estimates for the risk resulting from exposures to low levels of residues as might be in encountered in the diet is virtually non-existent.

The best example of the non-validity of the animal model extrapolations as applied to dietary exposures lies not with pesticide chemicals but with aflatoxin. Aflatoxins share one common trait with pesticide chemicals in that dietary exposures are generally at very low levels, typically in the parts per billion range. However, for aflatoxin we have the benefit of epidemiological studies to compare with the animal studies.

Potency estimates (*5*) for aflatoxin from animals studies range from about 30-26000 $(mg/kg\text{-}d)^{-1}$. The results from epidemiological studies of aflatoxin have been mixed with some studies failing to show a causal relationship between dietary intake of aflatoxin and cancer. Estimates from positive epidemiological data are generally less than 100 $(mg/kg\text{-}d)^{-1}$. Thus the animal studies can over estimate potency by a factor of two hundred.

There are at least two ways in which we can examine the aflatoxin results. Unlike most of the pesticide residues of concern, aflatoxin is both genotoxic and mutagenic. Thus it fits the classic models of carcinogenesis and extrapolation models, like the linearized multistage, much better than non-genotoxic and non-mutagenic substances like Ethylene Thiourea (metabolite of Ethylene Bisdithiocarbamate fungicides) and 1,1 Dimethyl Hydrazine (metabolite of the plant growth regulator Daminozide), where secondary mechanisms of carcinogenesis have been invoked. Yet even with a near perfect chemical model for carcinogenesis, the risk analysis exceeds the observed human incidence by several orders of magnitude. Regardless of the retrospective analysis which could be made regarding the source of the variance, the fact remains that even in this best case situation, the risk is over-estimated. If these models are not accurate for aflatoxin, how accurate can they be for substances for which the models were not designed?

There is a generally accepted rule in science that as the absolute level decreases, the uncertainty associated with a measurement increases. The Uncertainty Principle in Quantum Physics and Horowitz's Rule (*5*) in analytical chemistry demonstrate this in different ways. Horowitz has basically shown that the variability in analytical results increase exponentially with the decrease in the level of the chemical being tested.

If we make the assumption that the same kind of uncertainty exists in potency data, then some interesting parallels can be drawn. Thus, if the potencies estimated by the animal studies can be over estimated by a factor of 200 at 100 $(mg/kg\text{-}d)^{-1}$ (the calculated potency from epidemiological studies for aflatoxin), the error bars around estimates for chemicals like the ETU with an estimated potency of 0.6 $(mg/kg/d)^{-1}$, or Captan with an estimated potency of 2.3×10^{-3} $(mg/kg/d)^{-1}$, from animal studies, range probably from several thousand to several million fold with the error most probably being one of overestimation of potency.

Yet, we continue to talk about these numbers as if they are not only accurate but correspond to something in the real world. It appears that feeding studies at exaggerated levels with test animals are not a proper model for human dietary exposures. Thus, while the animal studies may be validated for occupational exposures they do not appear to be validated for dietary exposures.

One of the criticisms of epidemiological studies is that they typically lack the power to detect low level incidence of disease due to the small size of the study populations. The aflatoxin studies, however encompass regional population groups and appear to overcome this criticism and serve as a useful test of the accuracy of the animal studies in terms of low level exposures.

One of the common misconceptions about pesticides and their perceived threat to food safety is that it is a fairly recent phenomenon. But most of the chemicals which are the targets of popular chemophobia were first registered twenty to forty years ago. Thus, any threats to public health through the presence of residues in the diet should have been detectable by now. Yet the death rate from cancer in the United States over a forty five period, except for lung cancer, the have either been constant or are declining. For many chemicals, the liver, due to its role in the detoxification of chemicals, is a primary site for cancer, but the death rate from liver cancer has continued to decrease in the years since 1955. In fact, in those situations where cancer death rates have increased, e.g., breast, prostate, and pancreas, they have showed the greatest amount of increase in the years 1930-1950 and thus have greatly preceded the wide spread use of modern pesticides.

Some question the use of cancer death rate statistics, since it can be argued that reduced rate of mortality can be attributable in many situations to improvements in medical treatment. Therefore the incidence of cancer has often been cited to indicate increased health risks associated with pesticide residues in the diet. But historical trends in the incidence of cancer are difficult to interpret for several reasons. Just as advances in treatment can affect mortality statistics, advances in detection and increased access to routine medical examinations can affect the statistics of incidence. For example, the incidence of non-Hodgkin's lymphomas has increased about 30% between 1950 and 1984, but this appears to be due to better diagnosis among the older population groups, since the incidence rate has decreased among the younger groups (*7*).

Even in circumstances where the cancer incidence has apparently increased, a detailed analysis of the data reveals the possibility of many contributing factors. For example, Hawaii shows an incidence of thyroid cancer (*8*) which is at least double the national average. This might be suggestive of a causal relationship between the disease and an environmental factor unique to Hawaii, but when the rate of cancer incidence is viewed by ethnic background and sex, significant differences are observed. It is notable that the incidence rate decreased or remained constant for all women except for Filipino but increased for all men except for whites. It has been speculated that dietary factors involving concentrations of iodine in the diet and the consumption of certain foods, by specific ethnic groups, high in goitrogenic compounds may be important components of the high incidence of this disease in Hawaii. It has often been said that the public is at an increased risk from cancer due to the presence of pesticide residues in foods and often some reference is made to an observed increase in the incidence of some cancers. As the above example illustrates, the gross statistics can hide subtleties which belie simplistic interpretations. Diet may play a role in the increase incidence of some cancers but these dietary factors are more likely to be related to the increased consumption of fat and the reduced consumption of fiber than to be related to trace amounts of residues in foods. Some medical treatments, such as radiation treatment, have also inadvertently lead to the increased incidence of certain cancers. Other risk factors which have been identified include smoking and alcohol consumption.

Scheuplein (*9*) has used a quantitative approach to illustrate the relative contribution of various dietary factors to the total cancer risk from the diet. His analysis suggests that only about 0.01% of the cancer risk is due to pesticide residues and contaminants in foods. Nearly 99% of the risk comes from traditional components of the diet, fat and naturally occurring toxicants.

Scheuplein's analysis does not validate nor contradict the current approach used in evaluating food safety, but it does place the risks in perspective by examining the including all aspects of the diet.

It remains to be seen whether the development of better biochemically based models, or models incorporating pharmacokinetics will provide a more appropriate basis upon which to determine human risks of cancer.

Conclusion

The intention of this paper is not to excuse the use of pesticides and the presence of residues in foods but rather to provide some perspectives in the debate about food safety based upon an examination of parts of the risk assessment process. If a chemical represents a true threat to food safety, then in should be removed. But it is a disservice to consumers, and undermines the confidence of the public not only in the safety of the food supply but also in science, to raise unwarranted concerns about the food that they eat when those concerns are based upon unvalidated assumptions and flawed analysis of data.

While the public has been convinced that pesticide residues and food safety represent a real hazard, the majority of experts do not share this concern. EPA's Science Advisory Panel and the American Medical Association have stated that while there may be hazards to farmers and agricultural workers associated with direct exposure to pesticides, there is no evidence that indirect exposures such as that which occurs in the diet pose a threat the public health. The Government Accounting Office (GAO) has estimated that there are 6.5 to 33 million cases of illness a year due to bacterial contamination of foods. The GAO estimates that there are 9,000 deaths per year due to those factors. When examined against the demonstrable incidence of disease due to bacterial contamination of food, it is well to remember the admonition that is attached to the interpretation of the calculated risks. They are "plausible" upper bounds on the risk, the true risk could be between those numbers and zero and are more likely zero.

Literature Cited

1. Roberts, Leslie. *Science* **1990**, *249*, p. 616
2. Shank, Fred R., Carson, Karen L., and Willis, Crystal A. In *Pesticide Residues and Food Safety: A Harvest of Viewpoints*; Tweedy, B.G., Dishburger, Henry J., Ballantine, Larry G., and McCarthy John; Eds.; American Chemical Society Symposium Series No. 446, 1991.
3. Morgan, Karen J., Zabik, Mary E., Cala, Rosemary, and Leveille, Gilbert A. "Nutrient Intake Patterns for Children Ages 5 to 12 Years Based on Seven Day Food Diaries"; Michigan Agriculture Experiment Station Journal Article No, 9093, 1980.
4. Pao, Eleanor M.; Fleming, Kathryn H.; Guenther, Patricia M.; and Mickle, Sharon J. "Foods Commonly Eaten by Individuals: Amount Per Day and Per Eating Occasion"; Home Economics Research Report Number 44, Human Nutrition Information Service, United States Department of Agriculture, 1982.
5. Department of Health Services, State of California. *Proposition 65 Risk Specific Intake Levels: Aflatoxins*. October 1989.
6. Horwitz, William. *J. Assoc. Off. Anal. Chem.* **1983**, *66*, p. 1295.
7. Devesa, Susan S., et al. *JNCI* **1987**, *79*, p. 701.
8. Goodman, Marc T., Yoshizawa, Carl N., and Kolonel, Laurence N. *Cancer* **1988**, *61*, p. 1272.
9. Scheuplein, R. J. "Perspective on Toxicological Risk-An Example: Food Borne Carcinogenic Risk" In *Progress in Predictive Toxicology*; Elsevier Science Publishers, BV: The Netherlands, 1990.

RECEIVED August 15, 1991

Laboratory Testing of Ingredients

Chapter 7

Liver Cell Short-Term Tests for Food-Borne Carcinogens

Gary M. Williams

American Health Foundation, 1 Dana Road, Valhalla, NY 10595

Carcinogens occur in food from a variety of sources, including natural-occurring contaminants, such as mycotoxins, additives, such as saccharin, and chemicals formed from food components, such as pyrolysates (*1-4*). As with carcinogens of other origins, these elicit cancer in experimental models through a variety of mechanisms (*5*).

Carcinogens exert effects in two distinct sequences of carcinogenesis, the conversion of normal cells to neoplastic cells and the development of neoplastic cells into tumors. Carcinogens that form reactive species, such as electrophiles, that bind to DNA produce neoplastic conversion through alteration of gene function, especially in oncogenes and tumor suppressor genes. Such carcinogens have been designated as genotoxic or DNA-reactive. Other types of carcinogens that produce epigenetic cellular effects that may give rise to reactive moieties, such active oxygen species, possibly also cause neoplastic conversion., In addition, some epigenetic carcinogens do not seem to be capable of altering DNA, even indirectly, but rather increase cancer through enhancement of neoplastic development. Enhancement of cell proliferation seems to be an important effect of carcinogens that act in this sequence. Carcinogens have been classified by Williams and Weisburger according to their mechanism of action, DNA-reactive or epigenetic (Table 1).

A food-borne carcinogen that is DNA-reactive and, hence, is genotoxic in short-term tests, is aflatoxin, and an example of an epigenetic agent is saccharin. Food contains numerous agents of both types (*1*, *3*). Because of the different mechanisms of action of carcinogens, different types of test systems for identifying potentially carcinogenic food-borne agents are needed.

Short-term Tests for DNA-reactive Carcinogens

For the initial in vitro screening of potential carcinogen, the battery of tests incorporated in the decision point approach to carcinogen testing (Table 2) (*6*) has proven useful (*7*). This approach begins with evaluation of the structure of the chemical. The structures leading to electrophilicity and, hence, DNA-reactivity of chemicals have been thoroughly elucidated (*5*, *8*) and accordingly, structure provides important informa-

0097–6156/92/0484–0060$06.00/0

Table 1. Classification of Carcinogenic Chemicals

Category and Class	*Example*
A. DNA-reactive (genotoxic) carcinogens	
1. Activation-independent	propylene oxide
2. Activation-dependent	aflatoxin B_1
3. Inorganic[a]	nickel
B. Epigenetic carcinogens	
1. Promoter	butylated hydroxyanisole
2. Cytoxic	nitrilotriacetic acid
3. Hormone-modifying	amitrole
4. Immunosuppressor	purine analog
5. Peroxisome proliferator	phthalate esters
C. Unclassified	
1. Miscellaneous	dioxane

SOURCE: Reference 5.

[a]Some are categorized as DNA reactive because of evidence for damage of DNA; others may operate through epignetic mechanisms such as alterations in fidelity of DNA polymerases.

Table 2. Decision-Point Approach to Carcinogen Testing

Stage A.	Evaluation of structure
Stage B.	Short-term tests in vitro
	1. Mammalian cell DNA repair
	2. Bacterial mutagenesis
	3. Mammalian mutagenesis
	4. Chromosome tests
	5. Cell transformation
	Other short-term tests
Decision Point 1:	Evaluation of all tests conducted in stages A and B
Stage C.	Tests for epigenetic effects
	1. In vitro
	2. In vivo
Stage D.	Limited bioassays
	1. Altered foci induction in rodent liver
	2. Skin neoplasm induction in mice
	3. Pulmonary neoplasm induction in mice
	4. Breast cancer induction in female Sprague-Dawley rats
Decision Point 3:	Evaluation of results from stages A, B, and C and the appropriate tests in stage D
Stage E:	Long-term bioassay
Decision Point 4:	Final evaluation of all results and application to health risk analysis. This evaluation must include data from stages A, B, and C to provide a basis for mechanistic considerations. Dose-response information may be crucial.

SOURCE: Reference 6.

tion on potential genotoxicity and carcinogenicity (*9*, *10*). Carcinogens of the DNA-reactive type can usually be identified in short-term tests for genotoxicity, providing that appropriate bioactivation is represented (*11*), and hence, an essential component of the battery of short-term test systems in the decision point approach is tests with intrinsic bioactivation.

A system with broad biotransformation capability is the hepatocyte/DNA repair test (*12*,*13*). This test has been shown to be reliable for the detection of food-borne genotoxins (*3*,*13*,*14*). A wide variety of plant-derived (Table 3) and microbe-derived agents (Table 4) are positive in this system using rat hepatocytes (*13*). Safrole was negative in rat hepatocytes, but did elicit DNA repair in hepatocytes from mice and hamsters (*15*). For most microbial products, DNA repair was also elicited in mouse hepatocytes, although the response with aflatoxin B_1 was weaker than in rat hepatocytes (*16*, *17*).

Chemicals that are formed during the processing or cooking of food have also elicited DNA repair in rat hepatocytes (Table 5).

In contrast to natural-occurring substances, food contaminants and additives generally have been non-genotoxic in hepatocytes (Table 6). The negative results with the food preservatives butylated hydroxyanisole and butylated hydroxytoluene, suggest an epigenetic mode of action for their carcinogenic effects (*18*, *19*). A variety of food dyes have been shown to be inactive in the DNA repair test (*20*).

Another system in which intrinsic metabolic capability is retained is proliferating rat liver epithelial cells. In this system, mutagenesis (*21*), chromosome effects (*22*) and transformation (*23*) can be assayed. With mutations in the hypoxanthine-guanine phosphoribosyl transferase gene as the end point, several food-borne agents have been tested (Table 7). The results correspond to those in the hepatocyte/DNA repair test with the genotoxin aflatoxin B_1 being positive and antioxidants and organochlorine compounds being negative.

In vivo approaches are also available for identification of DNA-reactive agents (*5*). These include tests for genotoxicity, one of which is the in vivo/in vitro hepatocyte/DNA repair test (*24*, *26*), and limited in vivo bioassays (*6*) for preneoplastic or early neoplastic lesions, including induction of altered foci in rodent liver, skin tumor or induction in mice, lung adenoma induction in mice and breast tumor induction in rats.

Short-term Tests for Epigenetic Carcinogens

Short-term tests for epigenetic agents are not as well established as tests for genotoxicity. Moreover, since epigenetic carcinogens can operate through a variety of mechanisms, different types of tests will be required for these agents.

An approach to the detection of promoters emerged from the work of Yotti et al (*27*) and Murray and Fitzgerald (*28*) which demonstrated that the plant-derived skin neoplasm promoter, tetradecanoylphorbol acetate inhibited metabolic cooperation between cultured cells, A variety of agents has been positive in systems based on this phenomenom (*29*). This approach has been extended to a liver cell system (*30*, *31*) in which several food-borne agents have inhibited metabolic cooperation (Table 8).

Table 3. Results with Plant Products in the Hepatocyte/DNA Repair Test

Chemical	*Result*	*Carcinogenicity*[a]
D, L-amygdalin	-	
arecaidine		-
arecoline		-
carrageenan (degraded)	-	S
clivorine	+	
cycasin	+	S
flavone	-	
kaempferol	-	I
lasiocarpine	+	S
monocrotaline	+	S
petasitenine	+	L
pyrrole	-	
quercetin	-	L
safrole	-	S
senkirkine	+	L
tannic acid	-	L
viridefloric acid	-	
zearalenone	-	L

SOURCE: Reference 13.
[a] from IARCI=inadequate; L - limited; S = sufficient

Table 4. Results with Microbial Products in the Hepatocyte/DNA Repair Test

Chemical	*Result*	*Carcinogenicity*[a]
actinomycin D	+	L
aflatoxin B_1	+	S
aflatoxin B_2	+	S
aflatoxin G_1	+	S
aflatoxin G_2	-	
aphidicolin	-	
averufin	+	
azaserine	+	S
chrysophanol	-	
cytochalasin B	-	
doxorubicin (adriamycin)	-	S
duclauxin	-	
echinulin	-	
emodin	-	
flavoglaucin	-	
floccosin	-	
griseofulvin	-	S
luteoskyrin	+	L
luteosporin	+	
mitomycin C	+	S
ochratoxin A	-	I
patulin	-	I
penicillic acid	-	L
rugulosin	-	I
secalonic acid D	-	
skyrin	-	
sterigmatocystin	+	S
5,6-dimethoxy	-	
sterigmatocystin	+	
versicolorin A	+	
versicolorin B	+	
violaceol-1	-	
zanthomegrin	+	

SOURCE: Reference 13.

[a] From IARC; I = inadequate; L = limited; S = sufficient

Table 5. Results with Products Formed During Cooking of Food in the Hepatocyte/DNA Repair Test

Chemical	*Result*	*Carcinogenicity*[a]
2-amino-3methylimidazo [4,5-*f*]quinoline (IQ)	+	S
2-amino-3,4-dimethylimi-dazo[4,5-*f*]quinoline (MeIQ)	+	I
2-amino-3,8-dimethylimi-dazo[4,5-*f*]quinoxaline (MeIQx)	+	I
2-amino-3,4,8-trimethyl-imidazo[4,5-*f*]quinoxaline	+	
2-amino-3,7,8-trimethyl-imidazo[4,5-*f*]quinoxaline	+	
3-amino-1,4-dimethyl-5H-pyrido[4,3-*b*]indole(Trp-P-1)	+	S
3-amino-1-methyl-5H-pyrido[4,3-*b*](Trp-P-2)	-	S
2-amino-6-methyldipyrido-[1,2-*a*:3'-2'-*d*]imidazole(Glu-P-1)	+	S
2-aminodipyrido[1,2-*a*:3',2'-d] limidazole (Glu-P-2)		S
2-amino-9H-pyrido[2,3-*b*]-indole	+	
2-amino-3-methyl-9H-pyrido-[2,3-*b*]indole	+	S
benzo(a)pyrene	+	S

SOURCE: Reference 13.

[a] From IARC I = inadequate; L = limited; S = sufficient

Table 6. Results with Food-Borne Chemicals in the Hepatocyte/DNA Repair Test

Chemical	*Result*	*Carcinogenicity*[a]
Food Additives		
butylated hydroxyanisole	-	S
butylated hydroxytoluene	-	L
Food Constituents		
sodium fluoride	-	I
Food Contaminants		
carbadox	+	
chlordane	-	L
DDT	-	S
endrin	-	I
heptachlor	-	L
mirex	-	S
olaquindox	+	
polybrominated biphenyls	-	S

SOURCE: Reference 13.

[a] From IARC; I = inadequate; L = limited; S = sufficient

Table 7. Results with Food-Borne Chemicals in the Adult Rat Liver Epithelial Cell-HGPRT Mutagenesis Assay

Chemical	*Result*	*Reference*
Food constituents		
sodium fluoride	-	*Cell Biol. Toxicol* **1988**, *4*, 173.
Food additives		
butylated hydroxyanisole	-	*Fd. Chem. Toxic.* **1986**, *24*, 1163.
butylated hydroxytoluene	-	Fd. Chem. Toxic. **1990**, *28*, 793.
Food contaminants		
aflatoxin B_1	+	*Mutat. Res.* **1984**, *130*, 53.
aflatoxin G_2	-	*Mutat. Res.* **1984**, *130*, 53.
chlordane	-	*Adv. Med. Oncol Res. & Educ. I* **1979**, *273*.
DDT	-	*Adv. Med. Oncol. Res. & Educ. I* **1979**, *273*.
endrin	-	*Adv. Med. Oncol. Res. & Educ. I* **1979**, *273*.
heptachlor	-	*Adv. Med. Oncol. Res. & Educ. I* **1979**, *273*.
kepone	-	*Adv. Med. Oncol. Res. & Educ. I* **1979**, *273*.
Polybrominated biphenyls	-	*Environ. Res* **1984**, *34*, 310.

Table 8. Results with Food-Borne Chemicals in the Hepatocyte-Liver Epithelial Cell Assay for Inhibition of Intercellular Molecular Transfer

Compound	*Result*	*Reference*
Food additives		
butylated hydroxyanisole	+	*Fd. Chem. Toxic.* **1986**, *24*, 1163.
butylated hydroxytoluene	+	*Fd. Chem. Toxic.* **1990**, *28*, 793.
Food contaminants		
benzo(a)pyrene	-	*Carcinogenesis* **1982**, *3*, 1175.
chlordane	+	*Carcinogenesis* **1982**, *3*, 1175.
DDT	+	*Cancer Lett.* **1981**, *11*, 339.
heptachlor	+	*Carcinogenesis* **1982**, *3*, 1175.
polybrominated biphenyls	+	*Environ. Res.* **1984**, *34*, 310.

In addition to the in vitro systems, epigenetic agents can also be identified in rapid in vivo tests. Firstly, chemicals can be assessed for any of the biological effects e.g. peroxisome proliferation, known to underly the carcinogenicity of epigenetic agents. For identification of potential promoting agents, increased cell proliferation and induction of cytochrome P450 are highly predictive. Also, several rapid and efficient bioassays for promoters are detailed in the decision point approach (*6*). These include enhancement of genotoxin-induced altered foci in rodent liver, skin tumors in mice, lung adenomas in mice and breast tumors in rats. In a liver cancer system , the food additive butylated hydroxytoluene, was shown to be a promoter at high doses only (*32*) and a similar observation was made in a stomach cancer system for butylated hydroxyanisole (*33*).

Conclusions

Reliable methods are available for the detection of both DNA-reactive and epigenetic agents in food. The application of the approaches outlined above to evaluation of food packaging materials also has been described (*34*).

Many genotoxic natural products can be present in food, but relatively few synthetic chemicals that enter into food are genotoxic. DNA-reactive natural food products have been associated with human cancer (e.g. aflatoxins) whereas no synthetic chemical in food has led to cancer in humans (*35*). In fact, a substantial portion (i.e.30-40%) of cancer in the United States is believed to stem from nutritional imbalances and genotoxic natural foodborne carcinogens or carcinogen precursors (*1,36*).

The in vitro approaches described here provide information on the mechanisms of carcinogenicity of chemicals. Such information is assuming importance in assessment of human hazard from environmental chemicals (*37-39*). As just discussed, DNA-reactive carcinogens are distinct human cancer hazards, whereas epigenetic agents represent only quantitative hazards.

Literature Cited

1. Williams, G.M. In *Environmental Aspects of Cancer The role of Macro and Micro Components of Foods*; Wynder, E.L; Leveille, G.A. Weisburger, J.H.; Livingston, G.E.; Eds; Food and Nutrition Press: Connecticut, 1983, pp. 83-100.
2. Grasso, P. In *Chemical Carcinogens Second Edition*; Searle, CE Ed.; ACS Monograph 182; American Chemical Society: Washington, DC, 1984; pp 1205-1240.
3. Williams, G.M. In *Genetic Toxicology of the Diet*; Knudsen I., Ed., Allen R. Liss, Inc.: New York, 1986; pp. 73-81
4. Miller, E.C. and Miller J.A. *Cancer (Suppl)* **1986**, *58*: 1795-1803.
5. Williams, G.M. and Weisburger, J.H. In *Casarett and Doull's Toxicology,The Basic Sciences of Poisons*, 4th Ed.; Amdur, M.O.; Doull, J.; and Klaassen, C.D.; Eds.; Pergamon Press: New York, 1991, pp 127-200.

6. Weisburger, J.H.; Williams, G.M. In *Chemical Carcinogens*, Volume 2; Searle, C.E., Ed; American Chemical Society: Washington, D.C., 1984; pp. 1323-1373.
7. Williams, G.M.; Weisburger, J.H. *Mutat Res.* **1988**, *205*, 79-90.
8. Arcos, J.C., Argus, M.F., Wolf, G. *Vol. I-III B*; Academic Press: New York; 1968-1985.
9. Rosenkranz, H.S., Ennever, F.K. Chankong, V., Pet-Edwards, J., Haimes, Y.Y. *Cell Biol Toxicol* **1986**, *2*, 425-440.
10. Ashby, J., Tennant, R.W. *Mutat. Res.* **1988**, *204*, 17-115.
11. Williams, G.M. *Ann Rev Pharmacol Toxicol* **1989**, *29*, 189-211.
12. Williams, G.M., Laspia, M.F. and Dunkel. V.C. *Mutation Research* **1982**, *97*, 359-370.
13. Williams, G.M.: Mori, H.: McQueen, C.A. *Mutat Res.* **1989**., *221*, 263-286.
14. Williams, G.M. *Food Addit Contam* **1984**, *1*, 173-178.
15. McQueen, C.A. and Williams, G.M. In *Handbook of Carcinogen Testing;* Milman, H.A., Weisburger, E.K., Eds.; Noyes Publications: Park Ridge, NJ, 1985; pp 116-129.
16. McQueen, C.A., Kreiser, D.M. and Williams, G.M. *Environ Mut* **1983**, *5*, 1-8.
17. Mori, H., Kawai, K., Ohbayashi, F., Kuniyasu, T., Tamasaki, M., Hamasaki, T., Williams G.M. *Cancer Res.* **1984**, *44*, 2918-2923.
18. Williams, G.M., McQueen, C.A. and Tong, C. *Food Chem Toxicol* in press.
19. Williams, G.M., Wang, C.X., Iatropoulas, M. *Food Chem Toxicol* in press.
20. Kornbrust, D., and Barfknecht, T. *Environ Mut* **1985**, *7*, 101-120.
21. Tong, C., Telang, S. and Williams, F.M. *Mutat Res.* **1984**; *130*, 53-61.
22. Ved Brat, S., Tong, C., Telang, S. and Williams, G.M. *Annals New York Acad Sci* **1983**, *407*, 474-475.
23. Shimada, T., Furukawa, K., Kreiser, D.M., Cawein, A. and Williams, G.M. *Cancer Res.* **1983**, *43*, 5087-5092.
24. Mirsalis, J.C.Butterworth, B.E. *Carcinogenesis* **1980**, *1*, 621-625.
25. Ashby, J.: Lefevre, P.A.: Burlinson, B.: Penman, M.G. *Mutat Res.* **1984**, *156*, 1-18.
26. Hellemann, A.L.; Maslansky, C.J.; Bosland, M.; Williams, G.M. *Cancer Letter* **1984**, *22*, 211-218.
27. Yotti, L.P.; Chang, C.C., Troski, J.E. *Science* **1979**, *206*, 1089-1091,
28. Murray, A.W., Fitzgerald, D.J. *Biochem, Biophys, Res. Commun.* **1979**, *91*, 395-401.
29. Barrett, J.C., Kakunaga, T., Kuroki, T., Neubert, D., Trosko, J.E., Vasiliev, J.M., Williams, G.M. and Yamasaki, H. In *Long-Term and Short Term Assays for Carcinogens: A Critical Appraisal*; Montesano, R.; Bartsch, H., Vaino, J.; Wilbourn, J.; Yamasaki, H. Eds; IARC Scientific Publications: Lyon, France, 1986; No. 83, pp. 287-302.
30. Williams, G.M. *Ann, NY Acad. Sci* **1980**, *349*, 273-282.
31. Williams, G.M. *Food Cosmet Toxicol* **1981**, *19*, 577-583.
32. Maeura, Y.; Williams, G.M. *Food Chem. Toxicol* **1984**, *22*, 191-198.
33. Williams, G.M. Food *Chem Toxicol* **1986**, *24*, 1163-66.
34. Williams, G.M. *Reg Toxicol and Pharmacol* **1990**, *12*, 30-40.

35. IARC Mongraphs, Volumes 1 to 42, IARC Monographs Supplement 7, IARC. Lyon France, 1987.
36. Williams, G.M. and Weisburger, J.H. In *Surgical Clinics of North America, Vol 66, Nutrition and Cancer I*; Meguid, M.M.; Dudrick, S.J.; Eds; W.B. Saunders Co.: Philadelphia, PA, 1986; 66, 873-889.
37. Weisburger, J.H.; Williams, G.M. *Environ Health Perspectives* **1985**, *50*, 233-245.
38. Weisburger, J.H.; *Jpn, J. Cancer Res. (Gann)* **1985**, *76*, 1244-1446.
39. Williams, G.M., Banbury Report 25: Nongentoxic Mechanisms in Carcinogenesis, Cold Spring Harbor Laboratory, 1987,pp. 367-80.

RECEIVED November 1, 1991

Chapter 8

Bacterial Test Systems for Mutagenesis Testing

Johnnie R. Hayes

RJR Nabisco, Bowman Gray Technical Center, Reynolds Boulevard, Winston-Salem, NC 27102

Humans must have been interested in the safety of food for thousands of years. For the vast majority of this time, the major toxicological endpoint associated with this concern was acute toxicity. The methodology consisted of observations of symptomatic effects of consuming a new food. No thought was given to the long-term effects of consuming a specific food.

A major factor that sparked an interest in regulations to insure food safety was the adulteration of foods and inadequate food sanitation. Additional concerns resulting in the passage of regulations associated with food safety were the increased use of various substances added for technological purposes and an increased public demand for an essentially risk-free food supply.

In the United States, the regulation of food safety was left to the individual states until the beginning of the twentieth century. The national government took the lead from the individual states with the congressional passage of the Food and Drug Act of 1906. Notable expansions and amendments to this Act occurred in 1953, 1958 and 1962. Continuing changes resulted in the Federal Food, Drug and Cosmetic Act, as we know it today.

Paralleling the development of the current food and drug regulations was an explosion of new information in the areas of chemistry and the biological sciences, which continues today. Associated with this mass of new information was the evolution of a new scientific discipline – toxicology.

Toxicologists use the tools developed by other disciplines, as well as uniquely developed tools, to determine the potential adverse effects of chemicals and chemical mixtures on biological systems. Tools now exist that allow the toxicologist to investigate the interaction of chemicals from the molecular level to the whole animal, and even to ecosystems. In some cases, the ability to collect data using these tools outstrips the ability to confidently use these data to insure a risk-free environment and food supply for humans. This comes, in part, from the extreme complexities associated with the interaction of chemicals with biological systems and, in part, from the utilization of animal models to represent humans. Animal models may differ from humans in obvious, and also very subtle, ways. This complicates extrapolation of data from animal studies to the human population. However, even with current limitations, the toxicologist now has powerful tools to bring to bear on questions of food safety.

0097–6156/92/0484–0073$06.00/0

Role of Genetic Toxicology in Food Safety Assessment

One of the tools of modern toxicology is genetic toxicology. It is the goal of genetic toxicology studies to ascertain the potential for a chemical to interact with the genetic material of cells to produce hereditable changes in somatic and reproductive cells. Most recommendations for food safety assessment now contain genetic toxicology as an integral component. It is generally recommended that *in vitro* genetic toxicology studies be performed during the early stages of a food safety assessment. An example can be seen in the classical safety decision tree approach to food safety assessment, as recommended by the Scientific Committee of the Food Safety Council in 1980 and illustrated in Figure 1 (*1*).

After chemical characterization and acute toxicity testing of the test material, the decision tree splits into genetic toxicology testing and biotransformation and pharmacokinetic studies. These studies may complement each other because metabolism of the test material may yield metabolites of genotoxic potential and the genetic toxicology studies may indicate the potential for the production of genotoxic metabolites.

Current recommendations for food safety assessment, as recommended by the FDA in the current (1982) "Red Book", are illustrated in the next three figures (*2*).

The FDA divides their testing recommendations into three concern levels based upon a number of criteria, including exposure assessment and structure-activity relationships. The lowest concern level is Category I. The testing program for Category I is illustrated in Figure 2. If a battery of short-term genotoxicity assays indicates positive results, such as mutagenicity and/or DNA damage, a carcinogenicity bioassay is suggested. If the results are negative, this bioassay can be eliminated unless other data indicate otherwise. Concern Level II testing is illustrated in Figure 3. Positives in the genotoxicity battery require that carcinogenicity testing be conducted in one-to-two species. The testing requirements for the highest concern level, concern Level III, are illustrated in Figure 4. At this concern level carcinogenicity testing in two species is required, regardless of the outcome of the genotoxicity assays.

Genetic Toxicology Testing Strategies

As illustrated in Table 1, David Brusick has divided genetic toxicology assays into three main categories (*3*). Screening tests are used to determine the potential for a chemical or chemical mixture to interact with DNA to produce alterations in the genetic material. These tests have little, if any, ability to be directly extrapolated to human risks. Allegations of potential human health hazards are unjustified if these tests are based solely upon *in vitro* assays or utilize non-mammalian models. Hazard Assessment Tests generally provide more evidence of potential genotoxic effects but are not adequate for quantitative human risk assessment. Risk Analysis Tests are tests that produce quantitative estimates of transmissible mutation. These *in vivo* tests generally measure direct genetic damage to germ cells or inheritable effects in the offspring of treated animals.

Table 1. Categories of Genetic Toxicology Assays

• Screening Assays
• Hazard Assessment Test
• Risk Analysis Tests

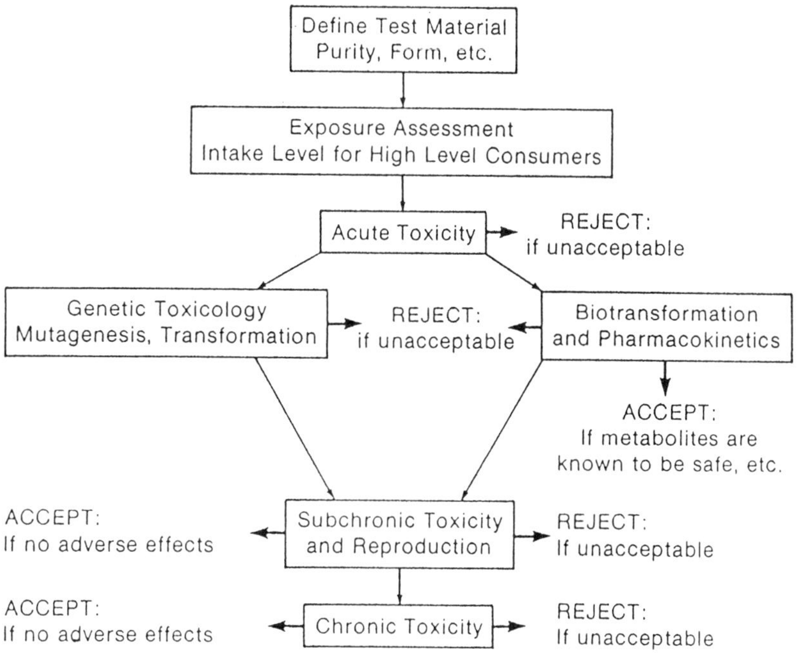

Figure 1. Safety decision tree recommended by the Scientific Committee of the Food Safety Council (*1*). (Reproduced with permission from ref. 16. Copyright 1989 Macmillan.)

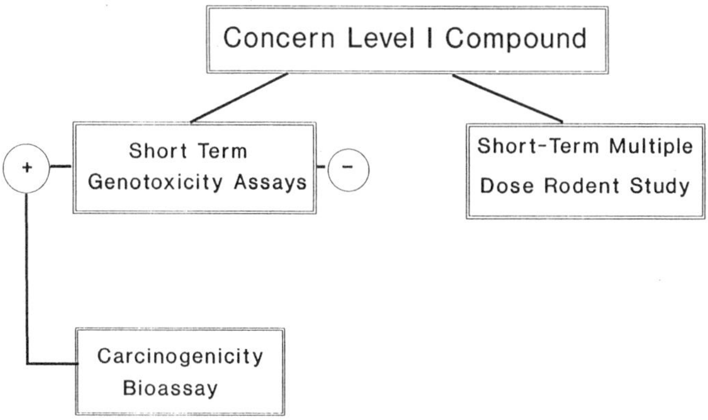

Figure 2. Decision tree for a U.S. FDA concern level I compound.

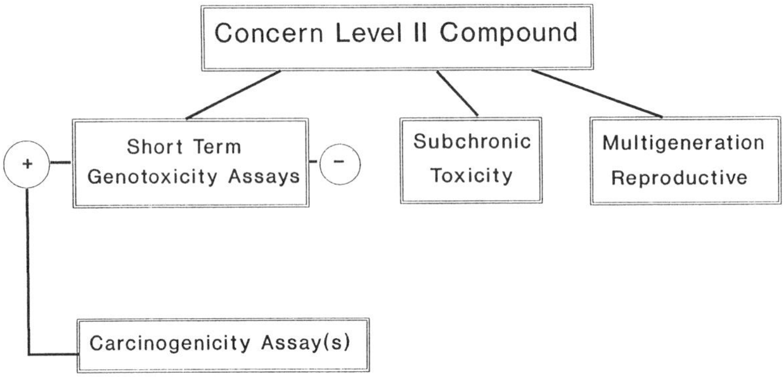

Figure 3. Simplified decision tree for a U.S. FDA concern level II compound.

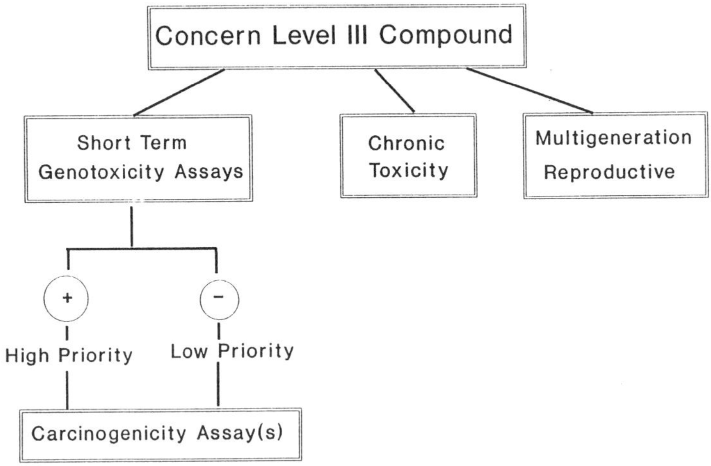

Figure 4. Simplified decision tree for a U.S. FDA concern level III compound.

Genetic toxicology testing strategies generally consist of two arms, one involves *in vitro* testing and the other *in vivo* testing. Recommendations for *in vitro* genetic toxicology generally consist of a battery of assays. Table 2 lists assays that may be commonly used in *in vitro* genetic toxicology testing.

As can be seen, these *in vitro* assays range from bacterial mutagenicity to mammalian unscheduled DNA synthesis. They cover a variety of genetic toxicology endpoints, including mutagenicity, clastogenesis and repair of damaged DNA. They also utilize two major categories of test subjects, bacteria and isolated mammalian cells.

Table 3 lists the common *in vivo* genetic toxicology assays. These tests generally parallel the *in vitro* tests with respect to genetic endpoints measured. They utilize mammalian species with the exception of the Drosophila sex-linked recessive lethal assay and the host mediated bacterial mutagenicity assay, which does use a mammalian species as the "host."

Bacterial Mutagenicity Assays

Bacterial mutagenicity assays are generally considered to be screening tests to determine the potential of a chemical or chemical mixture to interact directly with DNA to produce a mutation at a specific gene locus. As screening tests, data from bacterial mutagenicity assays are not directly useful for quantitative human risk assessment. However, they are useful for determining the potential for a chemical to interact with DNA to produce highly specific mutations.

The reason behind the development of bacterial mutagenicity assays and the high level of interest in these assays was the belief that they may predict carcinogenicity (*4*). If capable of predicting carcinogenicity, these assays would provide predictive data in weeks as opposed to years and provide an enormous savings of resources. A large effort has been devoted to determining the ability of these assays to reliably predict the carcinogenicity of a chemical.

There is still controversy associated with the ability of bacterial mutagenesis assays to predict carcinogenicity as measured by chronic animal bioassays. The percentage of accuracy for these short-term assays has been determined to be as high as 90%+ by some investigators and as low as 50% by others. There appears to be a consensus developing among most toxicologists that the lower percentages in this range are probably more accurate (*5, 6*).

The large variations in accuracy predictions arise from a number of factors that affect the predictability of short-term bacterial assays. Some of these are listed in Table 4. Non-genotoxic or epigenetic carcinogens are carcinogens that appear to not react directly with DNA. They are without activity in bacterial mutagenesis assays, do not produce evidence of DNA damage and have not been shown to interact covalently with DNA, yet they produce tumors in animal models. Various hypotheses related to their mechanism of action are under investigation. One of the more popular hypotheses is that these materials induce cellular proliferation. The increased DNA synthesis associated with proliferation may either increase the chances of normal mutations or the expression of preexisting DNA damage (*7*).

Another factor that influences assessment of predictability of short-term bacterial assays is that predictability is based upon comparison to chronic bioassays in animal models

Table 2. Short-Term In Vitro Genotoxicity Assays

Assay	*Endpoint*
Bacterial Cells	Gene Mutation
Ames/Salmonella	
E. Coli	
Yeast Mutation	Gene Mutation
Mammalian Cells	
Chromosome Aberration	Clastogenesis
Sister Chromatid Exchange	Chromosome Rearrangements
Unscheduled DNA Synthesis	DNA Damage
DNA/Xenobiotic Adducts	DNA Alteration
Cell Transformation	Transformation

Table 3. *In Vivo* Genotoxicity Assays

Assay	*Endpoint*
Mouse Coat Color (Spot Test)	Somatic Cell Gene Mutation
Drosphila Recessive Lethal Test	Sex-Linked Recessive Lethal
Rodent Micronucleus Assay	Chromosomal Aberrations
Bone Marrow Sister Chromatid Exchange	Sister Chromatid Exchange
DNA/Xenobiotic Adducts	Altered DNA
In Vivo/In Vitro Unscheduled DNA Synthesis	DNA Damage
Host-Mediated Bacterial Mutagenesis	Gene Mutation
Dominant Lethal Assay (Germ Cell)	Dominant Lethal Mutations
HeritableTranslocation (Germ Cell)	Translocation

(*8*). A material that is determined to be mutagenic in a bacteria assay but does not produce tumors in an animal bioassay is termed a "false positive." This is based upon the assumption that the animal bioassay was accurate. However, if the animal bioassay was inaccurate, a decision that the bacteria mutagenesis assay was a false positive would be incorrect. Because of the time and cost requirements of animal bioassays, it would be a significant amount of time before this conclusion could be corrected.

The unique biochemistry and physiology associated with the whole animal compared to the bacterial cell and the simplistic systems used to mimic mammalian metabolic activation systems in short-term bacterial mutagenicity test also affect the attempt to determine the predictability of the bacterial test. Most bacterial test systems use mammalian liver preparations to mimic oxidative metabolic activation of test materials to more reactive metabolites. These preparations neither have significant capability to mimic a number of other metabolic activation pathways nor most of the detoxification mechanisms. A test material that may be detoxified and thereby not yield tumors in animal models may or may not express its mutagenic potential in bacterial assays. Alternatively, a carcinogen that requires a metabolic activation pathway not present in the bacterial assay system will appear to be non-mutagenic.

Certain aspects of the bacterial mutagenicity assay systems that increase their sensitivity affect the ability to determine their predictability. For instance, the strains of *Salmonella* used in the Ames assay have been developed to have very poor DNA repair systems. These bacteria are highly sensitive to DNA damage because they cannot repair the damage. The animal models used in carcinogenicity tests have highly developed DNA repair mechanisms and may repair DNA damage before it is expressed. Therefore a test material that is positive in the bacterial assay may be negative in animal studies.

Although a number of other factors affect the ability to determine the predictability of bacterial mutagenesis assays (*see* Table 4), a major factor is a lack of knowledge of the mechanisms of carcinogenicity. Even though knowledge of these mechanisms has greatly expanded over the last 20-years, a greater understanding is needed before all the factors that influence our ability to use bacterial mutagenesis assays to predict carcinogenicity are understood. Overall, short-term bacterial mutagenicity assays have not lived up to the original expectations for them with respect to predicting carcinogenicity (*8*).

What is sometimes forgotten in the controversy over the ability of short-term bacterial assays to predict the carcinogenicity potential of a chemical is that genotoxicity is an endpoint unto itself. It may be too much to ask of a simple *in vitro* assay to predict the complex *in vivo* interactions associated with carcinogenicity. However, the bacterial mutagenicity assays will determine the potential of a chemical to interact with DNA to produce a mutagenic event under the conditions of the assay. Since certain of these assays can be evaluated on a semi-quantitative basis, they provide a method to compare the potential of two chemicals to interact with the bacterial genome to produce mutations. It is therefore possible to comparatively screen a number of materials to determine which has the lowest mutagenicity potential.

Bacterial mutagenicity assays provide a rapid and inexpensive method to screen chemicals for further research and to eliminate those that do not meet specific toxicological criteria. Therefore, even though these assays may not have a high percentage of predictability for carcinogenicity by themselves, they can serve an important function in screening chemicals for their potential ability to interact with DNA.

Genetic Toxicology Batteries

A number of genetic toxicologists believe that the addition of data from other short-term *in vitro* genotoxicity assays, especially mammalian cell assays, increases the ability to predict carcinogenicity. Although there is still controversy associated with this hypothesis, regulatory agencies recommend that a battery of short-term *in vitro* assays be used for genetic toxicology testing. Table 5 illustrates the battery suggested by the Food Safety Council in 1980 (*1*). As can be seen, these batteries represent a bacterial mutagenicity assay coupled with mammalian cell assays. The Food Safety Council recommendations include a mammalian cell mutagenesis assay to parallel the bacterial assay, an assay(s) to detect clastogenesis and a mammalian cell transformation assay. They also recommended *in vivo* genetic toxicology assays if positive responses were found in the *in vitro* assay. One of the recommended *in vivo* assays was a host-mediated bacterial cell mutagenesis assay. This type of assay generally involves injection of sensitive bacteria into the interperitoneal cavity of the "host" animal. The bacteria are harvested after the animal is exposed to a potential mutagen. After harvesting, the mutation rate, if any, is determined. These assays were developed to attempt to incorporate factors such as pharmacokinetics, mammalian physiology and mammalian biochemistry into a bacteria assay system. The number of problems associated with the host-mediated assays have resulted in this type of assay falling out of favor with many genetic toxicologists.

Short-term genotoxicity tests recommended by the FDA in the 1982 "Red Book" are listed in Table 6 (*2*). These, like the Food Safety Council's recommendations, include a bacterial and mammalian mutagenesis assay, an assay for DNA damage and a mammalian cell transformation assay. An *in vivo* assay, if necessary, is also included.

In both these recommendations, as well as others, the inclusion of mammalian cell assays is believed to increase the likelihood of detecting potential animal carcinogens whose mechanism of action would not be detected by the bacterial mutagenesis assays.

Utility of Bacterial Mutagenicity Assays

Even though there are a number of issues associated with the use of bacterial mutagenicity assays, they remain a component of food safety evaluations. A number of reasons account for their use and their advantages over certain other assays. The advantages are listed, in part, in Table 7. A major advantage in using the bacterial mutagenicity assays is their low costs and the short-term nature of the assay. Since data can be produced in a few days at a nominal cost, they are ideal for screening groups of test materials. This makes them useful in product development and determining the mutagenicity of various fractions of complex mixtures, such as food products.

Another advantage is their high sensitivity. The bacteria generally used in these assays have been developed to be extremely sensitive to mutagens. For instance, the *Salmonella* strains used in the Ames assay have cell walls that allow the passage of mutagens into the

Table 4. Factors Affecting the Predictability of Short-Term Bacterial Assays

Non-Genotoxic (Epigenetic Carcinogens)
Reliability of Chronic Animal Bioassays
Unique Metabolic Capabilities of Animals
Repair of Damaged DNA
Prokaryotic vs. Eukaryotic Cell Types
Sensitivity to False Positives/Negatives
Mutagenicity vs. Clastogenesis and other Mechanisms
Initiation vs. Promotion
Lack of Knowledge of Mechanisms of Carcinogenicity

Table 5. Assays Recommended by the Food Safety Council

In Vitro	
	Bacterial Cell Point Mutations
	Mammalian Cell Point Mutations
	Mammalian Cell Chromosomal Changes
	Mammalian Cell Transformation
In Vivo	
	Mammalian Cell Chromosomal Changes
	Host-Mediated Bacterial Cell Mutagenesis

Table 6. Short-Term Genotoxicity Testing Suggestions by the FDA

Test	*Endpoint*
In Vitro	
Ames Assay	Bacterial Mutagenesis
TK Locus Mouse Lymphoma Assay	Mammalian Cell Mutagenesis
Unscheduled DNA Synthesis	DNA Damage
Transformation	Mammalian Cell Transformation
In Vivo	
Drosphila Mutation Test	Sex-Linked Recessive Lethal

Table 7. Advantages of Bacterial Mutagenicity Assays

- Rapid
- Low Cost
- Sensitivity to Mutagens
- Large Database
- Availability
- Simple
- Semi-Qualitative
- Acceptable/Recommended

bacteria. As previously mentioned, the bacteria are deficient in DNA repair mechanisms, allowing increased expression of DNA damage. Bacteria are somewhat less sensitive to their environment than mammalian cells that originate from multi-cellular tissues whose environments are closely controlled. They are, therefore, less likely to yield either false positives or false negatives due to environmental conditions (*9*).

An additional advantage of bacterial mutagenicity assays, especially the Ames assay, is the availability of a large database of compounds and mixtures that have been tested (*10, 11,5,6*). This allows utilization of structure activity analysis and also aids in validating the assays for different chemical classes (*12*).

Bacterial mutagenicity assays are generally considered to be operationally simple to perform. They can be performed in a large number of laboratories, including contract toxicology laboratories. Therefore, the bacterial mutagenicity assays, especially the Ames assay, are generally available to investigators. A number of the more sophisticated genotoxicity assays may be available only in a small number of laboratories. Similar data from a large number of laboratories performing a particular assay, aids in the validation of the assay. For instance, interlaboratory variation is better understood and by using a common series of negative and positive controls, data from various laboratories can be more accurately compared.

An advantage of bacterial mutagenicity assays, especially the Ames assay, is that the dose response data appear to be semi-quantitative, at least within similar chemical classes. This allows the mutagenic potential of a number of chemicals to be compared and opens the possibility for greater understanding of structure activity relationships (*12*). The semi-quantitative nature of a number of other genotoxicity assays has not been adequately determined and some genotoxicity assays are not semi-quantitative.

The major advantage of the bacterial mutagenicity assays, especially the Ames assay, is that they are acceptable to the majority of genetic toxicologists as a screen for potential mutagenicity. They are also recommended and accepted by a number of national and international regulatory agencies as a component of a food safety assessment.

Methodology For Bacterial Mutagenicity Assays

As an example of the general methodology for bacterial mutagenicity assays, the following description is limited to the Ames assay. There are several variations of the Ames assay with the two most-used methods being the preincubation and plate incorporation assays (*13, 14*). The preincubation method preincubates the test material with and without the S-9 metabolic activation system and the bacteria tester strain at 37°C for approximately 20 minutes. Agar is added to the preincubation mixture and the resulting mixture distributed over agar plates containing limiting amounts of histidine. The plate incorporation method directly incorporates the test material with and without the S-9 system and bacteria tester strain directly into the plate without preincubation. With both methods, the plates are incubated for 48 hours at 37°C. Bacteria that have been reverted to the wild type phenotype, which does not require histidine in the medium, can grow into colonies. Bacteria that are not mutated and required histidine do not develop colonies. Mutagenicity is determined by counting, usually by automated methods, the number of colonies on the plate and the data expressed as revertants/plate.

Solvent controls are used when a solvent vehicle has been used to carry the test material into solution in the medium. Positive controls and sometimes negative controls are also tested. One positive control should be a compound which requires metabolic activation with the particular tester strain being used. An additional positive control which does not require metabolic activation in the particular tester strain is used with the assays without metabolic activation. These controls validate the assay by demonstrating the sensitivity of the tester strain to known mutagens under the conditions of the assay.

Various modifications can be used with the assay to address specific questions and to account for various assay conditions, such as when human urine is used as the test material.

Major Factors Affecting Bacterial Mutagenicity Assays

A number of factors can affect the use of bacterial mutagenicity assays in a food safety assessment, a few of which are listed in Table 8. One of these factors is the choice of bacterial strain. Table 9 list some of the *Salmonella* strains routinely used for the Ames assay. As can be seen, some strains are sensitive to frameshift mutations and others are sensitive to base-pair substitutions. Also, the specific target gene for each strain differs. Ames assays generally include two or more *Salmonella* strains and five strains are used in many cases.

Metabolic activation has been discussed above, but a few additional points can be mentioned. The classic metabolic activation system has been the supernatant from Arochlor 1254 induced rat liver homogenates centrifuged at 9,000xg, i.e. S-9. Preparations from species other than rats can be used to address specific issues. Because the quantitative and qualitative nature of the metabolically activated products may vary with S-9 preparations from animals pre-treated with a number of enzyme inducers and also vary with species, the results of mutagenicity studies can vary. This can be used to advantage to investigate the role of metabolic activation on potential mutagenicity.

The bacterial cytotoxicity of the test material is an important factor that can affect bacterial mutagenicity assays. If the test material is highly cytotoxic, the cytotoxicity may mask the potential mutagenicity. The threshold of mutagenicity may not be reached before too many bacteria are killed to produce a valid test. If the test material is not water soluble, it is necessary to use a solvent as a vehicle to carry the test material into the culture media. Use of a vehicle increases the number of factors that can influence the assay. For instance, the solvent will have its own inherent cytotoxicity and alter the physical and chemical nature of the assay media. Therefore, solvents used as vehicles in bacterial mutagenicity assays must be chosen with care.

Another major factor that can affect the results of bacterial mutagenicity assays is the physical and chemical nature of the test material. Table 10 list some of the factors associated with the test material that can affect bacterial assays. Water insoluble test materials complicate mutagenicity testing because of the aqueous nature of the assay media. This may necessitate the use of vehicles as discussed above. Even with a vehicle to carry the water insoluble test materials into the media, the test material may precipitate at the higher doses. This will limit the maximal dose that can be used in the studies. The pH and/or osmolarity of the media may be altered by either the test material or a solvent vehicle, although bacterial are less sensitive than mammalian cells (*9*). Such alterations can lead to cytotoxicity and produce false positives or negatives in the assay. When the assays are carried out with the inclusion of metabolic activation there are actually two biological systems that can be affected by either the test material or solvent vehicle; the bacteria and the metabolic activation system. If the test material or vehicle alters the activity of the metabolic activation system, this alteration could be expressed in the results of the mutagenicity assay. For instance, inhibition of metabolic activation could result in a false negative. From the above, it can be seen that the results of bacterial mutagenicity assays can be highly dependent upon a number of factors. It is imperative that these factors be carefully considered in the design of the studies to insure accurate, reproducible and meaningful results.

Disadvantages of Bacterial Mutagenicity Assays

Use of bacterial mutagenicity assays in food safety assessments has certain disadvantages associated with the assays. Several of these disadvantages are listed in Table 11 and have been previously discussed. If the assays are performed to determine carcinogenic potential, their reliability can be questioned (*8*). As noted before, they are more useful for predicting potential mutagenicity and DNA interactions than for carcinogenicity. Bacteria are prokaryote whereas mammalian cells are eukaryotic. Other differences, including the requirement to add mammalian enzyme systems that metabolically activate many mutagens, make extrapolation of results from bacterial mutagenicity assays to mammalian systems difficult.

Use of Bacterial Mutagenicity Assays in Food Safety Assessments

As noted, several expert groups and individuals have recommended bacterial mutagenicity assays, especially the Ames assay, as a component of a genetic toxicology battery in food safety assessments. A number of factors have been noted above that can influence the results of bacterial mutagenicity assays. These factors, among others, must be carefully considered in the design of the mutagenicity studies.

At times it is necessary to consider what to test in mutagenicity assays during a food safety assessment. If the test material is relatively chemically pure, such as a food additive used for technical purposes, the choice of what to test is straight-forward. If the test material is a complex mixture, the choice of what to test can be more difficult. For instance, one choice is to test the complete mixture in the form in which it is to be used in food. Alternatively, the major components of the mixture could be isolated and tested individually. Another choice would be to isolate and test various fractions of the mixture such as the organic solvent soluble phase, the aqueous phase, the acid soluble phase, as well as other phases. These could either be highly concentrated and the concentrates tested or the test could be done at concentrations reflecting their concentration in the original test material.

The exact approach as to how to test the material of interest is dependent upon the question to be addressed. If the results of the mutagenicity test are to be used to address the question of potential mutagenicity in the diet, it is preferable to test the material in the form in which it is to be used. This will allow any interactions that could occur in the mixture to be expressed. For instance, if the mixture were to contain several mutagens at concentrations below the threshold of detectability and their individual mutagenicities were additive, the

Table 8. Major Factors Affecting Bacterial Assays

- Bacterial Strain
- Metabolic Activation
- Cytotoxicity
- Solvents
- Physical/Chemical Nature of Test Material

Table 9. Bacterial Strains used in the AMES Assay

	Salmonella Strain	*Target Gene*	*Mutation Type*
*	TA-98	HIS D	Frameshift
* *	TA-100	HIS G	Base-pair substitution
* *	TA-1535	HIS G	Base-pair substitution
	TA-1537	HIS C	Frameshift
*	TA-1538	HIS D	Frameshift
	Others		
	Other species		

* Derived from parental strain D3052

* * Derived from parental strain G-46

Table 10. Physical/Chemical Factors of Test Material that Affect Bacterial Assays

- Water Solubility (Partition Coefficient)
- pH
- Osmotic Pressure
- Metabolic Activation

Table 11. Disadvantages of Bacterial Mutagenicity Assays

- Predictability of Chronic Effects Questionable
- Prokaryote
- Not Sensitive to Non-Genotoxic Carcinogens
- Method Dependent
- Requirement for Addition of Metabolic Activation
- False Positives/Negatives

resulting mutagenicity may be above the threshold of detectability. Conversely, the presence of antimutagens in the mixture or other inhibitory materials may result in no detectable mutagenicity.

Other questions to be addressed by mutagenicity testing may require alternative approaches. For instance, if a particular food preparation technique appears to result in the formation of mutagens, then it may be necessary to subfractionate the food to determine the chemical source of the mutagenicity. If it is expected that a particular manufacturing process may produce a mutagenic contaminate, it may be necessary to test subfractions of the mixture or individual components to determine the source of mutagenicity. Once discovered, it may be possible to modify the process to eliminate the mutagenic component.

A number of suggestions have been made as to how to interpret the data from mutagenicity test. These methods generally involve criteria based upon the occurrence of a dose response and the magnitude of change compared to the background control. One criteria used by many toxicologist is that a positive response is indicated by a dose response with at least one dose being twice the background.

Several statistical methods may be used to analyze Ames assay data (*15*). One method is the use of initial slopes determined from the dose response curves. Care must be used with this method that the slopes are determined from the linear portion of the curve because the revertants/plate have a tendency to plateau or even decline at high doses due to cytotoxicity and other factors. Various methods have been developed to aid in this type of analysis (*16*).

In a food safety assessment the results of bacteria mutagenicity assays must be used in context with data from the other genotoxicity assays and *in vivo* animal studies. Attempts are underway to develop rational methods of incorporating genotoxicity testing data into quantitative risk assessments. Additional work is required to develop risk analysis methods to integrate this type of data into quantitative risk analysis methods.

Literature cited

1. Scientific Committee of the Food Safety Council, Proposed System for Food Safety Assessment: Final Report of the Scientific Committee of the Food Safety Council, Food Safety Council, 1725 K Street, N.W., Washington, D.C. 20006, June 1980.
2. Bureau of Foods, U.S. Food and Drug Administration: Toxicological Principles for the Safety Assessment of Direct Food Additives and Color Additives Used in Food. U.S. Food and Drug Administration, Washington, D.C., 1982.
3. Brusick, D., Principles of Genetic Toxicology 2nd Ed., Plenum Press, New York, 1987, pp. 82-84.
4. Ames, B.N. et al. *Mutation Res.* **1975**, *31*, 347-364.
5. Ashby, J. et al. *Mutation Res.* **1989**, *223*, 73-103.
6. Travis, C.C. et al. *Mutagenesis* **1990**, *5*, 213-219.
7. Ames, B.N. *Environ. Molec. Mutagen.* **1989**, *14*, (Suppl. 16) 66-77.
8. Ashby, J. *Ann. N.Y. Acad. Sci.* **1988**, *534*, 133-138.
9. Scott, D. et al. *Mutation Res.* **1991**, *257*, 147-204.
10. Haworth, S. et al. *Environ. Mutagen.* **1983**, *5* (Suppl. 1), 3-142.
11. Mortelmans, K. et al. *Environ. Mutagen.* **1986**, *8* (Suppl. 7), 1-119.
12. Kalopissis, G. *Mutation Res.* **1991**, *246*, 45-66.

13. Maron, D.M. and Ames, B.N. *Mutation Res.* **1983**, *121*, 173-215.
14. Yahagi, T.M. and Ames, B.N. *Mutation Res.* **1977**, *48*, 121-130.
15. Mahon, G.A.T. et al. In *Statistical Evaluation of Mutagenicity Test Data*; D.J. Kirkland, Ed.; Cambridge University Press: Cambridge, 1989, pp. 26-65.
16. Hayes, J.R. and Campbell, T.C. In *Casarett and Doull's Toxicology: The Basic Science of Poisons,* 3rd ed.; Klaassen, C. D.; Amdur, M. O.; Doull, J., Eds.; Macmillan: New York, 1989; pp 771-800.

RECEIVED December 16, 1991

Chapter 9

Current Trends in Animal Safety Testing

John C. Kirschman

FSC Associates, P.O. Box 718, Lewisville, NC 27023

Since enactmant of the 1958 Food Additive Amendment, toxicological sciences have advanced faster that the statutes, particularly in relation to carcinogenicity testing. For the past three decades, other than for microbiological issues, food safety activities have generally focused on food additives rather than foods themselves. The knowledge collected during this period with improved methodologies, in analytical chemistry as well as biological sciences, along with the advent of biotechnology is now bringing focus to questions about the adequacy of present means of testing and evaluating the safety of complex mixtures known as foods. These developments, remaining issues and current trends will be discussed.

I plan to review the current trends that I perceive crossing the full spectrum of test segments used in the safety evaluation of foods and food components. Let me start with a bit of perspective regarding the numbers of chemicals we're dealing with. Of the 5,000,000 plus known chemicals in our universe, somewhere between 5,000 and 10,000 are being used worldwide as food additives (*1*).

In the U.S. under the FD&C Act and its 1958 Food Additive Amendment, premarket testing of food additives has been clearly codified.

While such premarket testing is not required for foods themselves, it is estimated that they are made up of several hundred thousand natural components. Compositional documentation is, however, actually extremely meager.

From the international perspective then, what is food? One could say that food is anything sold as such.

Food is defined by the U.S. FDC Act as articles used for "food or drink for man or animals, chewing gum, and articles used for components of any such article" which pretty well fits the above definition. Premarket testing and governmental approval is not required for foods. However, the law places responsibility on the person who introduces any food into commerce for assuring that it complies with all applicable safety standards and does not cause harm to the consumer.

0097–6156/92/0484–0088$06.00/0

Please keep these things in mind as I proceed now to discuss toxicological testing methodologies as designed for and applied to single chemicals. I'll then move on to issues we face in evaluating the safety of complex mixtures of unknown chemicals we consume as food.

If we had available tests for toxicity that were as reliable, reproducible and relevant to man, as for example pH paper is for determination of hydrogen-ion concentration, we would already be far along into toxicological testing of all chemicals. However, since testing will be possible in the foreseeable future on only a small portion of the universe of chemicals, testing must be based on carefully set priorities.

In essence the 1958 Food Additive Amendment, calling for toxicological testing of chemical additives to foods, but not for prior-sanctioned or food materials generally recognized as safe (GRAS) by recognized experts in the field, was a very pragmatic prioritization step for food chemicals.

The list of core toxicity tests called for on any new food additive includes those presented in Table 1. Shown as well are typical price and time estimates for performing such tests. Be assured that if everything proceeds perfectly well, without any hitches which require additional or repeat research and testing, one cannot hope to complete such a testing program on a single food chemical in less than 5 years. Even seven years is considered by most to be overly optimistic. I shall now briefly characterize these tests individually and mention trends and changes occurring recently within each of these areas.

Table 1. Toxicology Tests

Test	*(months)*	*(dollars)*
Acute	<1	2,500
Short Term	1	30,000
90 day Subchronic	3	75,000
Teratology	6	85,000
Reproduction	12	280,000
General Metabolism	12	350,000
Chronic/Carcinogen. (Rats)	24	775,000
Carcinogen. (Mice)	24	650,000
	Total	$2,247,500

Acute toxicity tests define the range of single oral doses that induce toxic and lethal responses. Ages ago, the need to compare the therapeutic potency of different lots of plant or animal extracts for use as medicinals led to development of bioassay procedures which estimated the ED_{50} (effective dose/50% response or median effective dose) of similar materials. If the effect measured was death, the ED_{50} became the LD_{50}. Later, this LD_{50} became a measurement for use in comparing the toxicities of different materials. It has been used as a basis for design of rational treatments of human poisonings and is helpful in designing longer multidose toxicity tests. Properly used, the LD_{50} provides information of the types of toxic effects, the onset of acute toxicity, and a quantitative estimate of the lethal dose.

The traditional LD_{50}, testing for 50% lethality, is now being replaced with procedures using significantly fewer (e.g. 5 animals/group or as few as 6 animals total) than the 10 animals per group used classically. A number of the newer regulatory guidelines permit or even encourage the use of studies yielding only an estimation of the lethal dose.

Based on the results of such lethality tests, the rating scale presented in Table 2 has been used for categorizing chemicals by toxic potency.

Table 2. Toxicity Rating

Toxicity Class		*Probable Lethal Dose for a 70 kg man (mg/kg)*
6.	Supertoxic	< 5
5.	Extremely toxic	5-50
4.	Very toxic	50-500
3.	Moderately toxic	500-5,000
2.	Slightly toxic	5,000-15,000
1.	Practically nontoxic	> 15,000

The next level of toxicity testing includes short term tests with repeated challenges given over 14-30 days. Such tests begin to provide the first meaningful indication of the tolerance animals have to reasonably expected exposure in use levels of the test material. With such information one achieves a rather dependable indication as to whether or not the compound is safe enough to warrant further pursuit with additional evaluation and use.

Subchronic studies are subsequently used to define the impact of repeated dietary doses over the greatest portion of the animals' growth and maturation period. They generally involve one rodent (e.g. Rat) and one non-rodent (e.g. Dog), 10-20 animals per group at each of three test dose levels, plus controls, fed for 90 days. The 90 day subchronic test provides the most substantial information and will undoubtedly be the workhorse test of the next decade. The indices, as listed in Table 3, are similar for both subchronic and chronic studies.

At this stage of a testing program, just before or during the subchronic studies, it can be extremely meaningful to start the absorption, distribution, metabolism and

Table 3. Subchronic/Chronic Indices

Food Intake
Growth
Mortality
Hematology
Clinical Chemistry
Urinalysis
Organ Weights
Body Weights
Gross and Histopathology

excretion (A,D,M & E) studies. They are very helpful in establishing actual target organ dosimetry by determining the stability and chemical character of the material absorbed from the G.I. Tract. Such information can be quite important in determining the direction of further research on the compound.

During or following the subchronic tests one should embark on testing for possible teratological and reproductive effects.

Teratology is the study of the potential effects of the test compound during *in utero* development. Generally performed as separate tests in the rat, mouse, hamster and rabbit, teratological potential may also be determined as an adjunct to multigeneration reproduction studies. A key issue in teratology testing and evaluation has been differentiating direct teratogenic effects on the fetus from those effects resulting secondarily from intoxication of the mother. The indices unique to a teratology study are shown in Table 4.

Table 4. Teratological Indices

Body Weights Over Test Period
Resorptions
Toxic Response Data
Time of Death
Pregnancy and Litter Data
Organ and Soft Tissue Abnormalities
Skeletal Abnormalities

Reproduction studies are generally done with rats and mice over two generations. Table 5 presents the key observations made in such tests.

Table 5. Reproductive Indices

Fertility Indices
Length of Gestation
Litter Size
Toxic Effects/Survival
Body Weight Data
Necropsy Findings
Histopathology

Chronic studies are used to determine the adverse effects of regular exposure to substances over periods of at least 12 months. Their purpose is to characterize the test material's chronic toxicity and to define the dose at which no adverse effects are observed. Rats and dogs are the primary species of choice for this test.

This brings us now to carcinogenicity tests the toughest part of the safety evaluation and regulatory science conundrum because of several test characteristics including those shown in Table 6.

Twenty-four month carcinogenicity studies are called for in both rats and mice. This length of treatment covers the greatest portion (90+ %) of the animal's lifespan. The objectives of both carcinogenicity and chronic testing can be achieved in a single

Table 6. Carcinogenicity Bioassays

a) High Dollar Cost
b) High Time Cost
c) Propensity for Equivocal Results
d) Questionable Relevance to Man
e) Statutory (Delaney) Zero Risk

study by accommodating the needs of both in the study design. Such a combined study will however require more animals than would either one singly.

A relatively recent review of carcinogenicity testing issues appears in the report of the "NTP Ad Hoc Panel on Carcinogenicity Testing and Evaluation" (*2*). This is recommended reading for one interested in thorough discussions on issues relating to carcinogenicity bioassays.

Subsequent to recommendations in this report, the NTP program has been undergoing changes to bring increased depth of science and understanding of mechanisms into the performance and interpretation its carcinogenicity bioassays. Nevertheless, the rodent chronic bioassay is at best a very crude tool for estimating cancer risk for man.

In 1981 Dr. Robert Squire (*3*), former director of the NCI Bioassay Program, had called for the weight of evidence to include the considerations given in Table 7.

Table 7. Carcinogenicity Weight of Evidence Indices

- Chemical similarity to other known toxicants
- Binding to dna, rna, protein
- Genotoxicity
- Metabolic and pharmacokinetic data
- Physiological, pharmacological and biochemical properties
- Number of species effected
- Number of tissue sites effected
- Latency periods
- Dose response relationships
- Nature and severity of lesions induced

As NTP gathered more of its own data on non-carcinogens and reviewed the control data, its position has also shifted from the early 1980's to include greater use of weight of evidence in their evaluations. Indeed in a recent 1990 article (*4*) Dr. Haseman of the NIEHS states "In particular, a decision rule that routinely labels a carcinogen whenever a single tumor increase is significant at the 5% level for any exposed group can result in a false positive rate considerably greater than the nominal 5%." He went on the say, "also, statistical decision rules should not be employed as a substitute for sound scientific judgement in the overall evaluation of these experiments."

The most provocative and perhaps significant new test to appear in the past two decades was the salmonella test for mutagenicity introduced by Dr. Bruce Ames in the early 1970's. For a while some people thought a "litmus test" had been found to replace the lengthy and costly carcinogenicity bioassays.

In the mid-1970's the FDA commissioned a FASEB Special Committee on Flavor Evaluation Criteria (SCOFEC) whose report recommended that all flavors should be tested with the Ames test. Those flavors testing positive were then to be tested further via the full gauntlet of toxicity tests then required for food additives, including carcinogenicicity testing. Estimates of this dollar cost to the food and flavor industry ran as high as three hundred million dollars. This recommended course of action was fortunately not pursued by FDA. Basically the reliability and relevance of the Ames test to human safety had not yet been adequately established.

Since that time, as more short term mutagenicity tests were developed, their reliability and relevance continued to be evaluated. In 1987 Tennant et al reported (*5*) that evaluation of results for 73 chemicals tested in four short-term tests (STT) for genetic toxicity and in two rodent bioassays for carcinogenicity demonstrated only about 60% qualitative agreement (i.e. concordance) between STT and bioassay results. The authors concluded that "no single in-vitro STT adequately anticipates the diverse mechanisms of carcinogenisis; and, more important, the advantage of a battery of in-vitro STTs is not supported by results of the present study." It was emphasized, however, that it would be prudent not to dismiss their importance for detecting genotoxic chemicals because of health concerns aside from cancer.

Brockman (*6*) pointed out that it is curious that the equally low concordance (67%) between rat and mouse carcinogenicity bioassay results for these 73 chemicals has not received equal attention. It is also true that a carcinogenicity assay in one rodent species does not adequately anticipate the diverse mechanisms of carcinogenisis in the other rodent species.

Since my topic is trends, I must point out that previous studies had shown concordance values as high as 90% for STTs and rodent bioassays, and as high as 85% for the two rodent bioassays. Indeed Tennant et al (*5*) point out, that the 73 NTP chemicals and their 60% incidence of carcinogenicity are probably not representative of the universe of chemicals but rather reflect the present chemical selection process for the NTP carcinogenicity assay.

Consider now the potential adverse impact of having had to test the 40% false Ames test positive flavors in chronic rodent bioassays which in turn yield equivocal results of questionable relevance to product safety for human use.

It is now felt that a more appropriate approach to prioritization and safety review of flavors is that proposed by the Flavor Extract Manufacturers Association (FEMA) (*see* Chapter 16, by Dr. Otho Easterday).

I'm sure the chapter by Dr. Gary Williams, will provide a more up-to-date assessment of the value and use of genetox tests. Research continues towards developing methods for effectively evaluating potential for additional toxicological endpoints including immunotoxicity, neurotoxicity and behavioral toxicity. Such methodology has not yet become a routine part of requried toxicity testing programs.

While the existence of many natural toxicants had been documented earlier (*7*), it was indeed interesting to read the 1983 science article (*8*) by Dr. Bruce Ames in which he described the presence of a great variety of natural mutagens and carcinogens, as well as anti-mutagens and anti-carcinogens in the human diet. Dr. Ames concluded that "characterizing and optimizing defense systems (e.g. natural anti-oxidants) may be an important part of a strategy of minimizing cancer and other age-related diseases". This is certainly a point to be considered extremely important as we consider whole foods.

This brings us to our future challenges regarding food safety. As prefaced by the considerations just described for flavors, attention is now turning to other natural components, chemicals introduced by processing (e.g. heat) and to foods themselves. Development of new foods via biotechnology is adding significant impetus for this new focus on natural food components.

Dr. Scheuplein's paper (Scheuplein, R., FDA, May, 1990 Conference on New Food and Food Chemicals at the National Academy of Sciences) is a fine treatise on this issue. Please keep in mind that most of the food related toxiciy issues faced since 1958, other than microbiological, have been concerned with single characterizable chemicals. We must now address mixtures of unknown chemicals eaten in large quantities (> 1% of diet) as we deal with foods themselves. Caution, yet due diligence, is needed in approaching safety testing and evaluation of complex matrices we know as foods.

In anticipation of challenges to the food safety sciences by new food biotechnologies a consortium of food and biotechnology companies assembled a group of expert scientists that undertook the development of a guideline document in which scientific criteria are recommended for use in evaluating new foods developed via biotechnology. This International Food Biotechnology Council's report on "Biotechnologies and Food: Assuring the Safety of Foods Produced by Genetic Modification" (*9*) has received a very broad international peer review, and was published in December 1990 as a special issue of Regulatory Toxicology and Pharmacology (RT&P). The executive summary of the IFBC document has been published as a separate article in the August, 1990, issue of RT&P.

I had the privilege to be one of the authors of the IFBC document and with the full endorsement of IFBC I now share with you the key findings and recommendations relative to product safety evaluation contained in chapters 5 & 6. By so doing, I do believe we'll be getting a close look at the future food safety issues and activities.

Before continuing however, I'd like to remind you again the tremendous issues raised by any attempt at straight forward application to foods of all the traditional toxicological approaches that have been applied to food additives.

Remember that our present safe and wholesome foods, variable mixtures from over 200,000 components, have never been thoroughly analyzed nor seen a laboratory animal toxicity test. We must rely heavily on the breeding and agricultural practices that have so successfully produced our wholesome and abundant foods and then integrate with them any new safety assessment procedures found necessary.

IFBC recommends that:

1. A decision tree approach be used for assessing the safety of whole foods and food components.

In evaluating genetically modified products via the decision tree it will be noted that their safety evaluation is geared principally to an evaluation of their inherent constituents as a means of ensuring the safety of the whole food as consumed.

Accordingly, the decision tree does not include a formal requirement for safety/biological testing of the final product (not unlike the introduction of new cultivars by traditional plant breeding practices). Nevertheless, a prudent manufacturer, who has the ultimate rersponsibility for product safety may, depending on the particular product being dealt with, undertake some degree of testing of the final product in animals and/or humans prior to placing the product on the market. Whenever such testing is considered, the specific approach, type and methods of test must be very carefully customized to the particular product keeping in mind the rationale of this overall document.

It makes the point also that following any appropriate, and if necessary, animal tests and setting a tentative exposure level, human volunteer studies to test for human tolerance should be designed. Following simple organoleptic evaluation, the first human study should involve the feeding of a single meal containing the macroingredient at a known dose level to one volunteer at a time. If no harmful effects are observed with several volunteers, studies involving the feeding of the novel food for a short period (initially about four weeks with follow-up studies of longer duration) should be performed.

Of course the less experience, knowledge and data one has about a new product's origins and composition the greater the effort needed to reduce the level of concern to an equivalent of that of the competing traditional product.

2. The safety evaluation of single chemical entities and simple chemically-defined mixtures used at low levels continue to be based on the conventional toxicology and safety evaluation practices presently being used.

It goes on to point out that complexity arises because some simple substances, such as sucrose and high-fructose corn syrup, are used at high levels in food, and therefore encounter many of the same safety evaluation problems as foods and complex mixtures. Conversely, there are many complex mixtures, such as spices, essential oils, or papain that are only used at low levels. The safety evaluation of such food components becomes a blend of the problems and opportunities that accompany traditional natural foods, and those that are associated with single ingredients used at low levels.

3. The initial basis of the safety evaluation of a genetically modified food should begin with consideration of the lineage of all genetic materials present in the final food product.

4. The principal feature of the safety evaluation of genetically modified food products should be a comparison of the composition of the new product with its traditional counterpart in regard to the levels of inherent constituents.

5. A food product should be considered to present no safety concern if analytical studies indicate that the concentration of inherent constituents does not differ significantly from the concentration range typical of the traditional food, and any new constituent(s), if present, is already accepted for use in food under the anticipated conditions of use.

6. Procedures for safety evaluation of whole foods and other complex mixtures should be closely linked to existing agricultural and food processing practices as well as to the regulatory status of comparable traditional foods and ingredients.

7. A food product should be considered to present no safety concern if use of the food would not be expected to alter significantly the present intake of it or its constituents in comparison with the traditional product, and the proposed conditions of use of the new product would not reasonably be expected to lead to such an intake of the food that the total intake of any constituent would exceed the amount acceptable under the standard of safety appropriate for that constituent.

8. Further safety evaluation of a food product should be required if:

(a) analytical studies demonstrate a significant change in the levels of inherent constituents of the food or;

(b) the new constituent(s) is not an accepted food ingredient and its safety under conditions of use requires further evaluation.

9. The standard for a significant nutrient (one that food supplies, in the average diet, 10% or more of the dietary need) should be the mean value reported in the literature plus or minus 20%. If the food is not extensively pooled, as for example potatoes, IFBC recommends that the standard should be the mean reported in the literature plus or minus 2 standard deviations or 75% of the reported range, where a standard deviation is not available.

If a nutrient in a food supplies less than 5% of the average dietary need, the nutrient may be considered non-significant for the purpose of this evaluation. The range from non-significant (less than 5%) to significant (more than 10%) is a judgmental area.

Depending on the nature and intended uses of macroingredients, studies in animals may be needed to supplement the chemical studies. It must be recognized that if animal studies are employed in the safety evaluation of whole foods and macroingredients that the traditional 100-fold safety factor approach to establishing acceptable human exposures will have limited validity. Indeed, one should not apply any biological test, or analytical test for that matter, to a new food product until it has been shown to work effectively with the traditional counterpart product.

Challenges facing us in any attempt to implement according to these IFBC recommendations include:

Food Composition. Relative to the task ahead, our knowledge and data base of composition of foods is grossly deficient to non-existent. In addition, the very fact that natural foods demonstrate extreme variabilities in composition due to genetic and environmental stress factors many analyses are necessary to obtain representative data.

Methods of Food Analysis. While a wide variety of sophisticated analytical chemistry techniques are available, they generally have not been adapted and applied to food systems. Also, there are very few chemists available or being trained in this field.

Animal Feeding Studies. As customarily performed and evaluated toxicology tests are inappropriate and will require significant revision before they can play an effective part in evaluating the safety of whole foods.

Human Clinical Testing. while the need for human trials with new foods is becoming more and more clear, there is yet no rubric under which such studies should be performed. Indeed, this subject is getting increased attention (*see* Chapter 11, by Dr. Glinsmann).

In essence we're faced with building a scientific bridge between the procedures used in the past for developing and introducing new cultivars of food plants and those becoming available as a result of incorporating recent biotechnological procedures in one step of the overall process.

There's a similarity between this bridge we need and the Tappan Zee bridge across the Hudson River at Tarrytown, New York. While the West end of the bridge rests on bed-rock, the East end rests on a multi-story concrete and steel caisson floating on the riverbottom. In order to bridge our new biotech foods with the traditional plant breeding and agricultural system, greater amounts of "bed-rock scientific data" will be necessary the more novel and different the new product is from the traditional products for which assurance of safety rests on centuries of experience. Enough data must be collected to demonstrate that there is no more reason to be concerned about the safety and wholesomeness of the new product than there is for the traditional counterpart product.

In summary, while we see some maturation in the safety and regulatory assessments of single ingredients over the past several decades, we presently find that:

* **Testing Costs** - The numbers and varieties of toxicity tests continue to rise.

* **Carcinogenic bioassays** - These are not as sensitive and reliable for determining carcinogenic potential for man as once believed.

* **Genetox Tests** - These tests, while useful for research quidance, by themselves are not reliable predictors for carcinogenicity potentials in mammals.

* **The 30 day and 90 day Feeding Studies** - These will become the workhorse toxicity tests for food materials in the next decade.

* **Metabolism and Pharmacokinetic Data** - Such data for single chemicals is increasing rapidly in importance.

* **Human Clinical Studies** - Use of human clinical studies will become an ever increasing part of the evaluation of new foods and their ingredients.

* **Immunotoxicity, Neurotoxicity, and Behavioral Toxicity** - Research continues towards development of appropriate tests in these areas.

* **Traditional Toxicity Tests** - The animal tests and approaches used traditionally by toxicologists are inadequate and often inappropriate for evaluating new food ingredients that are to be consumed at greater than 1% of the diet. While existing toxicological methodologies are designed to determine the levels at which exposure to a material leads to adverse effects, evaluation of new whole foods must be a matter of assuring that the level of concern of the new product is no more than its traditional counterpart with which it will compete.

There is really no reason to be more concerned about the safety of foods derived via use of biotechnology than there has been with traditionally derived foods. Nevertheless, the advent of biotechnology has brought a new focus onto the entire food system. Addressing the issues prompted by biotechnology focuses on the fact that much needs to be done for developement of methods for comparative food composition, nutrition, chemical and microbiological safety of foods in general.

Literature Cited

1. *Toxicity Testing - Strategies to Determine Needs and Priorities*, National Academy of Sciences, **1984**.
2. *Ad Hoc Panel on Carcinogenicity Testing and Evaluation*, National Toxicology Program Board of Scientific Counselors, **1984**.
3. Squire, R. *The Pesticide Chemist and Modern Toxicology*, 1981.
4. Haseman, J. K. *Fundam. Appl. Toxicol*, **1990**, *4*,637-648.
5. Tennant, R.W, Margolin B.H., Zeiger, E., Haseman J.K., Spalding, J., Caspary, W. Resnick, M, Stasiewicz, S., Anderson, B., and Minor, R., *Science* **1987**, *236*, pp. 933-941.
6. Brockman, H.E and D.M. DeMarini, Environ. *Mol. Mutagen.*, **1988**, *11*, 421-435.
7. *Toxicants occurring naturally in foods*, National Academy of Sciences, **1973**, 2nd Ed.
8. Ames, B.N., Magaw, R., and Gold, L. S. *Science*, **1987**, *236*, pp. 271- 280.
9. International Food Biotechnology Council, Regulatory Toxicology and Pharmacology. **1990** vol. 12, No. 3 Part 2 of 2 (S1-S196).

RECEIVED August 15, 1991

Chapter 10

Acute and Chronic Toxicity Testing in the Assessment of Food Additive Safety

David G. Hattan

Center for Food Safety and Applied Nutrition, U.S. Food and Drug Administration, 200 C Street, SW, Washington, DC 20204

This paper describes the history and present policy of the Food and Drug Administration regarding the use of the LD_{50} test procedure in support of the safety of food additives. Current issues affecting long-term testing such as the maximum tolerated dose and the survival of animals are discussed. The toxicological safety assessment of novel foods with low caloric density provides regulatory toxicologists with a special challenge.

In his classic paper (*1*), J.W. Trevan described the calculation and measurement of the acute toxicity of drugs, referred to subsequently as the LD_{50}, i.e., the single dose exposure of a test substance needed to produce lethality in 50% of the animals tested. Unfortunately, with the passage of time, the LD_{50} value became accepted as characterizing the acute toxicity of a substance. Actually, as will be discussed later, it is only an unrefined and extreme measure of the acute toxicity of a substance. Its very appearance of objectivity and discreetness led to its widespread utilization. Indeed, until recently, the LD_{50} was treated by some in the scientific community as possessing the characteristics of a biological constant (*2*).

The fact that the LD_{50} determination is not a constant value is well documented by a number of authors (*1-4*). Zbinden and Flury-Roversi (4) surveyed the acute toxicity literature for five substances to assess the consistency of results from several LD_{50} tests of these substances. The ratio of the largest dose to the smallest dose for LD_{50} determinations of these five substances in a number of tests varied from 3.7 to 11.3.

The significance of the LD_{50} value as a consistent indicator of acute toxicity has to be questioned when it can vary this widely. The reasons for this variability include differences in results of various studies with regard to species tested, age of animals, weight, sex, genetic influences, health and diet, degree of food deprivation, route of administration, ambient temperature, housing conditions, and season (*4*). In addition to these considerations, each test conducted to establish a value for an LD_{50} uses a large number of animals.

As described in the first edition (1982) and as anticipated in the revision of the FDA "Redbook" (now under way within the agency), it is not necessary to provide an exact LD_{50}. If the sponsor of a food additive wishes to provide the FDA with an LD_{50}, then an approximate value will suffice. The FDA has stated that the "classical" LD_{50} test is not an FDA-required procedure for determining safety and its use is not part of agency testing policy (*5*). There are several test procedures that have been developed recently to maximize information on acute toxicity without using large numbers of animals (*2,6*). Most of these newer study techniques emphasize studying each animal for additional information to assess the toxic effects on organ systems as well as to observe overall symptoms of toxicity and recovery (*6*).

Before deciding upon the actual doses to be used in testing a new substance, the intrinsic biological and chemical activity of the compound should be considered as well as factors such as chemical and physical characteristics, molecular weight, partition coefficient, and toxicity of related compounds. For example, the corrosiveness and local tissue toxicity of a highly acidic or caustic substance may bear little toxicological relevance to the degree of general systemic toxicity of the same substance (*7*).

If one is proposing to test a substance that appears to be relatively nontoxic, one might give a dose of 5 g (or mL)/kg body weight to a small number of animals (perhaps five) and monitor them closely for up to 14 days for toxicity and recovery. This period of observation could be followed by a pathological examination of the test animals to determine any organ system toxicity. If no test animals die at this dose, then the acute toxicity can be indicated as being "greater than 5 g/kg." This type of procedure to test first for low toxicity by administering a single, large dose is referred to as a "limit test." Practically speaking, this dose of 5 g or mL/kg is near the practical upper limit of what can be given by gavage in a single dose to a rodent.

Other types of dosing protocols are available, e.g., the so-called dose-probing design, in which one animal for each of three different widely spaced doses is tested. After a sufficient period of time (up to 14 days), one might decide based on these results whether and how to select other intermediate doses (*7*). Another type of acute toxicity design is often described as the "up-and-down" procedure. In this approach, one animal at a time is dosed; then another animal is dosed 1 or 2 days later with a different dose until an approximate LD_{50} is obtained (*6,8*). It is again appropriate to emphasize that a more complete examination of a few animals will provide more useful data than will a superficial examination of a large number of animals as in the traditional LD_{50} test.

Another general area of toxicity testing that has been the subject of intense and widespread discussion is that of long-term testing to assess the chronic toxicity and carcinogenic potential of food additives. Much of the recent debate has been concerned with the issues of "MTD" or "maximum tolerated dose" and animal survival.

The appropriate use of the MTD has been and continues to be one of the most intensely debated issues in toxicological testing. This is not surprising because the outcome of carcinogenicity testing and the ultimate fate of food additives are so dependent on the level of exposure (doses) of the test animals to the substance being tested. The so-called Delaney Clause of the Food, Drug, and Cosmetic Act prohibits a substance from being added to food if that substance produces cancer in either animals or man. On one hand, the dose to be tested should be relatively high to compensate for the inherent lack of sensitivity of the carcinogenicity bioassay, but on the other hand, the dose should not be so high as to be unrepresentative of the toxicity to be expected in humans at lower doses (*9*).

The MTD is determined (actually estimated) after a careful analysis of data from subchronic toxicity testing. Based on the National Cancer Institute guidelines of 1976, the definition of the MTD is as follows: the highest dose that causes no more than a 10% weight decrement, as compared to the appropriate control groups; and does not produce mortality, clinical signs of toxicity, or pathologic lesions (other than those that may be related to a neoplastic response) that would be predicted to shorten the animal's natural life span (*10*).

As time has passed and our experience with carcinogenicity testing has accumulated, it has become clear that a broader range of biological information is needed to select the MTD. For example, now it is possible to utilize data concerning changes in body and organ weight and clinically significant alterations in hematologic, urinary, and clinical chemistry measurements in combination with more definitive toxic, gross, or histopathologic endpoints to estimate the MTD (*11*).

Some of the advantages of using the MTD are (1) compensating for the insensitivity of the carcinogen bioassay, including the relatively small number of animals used for testing; (2) providing consistency with other models used in toxicology (high enough doses must be used in order to elicit evidence of presumed toxicity); and (3) allowing comparison of carcinogenic potencies of substances even when the data are collected from different studies (*9*).

Of course, there are also disadvantages to using the MTD concept. For instance, just using the word "maximum" allows some people, especially among the general public, to assume that these doses are impossibly high compared with those to which they as consumers are exposed. In reporting a carcinogenic response, it is rare that a full and balanced explanation is given of what the risks of exposure to a substance actually are in the context of the consumer's everyday experience. In addition, the definition of MTD is not consistent. Thus, it is possible for investigators looking at one set of data to conclude that their analysis shows that the MTD has been exceeded, whereas other investigators will conclude that the MTD has not been achieved. This disagreement might be over the interpretation of metabolism data or whether an organ alteration was adaptive or toxicologic. Finally, it is possible for a high dose that clearly exceeds the MTD to produce a carcinogenic response, while the next lower dose, which does not exceed the MTD, produces no carcinogenic response. Some investigators would claim that this substance should not be labeled an animal carcinogen (*9*).

McConnell (*9*) suggests that it is justifiable to use the MTD in carcinogenicity studies of substances for which there is little or no control over exposure, such as those in drinking water, food, air, or the work environment. Another method of deciding the dose might be to select a simple multiple of the human exposure, such as 1,000 or 10,000.

A second general topic that has attracted discussion with regard to its potential influence on the conduct of carcinogenicity bioassays has been the proportion of test animals surviving at the termination of chronic studies, also referred to as survival. Many toxicological guidelines have standards for valid negative carcinogenicity bioassays that require at least 50% survival of rats until 24 months of age (*12*).

This particular standard for valid negative carcinogenicity bioassays is included to help assure regulatory agencies that when a substance is tested for carcinogenicity, there have been a sufficient number of animals on test for a sufficient period of time for any tumorigenic potential to be adequately assessed. Until recently, there had been little or no indication that commonly used rat strains presented any problem of survival. Within the past year or so, however, industry (Burek, J. D., Merck, Sharp & Dohme Research Laboratories, West Point, PA, personal communication, 1990) and the National Toxicology Program (Rao, G. N., National Institute of Environmental Health Sciences, Research Triangle Park, NC, personal communication, 1990) are apparently having difficulty in reliably achieving a 50% level of survival at 24 months.

Industry is naturally concerned about this trend and is trying to monitor and solve this problem before it starts to impair their ability to conduct adequate chronic toxicity and carcinogenicity studies. The FDA and other regulatory agencies will be closely watching developments in this particular area of toxicological testing. It may be that if this is a definite trend across time rather than a short-term difficulty, serious consideration will have to be given to developing means of addressing this problem. One general suggestion is that animal breeders include adequate longevity as one of the desirable characteristics in selecting for their future generations of breeding colonies. At this point, breeders select mainly for fecundity and rapid growth in their breeding stocks.

The final topic for discussion is that of toxicity testing to support the safety of a class of substances referred to as "novel foods" or "low nutrient density foods." Subsequently in this paper, these materials will be referred to as NFs.

In the U.S. there is a keen interest in the development of public health policies and technological advances that will assist in the reduction of morbidity and mortality associated with ischemic heart disease (IHD). In 1985 coronary heart disease, cancer, and stroke resulted in 783,000, 330,000, and 210,000 deaths in the U.S., respectively, and resulted in costs of more than $100 billion in morbidity and hospital care (*13*).

One of the most useful approaches in lowering the toll from IHD would be to lower body weight and/or reduce the proportion of fat being eaten by the typical U.S. consumer. The U.S. Surgeon General's Report on "Nutrition and Health" recommends that the present proportion of fat in the diet (on average 37%) be reduced to no more than 30% and that the 25% of Americans who are now overweight could reduce their long-term health problems (obesity, diabetes mellitus, and atherosclerosis) by attaining their ideal body weights (*14*).

Given the persistence of dietary habits and the great difficulty in changing them, it may be that an entirely new approach will be required to produce the desired reductions in dietary fat and/or caloric intake. Toward this end, the research and development departments of international/multinational companies, including pharmaceutical and food and chemical companies, have been engaged in a search for food additives or food-like materials that may be used as complete or partial replacements for fats, oils, bulking materials (intended to act like flour or sugar), and artificial sweeteners in the foods we eat. These NFs are intended to fulfill all the technological requirements of fats, oils, bulking materials, and natural sweeteners as used in foods, but once in the gastrointestinal tract, they are neither digested nor absorbed. Such properties make these NFs essentially noncaloric.

Although NFs are interesting in and of themselves, they provide regulatory toxicologists with a special challenge to define exactly how they are to be tested to assure their safety as a common and chronic part of the American food supply. Some of the characteristics inherent to NFs complicate toxicological testing. For example, because NFs are food-like, they are consumed in food-like quantities. This means that the average consumer could, depending on how widely the material is spread throughout the food supply, consume substantial amounts each day.

The ingestion of relatively large quantities means that the usual dose exaggeration found in toxicological testing for food additives, which are used in much smaller amounts, will be unavailable for these substances. This is true because as the percentage of normal daily intake of a food additive reaches 1% or above, it is not possible to get dietary exaggeration of more than 10- to 20-fold. In many dietary feeding studies with other more "potent" food additives, it is not uncommon to have a 1000- or 10,000-fold exaggeration in the dietary mix used in a toxicological study, compared with the level of expected human exposure in the diet. With these latter "potent" food additives, it is entirely possible to apply safety factors of 100-fold, while with the food-like NFs, it is not possible.

This smaller margin of safety between the dose fed to animals and the level of expected human exposure means that the toxic endpoints in animal studies will have to be very clearly and unambiguously defined. In addition, it implies that even after the safety testing in animals is complete, additional careful studies with humans will have to be conducted to confirm that the human responds in much the same manner as the animal model used for testing.

Another potential complication of toxicological testing for these substances is that of obviating the possibility of nutritional interactions leading to seemingly substance-mediated toxicity. For example, if the new NF signifcantly reduced vitamin or mineral absorption, it might elicit adverse effects in the animal model simply because of this nutritional effect. Therefore, early studies with these materials will have to determine what, if any, nutritional effects these substances have and then modify the testing designs of the normal toxicological feeding studies to compensate for these nutritional effects.

As suggested above, these materials will probably pass through the gastrointestinal tract rather than be absorbed. This characteristic may require development of other data. For example, what might the influence of this substance be on the rate of gastrointestinal transit? Is there any laxation effect? If so, how much at what dose? Are there any adverse effects on disease states of the gastrointestinal tract, such as Crohn's disease or gastroenteritis? Are there any influences on the bioavailability of certain drugs, especially ones with a small therapeutic margin, such as the digitalis alkaloids? What about the long-term effect on human nutrition? Is there interference with the absorption of vitamins and/or minerals? If so, is there an effective way to counteract these effects, and if so, how?

In summary, although it is clear that novel foods with the capability of serving as low calorie food-like substitutes could provide a valuable contribution in altering and improving the macronutrient intake of those in the general population of the U.S. who need this assistance, it is also clear that there are major challenges in supplying answers to important animal and clinical toxicological questions. It is only after these difficult and, at times, unique issues have been satisfactorily resolved that these novel foods can be approved for addition to the U.S. food supply.

Literature Cited

1. Trevan, J. W. *Proc. R. Soc. London* **1927**, *101B*, 483-514.
2. Lorke, D. *Arch. Toxicol.* **1983**, *54*, 275-87.
3. Hunter, W. J.; Lingk, W.; Recht, R. *J. Assoc. Off. Anal. Chem.* **1979**, *62*, 864-73.
4. Zbinden, G.; Flury-Roversi, M. *Arch. Toxicol.* **1981**, *47*, 77-99.
5. Food and Drug Administration, "LD_{50} Test Policy," Fed. Regist. Oct. 11, 1988, 53, 39650-51.
6. Bruce, R. D. *Fundam. Appl. Toxicol.* **1985**, *5*, 151-57.
7. Gad, S.; Chengelis, C. P. *Acute Toxicity Testing Perspectives and Horizons*; The Telford Press: Caldwell, NJ, 1988.
8. Gad, S.; Smith, A.; Cramp, A.; Gavigan, F.; Derelanko, M. *Drug Chem. Toxicol.* **1984**, *7*, 423-34.
9. McConnell, E. E. *J. Am. Coll. Toxicol.* **1989**, *8*, 1115-20.
10. Sontag, J. M.; Page, N. P.; Saffiotti, U. *Guidelines for Carcinogen Bioassay in Small Rodents*, Department of Health and Human Services Publication (NIH 76-801), National Cancer Institute: Bethesda, MD, 1976.
11. Office of Science and Technology Policy, "Chemical Carcinogens; A Review of the Science and Its Associated Principles," Fed. Regist. March 14, 1985, 50, 10371-442.
12. *Toxicological Principles for the Safety Assessment of Direct Food Additives and Color Additives Used in Food,* Food and Drug Administration, Center for Food Safety and Applied Nutrition, Washington, DC, 1982.
13. Kinsella, J. E. *Nutr. Today* **1986**, *21*(6), 7.
14. The Surgeon General's Report on Nutrition and Health. Department of Health and Human Services Publication No. 88-50210, Washington, DC, 1988.

RECEIVED August 15, 1991

Chapter 11

Usefulness of Clinical Studies in Establishing Safety of Food Products

Walter H. Glinsmann

Center for Food Safety and Applied Nutrition, Division of Nutrition, U.S. Food and Drug Administration, 200 C Street, SW, Washington, DC 20204

The safety assessment of a food or food additive is a risk assessment of a cumulative food or additive use. It is limited by current scientific knowledge and often involves considerable judgment about science. The safety standard which is used to support regulatory actions varies according to product type, history of use, and regulatory category.

Traditionally, food additives are considered safe for food use when a large safety factor or margin of safety exists between the amount the consumer ingests and the amount that causes no adverse effects in the animal used to test for food additive safety. This approach may not be sufficient for estimating safety when there are high intended use levels, significant inter- and intra-species variation in physiological effects or metabolism of an additive, or when adverse health effects are different in animals and humans. Also, foods are increasingly being tested for their role in disease prevention or disease management. Animal or *in vitro* models often are insufficient for assessing validity of health claims or estimating health risk that might accompany special dietary or therapeutic uses of food products.

The thesis of this chapter is that recent developments in food uses and considerations about food safety contribute to an increased need for human testing to assure safety and suitability of new uses, and that several issues surrounding such testing involve complex ethical considerations and a novel mixture of current food and drug law. The agency is developing guidelines for clinical testing of food and color additives; however, these guidelines will pertain to only a limited segment of the concerns about clinical testing of new foods and food components.

Changing Food Use and Safety Concerns

Traditional food products have been primarily assessed for their safe use in the general population. They were developed to deliver adequate energy and nutrients and achieve functional or technical effects that produced improved products in terms of preservability, lack of contamination, improved organoleptic or aesthetic quality, and cost effectiveness. Many innovations in macro food components were made by processing of ingredients which had a long history of safe use and the resulting "new" food component was in fact a product that could be assessed predictably in terms of safety of use by estimating exposures to the component ingredients. A recent example of action on this type of product is the approval of the microparticulated protein product Simplesse for use as a fat substitute in frozen desserts. Protein quality was shown to be equivalent to that of the original ingredients and estimates of effects of increased exposure to protein and other components were easily assessed. Novel ingredients, such as artificial sweeteners with intense sweetness, are quantitatively minor components of foods and thus can be subject to safety assessments involving a large exposure safety factor based on traditional animal testing. Examples are the safety assessments supporting the approval of aspartame or acesulfam-K uses. This approach to product development and the ability to introduce a large exposure safety factor is applicable only when the amount of the new addition is small compared to the total diet.

Food products are now being developed to promote health and to prevent, mitigate, or manage disease, primarily chronic diseases and obesity, but also acute or subacute illness. A major focus of food product development is on the reduction of caloric density and on the reduction or addition of specific macro components such as saturated fatty acids and specific types of dietary fiber. In many cases, these products will have relatively high use levels as bulking agents, forms of dietary fiber, low intensity sweeteners, and fat substitutes. Many may be novel compounds without any history of use. In terms of safety assessments, animal models may be limited in their usefulness because of differences compared with humans in tolerance, metabolism, physiological responses, nutritional requirements, or susceptibility to the influence of how humans versus the animal model may consume food.

The development of new food products to promote health is in part supported by our greatly expanded knowledge of basic biological mechanisms involved in the development of diseases. Paradoxically, this expanded knowledge also leaves us with questions about the adequacy of the basic paradigms we have used to judge safety. As knowledge expands, so does the number and type of questions related to safety of use of food components or of the adventitious effects of manufacturing and preservation processes. These circumstances then reflect on the applicability of animal or *in vitro* models to predict safety. This places the FDA and others concerned with food safety assessments at a difficult juncture. Should we now develop and validate a new generation of animal and *in vitro* models to address a new generation of concerns about food safety? Alternatively, should we move in the direction of increased human testing? Although human testing would proceed in a graded fashion to minimize risk and would be performed with full informed consent, it raises questions of an ethical nature. Foods are mandated to be safe and until demonstrated to be safe, human testing is not easily justified as being appropriate.

I have drawn an oversimplified dichotomy to make a point, i.e., if we are to approve a number of new food additives for use for health promotion and disease prevention, we may need to consider some form of risk vs. benefit analysis in the approval process before we encourage expanded human clinical testing to assess safety of use. We also may need to consider greater use of restricted marketing and approval of manufacturing processes. The changing perceptions about how foods or food additives should be developed for health promotion and disease prevention may require a fundamental rethinking of the safety standards for approvals including dichotomous standards for marketing of foods for special dietary use as opposed to foods that are marketed for the general population for taste, aroma, and general nutritional content.

Finally, in the matter of law, it is important to note that, during the past year, Congress has passed the Nutrition Labeling and Education Act (NLEA). This act, which provides for mandatory uniform nutrition labeling, also alters our considerations andpotential need for human clinical testing in that it provides for health claims on food for the general population; it requires consideration of health claims for dietary supplements which are in a category of Foods for Special Dietary Use; and it defines "medical foods" in food law. The NLEA thus provides some additional stimulus to consider that efficacy to satisfy label claims as well as safety testing for foods and additives will need to be considered in the near future and that appropriate labeling will be defined to assure safety of use as well as potentially beneficial effects on health.

The Food–Drug Spectrum

Food safety evaluations are constrained by food law and regulations, by various policy decisions, and by previous regulatory actions and legal decisions. In this regard, it is important to recall that the definition of food is limited to articles used for food or drink for man or other animals, chewing gum, and articles used for components of any other such article. Foods do not have any particular beneficial health use associated with their definition; they are not approved on a benefit vs. risk paradigm; and there are no adequate directions for use associated with their labeling. Drugs, on the other hand, are articles intended for use in the diagnosis, cure, mitigation, treatment, or prevention of disease in man or other animals and articles intended to affect the structure or any function of the body of man or other animals. While it is possible to have a product that is both a food and a drug, the agency has not, as of now, defined a category of a food-drug hybrid. It is of interest that in the drug-cosmetic spectrum, the agency has taken such action in that tooth paste (a cosmetic) containing fluoride to prevent dental caries (a drug claim) is both a cosmetic and a drug and is regulated under both cosmetic and drug law. The NLEA now defines categories of foods eligible for health claims that may make the consideration of such a hybrid designation a viable consideration, namely by defining a medical food category as distinct from other foods for special dietary use.

In the past, when food law had been inadequate to assure appropriate marketing of a product which had beneficial health effects, such as total parenteral nutrition solutions or injectable vitamin products, such products were simply classified and regulated as drugs. Clinical testing for these products has been conducted under investigational new drug procedures. The area of "Foods for Special Dietary Use", which is defined as a food category, also could be logically thought of as an area to develop regulations for clinical suitability and safety testing under food law. Paradoxically this area was largely defined at a time when Congress wanted to limit the agency's ability to regulate vitamin and mineral supplements as drug products; hence, additional regulatory actions which would require clinical testing to document suitability of products for special dietary use were impractical. No action has been taken to implement clinical testing for foods for special dietary use with the exception of infant formulas.

Foods for special dietary use is a category that is principally defined by Section 411 of the Federal Food, Drug, and Cosmetic Act (FD&C Act) — often referred to as the Proxmire Amendment of 1976 — and Part 105 of Title 21, Code of Federal Regulations (21 CFR 105). Special dietary use is distinguished from general food use by (i) supplying particular dietary needs which exist by reason of physical, physiological, pathological, or other conditions, including but not limited to conditions of diseases, convalescence, pregnancy, lactation, allergic hypersensitivity to food, underweight, and overweight; (ii) supplying particular dietary needs which exist by reason of age, including but not limited to infancy and childhood; and (iii) supplementing or fortifying the ordinary or usual diet with any vitamin, mineral, or other dietary property. The Proxmire Amendment gave vitamin and mineral supplementation products special status and limited the FDA from establishing maximum limits on the potency of any vitamin or mineral dietary supplement unless for reasons of safety or from classifying a vitamin or mineral as a drug solely because it exceeds the level of potency which is nutritionally rational or useful. This amendment raises uncertainties about how to approach the area of health claims for food supplement products which are based on nutritional rationality. It effectively has prevented the implementation of a requirement for efficacy and safety testing data for the marketing of various combination nutritional supplement products providing claims for efficacy are made only in advertising (under the jurisdiction of the Federal Trade Commission) and not in labeling (regulated by the Food and Drug Administration). In this area, it is the agency's responsibility to prove that a given use of a vitamin or mineral supplement is unsafe before taking any adverse action against products that are properly labeled. The agency relies heavily on epidemiological and survey data to monitor this area.

A category of special dietary use foods that now relies on limited human clinical testing is infant formulas. This includes "exempt" infant formulas, which are formulas that are specially formulated to meet the needs of infants with inborn errors of metabolism or special dietary needs. These products now have independent regulatory status defined by the 1986 amendments to the Infant Formula Act of 1980. Regulations governing the safety, nutritional adequacy, and labeling of these products have been developed. Some formulas are exempted from the compositional requirements of standard infant formulas because they are specially formulated and labeled for use by infants who have specific metabolic needs because of inborn errors of metabolism, low birth weight, or an otherwise unusual medical or dietary problem. The assessment of safety of these formulas relies on clinical testing to assure their suitability (efficacy) for use. In general, clinical testing verifies that the product will support growth and development and will mitigate the obvious effects of inherited diseases. Application of exempt infant formula regulations to the approval of products for managing inherited metabolic diseases other than those expressed in infancy is not appropriate under the Infant Formula Act; such products are currently considered as medical foods.

Medical foods constitute a conceptual category of food use that has changed over time. A brief understanding of the evolution of this category may help in understanding how it may be classified and regulated in the future with regard to requirements for human clinical testing. In 1972, the concept of "medical foods" was developed to efficiently bring to market specialized formulas to manage infants with inborn errors of metabolism (e.g., low phenylalanine formulas for infants with phenylketonuria). These products, considered to be drugs before 1972, were in fact "orphan" products in that they were developed for the nutritional management of diseases that occurred very rarely. Thus, there would normally be no economic incentive to manufacture these products. Since early products in this category simply depended on minor nutritional modifications (e.g., deletion of a single amino acid) to make them safe and for target populations, full review for suitability and safety under investigational new drug requirements was not deemed necessary. Furthermore, it was assumed that recipients were being closely monitored by physicians. While these products subsequently became regulated as exempt infant formulas, the need remained for a medical food category to address the many newly formulated nutritional products that were being developed for disease management not limited to inborn errors of metabolism expressed early in development.

The Orphan Drug Act of 1982 subsequently defined a special procedure and incentive for developing orphan products, and in 1988 this Act was amended to include medical foods. By that Act, "The term 'medical food' was defined as food which is formulated to be consumed or administered enterally under the supervision of a physician and which is intended for the specific dietary management of a disease or condition for which distinctive nutritional requirements based on recognized scientific principles are established by medical evaluation."

Since the concept of medical foods originated, many products have been developed which have greatly expanded the applications for their uses. Most of these products have not been clearly distinguished as medical foods because, as yet, no regulations have been promulgated for this category of products. Medical foods have not been clearly delineated from other foods for special dietary use. Their unique features and requirements for clinical testing have not been defined by regulation or agency review. However, many are referenced in information targeted to physicians, e.g., they are listed and described with directions for disease related uses in medical references such as the Physician's Desk Reference.

In recent years, major expansion has occurred in the types of prevention and therapeutic claims made for "medical foods". At the same time there has been a major shift in the way in which the general food supply is perceived and marketed. Increasingly, we are seeing health messages on foods for the general population which suggest a role of food or food components in health promotion and disease prevention. Thus, lack of clarity as to differences between medical foods and foods for the general food supply with diet and disease prevention claims has created much confusion in the marketplace. Clinical testing to support claims or safety of use has not been defined.

The need to define medical foods as a separate product class was brought into sharp focus with the passage of the NLEA. The NLEA requires significant changes in nutrition labeling for the general food supply and also for special dietary use foods.

Furthermore, the NLEA contains a directive that the Agency implement procedures for allowing health claims on foods within the general food supply; specifies that food supplements, subject to Section 411 may be considered for health claims using a separate standard than that used to assess health claims for foods in the general food supply; incorporates the orphan products definition of medical foods into food law; and specifically exempts medical food from nutrition labeling. Claims may still be appropriate for medical foods but according to standards appropriate for a food-drug hybrid.

In this regard, it could be argued that foods for special dietary use and medical foods are defined by their distinct characteristic of use and that a different standard for documentation, clinical testing, and labeling may be appropriate. Special dietary use foods which are dietary supplements of vitamins, minerals, herbs, or other similar nutritional substances may contain a health claim on the label that would be targeted toward the general population with appropriate language. They are used to provide segments of the general population with a positive benefit within the context of a total dietary pattern. In contrast, medical foods are intended to be used in the specific dietary management of a disease or medical condition under the care of physician. They may require complex directions for use. They may not be safe for general food use and they are excluded from general health claim labeling.

The criteria and procedures for regulating health claims for general foods are now being developed to implement the NLEA. As a companion activity, it would be useful to establish criteria and procedures for clinical testing for allowing appropriate claims for medical foods. As a step in this process, the FDA has recently sponsored an analysis of "Guidelines for the Scientific Review of Enteral Food Products for Special Medical Purposes." Currently, a review is in progress to distinguish general health claims for generally available food products and special dietary use foods from treatment claims for medical foods, and to define approaches for developing high quality products for disease management. Many medical food products are still orphan products and there is considerable Congressional interest in removing obstacles to their development and marketing.

Part of the nutrition labeling package the Agency is working on includes a proposal for "Dietary Supplements of Vitamins and Minerals." Specific regulations on the labeling of these products have not been developed since the 1976 amendments and this proposal will go a long way in providing for uniform labeling of these products. The requirement for clinical testing to establish labeling for suitability and safety of use remains to be defined.

Food Safety Assumptions

With regard to safety testing paradigms that are routine and form the backbone of approvals for new food product uses, we , as noted earlier, often rely heavily on animal and *in vitro* data. We make certain assumptions which allow us to consider them as appropriate surrogates for predicting adverse endpoints in humans. We assume that qualitatively similar events occur in animals and humans and that a safety factor will account for differences in overall species sensitivity to the food additives. As we learn more about species and individual variation, we need to validate these assumptions about comparative metabolism with human clinical testing. In addition to validating animal toxicological testing models, human clinical testing may be the only practical way to approach newer issues in safety assessments. When we begin to consider new endpoints for our safety evaluations that focus for example on neurobehavioral or immunological responses or factors that may alter the expression of chronic diseases, we are faced with the fact that appropriate animal models by and large have not been developed. There also is a paucity of information on human clinical assessments that predict disease risk, multisystem endpoints, food sensitivity, and complex interactions between nutrients and drugs. Perhaps in some of these cases, the only valid approach will be to develop some form of post-marketing surveillance of diet and health relationships and to create a database on clinical measures which predict future health risk. The incorporation of these measures into nationally representative surveys of diet and health could then be an ancillary measure to assess long term effects of food products and diet patterns on health outcomes.

The assumption that food safety testing is adequate when targeted toward the normal healthy population is also being challenged. Perhaps most important in this regard is that we are having increasing difficulty in defining normal and healthy. As our population ages and our knowledge of genetics and predisposing factors to disease risk increases, we view our population as being very heterogeneous in terms of risk to adverse health effects from food components. Also, as noted, the complexity of endpoints for safety assessments is increasing both in the perceptions of the scientific community and in public debate. It becomes increasingly difficult and costly to anticipate safety outcomes and grant new product approvals. In this regard, it may in the future be prudent to change our tack and to consider that more questions about safety can be asked than answered. Global and highly speculative concerns with the validity of current safety testing will simply lead to a paralysis in new food product approvals and can lead to an erosion in consumer confidence that current approved food uses are safe. This situation may be more effectively managed by closer post-marketing surveillance or limited marketing approvals, expanded labeling requirements, or other changes in the way we consider and regulate our food supply with regard to assuring safety of food product use. Clinical testing could become important in a post-marketing mode — to assess new safety concerns raised by increased scientific knowledge; to evaluate impacts of changes in food use which alters exposures; and to investigate subpopulations in which there is preliminary evidence that they have unanticipated adverse health effects associated with a food product use.

Clinical testing prior to marketing also may change in its character. The traditional "gold standard" has been well-controlled clinical trials focused on endpoints that are clearly interpreted in terms of health risk. Such trials become more difficult or impossible to perform as the complexity of endpoints (e.g., immunological or neurobehavioral outcomes) increases and when there are no validated surrogate measures for predicting outcome. In this case, we may anticipate potential populations at risk (e.g., to coronary heart disease, stroke, or cancer) and consider multicenter trials with endpoints more referable to long term food use. An example could be the safety assessment of new uses of fats and oils which modify thrombogenesis, atherosclerosis, and blood clotting. Clearly, some populations might benefit from food products that diminish clotting and vascular wall reactivity. Others would be at increased risk. In this case, well controlled trials could look at efficacy for an acute health benefit but they may be inadequate to predict long term benefit vs. risk. Multicenter trials of populations at risk may be more appropriate to assess competing benefits and risks of new food product uses.

In discussing food safety assumptions and strategies that may be used to assure appropriate new food uses, I have assumed that we need to take a fresh look at food product approvals using largely food law which has some expanded authorities that are generally associated with drug law and that general health related claims for foods will continue to be based on nutritional considerations. In this regard, pharmacological effects of nutrients for disease management would continue to be drug uses. We could, as an alternative, consider new products being developed for health-related effects as OTC (over-the-counter) drug uses. It is my sense that such a view would be out of touch with the intended thrust of the NLEA in that it would not optimally facilitate innovation in new food product development.

Guidelines for Clinical Testing

Draft guidelines have been prepared for a revised "Redbook" (Toxicological Principles for the Safety Assessment of Direct Food Additives and Color Additives Used in Food). As with any human testing, a particularly important consideration is the protection of human subjects detailed on 21 Code of Federal Regulations, Part 50. Protocols for clinical testing should first of all follow the general guidelines of having clearly stated objectives, appropriate controls, defined methodologies with regard to interpretation of endpoints, and mechanisms for quality control. Limitations of the study should be clearly stated and appropriate statistical analysis should define the power of the study to detect its endpoint.

Minimization of risk can be augmented by developing a logical testing sequence with earlier testing focused on short term exposure and normal subjects. Often such studies would be used to define product acceptance and tolerance. Analysis of product metabolism or effects on metabolic or physiological processes may be an early focus to validate the appropriateness of animal studies and to aid in predicting possible adverse reactions. As testing progresses, studies would involve more extended exposures; the appropriateness of study duration and amount of product being tested would be judged by a review of all available scientific data. At this stage, it is important to have a firm idea of specific product formulations that are intended for marketing and projected population exposures because these studies may be critical to the approval process and must use a representative product and level of use to be judged valid for a safety assessment. Further clinical studies can then move on to longer term exposures that may focus on estimating endpoints for potential adverse reactions. Such endpoints could relate to nutritional interferences, food product–drug interactions, food intake, or specific metabolic features of the product under consideration. Finally, after a firm safety data base is established in normal populations, it may be necessary because of product characteristics or intended use levels to move into testing of populations with increased risk to potential adverse effects. Target populations could be those with altered health status, special nutritional or metabolic requirements, or high anticipated levels of exposure. In these studies, the data generated may also help define how a product should be labeled and marketed.

In addition to providing guidance for clinical testing of new food and color additives, the Redbook guidelines can also be used for other food safety testing. In this regard, a cogent case can be made for continuing safety assessments of a number of food products currently marketed as substances either generally recognized as safe or approved as food additives when major changes occur in exposure because usage has increased. The relatively recent trend to consider foods as beneficial for health promotion and disease prevention may result in changes in macrocomponent composition of diets that are not easily assessed in traditional animal models.

The general area of food for special dietary use has also seen a rapid development of many products, including foods for therapeutic use under the guidance of physicians. The safety of such uses is not routinely reviewed by the FDA. The efficacy of some of these medical food uses are reviewed by expert panels. Some medical foods contain novel ingredients or uses which have not been approved. In this area, clinical testing is more appropriately considered as satisfying investigational new drug procedures and rather than compliance with general guidelines for food additive safety testing.

Comment

There is a pressing need to provide label information on the potential health benefits of specific foods in our diet—a need augmented by the passage of the NLEA. Initially, claims in this area will be supported by scientific consensus, relying on data from a wide variety of predominantly clinical and epidemiological studies. A process will be established to petition the agency for additional claims. In a number of cases, there will be increased pressure to allow more specific claims for special dietary uses that are not easily approved for general population use. In this regard, we still need to define appropriated clinical testing for nutritional products for the medical management of disease or for use by select subpopulations with adequate directions for use. Guidelines for clinical testing of food or color additives will be useful for certain new product approvals and for testing safety of products already marketed when a new question of safety arises, or to verify that animal models used in safety assessments are adequate predictors of human health impacts. These guidelines are not complete with regard to issues raised by testing therapeutic claims for food products or assessing suitability and safety for novel products or uses that may not be safe for the general population. A large potential area for clinical testing is in the gray area between foods and drugs. This area requires further definition. Such definition will only come when clear regulatory categories are defined for various product uses in this area and regulations for marketing are established.

RECEIVED October 24, 1991

Chapter 12

Good Laboratory Practice Regulations

The Need for Compliance

W. M. Busey and P. Runge

Experimental Pathology Laboratories, Inc., P.O. Box 474,
Herndon, VA 22070

Historical Perspective - A Brief Overview

Through the Federal Food, Drug, and Cosmetic Act, the government has placed the responsibility of establishing the safety of regulated products with the sponsors of those products, and has made the Food and Drug Administration responsible for reviewing their efforts and determining if, in fact, safety has been established.

Prior to the promulgation of the Good Laboratory Practice (GLP) Regulations, the agency operated on the premise that reports submitted in support of these products were accurate representations of study conduct and results. However, inspection of several studies conducted for a major pharmaceutical manufacturer revealed unacceptable or questionable laboratory practices and inconsistencies in data (*1*). Major concerns included:

1. Poorly designed and conducted experiments, inaccurately analyzed or reported.
2. Technical personnel unaware of importance of adherence to protocol requirements, accurate administration of test materials, accurate recordkeeping.
3. Management did not assure critical review of data, proper supervision of personnel.
4. Studies impaired by protocol designs that did not allow for evaluation of all data.
5. Inadequate assurance of scientific qualifications and training of study personnel.
6. Disregard for proper laboratory, animal care, and data management procedures.
7. Failure of sponsors to adequately monitor studies/procedures conducted by contract laboratory facilities.
8. Failure to verify accuracy and completeness of reports prior to submission to FDA.

Further investigations of the laboratory facilities involved resulted in eventual convictions on criminal charges of fraud and a great concern for the validity of studies

0097–6156/92/0484–0114$06.00/0

already completed. The need for better control of nonclinical laboratory studies was soon recognized by both government and industry.

In response to this need, the FDA pursued numerous possible approaches to more controlled and consistent study conduct. After much investigation, review, consideration, and comment the FDA Good Laboratory Practice Regulations were eventually codified as Part 58, Chapter 21 of the Code of Federal Regulations. All regulated human and veterinary drugs and devices and food/color additives must comply with their directives.

Purpose and Content of the Good Laboratory Practice Regulations

The primary purpose of the GLP Regulations is to assure the quality and integrity of the studies submitted in support of the safety of regulated products. In order to accomplish this, many "common sense" requirements were incorporated into the regulations. These included provisions for criteria of study design, i.e., protocols and standard procedures; facility and equipment considerations; identification and control of test materials; recordkeeping; and reporting. These requirements reflect the demand for basic experimental structure necessary for conducting any high quality, scientifically sound study; however, the regulations have gone beyond those basics by requiring certain additional securities, primary among them the need for each study to have a study director, who assumes overall responsibility for a given study, and a quality assurance unit responsible for monitoring study conduct, the test facility, and reporting of results (*2*).

The regulations have been divided into several subparts, each dealing with a specific aspect of study conduct. Brief descriptions of each subpart follow (*3*).

General Provisions (Subpart A). This section identifies what products are regulated [58.1], and defines various terms prevalent in the conduct of nonclinical studies [58.2]. These include such terms as "The Act," i.e., the aforementioned Federal Food, Drug, and Cosmetic Act; test and control articles; nonclinical laboratory study; sponsor; testing facility; test system; raw data; specimen; quality assurance unit; study director; study initiation and completion dates; and types of submission applications. The section also addresses services contracted by the testing facility, and inspections by agency personnel.

Organization and Personnel (Subpart B). Subpart B has been designed to define the responsibilities of management, the study director, and the quality assurance unit as they relate to facility operation and conduct of nonclinical laboratory studies. The regulations are specific as to organizational relationships and responsibilities.

Management [58.31] has the responsibility for providing qualified and appropriately trained study personnel, assignment of the study director, providing adequate resources, and support of the quality assurance unit by its policies and directives.

The study director [58.33] has the overall responsibility for a given nonclinical laboratory study, and serves as the single point of control for all aspects of that study.

The quality assurance unit [58.35] is responsible for "monitoring each study to assure management that the facilities, equipment, personnel, methods, practices, records, and controls are in conformance" with the applicable regulations. Among its many responsibilities are maintenance of the facility master schedule, maintenance of all study protocols, periodic inspections of ongoing studies, reporting of inspection findings to management, review of the final study report, and preparation and release of the inspection statement required by the regulations.

Facilities (Subpart C). Requirements for animal care and supply and test and control material facilities are addressed by this section. Additionally, requirements for adequate laboratory operation areas and specimen and data storage (i.e., archives) are addressed.

Equipment (Subpart D). This section requires that equipment appropriate to the task be used for any given procedure, and provides for the periodic maintenance, calibration, standardization, cleaning, and/or inspection of that equipment. Specific standard operating procedures defining these activities and related remedial actions are a primary directive, as is the required recordkeeping of all such procedures.

Testing Facility Operation (Subpart E). Standard operating procedures, reagent and solution identification, and animal care requirements are addressed in this section. Standard operating procedures must be established for all routine laboratory operations; specifics are identified by the regulation [58.81(b)]. This regulation also provides for changes to established standard operating procedures and availability of those procedures to laboratory personnel.

Reagent and solution identification requirements are outlined, as are the requirements for dating of reagents.

Animal care requirements reflect basic directives for the receipt, isolation, housing, maintenance, and feeding of laboratory animals used in nonclinical laboratory studies. These regulations supplement those established by the various animal welfare regulations.

Test and Control Article Handling (Subpart F). Subpart F addresses a key area in the conduct of a nonclinical laboratory study in that it provides directive for the characterization and identification of test and control materials, the handling and custody of these materials, and determination of homogeneity and stability of test mixtures.

Protocol and Conduct of a Nonclinical Laboratory Study (Subpart G). A primary concern of any experiment is the procedure by which that experiment will be conducted. Subpart G addresses that concern by defining the requirements for the study protocol, that is, the written document that defines the objectives and establishes the procedures to be performed to meet that objective. It must be remembered that the protocol is directive, not documentation. Adherence to the protocol directive is confirmed through the documentation of the conduct of the study. Recordkeeping requirements are also defined by this section, and provide for the consistent recording of results and data so that, if need be, the study can be literally recreated from its documentation.

Records and Results (Subpart J). Among the most important facets used to support the safety of a regulated product are the final reports produced for the myriad of studies required to establish that safety. Subpart J identifies what information must be addressed in those reports, and how the related data and materials must be retained. It is essential that the final report provide an accurate reflection of the method and procedures employed and of the raw data resulting from the study. In order that sponsor and regulatory review may be accomplished efficiently and in a timely fashion, it is essential that data records and residual materials be retained in an organized and easily retrievable manner.

Disqualification (Subpart K). The FDA's power to enforce the GLP regulations is supported by its power to disqualify a testing facility. Such action is not taken without due consideration, and may only be invoked if all three criteria for disqualification as defined by the regulations are present [58.202]. These include: (1) that the facility failed to comply with one or more of the regulations defined in 21 CFR 58; (2) that the noncompliance adversely affected the validity of the study(ies); and (3) that lesser actions, such as warnings or rejection of individual studies, had not or would probably not be adequate to achieve compliance.

Compliance. Compliance with the GLP regulations is not as easy as it looks, nor is it as difficult as it can be made. It has been noted that "compliance" appears right before "complicated" in the dictionary, but compliance need not be complicated if approached in a common-sense, standardized manner.

There are numerous ways by which compliance in a study or a testing facility can be established. However, all involve a thorough and complete understanding of the final objectives. In establishing guidelines for compliance, certain aspects of nonclinical laboratory testing must be considered. Some will necessarily be tailored to the needs of specific situations, but the general considerations should include at least the following:

1. In order to effect the most complete protocol for a study or procedure, full awareness and understanding of regulatory and testing requirements are essential.
2. Standard operating or study-specific procedures should be developed before they are needed and should be reviewed periodically to insure continued compliance with the appropriate regulations. Procedures should be updated and revised promptly.
3. Ensure that all technical personnel are knowledgeable of and properly trained in the procedures they will be required to perform, and that they are aware of how their performance may affect subsequent aspects of the study.

4. Anticipate potential problems and concerns and be ready with possible resolutions.
5. Acknowledge unforeseen circumstances or deviations from protocol and/or standard operating procedures with prompt and appropriate documentation.
6. Stress the importance of timely and accurate recordkeeping for all aspects of the study.
7. Periodic review of study conduct and data, not only by the study director and quality assurance unit, but by the operating area's internal quality control mechanisms.
8. Communication between the laboratory, the study director, and quality assurance unit should be continuous, not just limited to QA inspections.

The GLP's may define what is needed for compliance, but they cannot guarantee it. That is the joint responsibility of the study director, management, and the quality assurance unit. Neither can fulfill their respective responsibilities towards compliance without the full support of its partners. These relationships should not be viewed as antagonistic, but as supportive. In reality, such relationships are often hard to achieve and maintain given the complexities of the industry. However, the benefits of a cooperative and supportive relationship greatly outweigh the efforts that must continually be expended to create and maintain such a relationship. It is ultimately on the strength of this relationship that compliance is based.

Conclusions - Where Do We Go From Here?

As the industry has matured and developed within the regulatory structure, the concerns of the past havc diminished but by no means disappeared. The majority of recent inspections by the FDA have been classified as "No Action Indicated," "VAI-1," or "VAI 2" (voluntary action classifications); this reflects the favorable attitude of industry towards maintaining compliance. However, deviations from the GLP Regulations were still present, with the most significant departures being (*4*):

1. Discrepancies between the final report and raw data.
2. Improper corrections/changes to raw data.
3. Implementation of protocol revisions without proper amendments to the approved protocol.
4. Lack of appropriate standard operating procedures and/or failure to revise SOPs when needed.
5. Incomplete information on facility master schedules and study protocols.

Inspectors found archiving and record retention procedures, animal care facilities, and laboratory operations generally in compliance with the regulatory requirements. This is encouraging, since it may be recalled that some of the major concerns that precipitated the GLP's involved these areas. The findings cited, however, appear to indicate there is an industry-wide need for tighter control over the "paper" aspects of the nonclinical laboratory study.

The GLP's have given the industry the means by which high quality, consistent, and reproducible studies can be conducted; however, the industry cannot comply with those directives without a continued commitment to compliance. Findings from recent GLP inspections indicate that the advantages of these regulations continue to be valid, and that the industry must not only maintain but increase its commitment to compliance. Despite our shortcomings, we have made significant inroads to the goal of "total compliance" set by the FDA; with continued commitment, we as an industry can meet this objective.

Literature Cited

1. Good Laboratory Practice Regulations; Proposed Rule, Department of Health, Education, and Welfare, Food and Drug Administration, 21 CFR 3e, Federal Register, Vol. 41, No. 225, pp 51206-51209.
2. Taylor, J.M. In *Good Laboratory Practice Regulations*; Hirsch, A., Ed.; Marcel Dekker, Inc.: New York, 1989, pp 6-8.
3. Good Laboratory Practice Regulations; Final Rule, Department of Health and Human Services, Food and Drug Administration, 21 CFR Part 58, Federal Register, Vol. 43, No. 247, pp 59986-60020.
4. James, G.W. In *Good Laboratory Practice Regulations*; Hirsch, A., Ed.; Marcel Dekker, Inc.: New York, 1989, pp 210-212.

RECEIVED August 15, 1991

Chapter 13

Importance of the Hazard Analysis and Critical Control Point System in Food Safety Evaluation and Planning

Donald A. Corlett, Jr.

Corlett Food Consulting Service, 5745 Amaranth Place, Concord, CA 94521

The Hazard Analysis And Critical Control Point (HACCP) system was updated and standardized in 1989 by the National Advisory Committee on Microbiological Criteria For Foods (NACMCF) (*1*). The broad utility of HACCP was recognized by the NACMCF, although by nature of the committee the risk assessment portion was developed only for microbiological hazards. Recent extension of the NACMCF microbiological risk assessment procedures to potential chemical and physical hazard analysis was suggested by Corlett and Stier (*2*). The combined hazard analysis for microbiological, chemical and physical food hazards provides a powerful tool for food safety evaluation and planning guided by HACCP principles and the blueprint for direct application of the specific HACCP system for preventive food safety in a commercial manufacturing operation.

The seven HACCP principles developed by the NACMCF for food safety are listed as follows. A description of each principle and definitions are provided in the NACMCF guide to help the user (*1*).

1. Assess hazards and risks associated with growing, harvesting, raw materials and ingredients, processing, manufacturing, distribution, marketing, preparation and consumption of the food.
2. Determine critical control points (CCP) required to control the identified hazards.
3. Establish the critical limits that must be met at each identified CCP.
4. Establish procedures to monitor CCP.
5. Establish corrective action to be taken when there is a deviation identified by monitoring a CCP.
6. Establish effective record-keeping systems that document the HACCP plan.
7. Establish procedures for verification that the HACCP system is working correctly.

It is clear that the assessment of food hazards and risks via Principle 1. is a most critical beginning for application of the other principles. For this reason it is the basis for food safety evaluation and planning using HACCP. A review of the application of risk assessment for microbiological, chemical and physical food hazards is provided in the following section.

Risk Assessment (HACCP Principle 1)

Risk Assessment consists of a systematic evaluation of a specific food and its raw materials or ingredients to determine the risk from biological (primarily infectious or toxin-producing food-borne illness microorganisms), chemical and physical hazards. The hazard analysis is a two-step procedure: hazard analysis and assignment of risk categories.

The first step is to rank the food and its raw materials or ingredients according to six hazard characteristics (A-F). A food is scored by using a plus (+) if the food has the characteristic, and a zero (0), if it does not exhibit the characteristic. A six characteristic ranking system is applied for microbiological, chemical and physical hazard ranking, although the characteristics are somewhat different for microbiological and chemical/physical hazards, as described later in this section.

The second step is to assign risk categories (VI-0) to the food and its raw materials and ingredients based on the results of ranking by hazard characteristics. Table 1 illustrates possible combinations of hazard characteristic ranking and hazard categories. Potentially highest risk is denoted by the highest number in the hazard category (i.e., VI.). In addition, note that whenever there is a plus (+) for hazard characteristic A (a special class that applies to food designated for high-risk populations), the resulting hazard category is always VI, even though other hazard characteristics (B-F) may or may not be a plus (+).

Several preliminary steps are needed before conducting the hazard analysis. These include developing a working description of the product, listing the raw materials and ingredients required for producing the product, and preparation of a diagram of the complete food production sequence. The listing of raw materials and ingredients is the starting point for the hazard analysis. If the specific mode of preservation for an ingredient is not known (raw, frozen, canned, etc.), the ingredient may be assessed for each type of preservation technique that may be utilized in preserving the ingredient.

Microbiological Hazard Characteristic Ranking. Microbiological hazard analysis and the ranking of food by hazard characteristics is explained in the NACMCF HACCP guide (*1*). I have made several minor changes in Hazard F, to differentiate ranking for consumer products and raw materials and ingredients as received by the processor before any manufacturing steps. The microbiological hazard characteristics are given in Table 2. As indicated earlier, rank the product and its raw materials and ingredients according to hazard characteristics A through F, using a plus (+) to indicate that the food product or its raw materials or ingredients exhibit the characteristic, and a zero (0) when they do not.

Table 1. Possible Combinations of Hazard Characteristic Ranking and Hazard Categories for Food Products and Food Raw Materials and Ingredients

Food Ingredient or Product[a]	*Hazard Characteristics (A,B,C,D,E,F)*[b]	*Risk Category*
T	A+ (Special Category)	VI
U	Five +'s (B–F)	V
V	Four +'s (B–F)	IV
W	Three +'s (B–F)	III
X	Two +'s (B–F)	II
Y	One + (B–F)	I
Z	No +'s	0

[a]These letters merely indicate different types of foods having different hazard characteristics and risk catagories. Normally the name of a food, raw material, or ingredient would appear under this heading.
[b]Hazard characteristic A automatically is risk category VI, but any combination of B through F may also be present.

Table 2. Microbiological Risk Characteristics

Hazard	*Description*
Hazard A:	A special class that applies to non-sterile products designated and intended for consumption by at-risk populations, e.g., infants, the aged, the infirm, or immunocompromised individuals.
Hazard B:	The product contains "sensitive ingredients" in terms of microbiological hazards.
Hazard C:	The process does not contain a controlled processing step that effectively destroys harmful microorganisms.
Hazard D:	The product is subject to recontamination after processing before packaging.
Hazard E:	There is substantial potential for abusive handling in distribution or in consumer handling that could render the product harmful when consumed.
Hazard F:	There is no terminal heat process after packaging or when cooked in the home. (Applies to food product, as used by the consumer.) There is no terminal heat process or any other kill-step applied after packaging by the vendor, or other kill-step applied before entering food manufacturing facility. (Applies to raw materials and ingredients coming into a food manufacturing facility.)

SOURCE: After NACMCF HACCP system (USDA-FSIS, 1990); and by permission of D. Corlett (Copyright D. Corlett by license from ESCAgenetics Corporation, course manual, "A Practical Application Of HACCP," 1990).

A brief discussion on "microbiologically sensitive" products, and raw materials and ingredients is useful. When scoring foods for Hazard Characteristic B, give the product a plus (+) if it is sensitive or contains microbiologically sensitive ingredient(s). Give raw materials or ingredients a plus (+) if they are microbiologically sensitive or contain sensitive foods (e.g.., such as a cheese/starch flavor blend).

A "sensitive ingredient" is defined as "any ingredient historically associated with a known microbiological hazard." The term "ingredient" normally also applies to raw materials. "Sensitive ingredient" was coined for microbiological hazards (infectious agents and their toxins), but it is also now used for ingredients and raw materials that are historically associated with known chemical or physical hazards.

The original list of microbiologically sensitive foods was based on the potential presence of the *Salmonella* species. Now any type of hazardous microorganism may cause a food to be "sensitive," and the list of sensitive foods has grown, particularly with the recognition that *Listeria monocytogenes* is a known threat in many foods. A partial listing of sensitive raw materials and ingredients is provided to assist in scoring a food, or its raw materials and ingredients, for Hazard Characteristic B. If there is a question as to whether a food is sensitive, it should be considered sensitive until more information is available for purposes of clarifying its status.

Microbiologically Sensitive Raw Materials And Ingredients:

Meat and poultry
Eggs
Milk and dairy products (including cheese)
Fish and shellfish
Nuts and nut ingredients
Spices
Chocolate and cocoa
Mushrooms
Soy flour and related materials
Gelatin
Pasta
Frog legs
Vegetables
Whole grains and flour (secondary contamination)
Yeast
Dairy cultures
Some colors and flavors from natural sources

Compounded ingredients may be considered sensitive if they are combinations of sensitive and non-sensitive ingredients. For example, a fat coated on milk powder, or compounded cheese flavor having cheese coated on starch. It is best to list all components of a blended material to determine if the blend contains a sensitive ingredient and also determine if it has received a controlled processing step that destroys hazardous microorganisms. In some cases, it is important to determine if microbiological toxins may also be present in a "processed" product, if the product is

to be used as an ingredient (e.g. heat stable staphylococcus enterotoxin in canned mushrooms).
Many raw materials and ingredients are not considered microbiologically sensitive even though they may occasionally be contaminated with hazardous microorganisms. A partial list includes:

Foods Not Normally Considered Sensitive:
Salt
Sugar
Chemical preservatives
Food grade acidulents and leavening agents
Gums and thickeners (some may be sensitive depending on origin, such as tapioca and fermentation-derived gums)
Synthetic colors
Food grade antioxidents
Acidified high salt/acid condiments
Most fats and oils (exception is dairy butter)
Acidic fruits

These lists are intended as a guide and are not necessarily an exhaustive listing of all sensitive and non-sensitive ingredients. When in doubt, it is recommended that assistance be obtained from authoritative sources including universities, regulatory agencies, trade organizations, consultants and consulting laboratories.

Chemical and Physical Hazard Characteristic Ranking. The following protocol for hazard analysis of chemical and physical food hazards complements the existing microbiological hazard analysis scheme given in the NACMCF system. Hazard characteristics for chemical and physical agents were developed in 1990 for use in the ESCAgenetics Corporation training course, "A Practical Application Of HACCP," and were recently published by Corlett and Stier (*2*). They are designed so that both chemical and physical hazards in food may be assessed by using the same six hazard characteristics.

Generally, hazard analysis for chemical and physical hazards is conducted like the procedure for microbiological hazards provided in the NACMCF guide. Although the six hazard characteristics are somewhat different, the same plus (+) and zero (0) scoring system and hazard category assignment procedures are used.

Table 3 provides the hazard characteristics for ranking foods for both chemical and physical hazards. This Table also includes examples of chemical and physical agents that could potentially be present in a food relative to each hazard characteristic. The concept of "sensitive" products, raw materials and ingredients is also used in Hazard Characteristic B for chemical and physical hazards.

Table 3. Hazard Characteristics for Ranking Foods for Chemical and Physical Hazards

Hazard	*Description*
HAZARD A:	A special class that applies to products designated and intended for consumption by high-risk populations, e.g., infants, the aged, the infirm, or immunocompromised individuals. (Examples are foods intended for persons sensitive to sulfites, and for infants where glass is of particular concern.)
HAZARD B:	The product contains "sensitive" ingredients known to be potential sources of toxic chemicals or dangerous physical hazards. (Examples are aflatoxin in field corn, and stones in agricultural products.)
HAZARD C:	The process does not contain a controlled step that effectively prevents, destroys or removes toxic chemical or physical hazards. (Examples include steps for prevention of the formation of toxic or carcinogenic substances during processing; destruction of cyanide-containing compounds by roasting of apricot pits; and removal of toxic processing chemicals such as lye or dangerous foreign objects such as sharp pieces of metal.)
HAZARD D:	The product is subject to recontamination after manufacturing before packaging. (Example is where contamination may occur when a manufactured product is bulk packed, shipped and packaged in another facility.)
HAZARD E:	There is substantial potential for chemical or physical contamination in distribution or in consumer handling that could render the product harmful when consumed. (Examples are contamination of a food from container or vehicle compartments that previously contained toxic chemicals or foreign objects; selling food in open containers; or where the potential for product tampering is high.)
HAZARD F:	There is no way for the consumer to detect, remove or destroy a toxic chemical or dangerous physical agent. (Examples are presence of toxic mushrooms or paralytic shellfish toxins, or presence of sharp metal objects buried in a food.)

SOURCE: Reference 2.

Example of the Combined Hazard Analysis and Hazard Category Assignment for Cheese Dip

The complete hazard analysis consisting of ranking of potential microbiological, chemical and physical hazards and assignment of hazard categories is illustrated in the example of a hypothetical cheese dip product.

Table 4 lists the raw materials and ingredients and for purposes of this illustration, gives a listing of potential microbiological, chemical and physical hazards that may be expected in these foods.

Table 4. Types of Potential Hazards in Cheese Dip

Ingredient	*Microbiological*	*Chemical*	*Physical*
Raw Celery	*Salmonella* sp.	Pesticides	Metal
	Shigella sp.		Wood
	Listeria monocytogenes		Rocks
Dried Mushrooms	*Salmonella* sp.	Pesticides	Metal
	Shigella sp.		Wood
	Staphylococcus aureus		Rocks
Soft-Ripened	*Listeria monocytogenes*	Pesticides	Metal
Cheese	*Salmonella* sp.	Antibiotics	
	Staphylococcus aureus	Hormones	
	EP *Escherichia coli*		
Water	Microbial pathogens	Various	n/u
Salt	n/u	n/u	Metal
Stabilizer	n/u	n/u	Metal

NOTE: n/u = not usually

SOURCE: From ECS Agenetics Corporation course "A Practical Application Of HACCP."

Forms 5.0 (Microbiological) (Figure 1), 6.0-A (Chemical) (Figure 2), and 6.0-B (Physical) (Figure 3) illustrate the ranking of hazard characteristics and assignment of hazard categories for three modes of preservation for the cheese dip product, and the ranking of all raw materials and ingredients.

Once the risk assessment is completed, utilize the NACMCF HACCP guide for completion of the HACCP plan for a specific food product, and it's raw materials and ingredients (*1*).

Conclusion

The combination of the hazard analysis and the critical control points make the HACCP system the ideal choice for food safety evaluation and planning. Principle 1. concerning assessment of hazards and risks associated with a specific food and it's ingredients is key to further development of the HACCP system. It is essential that the hazard analysis and risk assessment always be conducted correctly before attempting

HACCP PRINCIPLE 1. **HACCP WORKSHEET** **FORM** 5.0

RISK ASSESSMENT WORK-SHEET FOR MICROBIOLOGICAL FOOD HAZARDS

PRODUCT:___CHEESE DIP________________PAGE_1_OF__1__PAGES DATE:_______________
(DON'S DELIGHT)

FOOD PRODUCT(S)............AS USED BY THE CONSUMER.........................

PRODUCT	MICROBIOLOGICAL HAZARD CHARACTERISTICS ASSOCIATED WITH THE FOOD (+ FOR "YES"; O FOR "NO")						HAZARD CATEG.
	A HIGH RISK SPECIAL POPULAT.	B SENSITIVE INGRED-IENTS	C NO KILL-STEP IN PROCESS	D RECONTAM. BETWEEN PROC/PACK	E ABUSIVE HANDLING DIST/CONS	F NO TERM. HEAT PROC BY CONSUM	
(1) REFRIG.	0	+	+	+	+	+	V.
(2) FROZEN	0	+	0	+	+	+	IV.
(3) CANNED	0	+	0	0	0	+	II.

RAW MATERIALS AND INGREDIENTS...AS RECEIVED, BEFORE ANY MANUFACTURING STEPS BY THE FOOD FACILITY (SUCH AS COOKING)......

RAW MAT. OR INGRE.	A	B	C	D	E	F:NO KILL STEP BEFORE RECEIPT*	HAZARD CATEG.
RAW CELERY	0	+	+	+	+	+	V.
DRIED MUSHROOMS	0	+	+	+	0	+	IV.
SOFT-RIPENED CHEESE		+	+	+	+	+	V.
SALT	0	0	0	0	0	0	0.
WATER	0	+	0	+	0	+	III.
STABILIZER	0	0	0	0	0	0	0.

* NO HEAT PROCESS OR ANY OTHER KILL-STEP APPLIED AFTER PACKAGING BY SUPPLIER; NO HEAT PROCESS OR OTHER KILL-STEP BEFORE ENTERING FOOD PLANT.

DONSMICR

Figure 1. Risk assessment worksheet for microbiological food hazards.

HACCP PRINCIPLE 1. HACCP WORKSHEET FORM 6.0-A

RISK ASSESSMENT WORK-SHEET FOR CHEMICAL OR PHYSICAL FOOD HAZARDS

IS THIS SHEET TO BE USED FOR CHEMICAL OR PHYSICAL HAZARDS?___"CHEMICAL"_____

PRODUCT:___CHEESE DIP (DON'S DELIGHT)____________________DATE:_____________

FOOD ITEM	HAZARD CHARACTERISTICS KNOWN TO BE ASSOCIATED WITH THE FOOD AND IT'S INGREDIENTS (+ FOR "YES"; O FOR "NO")						HAZARD CATEG.
(1) PRODUCT	A HIGH RISK SPECIAL POPULAT.	B INGREDS. CONTAIN HAZARD	C NOT RE-MOVED IN MANUFACT.	D RECONTAM. BETWEEN MAN./PAC.	E CONTAM. BY DIST. OR CONS.	F CONS. CAN-NOT DE-TECT/REM.	
REFRIGERATED	0	+	+	+	0	+	IV.
FROZEN	0	+	+	+	0	+	IV.
CANNED	0	+	+	+	0	+	IV.
(2) RAW MAT'S AND ING'S							
RAW CELERY	0	+	+	+	+	+	V.
DRIED MUSHROOMS	0	+	+	+	0	+	IV.
SOFT-RIPENED CHEESE		+	+	+	0	+	IV.
SALT	0	0	0	0	0	0	0.
WATER	0	+	+	0	0	+	III.
STABILIZER	0	0	0	+	0	+	II.

NOTES: (1) AS USED BY CONSUMER
(2) AS ENTERING THE FOOD FACILITY BEFORE PREPARATION OR PROCESSING

Figure 2. Risk assessment worksheet for chemical food hazards.

HACCP PRINCIPLE 1. HACCP WORKSHEET FORM 6.0-B

RISK ASSESSMENT WORK-SHEET FOR CHEMICAL OR PHYSICAL FOOD HAZARDS

IS THIS SHEET TO BE USED FOR CHEMICAL OR PHYSICAL HAZARDS?__"PHYSICAL"______

PRODUCT:___CHEESE DIP (DON'S DELIGHT)____________________DATE:____________

FOOD ITEM	HAZARD CHARACTERISTICS KNOWN TO BE ASSOCIATED WITH THE FOOD AND IT'S INGREDIENTS (+ FOR "YES"; O FOR "NO")						HAZARD CATEG.
(1) PRODUCT	A HIGH RISK SPECIAL POPULAT.	B INGREDS. CONTAIN HAZARD	C NOT RE- MOVED IN MANUFACT.	D RECONTAM. BETWEEN MAN./PAC.	E CONTAM. BY DIST. OR CONS.	F CONS.CAN- NOT DE- TECT/REM.	
REFRIGERATED	0	+	0	+	0	+	III.
FROZEN	0	+	0	+	0	+	III.
CANNED	0	+	0	+	0	+	III.
(2) RAW MAT'S AND ING'S							
RAW CELERY	0	+	+	+	+	+	V.
DRIED MUSHROOMS	0	+	+	+	0	+	IV.
SOFT-RIPENED CHEESE		+	0	+	0	+	III.
SALT	0	+	0	+	0	+	III.
WATER	0	0	0	0	0	0	0.
STABILIZER	0	+	0	+	0	+	III.

NOTES: (1) AS USED BY CONSUMER
(2) AS ENTERING THE FOOD FACILITY BEFORE PREPARATION OR PROCESSING

DONSPHYS

Figure 3. Risk assessment worksheet for physical food hazards.

to apply the successive HACCP principles. Failure to conduct the risk assessment may lead to omission of critical control points and result in serious gaps in a food safety assurance program.

Acknowledgment

Portions of this paper are licensed to D. A. Corlett, and copyrighted in the ESCAgenetics course manual, *A Practical Application of HACCP* (*3*).

Literature Cited

1. USDA-FSIS. 1990. HACCP Principles for Food Production. Pamphlet containing the HACCP system developed by the *ad hoc* NACMCF HACCP working group, chaired by D.A. Corlett, U.S. National Advisory Committee On Microbiological Criteria For Foods. FSIS Information Office, Room 1160, South Building, Washington, D.C. 20250.
2. Corlett, D.A. and R.F. Stier. 1991. Risk Assessment Within The HACCP System. Food Control, April, 1991. p. 71.
3. Corlett, D.A. and R.F. Stier. 1990. Course manual: "A Practical Application Of HACCP". ESCAgenetics Corporation, San Carlos, California. Licensed to D.A. Corlett, 5745 Amaranth Place, Concord, California 94521.

RECEIVED November 15, 1991

Evaluation Guidelines

Chapter 14

Threshold of Regulation

Options for Handling Minimal Risk Situations

Alan M. Rulis

U.S. Food and Drug Administration, 200 C Street, SW, Washington, DC 20204

The Food and Drug Administration (FDA), under the Federal Food, Drug, and Cosmetic Act, requires the premarket safety evaluation of new uses of food additives. The statute defines as additives even those substances that may inadvertently become components of food by migrating from food packaging, and provides no cutoff below which chemicals migrating in very low amounts may be considered exempt from petition requirements. It is clear, however, that at very low levels of migration the agency's expenditure of resources to regulate such materials may result in negligible public health gain. What is an appropriate level to define as a "threshold of regulation," below which no petition for a new use need be submitted and approved? FDA's development of a scientific basis for such a regulatory cutoff using risk assessment has spanned several years. One approach considered by FDA employs a statistical analysis of potencies of known chemical carcinogens. The present paper will examine options open to the agency in this potentially precedent-setting policy area.

Is there a basis for defining a "Threshold of Regulation" (T/R) to exempt substances under the Federal Food, Drug, and Cosmetic Act (the Act) from premarket regulatory requirements? Specifically, what situations involving extremely low exposure to food chemicals migrating to food from food-contact materials (e.g., components of food packaging materials, or food handling equipment, etc.) could be considered *de minimis* (*1*) under the statute, and thus would not require the submission of a food additive petition and a full-blown petition review by the Food and Drug Administration (FDA)? The agency has been developing answers to such questions over several years (*2-5*), and is now nearing a workable solution (*6*).

Since the passage of the 1958 Food Additives Amendment to the Act, FDA has often considered such questions in regard to so-called "indirect food additives" (food packaging and other food-contact materials that are not added directly to food but become components of food by virtue of unintended migration to food), particularly when potential human exposure to such additives is extremely low and thus unlikely to produce any possible public health concern. The statute defines food additives, including the migrating indirect additives, quite broadly[1]. In particular it provides no exposure "floor" below which substances are not considered to be food additives and thus are exempt from premarket petition review and safety evaluation. Furthermore, the Delaney anticancer clause of the Act prohibits the use of any additive that has been shown to induce cancer in man or animal. (It is not intended that a T/R policy would be applicable to chemicals demonstrated to be carcinogenic.)

Until now the agency has lacked a policy under which to make T/R decisions in a consistent manner. Instead it has used a case-by-case approach. Since the 1958 amendment to the Act, FDA has written many letters exempting situations from food additive petition review because of the specific facts of a case. There are many examples of situations where the agency has deemed the minuscule human exposure to a chemical in question to be of no consequence and not a food additive concern under the Act. Representative examples might be the use of an adjuvant in the matrix of a food processing conveyor belt; a material used in nonfood-contact fixtures in a food processing plant; a colorant, polymerization catalyst, or other adjuvant, used at exceedingly low levels in a plastic food packaging material; etc.

Need for a Threshold-of-Regulation Policy

Today there is an increasing need for FDA to make decisions of the type described above more routinely, with greater consistency, and on a firmer scientific basis. It is also becoming more important for the agency to focus its limited resources more on issues of major public health impact and not to allow resources to be disproportionately focused on a myriad of minimal risk situations that are of negligible public health consequence. Yet present trends indicate that greater effort is in fact not always able to be expended on issues in direct proportion to their public health importance. Most of the food additive petitions reviewed by the agency are for food-contact substances (indirect food additives) rather than for direct food additives. Since 1958, FDA has reviewed and regulated an average of about 60 petitions per year for indirect additives, but only about 15 on average for direct food additives, color additives, and "generally recognized as safe" (GRAS) food ingredients. Even though indirect additive petitions are typically smaller and simpler to process than direct food additive petitions, the agency devotes over 40 percent of its petition review resources to the processing of indirect additive petitions. Some petitions are for such low-exposure uses of indirect additives that it may be legitimately questioned whether the safety decision results in any net measurable gain in public health protection. Yet the agency's formal review

---

[1]As has been amply noted previously, the Federal Food, Drug, and Cosmetics Act requires the premarket safety evaluation of all new uses of food additives. The statute, however, defines as additives even those substances that may become components of food by migrating from food packaging, and provides no cutoff below which chemicals migrating in very low amounts may be considered exempt from petition requirements.

mechanism for these petitions, including environmental impact considerations, legal reviews, and Federal Register publication, all operating under statutory time frames, must be fully engaged.

A T/R policy could help to alleviate this problem by providing a simpler alternative mechanism, apart from the full-dress petition review process, under which the agency could grant approvals for limited uses of a substance. Under such a policy any person could seek a T/R exemption from petition review for the proposed use of a given substance. The requestor would supply adequate information to identify the chemical in question, describe its proposed use, and provide limited data from which agency scientists could estimate the likely incremental human intake resulting from the proposed use. If, after reviewing the information, FDA decided to grant an exemption, it would issue a letter to the requestor and maintain a record of the decision in agency files. If the process were structured so as to permit decisions to be made in a short time compared to the time required for a full petition review, then the process would free resources for other more important issues. Of course, whether the net effect of this process on resources is helpful would depend on many considerations including the amount of current petition work actually diverted to the less arduous path, as well as any increase in petition submissions or requests for advisory opinions that are not now being sent to the agency from industry.

Somewhat paradoxically, instituting a T/R policy for indirect additives may, in some ways, actually represent a tightening of FDA requirements for indirect additives. First, it is the agency that would offer T/R exemptions to requestors; this is not thought of as a "do-it-yourself" exemption process. A request for an exemption need not be granted. Even if all nominal conditions were satisfied, FDA might decide not to exempt certain substances on the basis of knowledge of the chemical structures involved and the likelihood that those structures might be associated with high toxicity. Furthermore, current toxicological requirements for petitioned indirect additives presenting less than 50 parts per billion (ppb) dietary exposure consist, at minimum, of simply an acute feeding study and a literature search. A T/R level on the order of 1.0 ppb or lower, for example, would focus regulatory attention on a range of human exposures lower than 50 ppb. Users of indirect additives in applications resulting in dietary exposures of 10, 5, or 2 ppb or lower, under their *assumption* of *de minimis* status, would be encouraged to seek an agency opinion as to whether their application qualifies for an exemption from regulatory requirements. (In the initial stages, this might even result in a temporary increase in workload for the agency.)

During 1989, FDA conducted a Pilot Study to examine practical approaches to implementing a T/R policy. (The Study and its outcome have been described by Borodinsky (*7*). In that study, 35 T/R cases were examined at a total expenditure of about 120 person-hours of deliberation, or an average of 3.4 person-hours to reach a decision in each case. This is a considerable saving compared to the usual agency effort required to process a typical indirect food additive petition, which, although highly variable, may range from 250 to 500 person-hours on average.

Selection of a Threshold Level

A difficult issue in creating a T/R policy is the selection of an appropriate migration level to food or human dietary exposure level for the threshold. A simple solution might seem to be to arbitrarily pick a conservatively low level of migration to food from food-contact materials, for example, 25 or 50 ppb, and to define any situation with less migration to be below the threshold. This approach, however, is far from optimal. First, potential risk is related more closely to human dietary exposure than to migration. Setting a migration-based T/R level does not recognize this fact. Another approach would be to simply set the T/R level low enough to preclude any potential risk of toxicity from any chemical migrating into food, including ones of known high toxicity such as dioxin (TCDD). This approach, however, is not only unnecessarily conservative, but it would also require that the dietary intake level chosen as the threshold be set so low (femtogram levels in the case of TCDD) as to make such a policy useless. Not only is it unlikely that any materials used at such low levels would actually produce a technical effect in a food-contact material, but today's analytical measurements are insufficiently sensitive to routinely demonstrate the presence of the material below that level. Thus, in practice, no chemical would be able to pass such a threshold requirement.

Conversely, a level arbitrarily set too high would undermine the effect of the statute and perhaps create the possibility of unnecessary risk if the substance granted the exemption were to possess considerable toxicity.

To be relevant to potential human risk, the T/R level must be based on likely dietary intake from food, and not on migration. It also must be relevant to known toxic endpoints of chemicals at the level of intake, and as Schwartz has shown (*5*), it must be in the realm of present-day analytical capabilities. Because carcinogenesis occurs in animals at exposure levels generally lower than for most other types of toxic effects, a policy based on that toxic endpoint would provide a conservative measure of protection from almost all types of presumptive toxicity. For this reason FDA has considered that its T/R policy should use carcinogenesis as the basis for establishing the threshold (*2-4*). Such an approach is also consistent with the agency's established precedent for using upper-bound estimates of risk from carcinogenesis as a standard of negligible risk, in both its Sensitivity of the Method regulation for animal drugs (*8*) and its policy regarding Carcinogenic Impurities in Food Additives (*9,10*).

Use of Carcinogen Potencies to Establish a Threshold of Regulation

One approach to establishing a T/R level is to base that level on the degree to which presumptive carcinogenic risk may be ruled out, in the unlikely event that the compound is a carcinogen. FDA's approach to precluding potential carcinogenic risk makes use of potency data compiled from substances that have tested positive for carcinogenesis in animal feeding studies. Both the FDA (*2-4*) and others (*11-14*) have discussed this approach.

Carcinogen potencies are known to be lognormally distributed (Figure 1). From this distribution it has been shown (*2*) that the choice of a given exposure level for a

T/R excludes a probabilistically defined proportion of the area under the lognormal curve of potencies from producing dietary risk to humans above any chosen "Target Risk" level, so long as human exposure to the substance of concern does not exceed the T/R exposure level. The Target Risk level is that upper-bound level of presumptive lifetime risk deemed commensurate with negligible or *de minimis* risk, and is typically chosen to be 1 X 10^{-6} (*8-10*). Given this Target Risk and the range over which carcinogen potencies are known to be distributed, one may select a threshold level that provides adequate protection from presumptive carcinogenic risk in excess of the Target Risk.

In an earlier paper on this subject, a T/R level of 50 parts per trillion (ppt) was proposed for illustration (*2*) using the method described above. It was shown how that choice of T/R level is consistent with an 85 percent probability that an upper-bound risk of greater than 1 X 10^{-6} would be precluded for each exemption at that level, should the substance unexpectedly be a carcinogen. (Coupled with an assumed one-in-five probability of an untested chemical being a carcinogen, the choice of 50 ppt leaves a better than 97% probability that cancer risk will not exceed 1 X 10^{-6}.)

In fact, for any *range* of selected T/R levels there exists a *corresponding range* of probabilities that presumptive carcinogenic risk at some target risk level is precluded. These "Target Risk Avoidance Probabilities" correspond to areas under the lognormal curve of carcinogen potencies excluded by any given choice of T/R level. The relationship between these two variables is portrayed in Figure 2, which shows Target Risk Avoidance Probabilities as a function of the T/R level chosen. The shape of the curves in Figure 2 depends solely on the parameters that define the shape and position of the lognormal distribution of carcinogen potencies. Two curves from this author's work are portrayed in Figure 2, one corresponding to 343 carcinogens selected from the original data base compiled by Gold et al. (*2,15*) and the other, a more recent one using 477 carcinogens chosen from an updated Gold et al. database (*16,17*). The choice of 50 ppt as a T/R level is designated by Arrow "A" in Figure 2.

Schwartz has proposed a range of possible T/R levels between 100 ppt and 1 ppb (*5*). The lower bound for this range was justified on the basis of known diffusion coefficients for migrating species from polymeric food-contact materials, and represents a practical limit to current analytical capability for indirect food additives. At the upper limit (1 ppb) the target risk avoidance probability begins to exceed 50 percent. The range proposed by Schwartz is shown as the span between arrows "B" and "C" in Figure 2.

Recently, Munro et al. published a table of Target Risk Avoidance Probabilities (*see* Reference 11, Table 2) as well as parameters defining the lognormal potency curves for each of four carcinogen data sets they studied (*13*). Using their parameters, we have plotted in Figure 2 the Target Risk Avoidance Probabilities for two of their data sets, including the one they state to be of most relevance to the T/R problem. As can be seen from Figure 2, the Munro et al. analysis is in substantial agreement with the present analysis. Munro et al. argue that a dietary intake level as high as 1 ppb provides adequate protection from presumptive cancer risk, and that the level may be even higher, possibly as high as 10 to 15 ppb, if adequate data are available to preclude the genotoxicity of the chemical in question (*14*). The T/R level suggested by Munro

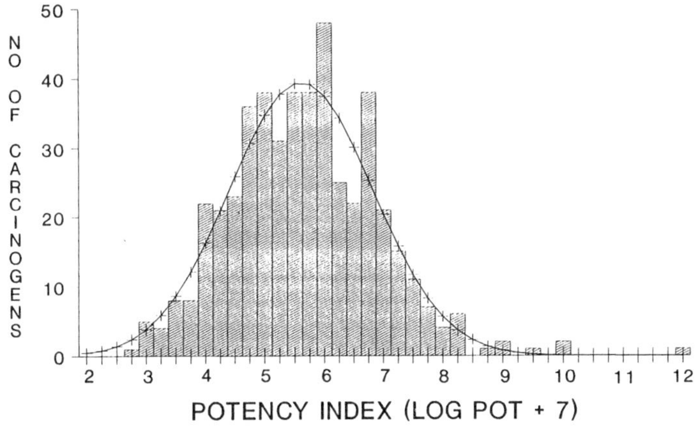

Figure 1. Probability distribution of carcinogen potencies based on the data base of Gold et al. *(15-17)* for 477 selected carcinogens.

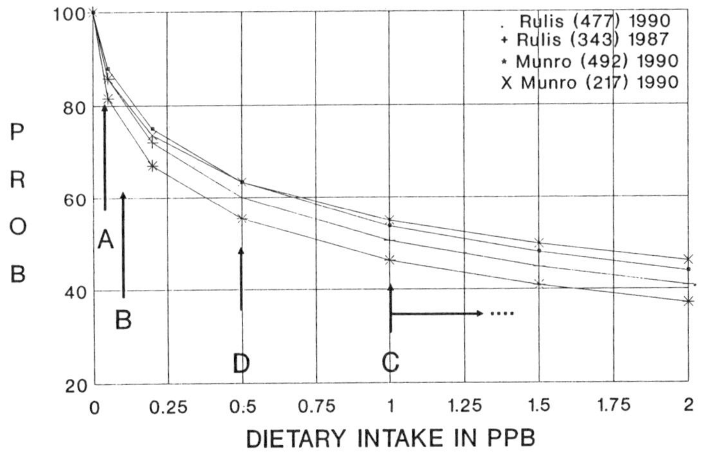

Figure 2. Target Risk Avoidance Probabilities as a function of human dietary intake. Data of Rulis (2) and Munro et al. *(11)*. **Ordinate** represents the probability that presumptive upper-bound risk of carcinogenesis would not exceed 1 X 10^{-6} upon lifetime ingestion at the dietary level indicated on the abscissa. **Arrow A** corresponds to a T/R level of 50 ppt (2). **Arrows B and C** represent a range of possible T/R levels described by Schwartz (*5*). **Arrow C** and above corresponds to the proposed T/R level of Munro et al. (*11-14*). **Arrow D** corresponds to a T/R level of 0.5 ppb. Data set 1 (•) of Rulis is from an unpublished analysis of 477 carcinogens selected from the Gold et al. data base. Data set 2 (+) of Rulis is from a previously published analysis of 343 Gold et al. carcinogens. Data set 3 (*) of Munro et al. is from a set of 492 Gold et al. carcinogens. Data set 4 (X) is from a set of 217 carcinogens selected by Munro et al. to be the best representative set for the purposes of establishing a T/R level.

et al. is shown in Figure 2 as arrow "C" at 1 ppb with an indefinite span to the right of that level.

Taken together, the data of Figure 2 show that an upper-bound presumptive risk of carcinogenesis from lifetime dietary ingestion of a carcinogen at a level of 1 ppb will be less than 1×10^{-6} with roughly a 50 percent probability (Arrow C), while at lower dietary intakes it becomes increasingly probable that the potential risk will not exceed that target risk level. Recall that these "risks" are conjectural and not actuarial in any sense. They are upper-bound estimates derived from a highly conservative linear extrapolation of data from animal studies. Furthermore, it has been presumed that the chemical in question is in fact a carcinogen. This is not likely to be true for more than about one in perhaps three to five randomly selected compounds.

At the present time it appears that a T/R level on the order of 0.5 ppb (arrow "D" in Figure 2) may represent a reasonable balance between necessary conservatism and practical utility. In the absence of any detailed toxicological information about a compound, including genotoxicity information, this level provides adequate protection from presumptive carcinogenic risk, and is also within the realm of analytical measurability.

Summary and Conclusions

Several impelling considerations currently point to establishment of a T/R for food packaging materials. The 1979 Monsanto vs. Kennedy decision of the United States Court of Appeals reminded the agency of the Commissioner's limited exemption authority under the present statute. FDA has yet to formally delineate its understanding of the scope and application of that exemption authority. The Monsanto court decision provides both an opportunity and an impetus to move forward with a T/R policy. Furthermore, industry petitioners for new food additives deserve to have consistent and expeditious decisions about their products under regulatory authority of FDA. These decisions must also protect the public health. Regulatory agencies need to find more ways to employ the "principle of commensurate effort," under which they systematically devote their limited resources to issues in proportion to the likely net public benefit. Expeditious handling of a larger number of trivial or near trivial issues would allow more attention to be focused on the less numerous, yet more important, issues. These are all major concerns related to the T/R policy for indirect food additives under development at FDA.

Scientific analyses suggest that even if the toxicological endpoint of carcinogenesis is selected as the key factor in setting a T/R level for indirect food additives, a level can be set that is both practical from the analytical standpoint and fully protective of public health. A level of the order of 0.5 ppb may be a reasonable starting point for such a policy, lying as it does, midway in a range bounded by analytical limitations on one end and by increasing probability of presumptive toxicity on the other. Of course, FDA has not yet settled on a specific T/R level, nor for that matter does it have the specific considerations of a policy fully laid out. When the agency develops its approach to a point where outside opinion and independent review will be helpful, we intend to publish a proposal in the Federal Register and request public comment.

Literature Cited

1. Monsanto vs. Kennedy, 613 F. 2d 947, DC Circuit, 1979.
2. Rulis, A. M. In *Food Protection Technology*; Felix, C.W., Ed.; Lewis Publishers, Inc.: Chelsea, MI, 1987.
3. Flamm, W. G., Lake, L. R.; Lorentzen, R. J.; Rulis, A. M.; Schwartz, P. S.; Troxell, T. C. In *Contemporary Issues in Risk Assessment, Vol. 2: De Minimis Risk*; Whipple, C., Ed.; Plenum Press: New York, 1987.
4. Rulis, A. M. In *Risk Assessment in Setting National Priorities*; Bonin, J. J.; Stevenson, D. E., Eds.; Plenum Publishing Corp.: New York, 1989.
5. Schwartz, P. S. Speech to the Food, Drug and Cosmetics Packaging Materials Committee of the Society of the Plastics Industry, Washington, DC, June, 1989.
6. Shank, F. R. Speech to the Food, Drug and Cosmetics Packaging Materials Committee of the Society of the Plastics Industry, Washington, DC, June 6, 1990.
7. Borodinsky, L. Speech to the Food Packaging Interactions Symposium, 200th National Meeting of the American Chemical Society, Washington, DC, August 28, 1990.
8. *Chemical Compounds in Food-Producing Animals — Criteria and Procedures for Evaluating Assays for Carcinogenic Residues, Fed. Regist.* March 20, **1979**, *44*, 17070-17114.
9. *Policy for Regulating Carcinogenic Chemicals in Food and Color Additives*, Advance Notice of Proposed Rulemaking, *Fed. Regist.* April 2, **1982**, *47*, 14464-14470.
10. Rulis, A. M.; McLaughlin, P. J.; Salsbury, P. A.; Pauli, G. H. In *Advances in Risk Analysis, Vol. 9*; Whipple, C., Ed.; Plenum Press, New York, in press.
11. Munro, I. *Regul. Toxicol. Pharmacol.* **1990**, *12*, 2-12.
12. Weisburger, E. K. *Regul. Toxicol. Pharmacol.* **1990**, *12*, 41-52.
13. Krewski, D.; Szyszkowicz, M.; Rosenkranz, H. *Regul. Toxicol. Pharmacol.* **1990**, *12*, 13-29.
14. Williams, G. M. *Regul. Toxicol. Pharmacol.* **1990**, *12*, 30-40.
15. Gold, L. S.; Sawyer, C. B.; Magaw, R.; Backman, G.; Veciana, M.; Levinson, R.; Hooper, N. K.; Havender, W.; Bernstein, L.; Peto, R.; Pike, M.; Ames, B. *Environ. Health Perspect.* **1984**, *58*, 9-314.
16. Gold, L.; Veciana, M.; Backman, G.; Magaw, R.; Lopipero, P.; Smith, M.; Blumenthal, M.; Levinson, R.; Bernstein, L.; Ames, B. *Environ. Health Perspect.* **1986**, *67*, 161-200.
17. Gold, L.; Slone, T.; Backman, G.; Magaw, R.; DaCosta, M.; Lopipero, P.; Blumenthal, M.; Ames, B. *Environ. Health Perspect.* **1987**, *74*, 237-329.

RECEIVED August 15, 1991

Chapter 15

Food Ingredient Safety Evaluation

Guidelines from the U.S. Food and Drug Administration

George H. Pauli

Division of Food and Color Additives, U.S. Food and Drug Administration, 200 C Street, SW, Washington, DC 20204

Procedures for the safety evaluation of food ingredients must take into account the legal authority for requiring safety testing, the capability of various scientific methodologies to address questions relevant to the safety of food, the risks to be encountered if safety questions are not addressed, and the societal consensus on what safety means. The societal value of committing scientific resources to address particular questions must also be considered. This consideration requires not only scientific knowledge of what may constitute a hazard, but also an understanding of how we have come to accept our present system of requirements.

It would be difficult to achieve a consensus on the best way to ensure safety if one had to design a safety testing system *de novo*. However, a remarkable consensus exists that the system which has evolved is effective at protecting public health and is achievable at an acceptable cost. This chapter presents an overview of how food ingredient safety assessments are made by the U. S. Food and Drug Administration (FDA). I will not presume more than a chemist's knowledge of toxicity testing, although I am sure that much of what I say will be well known by many in the audience. Several other chapters will provide more specific information on particular aspects of safety assessments.

Background

The FDA has had the authority to require safety testing of food ingredients only since 1958. Prior to that time, responsible companies tested ingredients for their own assurance that they were not selling a product that might be harmful, and government scientists occasionally tested ingredients that they thought might pose some risk. Industry testing often was done in consultation with government scientists to ensure wide acceptance of the results. Thus, when the Food Additives Amendment was passed in 1958, requiring premarket approval of all new ingredients, there was already a working consensus on the types of testing that would be needed. (The Food

Additives Amendment actually applies to the use of any substance that might become a component of, or otherwise affect the composition of, food. Thus, it also applies to packaging materials and food processing equipment. This chapter will focus on ingredients, however.) There are several principles that underlie the basis for such testing:

1. The dose makes the poison. Nearly any substance will be toxic at some dose. The objective is not to determine whether a substance can be toxic, but to determine whether some level of consumption can be considered safe.
2. In assessing toxicity, animal models can be used as surrogates for humans.
3. Different species, and individuals within a species, will vary in their sensitivity to a substance. Therefore, it is prudent to test in more than one species and to use sufficient numbers of animals to obtain statistically meaningful results.
4. Because of the variations described above, any extrapolation of data from one species to another introduces uncertainty. Therefore, safety factors have been applied to compensate for such uncertainty. Traditionally, a hundredfold safety factor has been applied to estimate a safe intake for humans based on a level producing no adverse effects during chronic feeding studies in animals; i.e., human consumption at a level 100 times less (in terms of the amount consumed in proportion to the body weight) than that producing no effect in animals is considered safe. Such a safety factor cannot be used for a substance that causes cancer at a higher dose, however, because a threshold for such an effect cannot be assumed.
5. The amount of testing required should be commensurate with the potential for risk posed by use of the ingredient.

When Congress passed the Food Additives Amendment, it allowed considerable discretion to government scientists on what testing should be required. The Food Additives Amendment does not require any specific testing to be done, although it requires sufficient data to conclude that the use of a substance is safe. Congress also recognized that complete certainty about the safety of any substance was impossible, a situation which has not changed with our substantially increased knowledge today. Correspondingly, FDA regulations define safety as a reasonable certainty in the minds of competent scientists that the substance is not harmful under the intended conditions of use.

Congress did include one provision restricting discretion: the Delaney Clause. This provision (which applies, in slightly different forms, only to food additives, color additives, and animal drugs) restricts the government from concluding that any additive is safe if it has been shown to induce cancer in man or animal when ingested or when applied by another route in an appropriate test. At the time of enactment, this provision was unlikely to affect decisions because in most cases scientists could not conclude to a reasonable certainty that use of a carcinogen would cause no harm. Its significance has probably been more symbolic than substantive. In recent years, however, we are seeing more examples in which a decision may depend solely on the Delaney Clause and efforts have been considered to amend or revoke it. The current

Secretary of Health and Human Services has stated that such a change would be appropriate.

In considering requirements for safety evaluation of ingredients, it is important to consider different categories of ingredients. Congress created three categories of exemptions from the requirement to demonstrate safety of an ingredient: (1) ingredients whose safe use is covered by other laws; (2) ingredients previously and explicitly found to be acceptable by the FDA or the U.S. Department of Agriculture; and (3) substances whose use is *generally recognized as safe* (GRAS) by experts qualified to make such determinations. This latter category is particularly significant in that a final decision on the safety evaluation is not reserved for the government. To ensure protection of public health, however, and to provide orderliness in decisions, FDA has involved itself in such GRAS determinations, establishing a petition process for positively affirming its agreement with independent GRAS determinations.

Under the Food Additives Amendment, GRAS determinations can be made on either of two bases: on a safe history of use prior to January 1, 1958, or on scientific procedures such as those used for food additives. The latter basis raises the question of how this differs from food additive approval. The intent of Congress is not clear but there is no evidence that Congress intended a standard weaker than that for food additives or that it intended to set up a dual system for premarket approval. FDA has concluded that the quality and quantity of data needed to demonstrate that a substance is GRAS by scientific procedures are the same as those needed to demonstrate the safety of a food additive, but that the data must be published in order to have general recognition. FDA expects that new substances would be evaluated as food additives.

Criteria for Safety Evaluation

An FDA review of the safe use of an ingredient is triggered by one of two circumstances: a petition to amend the regulations to permit a new use of an additive, or an agency initiated review stimulated by new data requiring a reconsideration of an earlier decision that use of an additive is safe. A reevaluation of earlier decisions depends primarily on the specific facts of a particular case and will, therefore, not be discussed here.

A sponsor petitioning for a change in the regulations to permit a new use of an ingredient bears the full burden of demonstrating that the requested use is safe. The petitioner should become expert on the safe use of the ingredient in question and the petition is the forum for demonstrating that expertise. FDA uses its general expertise to determine whether the petition provides an adequate demonstration of safety. FDA has issued several guidelines to aid the petitioner in preparing the petition, but the responsibility of providing adequate data is solely that of the petitioner. (These guidelines are available from the FDA upon request, with the exception of a more extensive set of guidelines on toxicity testing, discussed below, which is available from the National Technical Information Service for a nominal fee.)

A petition is, in effect, a scientific/legal document that must be sufficiently complete to allow any knowledgeable, objective observer to conclude that all reasonable safety questions have been addressed. Moreover, there must be sufficient detail for FDA scientists to reach their own conclusions on what the data demonstrate, independently of the conclusions of the original researchers.

Two important factors that govern the safety review are (1) that the safety of a substance being considered is safety under all conditions of use to be permitted and (2) that a determination that the requested use is safe applies to use by all possible companies, not just the petitioner. Thus, any controls needed to ensure safe use must be established before approval is granted so that all users of the ingredient are subject to the same controls.

Chemistry and Food Technology

The first criterion to be met is to establish an adequate identity for an additive. The common name for an additive usually defines either the major intended component of a commercial product or the source from which an additive is extracted. No commercial product is absolutely pure, however, so consideration must be given to the full range of components likely to be present in an additive under actual conditions of use. Possible source materials and manufacturing processes must be considered to determine the impurities likely to occur and multiple batches must be analyzed to determine actual composition and its variability. On the basis of such information, chemists and toxicologists can decide what specifications may be necessary to ensure safety. Any analytical method used to characterize a substance must be fully described and shown to be valid for the concentrations being determined.

The second criterion is whether the substance changes during use. This means that the types of food and the conditions under which the substance is intended to be used must be described. For example, does it decompose when heated or when present in acidic aqueous solutions? If so, what are the degradation products that will be consumed? Are restrictions needed for the types of foods in which the additive may be used? Does the petitioner want to request limits so as to avoid the need for addressing questions about conditions of use that may not be commercially important?

What is the technical effect to be achieved? In what amounts will an additive be used? In what types of foods? Is there a technologically self-limiting level of use such that the food would not be consumed if higher levels were present? This information is needed to assess how much of each component is likely to be consumed. If a tolerance is needed to ensure safety, that tolerance must be no higher than the amount needed to achieve the technical effect.

FDA uses consumer surveys to estimate portion sizes and frequency of eating so that, in combination with proposed use levels, it can reasonably estimate the amount of an additive likely to be consumed. Although consumer eating habits vary, FDA looks for the amount consumed by a person who eats relatively large amounts of the food in which the additive is used. This approach poses some problems when an additive is used in many different foods because the same person is unlikely to eat large amounts of each category of food. In such cases, a greater emphasis must be given to the average eater of the many different foods.

At the end of the chemistry evaluation, one should have a good idea of the amounts of each component of concern that are likely to be consumed if the requested permission to use the ingredient is granted. Sufficient information will have been presented to allow a chemist to verify the data, if necessary. One last important criterion must be considered. If the estimate of consumption depends on the establishment of limits to ensure purity or level of use, then the petitioner must present analytical methodology capable of verifying that such limits are being met. FDA may require samples of food containing the requested concentration of additive so that its analysts may evaluate the adequacy of the methodology in the laboratory, but in any case the petitioner must develop and validate the test methods.

Toxicology

FDA requires a core of toxicology data that depends on the substance and its use. Additional studies may be required to satisfy concerns raised during the initial, or core, testing. Feeding studies in laboratory animals are generally required. As noted previously, the extent of testing should be commensurate with the anticipated risks posed by the use of the substance. Therefore, FDA has devised a set of core requirements which considers both the chemical structure of the ingredient and the amount likely to be consumed to establish a "Concern Level" that gives guidance as to what studies are needed. Details for determining Concern Levels, their corresponding core test requirements, and guidelines for conducting tests are described in *Toxicological Principles for the Safety Assessment of Direct Food Additives and Color Additives Used in Food,* published by FDA in 1982 and available from the National Technical Information Service. This document, commonly known as the *Redbook,* is currently under revision and will be discussed later by other authors in this book.

FDA has established three structure categories in which all chemicals can be organized according to their functional groups. For each of these, ranges of dietary exposure are used to establish three Concern Levels, as shown in Figure 1. For the highest Concern Level, FDA normally requires carcinogenicity studies in two rodent species, a chronic toxicity study in a rodent species, a long-term (at least one year) study in a non-rodent mammal, and a two-generation reproduction study with teratology phase in a rodent species. Typically, the carcinogenicity, chronic toxicity, and reproduction/teratology studies are conducted as a combined study in a rodent. Generally, a 100-fold safety factor is applied to the no-observed-effect level to determine the maximum acceptable daily intake (ADI) for humans. (This approach is inapplicable, of course, to ingredients that are consumed in such large amounts that a 100-fold safety factor is impossible.) Observed toxic effects in any of the studies may indicate the need for more specialized studies to ascertain their significance for human health. For any petitioned ingredient, a thorough review of the toxicological literature is also needed to ensure that no relevant information is overlooked.

The *Redbook* also offers guidance on the design of specific studies to ensure that useful information will be obtained. Studies must be conducted according to good laboratory practices (as defined by regulation in 21 CFR 58) to ensure that results are credible. This requirement relates to the earlier discussion of ingredient identity and ensures that the lots of ingredient tested are representative of what would be consumed

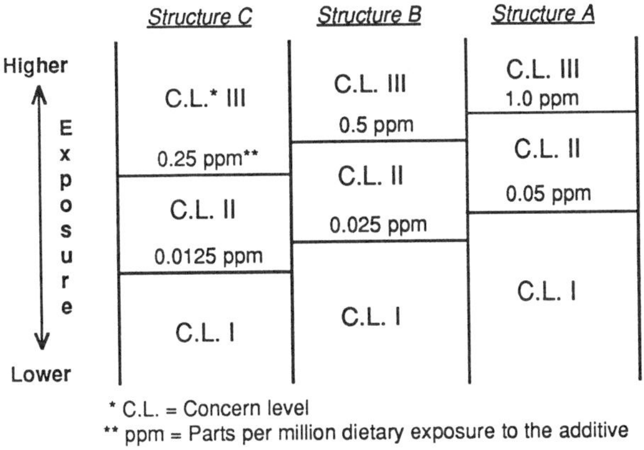

Figure 1. Concern level from exposure and structure. SOURCE: Toxicological Principles for the Safety Assessment of Direct Food Additives and Color Additives Used in Food; U.S. Food and Drug Administration, Bureau of Foods; Washington, DC, 1982.

by humans in the food supply. Also, good recordkeeping practices are essential to the usefulness of a study. FDA has required petitioners to provide more detailed information to resolve an issue because its significance was not clear from the original report. Resolution of such issues sometimes has required reevaluation of histopathology slides.

It is important to recognize that the *Redbook* is solely intended to provide guidance. The guidelines of the *Redbook* are not mandatory, because there is a need for judgment in making decisions on safety. There may be situations where other information is available that will help demonstrate safety without the need for full toxicological testing requirements.

Environmental Impact

All federal agencies are required by the National Environmental Policy Act to consider the environmental consequences of their actions, including the consequences of issuing rules permitting use of a food ingredient. Unless FDA can conclude that there will be no significant environmental impact from an activity to be permitted, and issues a public document stating the reasons, it must prepare an environmental impact statement. Although an environmental impact statement is rarely needed, there is still a need for data that would support a finding of no significant impact. Not surprisingly, FDA requires the petitioner to provide such information in an environmental assessment.

Data are needed to predict the environmental introduction, fate and effects of chemicals that would enter the environment through manufacture, use and disposal of a proposed ingredient. For the site of manufacture, FDA tries to avoid duplicating the environmental review of other governmental agencies by relying, to the extent possible, on a certification of compliance with federal, state, and local emissions requirements, including occupational exposure limits. The required data provide a basis for assessing the likelihood of an environmental impact. As with other aspects of premarket approval, foreign and American companies are treated alike. FDA is required to consider the environmental impact of its regulations anywhere in the world.

Although in most cases there will be little potential for a significant environmental impact from the use of a safe food ingredient, care is needed to ensure that situations where there could be a significant impact are not overlooked. An environmental assessment must be a complete document that, by itself, will show that there is no reasonable potential for an environmental impact. The FDA Center for Food Safety and Applied Nutrition's Environmental Impact Section makes available step-by-step guidance for petitioners preparing environmental assessments.

Nutrition

Ingredients may be nutrients or replacements for nutrients, or they may interfere with the utilization of nutrients. It is difficult to establish general guidelines for addressing nutritional concerns but one should be alert for potential effects. Animal feeding studies may reveal nutritional as well as toxicological information. The intended use will also provide information on the potential for adverse nutritional consequences. The need to generate new information is likely to vary with the individual case.

Microbiology

Finally, for some ingredients, the need for chemical information may be replaced, in part, by a need for biological information for one of two reasons. First, the technical effect to be accomplished by an ingredient may be biological. One needs data on efficacy to determine the amount of an ingredient needed to be effective as an antimicrobial agent. As stated above, if there is a need to set a tolerance on the amount of an ingredient to be used, the tolerance should be no higher than necessary. This limit is intended to prevent gratuitous use of ingredients. A corollary is that use of an antimicrobial agent at levels too low to be effective is also a gratuitous increase in one's consumption of the ingredient.

Second, in a growing number of cases, ingredients are being manufactured by biological rather than chemical means. The types of impurities that might be of concern will not be predictable from the laws of synthetic chemistry but will depend on the organisms used for manufacture. Biotechnology has been used in food manufacture for many years, but the new possibilities being made available through recombinant DNA technology present new issues to be addressed. Microorganisms used in food processing must be well characterized and understood to allow the design of a scheme that will provide an effective safety assessment. As a minimum, the following information is needed to evaluate the safety of their use.

1. Documentation and taxonomic identification of the specific strain of organism to be used.
2. Details of procedures used to guarantee cultural purity and genetic stability.
3. Quality control procedures to ensure use of a pure culture.
4. Description of methods to ensure absence of antibiotic formation by culture.
5. Evidence that microorganism isolates are neither toxigenic nor infectious.
6. Evidence of controls to ensure that viable cells of the production strain will not be present in food.

Summary

The data requirements for evaluating the safety of a food ingredient must be determined by specific case. In all cases, the ingredient must be adequately identified and there must be sufficient information on record to provide assurance that the amount consumed will not cause harm. The person intending to use the ingredient is responsible for ensuring that sufficient information is present in an official record for FDA to conclude that use of the ingredient is safe.

The cooperative efforts of government and industry scientists over several decades have led to a general acceptance of procedures for demonstrating the safety of food ingredients. Whether they are the best procedures is not answerable because, presumably, there may always be more efficient and effective procedures. FDA has issued many guidelines describing test procedures that it finds acceptable, and it will always consider proposals for more effective and efficient methods to determine the safety of food ingredients. The standard set by law is not that any particular tests be conducted, but that the use of the ingredients be demonstrated to a reasonable certainty to be harmless.

Bibliography

1. *Federal Food, Drug, and Cosmetic Act, As Amended*; U.S. Department of Health and Human Services, U.S. Government Printing Office, Washington, D.C.
2. *Title 21, Code of Federal Regulations, Parts 170-199*, U.S. Government Printing Office, Washington, D.C.
3. Rulis, A.M. *Food Drug Cosmetic Law Journal* **1990**, 45, 533-544.
4. Maryanski, J.H. *Food Drug Cosmetic Law Journal* **1990**, 45, 545-550.
5. Jones, D.D. and Maryanski, J.H. In *Risk Assessment in Genetic Engineering*; Levin, M.A. and Strauss, H.S., Eds.; McGraw-Hill, Inc.: New York, 1991, Chapter 4.
6. Kokoski, C.J.; Henry, S.H.; Lin, C.S.; Ekelman, K.B.; In *Food Additives*; Branen, A.L.; Davidson, P.M.; Salminen, S., Eds.; Marcel Dekker, Inc.: New York, 1990, Chapter 15.

RECEIVED August 15, 1991

Chapter 16

A Flavor Priority Ranking System

Acceptance and Internationalization

Otho D. Easterday[1], Richard A. Ford[2], Richard L. Hall[3], Jan Stofberg[4], Peter Cadby[5], and Friedrich Grundschober[6]

[1]International Flavors and Fragrances, Inc., Union Beach, NJ 07735–3597
[2]Research Institute for Fragrance Materials, Englewood Cliffs, NJ 07632
[3]Consultant, 7004 Wellington Court, Baltimore, MD 21212
[4]Consultant, 72 Wildwood Drive, Lake Monticello, Palmyra, VA 22963
[5]Product Safety Department, Firmenich SA, 1 route des Jeunes, CH–1227, Geneva, Switzerland
[6]International Organization of the Flavor Industry, 8 rue Charles-Humbert, CH–1205, Geneva, Switzerland

A system, accepted by the Codex Alimentarius Commission and others and developed by the combination of (1) the FEMA Decision Tree, (2) the FDA Prioritized Assessment of Food Additives (PABA — so called "Redbook", (3) the Consumption Ratio, and (4) the application of *Toxicological Adjustment Rules*, is briefly described. The system uses exposure estimates, structure–activity relationship elements, consumption ratio data, observed toxicity data, and adjustments to rank a large number of flavoring substances into seven levels of priority concern. We present information of the system's acceptance by National and Supra-National organizations. Sample exposure estimates for European and U.S. flavors are discussed in relation to the effect upon the priority level of concern. The current program with Supra-National organizations is described.

The number of flavoring substances known to be in use around the world, both as a result of intentional addition to food and as a result of their natural occurrence in foods, numbers in the thousands and may well exceed 10,000. While there is no evidence of any harm from these substances under normal conditions of use, the public is increasingly asking for evidence that these substances have been reviewed in a systematic manner using a scientifically sound, validated system.

Such a system or method of approach must be able to select out of the thousands of flavoring substances the few that, because of volume or use, chemical structure, etc., would be considered to have the highest priority for in depth evaluation. It must also provide confidence that those substances of lower priority present no significant risk as they continue to be used until such time as the review process can accommodate them.

A priority setting system (not identified as such) was called for as early as 1967 by JECFA [Food Agriculture Organization (FAO)/World Health Organization (WHO)

0097–6156/92/0484–0149$06.00/0

Joint Expert Committee on Food Additives] at their 11th meeting and in subsequent meetings. JECFA proposed consideration of consumption data and existing safety-in-use information of a substance as criteria for the selection of substances for further evaluation.

In responding to this request, a flavor priority ranking system was developed as described by Easterday, et al (*1*). The concept combined: (A) The Consumption Ratio-Food Predominant concepts (*2-10*), (B) The structure-activity- relationship "Decision Tree" (*11*, *12*) and (C) The computerized test data weighted method (*13*, *14*).

The concept was developed by an ad hoc group and applied to a set of flavoring substances derived from the Codex Alimentarius Commission's List B2 and an International Organization of the Flavor Industry (IOFI) subset. Working cooperatively with the Working Group on Flavors (a working party of the Codex Committee on Food Additives and Contaminants (CX/FA)) Codex List B and an IOFI subset was prioritized (*15*,*16*). A presentation to the CX/FA Meeting lead to the Working Group on Flavors' recommendation that one of the four methods examined be adopted by the CX/FA plenary session (*16*) (Rulis, A.M. et al., A Codex Flavor Priority Ranking System, Twentieth Session of the Codex Committee on Food Additives and Contaminants, Joint FAO/WHO Food Standards Programme, Codex Alimentarius Commission, 7-12 March, 1988, The Hague, unpublished data).

The prioritization of flavors was discussed further at a Strasbourg, France, Joint Council of Europe (COE)/European Community Workshop in 1987 by Rulis and Hall (*17*,*18*). Doctor Rulis, a member of the ad hoc working group, was invited to present the system and the results obtained to the JECFA (*19*). The system, as developed, is consistent with the principles discussed in the WHO monograph entitled *Principles for the safety assessment of food additives and contaminants on food* (*20*). The Codex Alimentarius Commission and JECFA have endorsed, or officially accepted the system. The Flavor and Extract Manufacturers' Association of the United States (FEMA) — Expert Panel uses the system as well. Further refinements were made and presented at the COE, 2nd International Consultation on Flavors (Hall, R. L., et. al., A Method for Prioritizing Flavor Substances for Safety Evaluation. 2nd International Consultation on Flavors, 27-28 April 1989, Strasbourg, Council of Europe, 1989, unpublished data.) and at the London Meeting on "Harmonization and Consistent Approach to Regulation of Flavors" (*21*). The latter two sessions further endorsed the use of the priority setting method for application to flavors.

The United States Food and Drug Administration (FDA) and IOFI use the system. The current data bank uses the FDA's mainframe computer. Other national industry flavor associations and/or national governments are supportive, participate or provide information. The work is coordinated by the International Committee on Flavour Priority Setting (ICFPS).

We briefly outline the historical developments within national and supra-national organizations. We summarize their endorsements, the official and unofficial acceptances and the progressive internationalization of the ad hoc working groups' effort to the current period.

This method is not intended and should not be used as a technique for safety evaluation. Clearly, however, any setting of priorities carries an implied estimate of what formal safety evaluation would likely confirm. It does this only in order to integrate and to apply systematically the available information and a sensible, broad, conservative judgement to insure that the more urgent risks are examined first, and in greatest depth, and that minor or insignificant risks are appropriately delayed, and not allowed to interfere or preempt limited resources. It should also provide confidence that those substances in a lower priority class may continue to be used with minimal risk until such time as in depth examination is possible.

The method used for establishing priorities among flavoring substances is composed of a set of simple procedures that, when applied to any inventory of flavor substances, produces subsets that can be designated for further review and evaluation. The work has been completed with the Codex List B and the chemically defined flavor substances approved for use in foods in the United States. However, the COE has discussed this technique for application to the flavors mentioned in its Blue Book (*22*), Pink Book (*23*) or subsequent revisions. The priority ranking of the flavoring substances submitted in the European Inventory to the European Community is currently being investigated.

The system is "risk-based". The highest priority is alloted to those flavors that have the greatest presumptive risks. The principal components of risk are inherent toxicity and exposure. The system is selective and can differentiate among substances thus creating subsets of substances of manageable size appropriate to the resources available for safety evaluation The system is structured for computerization, can incorporate a large range and number of substances and is flexible. It can easily incorporate new information for any substance. Thus, it can test the effect of and incorporate new data on existing priority lists.

Since classical toxicological data is lacking for many flavoring substances, the system can utilize other information in the absence of specific toxicity data for a particular flavoring substance. This method utilizes, in decreasing order, exposure (intake), structure-activity relationship information and natural occurrence in food. The method considers available toxicological data and past evaluations, utilizing information contained in current computerized data banks, rather than original study reports. The system does not underestimate risk. It is organized to be conservative. The priority setting method does not bias any special type of regulation. It is a rational and scientifically based priority setting system useful for focusing safety evaluation in any reasonable regulatory operation.

An inventory of any set of candidate substances to be ranked is prepared. The chemical structure of the substances must be identified with certainty. Following the construction of the inventory, a set of "hybrid priority levels" is formed. The initial assignment of "presumptive concern" for each flavor to be ranked is based upon two well established procedures. Both procedures use: (a) estimates of probable intake and (b) information about chemical structure. These procedures combine chemical structural information (in the form of assignments to discrete structural categories) with human intake (exposure) data to permit the allocation of all flavors into one of several "concern levels".

One method, the U.S. FDA Redbook procedure uses tables of chemical structures to assign flavors into structure categories (*24*). The second of the two procedures, the FEMA "Decision Tree", employs a decision tree composed of 33 questions to assign flavors into chemical structure categories (*11*). For both procedures, substances are assigned to one of three chemical structure categories corresponding to "low", "intermediate" or "high" presumptive toxicity. These structure assignments are combined with the estimates of exposure through intentional addition to food to categorize the flavors into one of several initial "concern levels", three for the FDA Redbook procedure and four for the FEMA Decision Tree procedure. The "concern levels" derived from the two procedures are combined into a single set of "hybrid priority levels". In this manner, the highest "concern level" by both procedures is correlated with the highest "hybrid priority level." Likewise, the lowest "concern level" is related to the lowest "hybrid priority level". Thus, by definition, the "hybrid priority levels" are the sums of the respective numbers characterizing the individual procedure's "concern levels", with the "hybrid priority level" 7 being the highest (Table 1).

Table 1. Merging Procedure For Hybrid Priority Levels

Hybrid Priority level	*RCL*[a]	*FCL*[b]
7	3	4
6	3	3
	2	4
5	3	2
	2	3
	1	4
4	3	1
	2	2
	1	3
3	2	1
	1	2
2	1	1

SOURCE: Reproduced with permission from ref. 19. Copyright 1989 World Health Organization.

[a]Concern level defined by the FDA Redbook method.

[b]Concern level defined by the FEMA Decision Tree method.

The next operation in setting flavor priorities is to adjust the hybrid priority level assignments by considering the quantity of the flavor substance's natural occurrence in food. This can be done by invoking the concept of "consumption ratio" developed by Stofberg (*8*). By definition, the "consumption ratio" is the ratio of the per capita intake (exposure) resulting from the flavor substance's natural occurrence in food to the per capita intake of the flavor from its intentional addition to food (*6, 26*). A large consumption ratio indicates that the human intake (exposure) from natural occurrence sources of the flavor in food is much larger than the intake that is derived from its intentional addition to food. Similarly, a consumption ratio of zero signifies that the flavor material does not have natural occurrence in food as currently known. Consumption ratios vary from 1×10^{-3} to 1×10^{5}.

For the assignment of flavors to priority levels, the effect of the consumption ratio depends upon whether the initial priority level is high or low. If the "hybrid priority level" is low, a high consumption ratio will lower the priority since it is extremely difficult to control a substance that is primarily consumed as a result of its natural occurrence in food. But, if the "hybrid priority level" for the flavor is high initially, a large consumption ratio suggests a larger level of risk that demands scrutiny despite difficulty of control. Because of these arguments, the priority setting method uses the consumption ratio to adjust the initial assignments to the "hybrid priority levels" as follows:

1. Any substance placed in "hybrid priority level" 7 should not have that priority reduced by the consumption ratio. The presumptive risk is in no way reduced by heavy intake of the flavour from natural sources, but is increased. However, there is no need for a higher priority than the highest already available. Thus, all substances in "hybrid priority level" 7 remain there.
2. Substances in "hybrid priority level" 6 with a consumption ratio higher than 100 should be moved to priority level 7, since intake of the substance from natural sources requires review. Consumption ratios of less than 100 should have no effect.
3. At "hybrid priority level" 5, the consumption ratio should be without effect.
4. At "hybrid priority level" 4 and below (and for consumption ratios ≥3.2), the hybrid level of concern should be reduced by the logarithm (base 10) of the consumption ratio rounded up or down to the nearest integer (Table 2).

Table 2. Effect Of Consumption Ratios On Hybrid Priority Levels Of 4 And Below

Consumption Ratio (CR)	*Log CR*	*Level Reduction*
CR ≥ 3200	≥4	-4*
320 ≤ CR ≤ 3200	3	-3*
32 ≤ CR ≤ 320	2	-2
3.2 ≤ CR ≤ 32	1	-1
CR ≤ 3.2	<1	0

SOURCE: Reproduced with permission from ref. 19. Copyright 1989 World Health Organization.

*The priority level should not be reduced below zero.

The next and last sequential operation for the priority setting system is an adjustment for toxicological data, if available, for a specific flavor present on the priority list. After adjustment for the consumption ratio, a set of guidelines are used to apply summaries of existing toxicological data and scientific judgments to determine a final priority assignment. It is important to use summaries, rather than original reports. The use of summaries not only expedites the process, but it also tends to counteract the all-to-easy temptation to evaluate, instead of simply to prioritize.

Unless otherwise stated, these guidelines are intended to apply only to oral data. The parenthesized references are to the definitions that follow the guidelines.

1. Seriously adverse data(a) not previously evaluated by the WHO/FAO Joint Expert Committee on Food Additives (JECFA) raise the substance to the highest priority level.
2. Adverse data weigh much more heavily than favorable data of equal quality.
3. Suggestively adverse data(b) not clearly overridden by substantially more data of higher quality raise the substance by three priority levels, or to the highest level unless guidelines 7 or 8, below, apply.
4. Data by non-oral routes or in non-mammalian species are given weight only in the absence of the oral data, unless there are data indicating relevance to ingestion.
5. Data from short-term (mutagenicity) tests have no weight unless, in the absence of chronic data, the results from two or more different tests are positive for mutagenicity.
6. A prior JECFA review that resulted in setting either a specific Allowable Daily Intake (ADI) (not a temporary ADI), or an "ADI not specified" reduces the priority level to zero unless 1, above, applies.
7. Data from chronic studies of at least moderate quality(c), showing no adverse effects at feeding levels 1000x probable daily intake, reduce the priority of a substance by three levels, but not below level one, unless guideline 1, above applies.
8. Data from subchronic studies of at least moderate quality(c), showing no adverse effects at feeding levels 1000x probable daily intake, from substances in priority levels five and below reduce the priority by three levels, but not below level one, unless 1, above, applies.
9. Data from LD_{50} tests have weight only in absence of data from repeated dose studies, and only if the LD_{50} is less than 100 mg/kg, in which case they raise the priority to the highest priority level.
10. Mixed data, generally favorable(e) but of poor quality and thus raising or leaving some questions, have no impact.
11. Data of poor quality(f) have no weight unless seriously adverse (*see* Guideline 1).

Definitions

(a) "Seriously adverse data" means data from other than single-dose acute studies that indicate the potential for proliferative lesions, necrosis, reproductive, or reproductive organ effects, developmental or teratological effects, or other irreversible effects.

(b) "Suggestively adverse data" means data that do meet the definition of "seriously adverse"(a), but that imply the presence of dose-related adverse effects that are very unlikely to be serious or irreversible, or to occur within reach of human exposure, but that nevertheless suggest a somewhat higher priority for future evaluation than if such effects had been absent.

(c) "Moderate (or credible) quality" means data from a study that does not meet guidelines for "high quality"(d) It does meet minimum standards, but has defects that render it inadequate for supporting persuasively the absence of adverse effects.

(d) "Higher quality data" are from studies that meet published guidelines for current scientific quality. (This term is not used in the toxicological guidelines, but is given here for clarity in connection with other definitions.).

(e) "Mixed data, generally favorable..." mean data from two or more studies that fall short of desirable minimum ("core quality") standards, and that do not agree in being entirely negative in their conclusions. Any adverse implications, in type and frequency of effects, and in the relation of the dose level(s) associated with such effects to possible human exposure, fall well short of "seriously adverse." Such effects are, at most, "suggestively adverse" and are outweighed in quantity and quality by data showing no effect.

(f) "Poor quality" means significantly short of desirable minimum quality standards in one or more critical respects, but not wholly lacking in some possible indicative value.

****

Almost regardless of priority, where there is certain knowledge of toxicological testing currently under way, it seems reasonable to suggest that it will generally be sensible to delay review until the results are available.

We regard the careful use of existing toxicological data, in a highly conservative way, to influence the final priority as only rational. If we wish to decide which of two substances to evaluate first, no reasonable toxicologist would choose a substance with data that show little or no risk over a similar substance on which no data are available. Likewise, if suggestively adverse data exist, even if not clearly relevant to the final safety evaluation, it is only logical to raise the priority so that such data can be reviewed in a timely manner.

The ad hoc Committee has applied the priority setting system to two flavor lists. They were the Codex List B - IOFI subset (602 flavoring substances) and a list of 1229 U.S. approved synthetic flavors. The data base was created in the FDA computer by merging files from the FDA computer with data files from FEMA and from the National Academy of Sciences. The resulting data file is in a d-Base III data base and is conveniently accessible. For each flavor ingredient, the database contains:

1. Chemical name;
2. FEMA number;
3. Chemical Abstract Service Registry Number (for certainty identification);
4. Exposure or intake (total number of pounds, which disappears yearly into the U.S. food supply);

5. FEMA chemical structure category;
6. FEMA concern level;
7. Redbook chemical structure category;
8. Redbook concern level; and
9. Consumption ratio (if available).

The system's computer printout is illustrated for Codex List B - IOFI subset priority level 7 substances in Table 3 as produced on April 30, 1990:

Table 3. Illustrated List Of Codex List B - IOFI Subset Flavors For The Codex Committee On Food Additives And Contaminants (1987 Poundage)

Final Priority Level 7					*R*		*F*					
CAS No.	*Name*	*ID*	*CRatio*	*Poundage*	*S*	*C*	*S*	*C*	*L*	*LC*	*TR*	*P*
000075070	Acetaldehyde	2003	12.10	112000.00	A	3	1	C	6	6	1	7
000123682	Allyl Hexanoate	2032		9620.00	C	3	2	C	6	6	3	7
000057067	Allyl Isothyocyanate	2034	5.306	38200.00	C	3	2	C	6	6	3	7
002835394	Allyl Isovalerate	2045		2.00	C	1	2	A	2	2	1	7
004180238	Anethole	2086	1.788	37700.00	C	3	3	D	7	7	1	7
000087296	Cinnamyl Anthranilate—Prohibited	2295		0.00	C	1	2	A	2	2	1	7
000119846	Dihydrocoumarin	2381	0.000	8270.00	C	3	3	D	7	7	10	7
000140670	Estragole	2411	1.230	300.00	C	1	3	C	4	4	1	7
000121324	Ethyl Vanillin	2464		1570000.00	C	3	2	D	7	7	10	7
000050215	Lactic Acid	2611	0.000	3180000.00	C	3	1	D	7	7		7
067633970	3-Mercapto-2-Pentanone	3300		1.00	B	1	1	A	2	2	1	7
000092488	6-Methylcoumarin	2699		840.00	C	2	3	C	5	5	1	7

Symbols: R = FDA Redbook; F = FEMA Decision Tree; S = Structural Category; C = Concern Level; L = Initial Hybrid Concern Level; LC = Initial Hybrid Concern Level corrected by Consumption Ratio; TR = Toxicological Adjustment Rule Applicable; and P = Final Priority Level after Toxicological Rule Adjustment.

These listings or printouts can be merged, sorted by CAS or FEMA numbers, by priority level, or produced alphabetically. Table 4 illustrates the Codex List B - IOFI Subset as produced on April 30, 1990 by priority level for the initial assignment, and as adjusted for consumption ratio and toxicological data (rule). A similar tabluation can be produced for the U.S. approved flavors.

The tabulation illustrates that the system assigns relatively few substances to the highest priority levels, while the largest number of substances are assigned to lower levels.

The method can distinguish a relatively small proportion of substances that may be considered to be of high presumptive concern in a large inventory of flavors.

The ad hoc committee working with the Codex organization has expanded its scope cooperatively with the Council of Europe and other interested national and supra-national organizations including the European Community and the International Organization of the Flavor Industry (IOFI). Recently, the name of the working group was changed to the International Committee on Flavour Priority Setting (ICFPS) to reflect its international activity. Additional internationally recognized individuals

Table 4. Number Of Flavors Assigned To Priority Levels Before And After Adjustments

Priority Level	*Number of Compounds in each Priority Level* *Initial Assignment*	*After Adjustment for Consumption Ratio**	*After Adjustment for Toxicological Data*
7	5	5	6**
6	30	30	17
5	17	17	10
4	70	61	42
3	94	94	87
2	385	365	313
1	1	19	86
0	—	11	31
TOTAL(S)	602	602	602

SOURCE: Reproduced with permission from ref. 19. Copyright 1989 World Health Organization.

*The substances in this table were chosen from the Codex List B - IOFI Subset which is primarily a list of artificial flavoring substances, so the effect on the consumption ratio is under represented

**The numerical difference between Table 3 and Priority Level 1 Substances is due to the dynamics of the data at the time each tabulation was prepared (a time-frame difference).

have been or will be invited to join as members of this Committee. The current committee is shown in Table 5.

In order to prepare the European Community's (EC) Flavor Directive, a decision at the Council of Ministers required the preparation of a flavor inventory. The EC's inventory results from the collection of those substances permitted for use by any of the EC member countries. This inventory is large. It contains several thousand flavoring substances and flavoring source materials. It includes artificial, nature identical and natural flavoring substances and flavor source materials. The European historical regulatory environment favored the use of nature identical and natural flavoring ingredients which tends to increase significantly the number of substances available for use in Europe as compared to the US or other governments. This situation causes the EC's inventory to be quite large. The distribution of substances and their volume of use has been illustrated by Cadby (*21*). These distributions are shown in Figures 1 and 2.

The question has been raised if the US consumption data obtained through poundage surveys by the National Academy of Sciences can be used as appropriate exposure data for priority setting purposes for the European lists since there has never been comparable surveys conducted in Europe. Flavor experts have tentatively responded by commenting that the US and European cultural and eating consumption patterns are converging. IOFI has conducted an exploratory survey for approximately 50 materials used in the EC which were included in the US and Codex priority setting system lists. This IOFI survey has indicated some consumption differences between the EC and the US data; however, the differences were not sufficient to significantly affect the priority level previously determined for these substances. The preliminary findings suggest that the US data can be used until better European consumption data can be obtained. At the request of the COE, a second set of approximately fifty substances are now being examined.

It is possible, in a very short time period, to organize the EC flavor inventory for priority setting purposes; that is, to assign the Redbook and FEMA chemical structure categories and to estimate the concern levels for creating the required input for the computer to determine the "hybrid priority levels". Experienced flavorists can provide reasonable estimates for exposure (intake). Since analytical data on natural occurrence are available as background information for the discovery of nature identical flavoring substances, utilized more in Europe, it is possible to make reliable estimates for the consumption ratio values. As Cadby has mentioned, the total number of nature identical substances is large; thus, the likelihood for high exposures (intake) among these substances will be individually low. The system is sufficiently exposure driven so the low exposures will tend to lower the substance's priority level.

These estimates will provide to the priority setting system all of the necessary input data for the operation of the system and the production of a preliminary priority level sort. From this sort, we can concentrate upon levels 7 through 4 (the substances of highest concern) for additional refinement. "Refinement" is: (A) The development of more precise concern levels; (B) Conducting surveys for exposure (intake) data; and (C) clarifying the consumption ratio estimates.

Table 5. Members of the International Committee on Flavor Priority Setting

Member	*Affiliation*	*Country*
Burdock, Dr. G.	FEMA	U.S.A.
Cadby, Dr. P.	IOFI	Switzerland
Dodgen, Mr. D.	FDA	U.S.A.
Ford, Dr. R.	RIFM	U.S.A.
Grundschober, Dr. F.	IOFI	Switzerland
Gry, Dr. J.	COE/EC	Denmark
Hall, Dr. R. L.	FEMA	U.S.A.
Hardinge, Ms. J.	EFFA	England
Herrman, Dr. J.	JECFA	Switzerland
Irausquin, Dr. H.	FDA	U.S.A.
Nally, Ms. R.	USDA	U.S.A.
Ronk, Mr. R.	FDA	U.S.A.
Rulis, Dr. A.	FDA	U.S.A.
Schwartz, Dr. P.	FDA	U.S.A.
Shenkenberg, Mr. D.	USDA	U.S.A.
Stofberg, Dr. J.	FEMA	U.S.A.
Liaison:		
Crawford, Dr. L.	USDA	U.S.A.
Emerson, Dr. J.	FEMA	U.S.A.
Newberne, Dr. P.	FEMA-FEXPAN	U.S.A.
Pisano, Mr. R.	FEMA	U.S.A.
Shank, Dr. F.	FDA	U.S.A.
Thompson, Mr. D.	FEMA	U.S.A.
Chairman:		
Easterday, Dr. O.	FEMA	U.S.A.
Vice Chairperson:		
Howell, Ms. J.	FEMA	U.S.A.

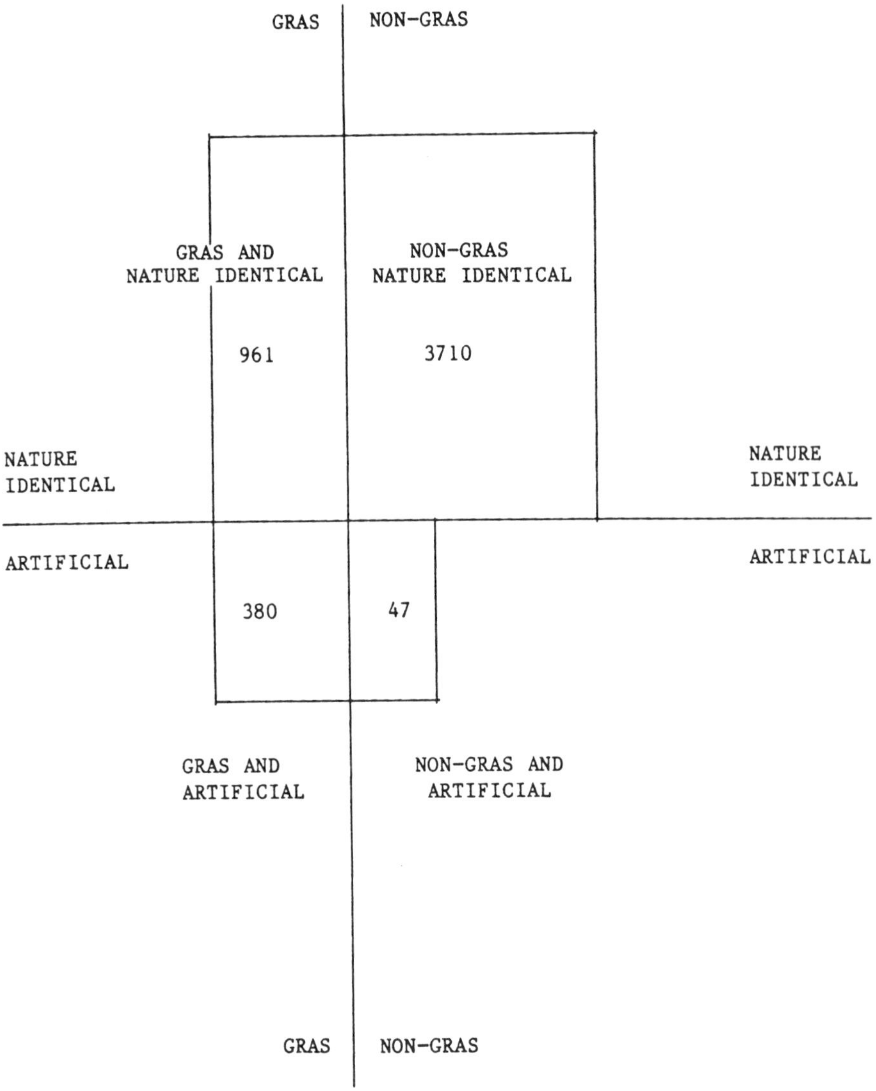

Figure 1. Numerical Distribution Into Four Sub-Categories of the Flavoring Substances Submitted to the EC Commission.

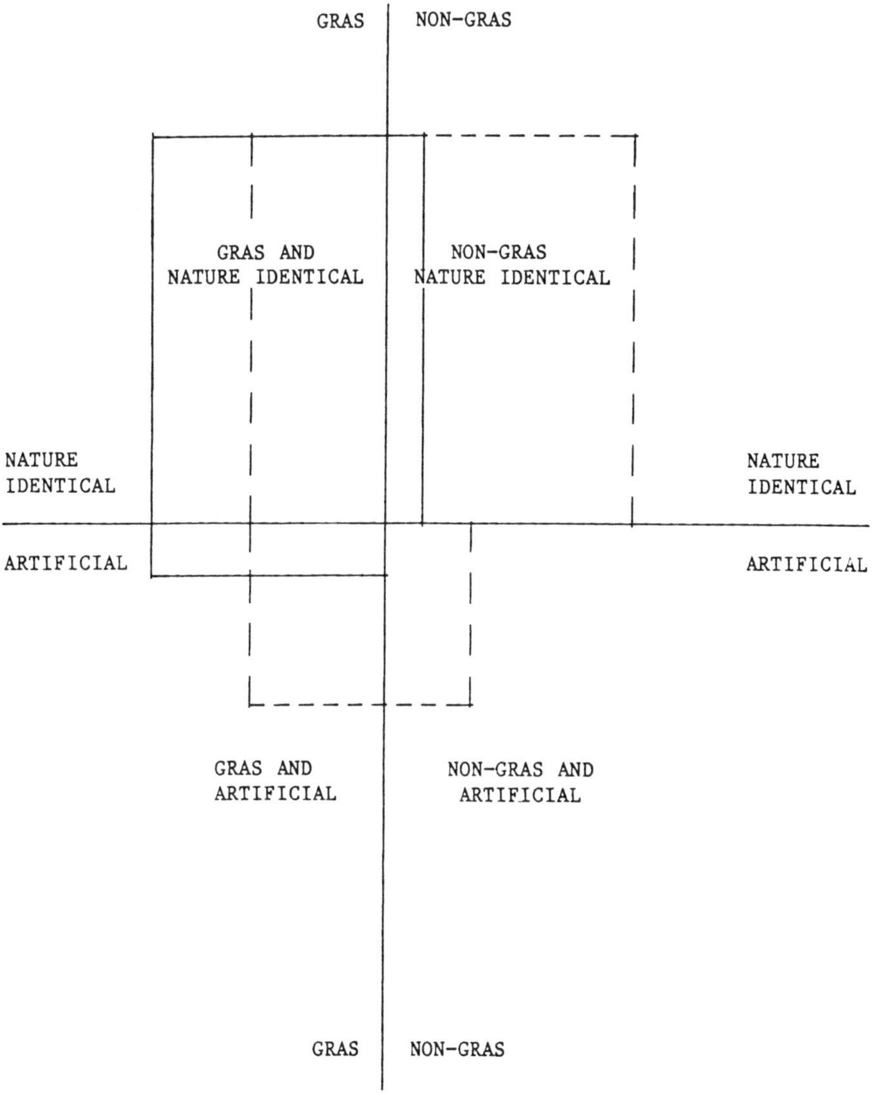

Figure 2. Relative Usage Volumes in Europe of the Different Sub-Categories of the Flavoring Substances Submitted to the EC Commission (Estimated)

An inquiry on volume of use of several hundred additional substances is now being undertaken in Europe for those substances having no exposure data (current or potential use in commerce).

Because many substances have currently little or no usage, they would be of low priority for further evaluation. They should, however, be maintained in the inventory and may achieve higher priority levels when their use increases. By applying the priority setting system on the existing inventory the high priority substances requiring safety evaluation will be identified.

Other regulatory harmonization matters have been examined by Cadby (*21*), such as the development of more uniform flavor regulations, the mutual acceptance or reciprocity of expert body evaluations. These considerations will globally decrease duplicating government and industry costs. It will reduce future testing, animal use and safety evaluation effort.

The International Committee on Flavour Priority Setting (ICFPS) is expanding its international membership and is now alternating its meetings in Europe and the United States. The Committee is composed of government and industry representatives working together for the common objective of preparing priority lists of flavoring substances. These priority lists can be used for subsequent safety evaluation purposes by national or supra-national organizations, for any set of flavoring substances. Joint discussions are underway for exchanging computer software and data banks and for providing the necessary training assistance for using the priority setting system. This participation with the ICFPS and elsewhere will assure the required experience and judgement making the priority setting system a valuable product for subsequent safety evaluation of flavors or, with some modifications, on any set of chemical substances, for example, pesticides, drugs, etc.

In summary, the priority setting method described can be used for prioritizing flavor substances for further safety evaluation. The system is focused to arrange substances in terms of their safety concerns and not to provide determinations of safety. The system is designed for use for any inventory, of any size, of discrete chemical substances and it uses simple procedures. The system provides a screening procedure, based upon chemical structure, human exposure, consumption ratio and toxicological adjustments which is operationally applicable even in the absence of complete toxicological data. It permits the use of summary toxicological information without the detailed examination of original data.

In addition the system provides also the rationale for a regulatory approach characterized by a list with the top priority flavoring substances and a temporary authorization for using flavoring substances of lower priority ranking which have not yet been evaluated.

This chapter has briefly outlined the development of the priority setting system, its acceptance by national and supra-national bodies. We have discussed the international role of the International Committee on Flavour Priority Setting with regard to future cooperation with the European Community, Council of Europe or other interested parties. The Committee's work is uniquely carried out cooperatively with government and industry participation through its membership in association with flavor organizations such as IOFI, FEMA and others to achieve a common goal for developing priority setting lists required for subsequent safety evaluation. The Committee's work is efficient, economical and productive.

Acknowlegments

The authors of this paper wish to thank the following staff members of the Division of Food Chemistry andTechnology, Center for Food Safety and Nutrition, Food and Drug Administration, 200 "C" Street, S.W., Washington, DC 20204, for their valuable contributions in the preparation of this paper and to the development of the Flavor Priority Setting System: Drs. Alan M. Rulis and Hiltje Irausquin.

Literature Cited

1. Easterday, O. D., et al. *A Combined Three-Method Safety/Risk Priority Ranking System*, Hommage au Professeur Rene' Truhaut, Jubile scientifique du Professeur Rene' Truhaut, Paris, 1984; p 338-342.
2. Stofberg J., *Perfumer & Flavorist*, **1977/1978**, 2 (7), 20-25.
3. Stofberg J., *Perfumer & Flavorist*, **1978/1979**, 3 (6), 62-64.
4. Stofberg J., *Perfumer & Flavorist*, **1980**, 5 (4), 17-22.
5. Stofberg J., *Perfumer & Flavorist*, **1981**, 6 (4), 69-72.
6. Stofberg J. and Stoffelsma J., *Perfumer & Flavorist*, **1980/1981**, 5 (7), 9-35.
7. Stofberg J., Flavor-Safety Evaluation, Food Engineering Int'l., 1983, 41.
8. Stofberg J., *Perfumer & Flavorist*, **1983**, 8 (3), 61-64.
9. Stofberg J., *Perfumer & Flavorist*, **1983**, 8 (4), 53-62.
10. Stofberg J. and Grundschober F., *Perfumer & Flavorist*, **1984**, 9 (4), 53-83.
11. Cramer G. M. et. al., *Fd. Cosmet. Toxicol.*, **1978**, 16, 255-276.
12. Ford R. A., and Hall R. L. *Proc., Annual Summer Meeting, The Toxicology Forum*, **1983**, p 258-272.
13. *Toxicological Principles for the Safety Assessment of Direct Food Additives and Color Additives Used in Food*, Bureau of Foods, Food and Drug Administration, 1982.
14. Rulis A., et al., *Reg. Toxicol., and Pharmacol.*, **1984**, 4, 37-56.
15. Proc., Up-dated Codex List B of Food Additives, Report of the Eighteenth Session of the Codex Committee on Food Additives, Joint FAO/WHO Food Standards Programme, Codex Alimentarius Commission, ALINORM 87/12, Appendix V, 76-90, 5-11 November 1985, The Hague.
16. Proc., Priority Ranking System for Flavors, Report of the Twentieth Session of the Codex Committee on Food Additives and Contaminants, Joint FAO-WHO Food Standards Programme, Codex Alimentarius Commission, ALINORM 89/12, Appendix VI, 51-58, 7-12 March, 1988, The Hague.
17. Rulis, A.M. Proc, Priority Setting of Flavors. In *Partial Agreement in the Social and Public Health Field* (P-SG (88) 9); Joint Council of Europe/Commission of the European Communities Workshop on a Priority Ranking System for Flavorings, 3-4 December 1987, Strasburg, Council of Europe, 1988.
18. Hall, R. L. Proc., A Suggested Priority Ranking System for Flavors In *Partial Agreement in the Social and Public Health Field* (P-SG(88) 9); Joint Council of Europe/Commission of the European Communities Workshop on a Priority Ranking System for Flavorings, 3-4 December 1987, Strasbourg, Council of Europe, 1988.

19. Proc., A Method for Setting Priorities for the Safety Review of Food Flavoring Ingredients. In *Evaluation of Certain Food Additives and Contaminants*, Thirty-third Report of the Joint FAO/WHO Expert Committee on Food Additives, Annex 4. World Health Organization (Technical Report Series 776), Geneva, 1989.
20. Proc., *Principles for the Safety Assessment of Food Additives and Contaminants in Food*, World Health Organization (WHO Environmental Health Criteria, No. 70) Geneva, 1987.
21. Cadby, P. Proc., Priority Setting and the Harmonization of Flavor Regulations, Accepted for publication. *Harmonization and Consistent Approach to Regulation of Flavors*, St. Mary's Hospital Medical School, 15-16 May, 1990, London, 1990.
22. Proc., *Flavouring Substances and Natural Sources of Flavourings*, 3rd Edition, Council of Europe, Strasbourg, 1981.
23. Proc., *Flavouring Substances Not Fully Evaluated*, Council of Europe, Strasbourg, 1981.
24. *Toxicological Principles for the Safety Assessment of Direct Food Additives and Color Additives Used in Food (the Redbook), Bureau of Foods*; U.S. Food and Drug Administration: Washington, D.C., 1982.
25. Stofberg, J. and Kirschman, J. *Food and Chemical Toxicology* **1985**, *23*, 857.
26. Stofberg, J., Proc., Priority Setting of Flavor Safety Evaluation Based on Relative and Total Consumer Exposure and Structure Activity Relationship. In *Partial Agreement in the Social and Public Health Field* (P-SF(88)9). Joint Council of Europe/Commission of the European Communities Workshop on a Priority Ranking System for Flavourings, 3-4 December 1987. Strasbourg, Council of Europe, 1987.

RECEIVED September 16, 1991

Computer Modeling of Risk Assessment

Chapter 17

Expert Systems and Neural Networks in Food Processing

George Stefanek and John M. Fildes

Illinois Institute of Technology Research Institute, Chicago, IL 60616

Expert system and neural network technology has recently matured to the point where it is often economically attractive to apply it in an industrial setting. This paper presents an overview of expert system and neural network technology, describes relevant applications in sensor fusion, diagnostics, and process control, and discusses factors which determine where to appropriately use each technology.

Recent studies have established the potential for application of new model-based process control strategies (*1*). These studies indicate that improved control systems will be developed by integrating artificial intelligence (AI) techniques with new and improved sensor technologies. These systems offer the promise of better productivity and more consistent control of product quality. Although these studies are not specifically focused on food processing, their conclusions are applicable to the food industry.

Process control will be increasingly performed by distributed control systems composed of programmable logic controllers for regulatory control and networked personal computers and workstations for supervisory control. Advances in control methodologies will increasingly come in the form of software, especially AI. Expert systems and neural networks are starting to be used for sensor fusion, process diagnostics, and model-based real-time process control. By implementing these technologies in appropriate applications, it has often been found that a 10:1 rate of return can be realized (*2*).

The food industry has not embraced some of the new software technologies to improve the process control environment. AI technologies are often not well understood and their cost justification in specific applications is unclear. The requirements for regulatory compliance present additional barriers to adopting flexible control strategies in food processing. AI-based control systems can provide greater efficiency, quality, and lower cost — but these benefits cannot be fully realized unless AI is applied in regulated aspects of processing.

0097–6156/92/0484–0166$06.00/0

This article will survey the application of AI technologies in food processing beginning with the concepts behind expert systems, and neural networks. Special emphasis will be placed on the use of these technologies for (1) sensor fusion to improve the safety and stability of food process operations, (2) diagnostics to minimize time in troubleshooting machinery and processes, and (3) statistical process control to improve the safety and efficiency of food process operations. The factors that govern the selection of an expert system, or a neural network, or a hybrid system, will be explained for each type of function.

Background

Food processors must balance productivity against requirements for quality and regulatory compliance. The corporate response to this challenge has been to employ techniques such as statistical quality assurance (SQA), statistical process control (SPC), and mechanistic modelling. SQA uses off-line measurements of a product's performance to ensure that the process is under control. The off-line measurements include chemical analysis and taste-testing. SPC is an extension of SQA that involves correlation of a product's performance with on-line measurements of controllable factors such as pH, temperature, pressure, and flow rate. Development of an SPC model has historically utilized regression techniques and these have proven useful only when there are a small number of controllable variables. Mechanistic models are an alternative to statistical techniques. A mechanistic model is based on a rigorous mathematical description of the process. It is difficult to establish and it is not easily moved from one installation to another.

A major difficulty with conventional measurement and control technologies is the inability to handle many of the situations that occur in food processing. Measurements in food processing involve noisy signals and the need to assess subjective quality factors such as taste, odor, texture, and color. Plus, control strategies have to accommodate non-linear regulatory control functions, supervisory control strategies for process optimization, and fault detection and diagnosis.

Utilization of artificial intelligence in process control systems answers many of these needs. Expert systems embody the experience of process engineers, so they facilitate fault diagnosis and process optimization subject to constraints. Neural networks allow the construction of complex empirical models of sensor transfer functions and control functions, so they handle non-linear situations. Neural networks and expert systems can both be used for pattern recognition, noise filtering, and data reduction. In this case, the choice between the two approaches will depend on the differing developmental requirements for each system. These distinctions will be explored further in this paper.

Some industries such as the chemical and power industries have already built and deployed prototype AI-based systems that do process control. In the chemical industry, much research has been done to produce prototype systems which validate the usefulness of AI technology in solving problems that are similar to those encountered in food processing. Some of these applications have included plant-wide control strategy planning (*3*), supervisory level real-time process control (*4*), and supervisory control for chemical reactor fault tolerance (*5*). Expert systems have been applied at the supervisory level to tune controllers, perform process and control system fault diagnosis, and to restructure the control system (*6*).

In the power industry, prototype expert systems have been developed to do boiler tube failure diagnosis, turbine condition monitoring, condenser and feed water heater diagnosis (*7*), scheduling of power units for large power plants (*8*), among others (*1*).

The following sections will examine the details of AI technologies and their application to problems in food processing. Expert systems will be described first, followed by neural networks. Finally, applications will be presented in the areas of sensor fusion, diagnosis, and real-time process control.

Expert Systems

Expert systems are computer programs that contain the expert knowledge of a specialist in a specific application domain, and the ability to reason through the knowledge to establish a concluding hypothesis. The concluded hypotheses may be decisions for diagnosis, control, scheduling, planning, or a myriad of other applications. The power of expert systems comes from the capability of being able to represent abstract knowledge symbolically. By using symbol representation, any kind of knowledge can be described and formalized, including high level heuristic concepts. These high level concepts can be used by a reasoning paradigm within an expert system program to emulate human-like decision making.

An expert system (Figure 1) consists of the following components: (1) a knowledge base for a particular application, (2) a structure to represent the knowledge, and (3) an inference strategy to reason through the knowledge.

Rule-Based Systems. The knowledge base contains the formalized knowledge of an expert in a narrowly defined application domain. The knowledge is represented as If-Then-Do rules or Objects with associated properties and methods (*10*). In rule based systems, the "IF" part of a rule consists of a set of premises corresponding to conditions. The conditions in the "IF" part of a rule can have a string symbolically represent a condition which can be in the form of a relational triplet (e.g. temperature > 200), a boolean expression (e.g. relay_valve_is_broken TRUE), or a check of set membership (e.g. Limit_Valve_5 is a member of working_LV_set). Conditional statements can be linked with either an "and" or "or" conjunction. The "THEN" part of the rule is a hypothesis that is set to TRUE or FALSE depending on the evaluated conditions. The "DO" portion of a rule invokes some actions to be taken, given the hypothesis is TRUE. These actions may include the execution of a program, setting a flag to some value, introducing or retrieving data, etc.

Object-Based Systems. Knowledge can also be represented within objects that belong to a class hierarchy. Representing data and knowledge in object form, communicating between objects using messages, and having objects take actions via prescribed methods is called object-oriented programming. Objects in a class hierarchy inherit data from other objects higher in the hierarchy. Each object

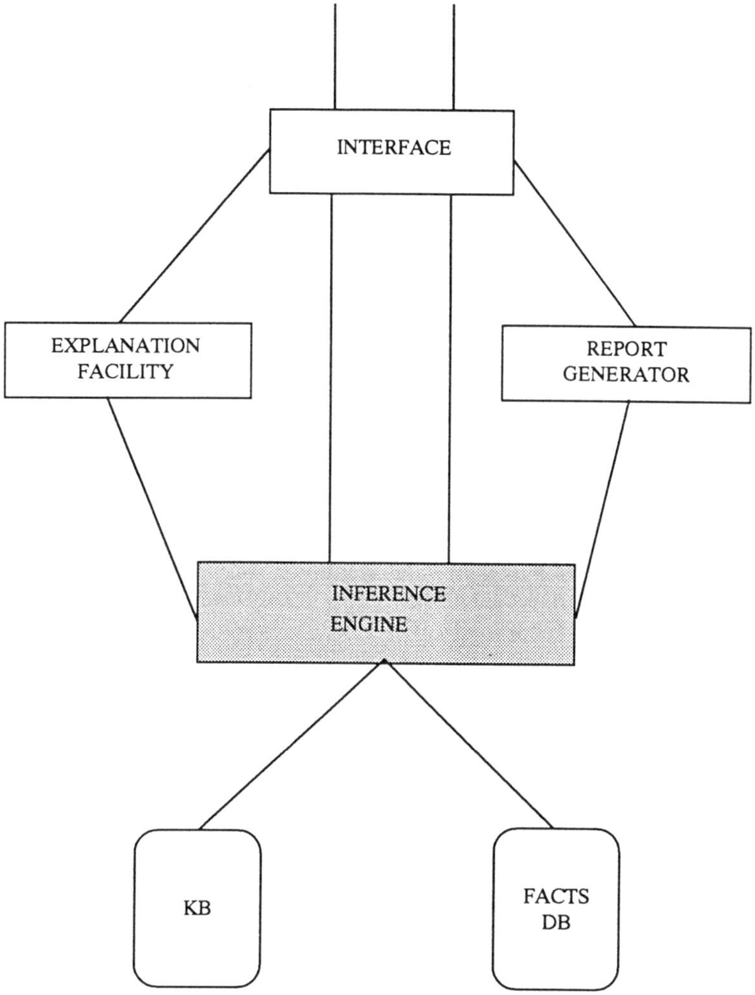

Figure 1: Expert System Modules.

encapsulates data and knowledge and stores it in specific object properties. Each object property or slot may have methods associated with it that retrieve data or take some action in response to a property access or change.

An example of the use of objects in process control is found in representing the operating state of a system at any given time, i.e. state estimation (*11*). Each state is specified by a set of events and each event is defined by a set of specifications. Each specification set identifies a sensor modelling as an object along with its properties such as sampled value, setpoint, comparison operator, and the minimum and maximum times that a condition must last. Another example where an object representation can be used is in representing a system topology in a process control environment. Objects can represent relay valves, limit valves, pipes, mixing tees, branch tees, actuators, controllers, and transmitters. The objects in the topology indicate their relationship in the topology and any detailed information about the object.

Figure 2 shows an example of a limit-valve object hierarchy containing properties showing each object's location within the topology.

Object representation and rule representation can be used together to form a hybrid system which more accurately describes the knowledge and reasoning processes within a domain. Typically, an object-oriented approach to representing data is used when data with many properties must be represented.

Generally, control and heuristic knowledge is better expressed in rules since it can be represented symbolically and is more easily and explicitly formalized using the IF-THEN-DO paradigm. Also, model-based approaches use object representation to describe the structure and behavior of a system. The model based approach doesn't need to use heuristic rules extracted from an expert, but can use structural and behavioral knowledge for diagnosis or control. These approaches tend to be computationally expensive therefore hybrid systems which use some rule-based heuristics are a more pragmatic alternative.

Inferencing. The interrelationships within a knowledge base can be complex, containing large interrelated decision trees with many levels. In order to traverse decision trees represented as rules in the knowledge base, an inference engine must be used. The inference engine is software that operates on the knowledge base using a methodology to solve problems that simulates human reasoning. The most common reasoning mechanisms are called forward chaining and backward chaining. Forward chaining works from a set of facts to try to establish all rules whose conditions satisfy the facts. It works from an initial state toward a goal state. The inference engine cycles through the knowledge base selecting rules whose conditions are met and putting the resulting true hypotheses in the facts database after each cycle until no more rules can be established as true. If many rules are selected during a single cycle, then the inference engine may use a conflict resolution strategy, set by the designer, to select the best rule or object. The concluding rule or object on the last cycle is the final conclusion.

Backward chaining starts with a hypothesis and tries to find support for that hypothesis. It works from a goal state to an initial state. This requires matching facts to conditions and finding support for the hypothesis in the knowledge base. The type of inference engine that is chosen is based on the application. For instance, in a diagnostic expert system a hypothesis is chosen and its conditions are evaluated for support. If support exists, then the actions associated with the goal hypothesis are executed. In contrast, a forward chaining mechanism would be used for state estimation in process control. In this case, sensor data would be supplied to the inference engine which would infer a conclusion about the state of the system from the forward propagation of known facts.

Knowledge Acquisition. In developing an expert system, a knowledge acquisition phase is conducted. The knowledge acquisition phase consists of an interview process where a knowledge engineer attempts to elicit knowledge from the domain expert. The knowledge acquisition process continues in cycles where each cycle consists of eliciting a comprehensive subset of the knowledge from the expert, formalizing the knowledge using one of the knowledge representation strategies, and presenting the formalized knowledge to the expert for review and verification. The knowledge acquisition phase is the most difficult and time consuming task in developing an expert system. Some experts are able to communicate their knowledge of a subject precisely and comprehensively while others have a difficult time expressing themselves. Often the expert does not realize all the fundamental knowledge that is used in order to make a decision and therefore omits some of this knowledge during the interviews. This will necessitate additions or changes to the knowledge base followed by reverification from the expert.

There is an on-going debate as to whether domain experts or knowledge engineers should develop expert systems. The accuracy and validity of an expert system correlates directly with the quality of the domain expert's knowledge and the ability to communicate that knowledge. Some companies have begun training process control engineers to use AI technology since they believe that it is more practical and important to have domain experience rather than AI experience. The difficulty in this philosophy is that AI technology can be complicated in itself. Superficial knowledge may lead to naive and poorly designed systems. Choosing the project team for designing an expert system thus becomes a very important aspect to the success of the project.

Additional complications in developing expert systems arise in process control. Frequently, the process evolves through empirical knowledge. In this case, the expert system only propagates the existing non-optimal procedures. An alternative approach has recently been suggested (*12*) to incorporate statistically designed optimization experiments into the knowledge acquisition process. This should produce an expert system that embodies optimized processing procedures.

Expert System Tools. Expert system development environments or shells are available from many vendors (*13*). Some are based on conventional languages such as C and others on the LISP programming language. There are specific shells for doing diagnostic expert systems, real-time systems, and others which contain implemented features which don't have to be designed by the developer. The recommended platform for development of expert systems is either a workstation, or a 386-class PC.

After development, the most likely run-time environment is a 286 or 386-class PC, or a workstation running a multi-tasking operating system.

Neural Networks

Neural network programs consist of a network processing architecture that functions similarly to the way the brain processes information. Neural networks, sometimes known as connectionist architectures, are parallel distributed systems that consist of the following components (*14*): (1) a set of nonlinear neural processing elements (nodes), (2) interconnections between the nodes, (3) weights associated with each interconnection, (4) a transfer function associated with each node, and (5) a learning algorithm to adjust the strength of connections.

Neural Network Architecture and Processing. The neural network processes incoming data in parallel where each of the input neural processing elements feeds the input vector data to the middle layer of the net as shown in Figure 3. Each neural processing element j sums the product of the inputs X_i and weights W_{ij} of the incoming interconnections from the previous layer, subtracts a constant bias B_j, and then applies an activation function F to generate an output Y_j:

$$Y_j = F(\Sigma(W_{ij}X_i - B_j) \quad (1)$$

The activation function can be a sine, cosine, etc. Non-linear activation functions are frequently needed and the sigmoid function is commonly used. In this case, the output from each node is expressed by:

$$Y_j = 1 \div (1 + \Sigma(W_{ij}X_i - B_j)) \quad (2)$$

The output from each node in each layer fans out to all other nodes in the next layer in the architecture and the process continues until the output layer is reached. The result obtained from each node k in the output layer of the network can be expressed solely in terms of inputs and weights:

$$O_k = F\ (\Sigma\alpha_{jk}\ F(\Sigma(W_{ij}X_i - B_j))) \quad (3)$$

where α_{jk} are weights for the interconnections between the output layer and the adjacent hidden layer. Studies have shown that the form of Equation 3 corresponds to specific families of non-linear regression curves (*15*). The inputs to the network correspond to the independent variables and the weights of the network correspond to the adjustable coefficients of the regression model. Unlike the regression model, the elements of the neural network are massively interconnected and processing works in parallel. The rate of convergence toward a steady state has been shown to be independent of the number of nodes in the network (*16*).

The strength of a connection is denoted by its weight which is adjusted by one of several learning rules, the most popular of which is the Generalized Delta rule or

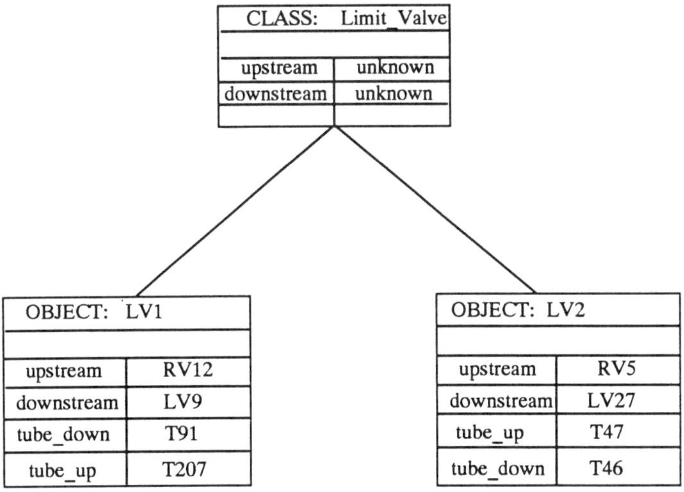
CLASS: Limit_Valve
upstream
unknown
downstream
unknown
OBJECT: LV1
upstream
RV12
downstream
LV9
tube_down
T91
tube_up
T207
OBJECT: LV2
upstream
RV5
downstream
LV27
tube_up
T47
tube_down
T46

Figure 2. Object Representation in Class Hierarchy.

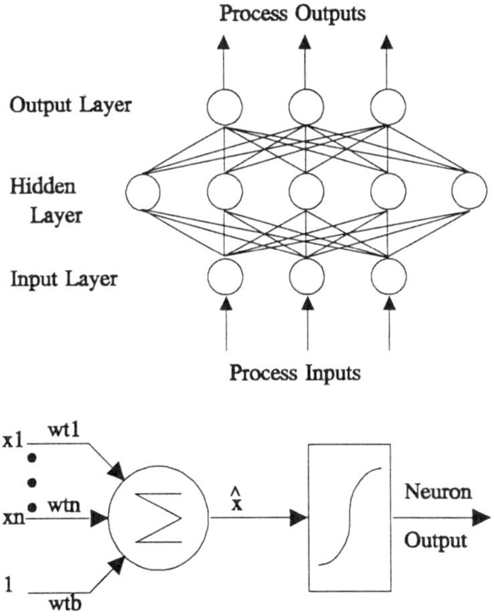

Figure 3: Neural Network Architecture

backward propagation (*17*). This learning rule was designed for multiple layer neural networks and works by modifying the weights in the hidden layers through the backward propagation of the derivative of the error. The network therefore computes for each weight used in the forward pass, the gradient of the output error with respect to that weight. The weight is changed in the direction that reduces the error. The weight changes proceed from the output layer to the next hidden layer until the input layer is reached. This learning scheme has become popular since it has been shown capable of learning many different types of representations in the hidden units such as learning sets of optimal filters for discriminating between similar noisy signals.

There are many neural network paradigms and they are classified by two modes of learning: (1) *supervised learning* and (2) *unsupervised learning*.

Supervised Learning. The most commonly used learning paradigm is supervised learning (*18*). Supervised learning involves training the neural network by showing it associated pairs of input and output pattern vectors. The input vectors describe the pattern that is to be classified and the output vectors describe the classification. The net adjusts its weights internally until it learns how to correctly classify the input vectors to the associated outputs. Many input patterns, possibly 10,000 or more, are introduced to the network for training and classification.

Unsupervised Learning. Unsupervised learning or competitive learning does not involve any a priori training of the net. Instead, as data is introduced into the network it is grouped into clusters in the net. An example of this type of learning is found in the ART model by Grossberg and Carpenter (*19*), and Kohonen's self organizing maps (*20*). These networks have not been used as extensively as the supervised learning systems because they are more difficult to implement and have been shown to have unstable learning characteristics. That is, the network's adaptability enables prior learning to be washed away by more recent learning. Also, if too many clusters develop in the net it becomes more difficult for the net to stabilize. These networks, however, have the great advantage of adapting on-line to gradual changes in the input set and responding to those changes shortly after they occur.

Applications of Expert Systems and Neural Networks

Expert systems have not been extensively used in the food processing industry, but they can be applied to sensor fusion, process diagnostics, and real-time process control. Expert systems are usually used for problems that are well defined (*21*). Current prototype applications have included fuzzy predictive control of corn quality during drying (*22*), a system for carcass beef grading (*23*), and various diagnostic systems.

Neural networks are usually used when a generalized classification scheme is needed. Compared to expert systems, neural networks are a more recent development and they have been applied less extensively. A recent review uncovered 181 applications of neural networks in 56 companies (*24*). Most of these studies were in the investigative phase and very few were in process control. Current applications include the use of neural nets for interpretation and management of sensor data (*25*), sensor calibration (*26*), and for adaptive control (*27*). Neural networks excel in categorizing patterns under noisy conditions and can adapt themselves to changing conditions by unsupervised learning techniques. The ideal applications in process industries are for adaptive control and sensor interpretation (*28-29*). The advantage of neural networks over expert systems is that the neural network can learn, and deal with noisy data. Also, exact decision rules do not have to be known to build the network. Neural network software is slowly maturing and is now available through several companies for PCs and workstations (*30*). Also, neural network hardware is available from several vendors which greatly increases the speed of processing for complex networks.

As described previously, expert systems and neural networks have common and unique uses. A survey of applications will exemplify the distinctions. The survey will be divided into sensor fusion, diagnostics, and real-time process control. A concluding example in thermal processing will show how expert systems and neural networks can be used together for advanced process control.

Sensor Fusion. Sensor fusion is a technique of reducing sensor data from multiple sensors to a smaller and more useful representation of a process. It also provides access to subjective product characteristics such as taste, odor, texture, and color. The techniques involved in sensor fusion include data reduction, data reconciliation, and data interpretation. Pre-processing of sensor data may be required prior to using these techniques. Filtering of noise is one of the most common applications for pre-processing. In addition to conventional analog and digital techniques, neural networks can be used for filtering because they excel at classifying data under noisy conditions.

Data reduction can improve the usefulness of data. One technique for data reduction is hierarchical analysis which can be achieved with a multi-level neural network architecture. Feature subsets can carry the same kind of information except at different degrees of resolution. For instance, when sampling a signal, the sampling can be course or fine. Neural networks can be arranged in a hierarchy starting with the coarsest representation in the uppermost level and successive levels carrying more detailed representations. The signals which can be categorized by the coarse neural net classifier will be handled at the top level in the hierarchy, signals which can only be distinguished by very detailed information will be handled by the lower levels. The processing will be started at the top layer, if it cannot classify the signal, the data will successively be passed to lower levels until a classification is reached (*31*). By having a hierarchical analysis scheme, classification can be done quicker.

Data reduction can also be achieved by feature extraction. Both expert systems and neural networks can be used for this purpose. Features for modelling, such as rise and fall times of signals, pulse durations, energies in defined frequency bands of the power spectrum, zero crossings, location and values of maxima and minima, etc. can be used by the expert system or neural net to extract the most appropriate data for use by the control system. Neural networks are better for large sensor systems, or when the characteristics that classify the data are not completely specified.

Sensors suffer from non-linearities, coupling, and noise, thus leading users of the sensors to calibrate, compensate, and filter the output in order to obtain the most significant information. Neural networks have been used for sensor calibration. For example, a displacement sensor has been calibrated by a neural network by presenting the network with the sensor's output for a known input (*26*). The transfer function of the sensor is highly non-linear. The mapping of data compensates for environmental differences between factory standards and the operating environment. During operation of the sensor, the neural net maps input data to the proper classification to provide calibrated readings. This method was compared to sensor calibration using curve fitting and was found to have similar results. The advantage of the neural network approach was that a model for the mapping did not have to be assumed as is necessary for the use of curve fitting.

Sensors are important to process control since they monitor the status of critical parameters. This data is then used by the process control system to optimally control and time the process. Expert systems can be used for sensor validation to increase reliability by checking sensor data against sensor and process dynamics. Sensor data out of the expected range may indicate that a problem has developed and the expert system can sound an alarm and suggest appropriate corrective action. In addition, data reconciliation techniques allow the accuracy of sensors to be determined by using redundant data from other measurements to estimate probable sensor responses. Expert systems have been used for heuristic-based reconciliation (*2*).

Neural networks can be used to construct models that assess subjective quality factors such as taste, odor, texture, and color. Although these characteristics are determined by measurable quantities such as composition, the relationships are complex and difficult to establish. For example, a neural network can be used to analyze orange-grapefruit juice samples for the purity of orange juice in the blend. First samples with a known origin having a specific feature vector are used to train a neural network in a supervised fashion. The feature vector may contain data on trace elements in the juice such as Ca, Cu, Fe, Mg, etc. Random samples can be then introduced to the net for classification. This analysis is currently accomplished by conventional pattern recognition techniques (*32*). The advantage of using a neural network is that pattern recognition rules don't have to be explicitly developed, rather the supervised training phase is used.

Diagnostics. Diagnostics may involve many different problems from fault diagnosis in circuits to diagnosis of faults in large process control systems. Either conventional expert systems or neural networks can be used. Usually, the best approach is the traditional expert system approach since diagnostic rules can be formalized by interviewing experts and heuristics can be added to bypass steps in the diagnostic process. An emerging methodology is the model-based approach which requires a structural description of the system and knowledge of the behavior of individual components. By using this approach knowledge acquisition from an expert is minimized and system performance is less brittle. Diagnostic systems are the most popular area for applying expert system technology. Diagnostic systems have been built that diagnose problems of machinery in manufacturing environments such as digital circuits, PLC controllers (*33*), pneumatic circuits (*34*), sensor systems, and process control topologies (*35*). For example, in the food processing area a system has been built to diagnose pneumatic circuit problems for a food processing machine. The purpose of this system was to reduce the time it takes to diagnose problems and enable technicians on the factory floor to try to troubleshoot the problem without calling in an engineer.

Real-time Process Control. A control system can be as simple as a single loop, or it can involve multiple loops that are interconnected in a distributed architecture. Within each loop, regulatory control of process parameters occurs at the lowest level. Measurements at this level include temperature, pressure, and flow. Controlled parameters are heat flux, and flow rates of components. Safety interlocks are also incorporated at this level because rapid response and reliability are essential. Expert systems and neural networks are useful at the regulatory control level to filter noisy sensor data, handle non-linear transfer functions, and provide alarm response sequences. An example is provided by application of expert system technology to the commonly used proportional-integral-derivative (PID) controller (*33*). The PID controller is a procedure that can be implemented by a simple software program. Normally, the operator tunes the response constants of the controller based on operating experience. This experience can be embodied in an expert system so that the response of the PID controller can be dynamically tuned to adjust to changes in the response of the process to thermal variation.

At the supervisory level of process control, sensors are integrated with computational functions to achieve sensor fusion and tighter control. For process control, expert systems are useful for fault detection, flexible start-up and shut-down sequencing, and process simulation and optimization. Continuing with the PID example, an expert system could augment the PID controller. If the process temperature goes out of range, the expert system would decide whether to adjust the heat flux by use of the PID controller. Alternatively, the expert system could decide to adjust reactant flow rates to prevent scale formation or run-away conditions that might occur if the heat flux becomes excessive, or the flow rate too slow. Constraints imposed by required production rates and process economics can also be factored into the expert system.

At the supervisory level, neural networks are useful for predictive control, especially when there are long time constants and important uncontrolled factors in the process. Neural networks have been shown to be capable of acquiring an interpretation of a process. After the network, future behavior may be predicted (*25*). In other words, the net can learn a representation of the underlying process. Neural networks can also replace taste-testers to provide real-time model-based control of product quality (*37*). Based on the outcome of the neural network analysis, an expert system could tune the setpoints in the process.

The final application involves research in our laboratory in the area of thermal processing. This example will focus on the High Temperature Short Time (HTST) pasteurization section of a dairy processing plant. A simplified description of a typical HTST process stream follows. Raw milk is supplied from a feed tank to a plate heat exchanger where it is heated by pasteurized milk. The raw milk then goes through the heater section of the heat exchanger where steam heats it to the desired temperature. It then passes through a holding tube where the residence time is more than 15 seconds.

The temperature at the outlet of the holding tube must exceed the legal requirement, else a flow diversion device directs the under-pasteurized milk back to the feed tank.

Control of HTST operations is mostly by hard wired systems, such as a sealed timing pump in the feed line, because this ensures that the process cannot be operated in an improper manner. Programmable logic controllers (PLCs) are utilized in dairy operations, but they have not been used on public health functions. PLCs can provide greater efficiency, quality, and lower cost — but their full benefit cannot be realized unless they are applied in regulated aspects of processing. Better control of HTST processing requires prediction of the temperature of the pasteurized milk at the end of the holding tube. The lethality of HTST processing is such that a variation of ±18°F requires a 100 times change in the flow rate to maintain equivalent lethality. If a minimal over-temperature is to be used, the system must predict unacceptably low temperatures far enough in advance to allow corrective action in the form of adjustments to the steam valve setting and flow rate of the milk. Attainment of this goal is complicated by uncontrollable factors, such as steam pressure, and by slow process dynamics that include a 15-second delay in the holding tube.

Work being organized at our laboratory is examining an improved control system that will use a neural network model of the process to analyze variation of the temperature at the end of the holding tube in terms of controlled (i.e. milk flow rate and steam valve setting) and uncontrolled (i.e. steam pressure and heat exchanger efficiency) factors. The prediction can then be used by an expert system for revision of the settings of the steam valve and the milk pumping speed. The expert system will embody the knowledge that is required to optimize the settings of the steam valve and milk flow rate, subject to constraints inherent in the system such as heat exchanger fouling and energy utilization. The system will operate only slightly above the compliance lethality with the objective of not using the diversion device. The expert system will also be used for sensor validation, data reconciliation, and fault diagnosis.

Summary

The utilization of expert systems and neural networks in other processes indicates that these technologies offer great promise to improve the safety, quality, and productivity of food processing. The ability to introduce this new technology into a company will require the backing of management, but this should be justifiable on the basis of productivity improvements. It is also important to start with smaller, prototype projects that can actually be deployed within the company.

Literature Cited

1. Fildes, J. GRI Industry Workshop — Advanced Combustion and Process Control, Final Report to Gas Research Institute, GRI 90/002, IIT Research Institute, Chicago, IL, 1990.
2. Rowan, Duncan A., *AI Expert* **1989**, *August*, pp 30-38.

3. Stephanopoulos, G.; Johnston, J.; and Lakshmanan, R. *An Intelligent System for Planning Plant-wide Control Strategies*, IFAC 10th World Congr. on Automatic Control, Munich 1987.
4. Astrom, K.; Anton, J. J.; and Arzen, K. E. *Automatica* **1986**, *22*, pp 277-286.
5. Basila, M. R.; Stefanek, G.; and Cinar, A. *Computers in Chem. Eng.* **1990**, *14*(4), pp 551-560.
6. Bartos, Frank J. *Control Engineering* **1990**, *July*, pp 63-66.
7. Valverde, James L.; Gehl, S. M.; Armor, A. F.; and Scheibel, J. R. *The Role of Expert Systems in the Electric Power Industry, Expert Systems: Planning/Implementation/ Integration*, 1990, Vol. 1, No. 4.
8. Ouyang, Z. and Shahidehpour, S. M. *Power Generation Scheduling Using an Expert System*, Proc. of Sixth IASTED Conf. on Expert Systems, December 1989.
9. Fildes, J. *Intelligent Systems Review* **1990**, *2*.
10. Winston, Patrick In *Artificial Intelligence*; Wesley Inc.: Reading, Mass., 1984.
11. Raeth, P. *AI Expert* **1990**, *September*, pp 54-59.
12. Candy, D. *Pro. Conf. Food Processing Automation, Am. Soc. of Ag. Eng.* **1990**, pp 119-124.
13. Neural-Network Vendors and Features *AI Expert* **1989**, *December*, pp 60-61.
14. Willard, J. *Proc. Conf. on Food Processing Automation, Am. Soc. of Ag. Eng.* **1990**, pp 167-171.
15. White, H. *AI Expert* **1989**, *December*, pp 48-52.
16. Hopfield, J. J. and Tank, D. W. *Bio. Cybern.* **1985**, *52*, pp 1-12.
17. McCelland, J. L. and Rumelhart, D. E. In *Parallel Distributed Processing*, Vol. 2, MIT Press: Cambridge, Mass., 1985.
18. Hopfield, J. J. *Proc. Nat. Acad. Sci. U.S.* **1982**, *79*, pp 2554-2558.
19. Carpenter, Gail A. and Grossberg, S. *IEEE Computer* **1988**, *March*, pp 77-88.
20. Kohonen, T. *Self-Organization and Associative Memory*; Springer-Verlag: New York, 1984.
21. Hillman, D. *AI Expert* **1990**, *June*, pp 54-59.
22. Zhang, Q.; Litchfield, J. B.; and Bentsman, J. *Proc. of Conf. on Food Processing Automation, Am. Soc. of Ag. Eng* **1990**, *May*, pp 313-320.
23. Chen, Y. R. and McDonald, T. P. *Proc. Conf. on Food Processing Automation, Am. Soc. of Ag. Eng* **1990**, *May*, pp 244-255.
24. McCusker, T. *Control Engineering* **1990**, *May*, pp 84-85.
25. Pao, Y. and Sobajic, D. J. *Neural-net Technology for Interpretation and Management of Sensor Data, Sensors Expo International* **1989**, pp 206C-1-206C-8.
26. Masory, O. and Aguirre, A. L. *Sensors* **1990**, *March*, pp 48-56.
27. Guez, A.; Eilbert, J. L.; and Kam, Moshe *IEEE Control System Magazine* **1988**, *April*, pp 22-25.
28. Bavarian, B. *IEEE Control Systems Magazine* **1988**, *April*, pp 3-7.
29. Whil, Manfred G. *Sensors* **1988**, *September*, pp 11-18.
30. Expert System Vendors and Features *AI Expert* **1989**, *September*, pp 53-57.
31. Nestor Inc. *Neural Network Systems: An Introduction for Managers, Decision-Makers and Strategists*, 1988, pp 15.
32. Nikdel, S. *Scientific Computing and Automation* **1989**, *August*, pp 19-23.

33. Myers, Douglas R.; Davis, J. F.; and Hurley, C. E. *Application of Artificial Intelligence to Malfunction Diagnosis of Sequential Operations which Involve Programmable Logic Controllers*; Extended Abstracts of AICHE Annual Meeting, Abstract 37B, April 1989.
34. Stefanek, George *An Expert System for Pneumatic Circuit Diagnostics*, IITRI Technical Report, 1990.
35. Davis, Randall; Shrobe, H.; Hamscher, W.; Wieckert, K.; Shirley, M.; and Polit, S. *Proc. of the National Conference on Artificial Intelligence* **1982**, pp 137-142.
36. Joseph, B. *Control* **1990**, *August*, pp 58-61.
37. Gould, L. *Sensors* **1989**, *September*, pp 75-84.

RECEIVED August 15, 1991

Chapter 18

Predicting Chemical Mutagenicity by Using Quantitative Structure–Activity Relationships

Alan J. Shusterman

Department of Chemistry, Reed College, Portland, OR 97202–8199

Molecular orbital calculations have been performed for several classes of chemicals that are mutagenic in the Ames test in *Salmonella typhimurium*. The results of these calculations, along with the hydrophobicity of the mutagens, allows construction of quantitative structure-activity relationships (QSAR) for mutagenicity and prediction of mutagenicity over a very wide range of structure and activity. The QSARs also provide insight into the mechanisms responsible for the metabolic activation of these mutagens. QSARs for the action of nitro-polycyclic aromatic hydrocarbons, phenyl- and heteroaromatic triazenes, and aminoimidazoles such as IQ and MeIQ will be presented.

The public has become increasingly aware of the chemical basis, both real and perceived, of many diseases. Consequently, the public has also begun to demand information regarding the toxicity of chemicals found in the public domain. The desire to avoid unnecessary chemical contamination of basic substances such as food, water, and air is especially pronounced. Scientists are in a poor position, however, to provide toxicity information because of the huge number of untested chemicals, both synthetic and naturally occurring, the lack of unambiguous biological tests for establishing toxicity, and the expense of performing these tests. Mathematical models that can predict several types of toxicity on the basis of chemical structure without the need for biological testing are, therefore, highly desirable.

Quantitative structure-activity relationships (QSAR) for biological systems have had a long and successful history as tools for the study of biochemical reaction mechanisms, and for the rational design of therapeutic drugs. QSARs have also been used to study several types of toxicity, and this work has been the subject of a recent review (*1*). This paper describes recent work devoted to the development of QSARs for chemical mutagenicity in Ames bacteria, *S. typhimurium* (*2-4*). Most of this work has been carried out in collaboration with Dr. Corwin Hansch and his associates at Pomona College.

0097–6156/92/0484–0181$06.00/0

QSAR Methodology

Before one can construct a quantitative structure-activity relationship for a set of chemicals the following must exist: a quantitative test for the biological activity of interest, a quantitative description of chemical structure or some structure-dependent property, and a mathematical formalism for relating structure and activity. While each of these areas continues to be the subject of vigorous research, the approach taken here is a traditional one.

Activity is defined using the experimental dose-response curves observed for different chemicals acting on various strains of *S. typhimurium* as measured by the Ames test (*5*). The test has the advantage of being readily quantifiable, and reasonably reproducible from laboratory to laboratory. On the other hand, the results of the Ames test should not be construed as representing carcinogenicity, since the formation of cancer involves a significantly more complicated chain of events.

The mutagen families described here all require metabolic activation of some sort. It is also anticipated that certain types of metabolic transformations may render the molecule inactive. Therefore, a reasonable set of structure descriptors would be those properties most closely related to chemical reactivity, particularly the processes involved in (de)activation. Such descriptors include the electronic and steric/shape properties of the molecule. Another key factor, often overlooked by chemists, is the relative hydrophobicity of the molecule. The hydrophobic properties of a mutagen affect its penetration of biological membranes, and its binding to metabolic enzymes. Following accepted practice, log P has been employed as a measure of relative hydrophobicity where P is the mutagen's octanol–water partition coefficient.

Finally, multiple linear regression equations are used to correlate activity with the relevant structural properties. As described below, great care must be taken to guarantee that each term occurring in the regression equation is justified both statistically and chemically. There is a great potential for confounding variables, which accidentally parallel the behavior of more meaningful structural parameters (at least for the limited set of compounds under consideration), to lead to QSARs that either lack generality or whose apparent chemical interpretation is unrealistic (*see* below).

Quantum Chemical Calculations

Hammett substituent constants, σ_X have traditionally been used to describe the electronic properties of different chemical structures. This approach has restricted the range of chemical structures that can be studied to those that can be described by the interchange of different substituent groups on a common skeleton. This limitation on structure also affects one's ability to manipulate other properties such as hydrophobicity, and limits the range of biological activity that can be studied. We have attempted to overcome these problems by undertaking the use of quantum chemical calculations, such as the semi-empirical MNDO[6] and AM1[7] methods, to describe electronic properties related to chemical reactivity.

We have made use of several quantum chemical indices including molecular orbital energies (ε_i), electron densities associated with particular atoms (q_i), and calculated reaction enthalpies (ΔE_i), as measures of a given mutagen's electronic properties. As shown below, it is quite common for several parameters to be collinear, and QSARs can be derived for a given family of mutagens using a variety of electronic descriptors. However, given this high degree of collinearity, it is essential that mechanistic conclusions not be based solely on the presence of a given electronic descriptor in the QSAR.

Triazenes

Venger et al., initially reported a QSAR (Equation 1) for 17 aryltriazenes, **1**, acting on *S. typhimurium* strain TA92 (*2c*). The triazenes do not exhibit any activity except in the presence of the S9 fraction obtained from rat liver microsomes. Thus, it was concluded that cytochrome P-450 activation of the triazenes was essential.

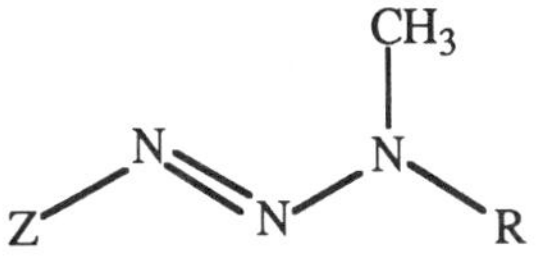

1

Mutagenic activity was defined as log 1/C, where C is the molar concentration of triazene causing 30 revertants above background/10^8 TA92 bacteria. A low mutation rate was chosen so that the cytotoxicity of the mutagen would not interfere with the test results. The statistical parameters associated with this equation are n, the number of data points; r, the correlation coefficient; and s, the standard deviation. Figures in parenthesis are for construction of the 95% confidence intervals.

$$\log 1/C = 1.04\ (\pm 0.17) \log P - 1.63\ (\pm 0.35)\ \sigma^+ + 3.06 \qquad (1)$$
$$n = 17, r = 0.974, s = 0.315$$

Several points concerning this equation are noteworthy, the first being that the quality of the correlation is unusually high. This can be attributed, in part, to the fact that all of the data were collected in a single laboratory where the testing of each mutagen could be carried out in a reproducible fashion. Also interesting are the appearance of both hydrophobic and electronic terms in the equation, and the coefficients associated with each. Triazene mutagenic activity is increased by attaching substituents to the benzene ring that either render the triazene more hydrophobic (larger log P), or more electron-rich (smaller σ^+). Cytochrome P-450 is known to activate more hydrophobic substrates preferentially, and so the coefficient with log P may reflect this selectivity, or it may reflect the ease with which more hydrophobic substances can penetrate the bacterium. Likewise, electron donor substituents are expected to facilitate oxidation of the triazene by P-450 (*2a*).

We have followed up on this study by preparing several more triazenes in which Z is a heterocyclic ring (*2a*). Since σ^+ constants are not available for most of these rings, we have also performed MNDO calculations on the parent triazenes in order to describe their electronic properties. One triazene (Z = 1,2,4-triazole) is inactive, while another triazene (Z = 2-dibenzofuran) is more active than any of the aryltriazenes originally studied by Venger et al. Two QSARs, Equations 2 and 3, were found that correlated the behavior of the 17 aryltriazenes along with 4 of the heterocyclic triazenes.

$$\log 1/C = 0.95\,(\pm 0.25)\log P + 2.22\,(\pm 0.88)\,\varepsilon_{HOMO} + 22.69 \qquad (2)$$
$$n = 21,\ r = 0.919,\ s = 0.631$$

$$\log 1/C = 0.97\,(\pm 0.24)\log P - 7.76\,(\pm 2.73)\,q_{HOMO} + 5.96 \qquad (3)$$
$$n = 21,\ r = 0.931,\ s = 0.585$$

ε_{HOMO} is the energy of the highest *occupied* molecular orbital (HOMO) in eV. q_{HOMO} is the electron density on the alkyl N in the HOMO. The relationship between hydrophobicity and activity defined by these equations is essentially identical to that discovered earlier by Venger et al. At the same time, the electronic terms in these equations can be interpreted in a fashion consistent with that given for Equation 1. As the HOMO rises in energy, i.e., ε_{HOMO} becomes *less* negative, the molecule becomes easier to oxidize and activity increases. Equation 3, on the other hand, would appear to indicate that higher activity is associated with lower electron density on N. A closer inspection of the HOMO, however, shows that this orbital is largely concentrated on Z, i.e., the aromatic or heterocyclic ring, and not on the triazene. Electron-donor groups, while making the triazene easier to oxidize, also concentrate more of the electron density of the HOMO on Z, hence the paradoxical correlation between activity and q_{HOMO}. It is important to remember that the electron density on an atom in a particular orbital is not necessarily indicative of the chemical reactivity of that site in the molecule.

Equations 2 and 3 predict the 1,2,4-triazole to be 10^6–10^7-fold less active than the 2-dibenzofuran in accord with the observed inactivity of the triazole. The activity of one heterocyclic triazene, Z = 2-thiazole, is greatly underpredicted by Equations 2 and 3. In this case, a σ^+ value for the thiazole ring is available, and Equation 1 also underpredicts the activity of this triazene. Since the MNDO and Hammett parameters give a consistent prediction for this triazene it is reasonable to believe that an additional, unknown mutation mechanism is acting in this case.

Nitro-Polycyclic Aromatic Hydrocarbons

Nitroarenes are mutagenic in both *S. typhimurium* strain TA98 and TA100, and do not require the presence of S9 microsomes. QSARs describing the mutagenic activity of these two systems are given in Equations 4 (*3a*) and 5 (*3b*).

$$\log TA98 = 0.65\ (\pm 0.16) \log P - 2.90\ (\pm 0.59) \log (\beta 10^{\log P} + 1) - 1.38\ (\pm 0.25)\ \varepsilon_{LUMO} + 1.88\ (\pm 0.29)\ I_L - 2.89\ (\pm 0.81)\ I_a - 4.15\ (\pm 0.58) \quad (4)$$

$$n = 188,\ r = 0.900,\ s = 0.886,\ \log P_o = 4.93,\ \log \beta = -5.48$$

$$\log TA100 = 1.36\ (\pm 0.20) \log P - 1.98\ (\pm 0.39)\ \varepsilon_{LUMO} - 7.01\ (\pm 1.2) \quad (5)$$

$$n = 47,\ r = 0.911,\ s = 0.737$$

TA98 and TA100, the mutagenic activity of the nitroarene in each strain of the Ames bacteria, are defined as revertants/nmol mutagen. ε_{LUMO} is the energy of the lowest *unoccupied* molecular orbital in eV (AM1 for TA98, MNDO for TA100). The negative coefficient with ε_{LUMO} in each equation indicates that the nitroarene becomes more active as this empty orbital falls in energy, i.e., the compound becomes a better electron acceptor. This relationship is consistent with the generally accepted activation mechanism for nitroarenes which postulates initial reduction of the nitro group to give a hydroxylamine intermediate, which ultimately reacts with DNA.

The TA98 QSAR shows a two-term dependence of activity on log P. These two terms describe a bilinear function; activity rises as 0.65 log P for $\log P < 4.93$ and then falls as -2.25 log P for $\log P > 4.93$. The simple linear relationship seen for TA100 and log P may simply be due to the lack of more hydrophobic compounds in the smaller TA100 data set. The appearance of a bilinear activity-hydrophobicity relationship indicates that there is an optimal log P for mutagenic compounds ($\log P_o$). More hydrophilic compounds are less likely to penetrate the bacterial membrane, while more hydrophobic compounds are likely to become sequestered in lipid phases before they can react with DNA.

Equation 4 also contains two indicator variables. These variables are set equal to either 0 or 1 depending on the absence or presence of a particular structural feature. $I_L = 1$ signifies the presence of 3 or more aromatic rings in the nitroarene. Compounds containing 3 or more rings are 76 times more active than 1 or 2 ring compounds other factors being equal. This jump in activity may be due to an increased ability for larger compounds to intercalate into bacterial DNA. $I_a = 1$ indicates the arene belongs to the acenthrylene family. The negative coefficient with I_a shows that these compounds are all much less active than would be predicted on the basis of their hydrophobic and electronic properties alone.

Equations 4 and 5 span a tremendous range in structural types and in mutagenic activity (ca. 10^8 revertants/nmol). Thus, these equations are powerful predictors of nitroarene behavior. Inspection of the TA98 data set shows that there is also a significant variation in each of the structural properties; variation in ε_{LUMO} spans a 2.84 eV range accounting for an 8000-fold range in activity, while log P varies from -0.02 to 7.84 accounting for 1700-fold ($P < P_o$) and 3.5×10^6-fold ($P > P_o$) ranges in activity.

IQ-type Mutagens

The IQ-type mutagens, i.e., 2-amino derivatives of imidazo[4,5-f]quinoline, **2**, are especially potent mutagens which occur as pyrolysates in various types of cooked food (*4, 8-12*). The compounds all require the presence of the S9 microsomal fraction, and lose most of their activity when either the 2-amino or 3-methyl group is missing, indicating that the key mutagenic processes involve this part of the skeleton.

Unfortunately, the compounds isolated from food make a poor data set from the standpoint of QSAR development since their mutagenic activities are all nearly the same, and their hydrophobic and electronic properties span a relatively narrow range ($-8.60 \leq \varepsilon_{HOMO} \leq -8.22$ eV, $1.01 \leq \log P \leq 2.62$). Expansion of the data set is essential in order to sort out the structural factors responsible for *relative* mutagenicity. Debnath et al. have synthesized 5- (Y = H) and 6-substituted (X = H) derivatives of 1-methyl-2-aminobenzimidazole, **3** (*4*). These compounds, like the IQ derivatives, require S9 microsome activation, and are inactive when either the 2-amino or 1-methyl group is absent. It is reasonable, therefore, to assume that these compounds act by the same mechanism as the IQ-type mutagens.

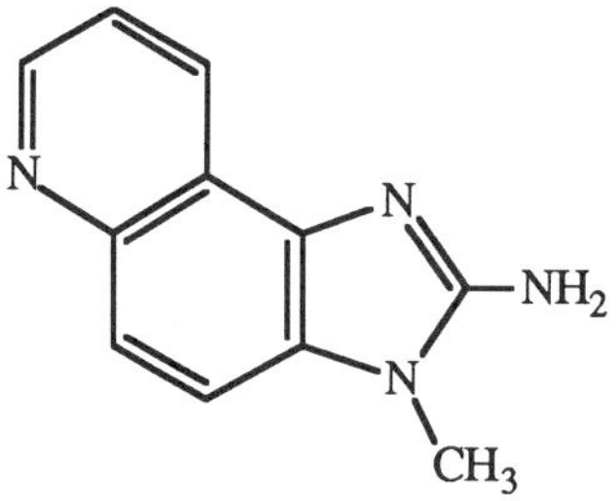

2

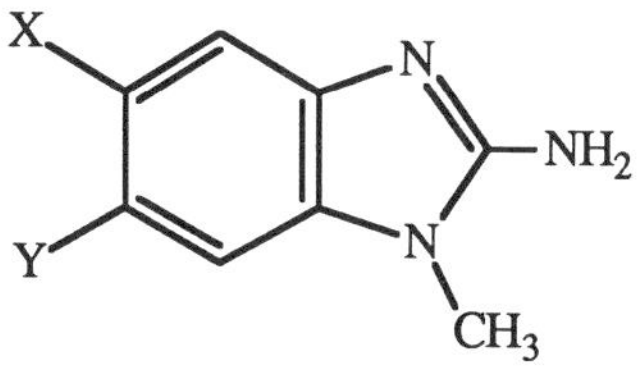

3

In order to facilitate the study of this enlarged data set we have excluded several compounds whose mutation rates in TA98 are known. Compounds with 1 or 2 methyl substituents on the exocyclic N were excluded, since it was assumed that cytochrome P-450-catalyzed N-demethylation would occur, and the observed mutation rates would not be representative of the parent compound. Similarly, two imidazonaphthalene derivatives were found to be much less active than predicted and were excluded. A QSAR describing the remaining compounds in our data set is given in Equation 6.

$$\log TA98 = 1.31\ (\pm 0.74) \log P - 0.30\ (\pm 0.09)\ \Delta E_{NH+} + 64.92 \qquad (6)$$
$$n = 22,\ r = 0.906,\ s = 1.011$$

In this equation, log TA98 is the number of revertants/nmol mutagen, and ΔE_{NH+} is the reaction enthalpy (in kcal/mol) for conversion of a hydroxylamine intermediate into a putative nitrenium ion, Equation 7 (geometries and energies calculated using AM1).

$$\text{Imid-NHOH} \rightarrow \text{Imid-NH+} + \text{OH-} \qquad (7)$$

A positive relationship between hydrophobicity and activity is observed, in accord with the need for P-450 activation and membrane penetration. Activity also increases with increasing ease of nitrenium ion formation (smaller ΔE_{NH+}).

Several points concerning this QSAR are noteworthy. First, all of the terms in Equation 6 are significant at the 99% confidence level according to the t-test. On the other hand, the confidence interval for the log P coefficient is rather large. Indeed, removal of one compound from the data set, 2-amino-5-hydroxy-1-methylbenzimidazole, slightly improves the quality of the correlation ($r = 0.914$, $s = 0.853$), but at the expense of the log P term, which is only significant at the 87% confidence level. If log P is removed from the equation altogether, then $r = 0.901$ for this smaller data set. Since this one compound is significantly more hydrophilic than any of the others in the data set not much trust can be placed in the log P term until its role is validated by the testing of more compounds.

Second, since nitrenium ion formation is believed to be essential before reaction can occur with DNA, it is very interesting to find a correlation between activity and ΔE_{NH+}. On the other hand, it is important to keep in mind the possibility of collinear variables. In this case, the calculated ionization potential of the parent amine is strongly correlated with ΔE_{NH+} ($r = 0.887$). Thus, the dominant electronic factor may be ease of amine oxidation to generate a hydroxylamine, conversion of the hydroxylamine to a nitrenium ion, or even some still undetected process. Nevertheless, we believe that Equation 6 is a good starting point for designing and testing more compounds belonging to this critical family of mutagens.

Inspection of the molecular and electronic structure of the nitrenium ions related to **2** and **3** is also revealing. The AM1 geometry for the imidazole and adjacent benzene ring of the nitrenium ion clearly shows a pattern of alternating single and double bonds, similar to the imine resonance structure, **5**. Inspection of bond orders shows that pi bond localization is predicted to occur for the nitrenium ion.

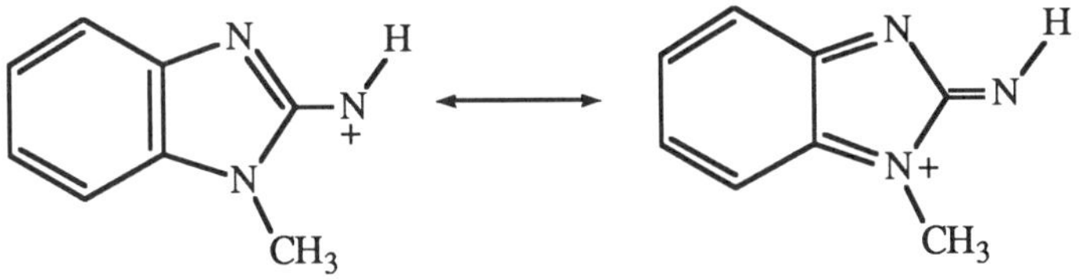

4 minor — **5** major

While it has been generally assumed that such a nitrenium ion would react with DNA exclusively at the exocyclic N, it would appear that other sites for nucleophilic attack are also available. For example, the nitrenium ion might transfer a methyl group to the DNA (Figure 1). This might serve to explain the need for a methyl group on the imidazole ring.

Another intriguing possibility is nucleophilic attack on the benzene ring at C-6. If the nucleophile is water, rearrangement would yield a much less active phenol

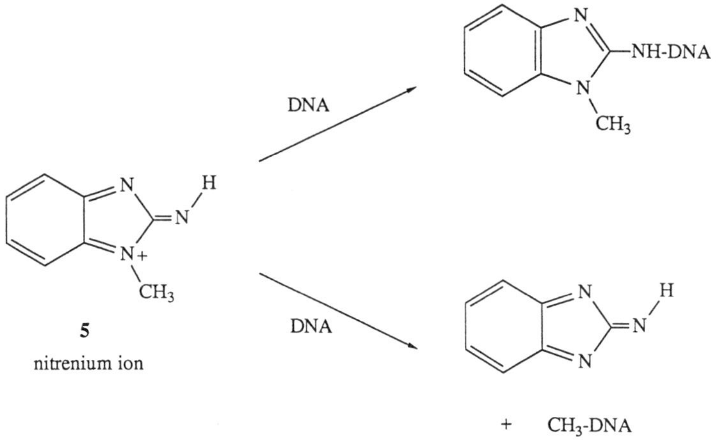

Figure 1. Alternative modes of nitrenium ion capture by DNA.

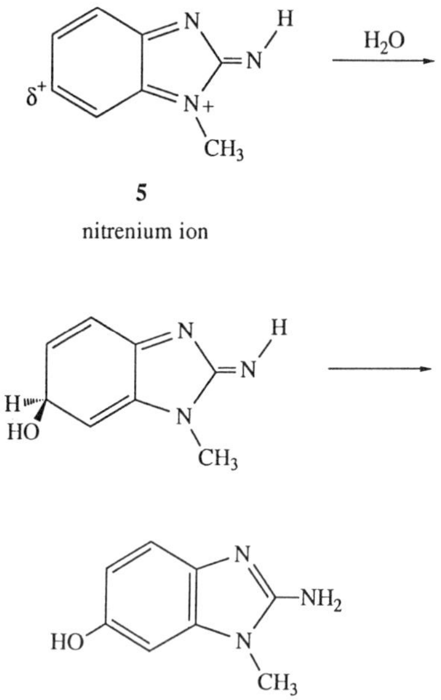

Figure 2. Mechanism of phenol formation from a nitrenium ion.

(Figure 2). Some support for a deactivation reaction occurring at C-6 can be found by comparing the mutagenic activities of the 5-CN and 6-CN derivatives of **3**. The 6-CN derivative is 41 times more mutagenic, possibly due to a steric effect; the substituent blocks attack of water on the benzene ring of the nitrenium ion, preventing phenol formation and deactivation, and enhancing mutagenicity. Metabolic conversion of food mutagens to analogous phenols appears to occur *in vivo* as well.

Conclusion

The Ames test provides useful quantitative data which can be correlated using traditional QSAR techniques. The resultant QSAR equations can be used to correlate, and potentially, to predict mutagenic activity over a very wide activity range, and can accommodate a broad range of chemical structures. Consideration of relative hydrophobicity, as reflected in log P, is often essential in order to correctly account for the influence of structure on activity. Variations in chemical structure that can not be treated using Hammett substituent constants appear to be well treated using a variety of parameters derived from semi-empirical quantum chemical calculations. However, care must be exerted in the interpretation of the resultant QSARs due to the great potential for collinear variables.

Acknowledgment

This research was supported by a grant for fundamental studies in toxicology from the R. J. Reynolds Company. The support and advice of Corwin Hansch is gratefully acknowledged. The assistance of Asim Debnath in communicating unpublished results concerning the mutagenic activities and log P values of substituted 1-methyl-2-aminobenzimidazoles is gratefully acknowledged.

Literature Cited

1. Hansch, C.; Kim, D.; Leo, A.J.; Novellino, E.; Silipo, C.; Vittoria, A. *CRC Crit. Rev. Toxicol.* **1989**, *19*, 185-226.
2. (a) Shusterman, A.J.; Debnath, A.K.; Hansch, C.; Horn, G.W.; Fronczek, F.R.; Greene, A.C.; Watkins, S.F. *Mol. Pharm.* **1989**, *36*, 939-944. (b) Shusterman, A.J.; Johnson, A.S.; Hansch, C. *Int. J. Quant. Chem.* **1989**, *36*, 19-33. (c) Venger, B.H.; Hansch, C.; Hatheway, G.J.; Amrein, Y.U. *J. Med. Chem.* **1979**, *22*, 473-476.
3. (a) Debnath, A.K.; Compadre, R.L.L.; Debnath, G.; Shusterman, A.J.; Hansch, C. *J. Med. Chem.* **1990**, in press. (b) Compadre, R.L.L.; Debnath, A.K.; Shusterman, A.J.; Hansch, C. *Environ. Mol. Mutagen.* **1990**, *15*, 44-55. (c) Compadre, R.L.L.; A. J. Shusterman, C. Hansch, *Int. J. Quant. Chem.* **1988**, *34*, 91-101.
4. Debnath, A.K.; Shusterman, A.J.; Raine, G.P.; Hansch, C. unpublished results.
5. Maron, D.; Ames, B.N. *Mutat. Res.* **1983**, *113*, 173-215.

6. Dewar, M.J.S.; Thiel, W. *J. Am. Chem. Soc.* **1977**, *99*, 4899-4907.
7. Dewar, M.J.S.; Zoebisch, E.G.; Healy, E.F.; Stewart, J.J.P. *J. Am. Chem. Soc.* **1985**, *107*, 3902-3909.
8. Kaiser, G.; Harnasch, D.; King, M.T.; Wild, D. *Chem.-Biol. Interact.* **1986**, *57*, 97-106.
9. Sugimura, T. *Science* **1986**, *233*, 312-318.
10. Jagerstad, M.; Grivas, S. *Mutat. Res.* **1985**, *144*, 131-136.
11. Nagao, M.; Wakabayashi, K.; Kasai, H.; Nishimura, S.; Sugimura, T. *Carcinogenesis* **1981**, *2*, 1147-1149.
12. Alexander, J.; Wallin, H.; Holme, J.A.; Brunborg, G.; Soderlund, E.J.; Becher, G.; Mikalsen, A.; Hongslo, J.K. *Mutation in the Environment, Part E*; Wiley-Liss: New York, 1990, p. 159.

RECEIVED November 18, 1991

Chapter 19

HazardExpert

An Expert System for Predicting Chemical Toxicity

Michael P. Smithing[1] and Ferenc Darvas[2]

[1]CompuDrug USA, Inc., Austin, TX 78720
[2]CompuDrug Chemistry Ltd., Budapest, Hungary

HazardExpert is one of several advanced expert systems developed during the past decade by CompuDrug Chemistry Ltd., Hungary and marketed by both CompuDrug Chemistry Ltd. and their US subsidiary CompuDrug USA, Inc. This paper will first explain the general ideas behind CompuDrug's expert systems and what these systems, as a group, are meant to achieve. From this point, the paper goes on to HazardExpert as an example of an expert system, examining the specific objectives of the systems and how these objectives are met. Finally, examples of possible uses of the program are detailed.

What Is An Expert System And Why Is It Not A Database?

An expert system is quite different from a traditional database, although the typical user will often confuse the two. Databases consist of facts or items which have been collected from somewhere and are stored in a logical manner. Expert systems contain not only facts or items, but also rules which the system applies. The organization of these rules is also orderly, and in many cases is quite similar to that of a database; however, an expert system is different from a mere database in that the system applies the information rather than just storing it. In this way, the system appears intelligent, although never as intelligent as a human expert. By storing rules and utilizing methods to apply them, an expert system can occupy less space and operate more quickly than a simple database. In addition, the system does not know, or care, if the requested information has ever been assembled before - the rules are applied in the same manner to both known data (often used for justification of the system) and unknown data, where the true power of an expert system becomes apparent in generating new ideas and insights into new problems.

0097–6156/92/0484–0191$06.00/0

Expert Knowledge, Not Published Reports

The secret behind any good expert system is the expert. It is the expert who, from his or her own knowledge of the subject matter, formulates the rules which comprise the system. These rules must be not only exact, but also well formulated to anticipate the wide variety of possibilities which the computer may pass to the system. In addition, knowledge is continually being updated as advances are made. Of course, no one would think of buying a database and never adding anything new - this is not the accepted use of a database. The same applies to an expert system. Just as facts need to be updated, changed and added, so does knowledge. Each expert has their own way of understanding a certain problem, and the formulation of a question often severely modifies the answer. Thus, every expert user should modify the knowledge in an expert system to fit their expectations and experiences. The process requires continual updating as new experiences are gained, the modification of existing knowledge to fit personal experiences, and even the deletion of irrelevant knowledge which is unused and adversely effects the operation of the system, i.e. taking extra time to produce useless results.

Two Levels Of Use, Expert And Novice

Most expert systems are used at two widely different levels. The first of these is the local expert. Expert users of expert systems have two main duties within the system. The first is to update the knowledge of the system and keep it current. This allows all users of the system to obtain satisfactory results. Their second duty is to actively use the system to their own benefit. While this also provides verification of the system, regular use will benefit the expert by allowing him or her to focus his or her time and energy on areas which are more complex than those handled by the system. The expert system will, in effect, codify and apply the expert's knowledge just as he or she would, and often will produce results in a shorter period of time, depending on the difficulty of the task.

The second group is comprised of novice users. While not experts themselves, the novice group consists of those who have (more or less) occasional need for or interest in the results provided by the expert. Users in this class often are able to replace the expert with the expert system, thus saving the expert's time for more useful projects than answering (often mundane) questions. However, as users in this group are not experts themselves, they should not be encouraged to base critical decisions on this information alone. In such cases, the expert system can significantly reduce consultation time by providing both the expert and the novice with a common basis upon which the discussion may be based.

HazardExpert - An Introduction To The Idea

HazardExpert is an expert system which predicts the overall toxicity of an investigated organic compound in a variety of living systems. To obtain as complete an assessment as possible, HazardExpert examines the investigated compound and its potential metabolites for both toxicokinetic and toxicodynamic effects. As the basis for its toxicodynamic predictions, the system utilizes a complete set of rules derived from the Environmental Protection Agency's report (*1*) on toxic segments, where each rule details a toxic segment as well as the sensitivity of the different classes of species to the segment given the dosage and means of exposure. Metabolic predication is accomplished by the "MetabolExpert" module of the program which is based on CompuDrug's MetabolExpert (*2*) expert system and draws upon the standard text, *Drug Metabolism* (*3*), by Testa and Jenner. The calculation of toxicokinetic effects is basedon molecular weight, obtained through simple calculation; pKa, claculated by a mini-expert system developed by CompuDrug; and on hydrophobicity calculated by the "Pro-LogP" module of HazardExpert, which is based on CompuDrug's Pro-LogP (*4*) expert system utilizing the calculation algorithm described by R.F. Rekker (*5*).

Custom Work For The EPA

The HazardExpert project started as a piece of custom software developed for the United States Environmental Protection Agency under a contract to Dynamac Corporation, and funded through Superfund and the Resource Conservation and Recovery Act. The stated goal was to produce a computerized expert system to facilitate the prediction of the health and ecological effects of organic chemicals based on certain sub-structural molecular features and selected properties. The system was to make predictions for both human and non-human biota, and provide qualitative estimates of toxicokinetic activity and toxicity levels for the investigated compound. Further, the system was to contain the ability to identify expected metabolites of an organic structure and provide an indication of their expected toxicity.

The final version of the system presented to the EPA summarized both toxicodynamic and toxicokinetic effects. The evaluation of toxicodynamic effects was based on a substructure search of the investigated compound, utilizing the substructures set out in the EPA interim report. Toxicokinetic effects are estimated primarily on the basis of the hydrophobicity of the compound, but molecular weight and pKa are also considered. These effects are further modified by considerations of dosage and means of administration, and fuzzy logic is employed to estimate the validity of the final prediction. This means, in effect, that each substructure in the knowledge base has been graded by a panel of experts for both the level of expected toxic effect (high, medium, or low) and for the validity of the initial prediction of toxicity (the accepted values are surely true, possibly true, uncertain, less probable, not probable) and each of these values has been set for each species group for which evaluation can be carried out.

Final results are displayed graphically on two different screens. The first of these emphasizes the actual prediction, on the left detailing the toxicokinetic effects and on the right outlining the specific toxicodynamic effects anticipated. The second screen displays the integrated prediction, with the final qualitative prediction (using the terminology set down in the EPA report), as well as the display of the arguments for and against the prediction. (True positive, false positive, true negative, false negative.)

HazardExpert - A Commercial Product

CompuDrug intends HazardExpert to be used as a quick and easy evaluation of expected toxicity. In-depth toxicological studies will never be replaced by the system, however we believe that, by continually updating the knowledge of the system based on toxicological investigations, reasonable results can be obtained at minimal cost, both monetarily and in terms of time.

The commercial product contains some key advances over the original EPA product. The knowledge bases have been expanded, and individual modules, including the metabolic prediction module and that for the calculation of hydrophobicity have been improved upon. As the product is still early in the development cycle, continuous improvements to the system are being made, and feedback from users is continually bringing the system closer to the demands of industrial users. Recently, user feedback resulted in the addition of an option permitting the user to override the logP and pKa data calculated by HazardExpert with the user's own experimental data. Current development involves a complete restructuring of the calculation of pKa, as well as the ability of the user to modify and/or rewrite the algorithms used in the final analysis of the data. Thus, while the scientists from the EPA had a solid idea of the algorithms which should be employed in the interpretation of their data, most industrial users require the flexibility to reinterpret data as well as to simply modify it. Current developments should lead to a situation in which the system as a whole could be tailored in-house to fit the needs of a variety of users across industrial boundaries.

The Heart Of HazardExpert - The Knowledge Bases

All of CompuDrug's rules are based on the principal of QSAR - Quantitative Structure Activity Relationship. Thus, each rule -- whether toxic, metabolic, or hydrophobic -- contains both a sub-structure and an activity. For convenience, these rules are coded in CompuDrug's linear notation format MolNote, based on the following simple rules:

1. Heavy atoms are entered with unique indices written after their symbols (i.e., C1, Br13). Hydrogen atoms are generally added automatically.
2. The symbol X may stand for any heavy atom, XA, XB, etc. are atomic variables, and must be followed immediately by a list of possible values. (i.e., XA = C,N,O)
3. Carbon atoms may be written as an index without chemical symbol (i.e., 2 or C2)
4. The following are the bond types:

-	Single
=	Double
#	Triple
*	Aromatic
?	Any bond

5. Bonds are placed between atoms. An unbroken chain containing different bonds may be written on one line, or on several consecutive lines.

6. An unbroken chain of similar atoms connected by the same type of bond is represented Ai.<b>.Aj where Ai is the first atom in the chain, Aj the last, and b the bond type (i.e., N1.-.N3 = N1-N2-N3). For rings, this notation is preceded by an "R". Thus, R1.*.6 is benzene.

Metabolism And Its Role In Predicting Toxicity

One of the major advantages of HazardExpert over any other system for toxicity assessment is the ability of HazardExpert to offer a prediction not only for the suspected toxin, but also for its metabolites. With the metabolic generation module of HazardExpert, it is possible to assess the risk of an entire system, from the time a compound enters the organism until it is excreted. Without the aid of metabolic prediction, potentially dangerous reactions may be overlooked, and the danger of a compound severely underestimated.

Metabolic transformation rules contain the following four main parts:

Active	Bonds which are broken.
Replacing	Bonds which are created.
Positive	Conditions necessary for the reaction to occur.
Negative	Conditions which inhibit the reaction.

In the example in Figure 1 (Para-hydroxilation), the benzine ring and the bond binding it to the rest of the molecule form the positive condition (Structure). Oxygen bonded to the para position forms the replacing substructure (Activity). Bonds at any position except the primary position would inhibit the reaction, as outlined in the negative conditions.

Toxicokinetic Effect Calculation Rules

In comparison with metabolic or toxic rules, these are rather simple, with the structure being directly related to a specific additive value (Activity). When examining the two rules presented in Figure 2, make note of the following:

- Segment type may be Basic or Acidic for pKa calculations and Fragment or Interaction for logP calculations.
- Standard deviation is given for logP, whereas separate positive and negative deviations may be entered for pKa.
- Hydrogen atoms may be entered just as heavy atoms in logP segments, as required by Rekker's fragment set.

Toxicodynamic Effect Calculation Rules

While the rules governing toxicodynamic effects contain more information, they are really no more complicated than those for the calculation of toxicokinetic effects.

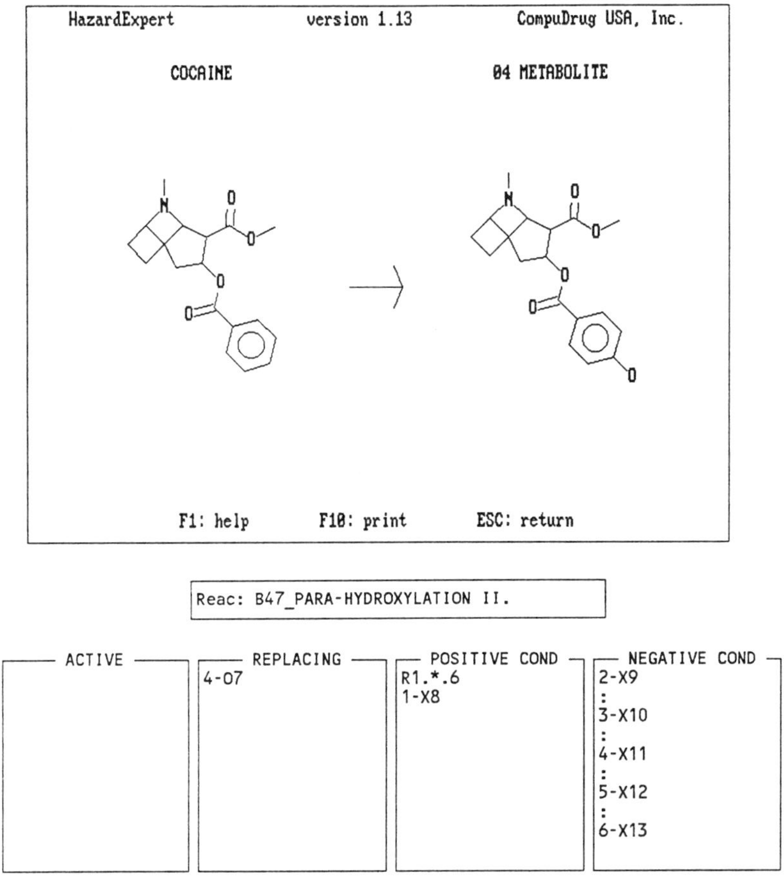

Figure 1. Metabolic transformation.

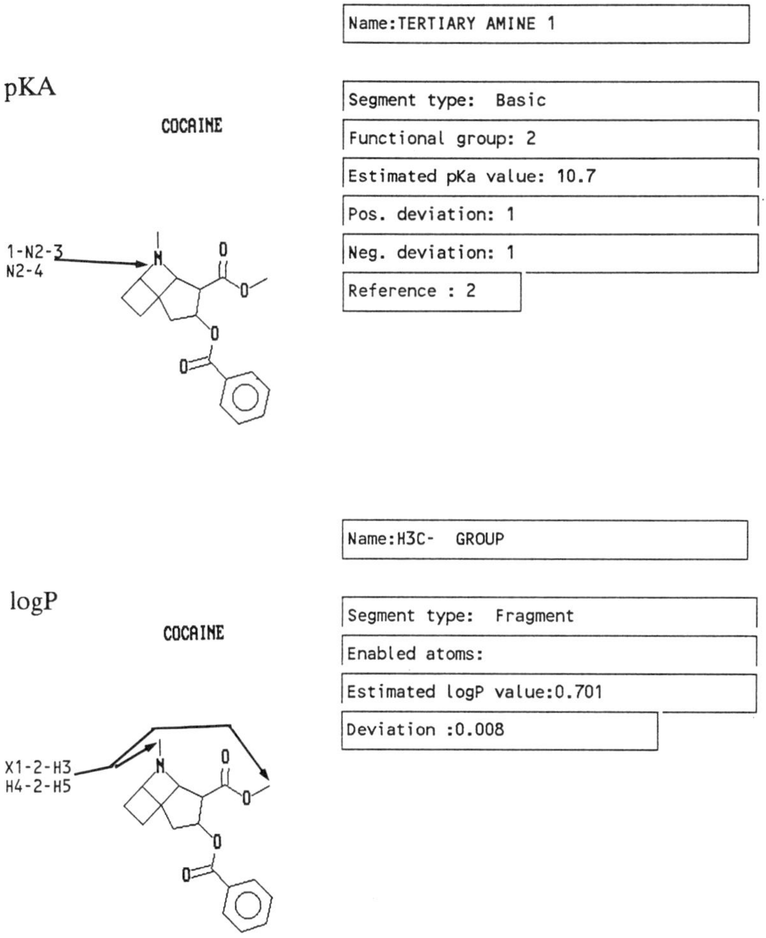

Figure 2. pKa and log P segment rules.

Toxicodynamic rules consist of a segment (the structure) and two "effects groups". The first of these sets the relative specific toxicodynamic effects, while the second governs species sensitivity to the segment.

In the example in Figure 3, the diagram presents both the raw toxicodynamic effect (solid lines), as well as the weighted version which has been altered to account for species sensitivity and dosage (filled area).

Lock Into The Future, Not Into The Past.

In comparison with traditional toxicity studies, HazardExpert is both fast and inexpensive. However, the main benefit of HazardExpert, and indeed of any of CompuDrug's expert systems, is not simply speed and ease of use but predictive ability. By its very nature, a database is focused on the past. All of the information is set, unchangeable facts uncovered in the past. If one's needs are primarily the retrieval of information previously available, a database will certainly be a better choice than an expert system. Expert systems, on the other hand, focus on the future. The true application of HazardExpert and any of CompuDrug's other expert systems is to predict future results on the basis of the knowledge available today. Thus, today's expert system predictions may become tomorrow's database entries. Scientists working on advancing the limits of our current knowledge - those adding to our existing databases - can utilize expert systems as valuable tools.

HazardExpert - The Range Of Possible Uses

Product Development. HazardExpert provides a means for research chemists working in the early stages of product development in many industries to give potential products a "trial run" before significant resources are devoted to it. While still in the early stages of design, not only the toxicity of the actual product can be assessed, but also the toxicity of any potential metabolites. Comprehensive use of HazardExpert can significantly decrease the amount of resources devoted to "dead end" products, freeing these resources for more profitable ideas.

Environmental Auditing. The predictive abilities of HazardExpert make it a valuable tool in the initial stages of any environmental audit. The quick and inexpensive estimates provided by the system can be used as the basis of bids submitted as well as in the formulation of the initial hypotheses of the audit once a contract is awarded.

Environmental Protection Groups. In general, environmental protection groups lack the resources to undertake comprehensive environmental impact studies of their own. The low price per estimate of HazardExpert provides a means through which these groups can accurately assess toxicity in the environment - an ability which is in the best interests of everyone.

		Toxicity based on the EPA report:	Toxicity modified by dose, partial effect of exposition and administration:
Oncogenicity	:	3	1
Mutagenicity	:	3	1
Teratogenicity	:	3	1
Membrane Irritation	:	40	20
Other Chronic Toxicity	:	41	16
Acute Toxicity	:	3	1
Neurotoxicity	:	3	1

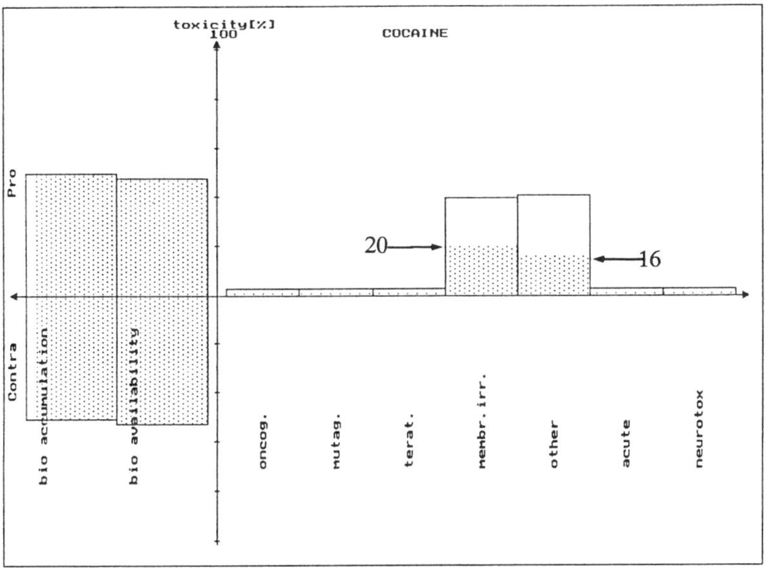

Compound	:	COCAINE
Species	:	Mammals
Administration	:	Oral
Toxic segment	:	BENZOIC ACID
Molecular weight	:	315.35
logP	:	1.111+-0.484
pKa	:	Basic:10.7

COCAINE

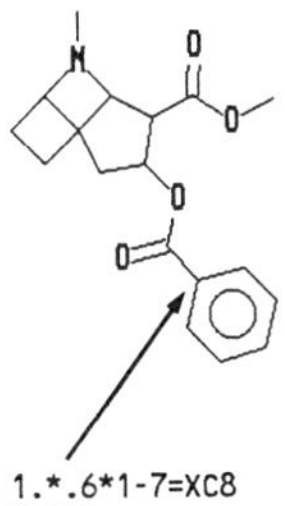

1.*.6*1-7=XC8
7-XB9
2-XA10
3-XA11
4-XA12
5-XA13
6-XA14

XB: O,N,CL,BR,I

Name:BENZOIC ACID

Enabled atoms: 7,1,6,5,4,3,

Reference: 1

Tox. dinamial eff.	
Oncogen	3%
Mutagen	3%
Terratogen	3%
Membran irritation	40%
Other chronic	41%
Acute	3%
Neurotoxic	3%

Species sensitivity	
Microbes	100%
Algae	100%
Plants	100%
Aquatic invertabres	40%
Soil invertabres	40%
Fishes	40%
Birds	40%
Mammals	40%

Figure 3. Toxic segment rule and application.

Conclusions

HazardExpert is one of several expert systems developed by CompuDrug Chemistry Ltd. for use in the prediction of the results of potential scientific studies. The program is not a database based on past results, but rather a knowledge base system drawing upon knowledge available today to predict the results of tomorrow, in a manner similar to that of an expert in the related field. With proper maintenance of the knowledge contained in the system, HazardExpert is a valuable tool to a variety of scientists and laymen with differing needs for similar information across a wide variety of industries.

Acknowledgment

I would like to thank Dr. Sam Clifford for his role in convincing me to write this paper, the entire crew at CompuDrug Chemistry Ltd. in Budapest for creating the programs and teaching me all I know about them, H. Dwain Lovett for his assistance in creating the finalized version of the text, and the American Czar for correctly identifying one toxic compound.

Literature Cited

1. Brink, R.H.; Walker, J.D. ***EPA TSCA ITC Interim Report***; US EPA: Rockville Md, June 1987.
2. ***MetabolExpert version 9.12***; CompuDrug USA, Inc: Austin TX, 1990.
3. Testa B.; Jenner P. ***Drug Metabolism: Chemical and Biochemical Aspects***; M. Dekker, Inc.: New York, 1976.
4. ***Pro-LogP version 4.1e***; CompuDrug USA, Inc.: Austin TX, 1991.
5. Rekker, R.F. ***The Hydrophobic Fragmental Constant. Its Derivation and Application. A Means of Characterizing Membrane Systems***; Elsevier: Amsterdam, 1977.

RECEIVED August 15, 1991

Chapter 20

Using the Menu Census Survey To Estimate Dietary Intake

Postmarket Surveillance of Aspartame

I. J. Abrams

MRCA Information Services, 2215 Sanders Road, Northbrook, IL 60062

MRCA has been tracking the actual consumption of Aspartame by individuals from their diets since its approval in 1981. This monitoring was mandated by The Food & Drug Administration (FDA) as a condition of its approval of Aspartame as a food additive, as published in the Federal Register on July 24, 1981. One objective of this tracking was to compare the actual use with the maximum projected consumption level of 34 milligrams per kilogram of body weight of the eater, per day, at the 99th percentile for the total sample. This estimate was computed by MRCA in 1975, using food consumption data from the Menu Census Survey of 1972-73, and the concentrations of Aspartame that were proposed by General Foods Corporation in packets for table-top use, in various powdered mixes, and in several liquid products including carbonated soft drinks. It was included in the petition which was submitted to FDA by General Foods Corporation in March 1976. Thus, MRCA's work in estimating the intake of Aspartame began over fifteen years ago.

The methodology for using food consumption data from Menu Census Surveys to estimate the intake of substances from the diet was originally developed, in the early 1970's, by the National Academy of Sciences GRAS Review Committee Phase I. It has been refined substantially since then, in continued work with The Food & Drug Administration, with the National Academy of Sciences GRAS Review Committee Phase II and Phase III, and with many commercial organizations. Over these years, the Menu Census Surveys were used to estimate the frequency distributions of the intake of close to 2,000 substances from the diet. In order to understand this methodology, it is necessary to describe in detail the Menu Census Surveys.

The Menu Census Surveys

The Menu Census Surveys provide an in-depth continuing record of food and beverage preparation and consumption by U.S. households and individuals. Although the primary use of these data is for marketing and product development, a growing use has developed in the past 20 years for estimating the intake of direct and indirect food additives, by both The Food and Drug Administration and by commercial organiza-

0097–6156/92/0484–0201$06.00/0

tions, in order to meet FDA's regulatory and review requirements. In addition, these data have been used by commercial organizations to estimate the intake of vitamins and minerals from the diet, in order to support nutritional claims, or to explore new product opportunities. Some of their key applications, and typical user groups, are shown in Table 1.

MRCA has been tracking all food preparation and consumption at-home and away-from-home through the Menu Census Surveys since 1957. These surveys were conducted once in five years from 1957 to 1977, and then continuously from 1980 forward. All foods are reported, except the use of table salt, pepper, and tap water. The surveys are currently based on nationally representative rotating samples of 500 households per quarter, or 2,000 households per year, containing about 5,500 members. Each household reports all food preparation and consumption, daily, by mail, in 14 consecutive daily diaries as summarized in Table 2. The households are distributed uniformly throughout the year, with about five or six new households starting their two-week reporting period each day of the year. All diaries are completed by the homemakers, who are also long-term members of MRCA's National Consumer Panel (NCP), the Weekly Purchase Diary Panel, who are therefore experienced in reporting their purchases of food products, in great detail, via diaries by mail.

Diets, Psychographics, and Household Demographics. A separate questionnaire, administered following the 14th day of reporting, provides, for each household member, detail information on age, sex, pregnancy status, weight, height, special diets followed, reason for the diet, foods encouraged or discouraged from eating, use of table salt, and the consumption of vitamin and mineral supplements, including kind, potency, amount and frequency. Another questionnaire, completed only by the homemaker, covers attitudes, awareness, and interests in a wide range of subjects dealing with lifestyle, food preparation, cooking skills, nutrition, food additives or preservatives, low-calorie products, sugar substitutes, and the use of information shown on the labels. In addition, an extensive set of demographic classifications is available for each household, obtained as part of their participation in NCP, the Weekly Purchase Diary Panel.

Intake Surveys. An intake study for a given substance usually includes Frequency Distribution Reports of the Daily Intake of the Substance by Individuals in several age groups, in milligrams (MG) and in milligrams per kilogram of body weight (MPK) of the eaters. Also provided are the corresponding Sources of Mean Intake Reports, which show the contribution of each specific food to the total overall mean intake of the substance, by the same age groups.

The intake study for a given substance is based on a detailed listing of all the foods which contain the substance, or which the manufacturer plans to include in the petition to FDA for the use of this substance. This list is prepared using the detail Menu Census food classification code book. At the same time, the manufacturer also provides the actual or the proposed concentrations of the substance in each food item on the list. The amount of the substance consumed by eating any food item on the list is computed by multiplying the concentration of the substance by the quantity of the food eaten (Table 3).

Table 1. Menu Census Survey: Typical Applications and User Groups

Applications	*Users*
National Food Trends	Top Management
	Mergers & Acquisitions
Defining the Marketplace:	
• Market Size Estimates	Product/Marketing Management
• Competitive Framework	Advertising Management &
• Product Positioning	Agencies
• Copy Theme Development	Legal Department
Managing Brands:	
• Line Extension	Product/Marketing Management
• Recipe Ideas	Advertising Management &
• Tie-In Promotions	Agencies
• Ad Themes	Nutritionists/Home Economists
Evaluating New Product Opportunities:	
• Define Markets & Segments	New Product Development
• Estimate Potential	
• Target Audiences	
Forecasting:	Strategic Planning Groups
• Long & Short Range	Marketing Management
• Detect Trends - Early On	Product Management
Intake Studies:	
• Nutrition	Food Development &
• Food Additives	Technology
	Regulatory Affairs
	Government Agencies
	Nutritionists

Table 2. The Daily Diary

Each daily diary provides the following information:

1. A detail description of each dish eaten, and items added to it at the table
2. At-home or away-from-home
3. At breakfast, lunch, or dinner meals; or at morning, afternoon, evening, or bedtime snack eating occasions
4. The position of the dish in the meal
5. Which household members ate the dish, and each item added to it at the table

For all dishes eaten at-home, information is provided on:

1. The number of guests who ate it, by children vs. adults
2. First time vs. leftover serving
3. Method of preparation and appliance used
4. Brand names of commercially prepared products
5. Form as obtained, such as fresh, frozen, canned, etc.
6. Packaging material in contact with the food
7. For homemade dishes, the reporting includes a detail description of every product used as an ingredient, fats and oils used as agents for frying, or flour for dusting breadboards. For each ingredient, the diary provides the brand name, form as obtained, packaging material, and whether the ingredient was itself a leftover.
8. Who ate the meal together, at what time, and if it was a special occasion

For foods eaten away-from-home, information is provided on:

1. The type of place, such as at friends, school, restaurant, lunch counter, etc.
2. The name of the food service facility
3. If from a vending machine
4. If eaten at the place where it was obtained.

Quantities of Food Eaten. Note that the 14 daily diaries provide only the incidence of eating each food product by an individual, but not the quantity eaten by each person, since reporting quantities for 14 days would be too burdensome to the homemaker. Instead, the average grams per eating occasion have been calculated from the most recently available USDA National Food Consumption Survey (NFCS) of 1977-78, for persons grouped by age and sex, using a linear "smoothing" procedure on these estimates when needed.

Since the USDA Survey provides grams of foods for end dishes as-eaten, these estimated average amounts per eating occasion are used only to quantify dishes as-eaten, such as milk when consumed as a beverage, or sugar when added to coffee or tea, or oil used as a salad dressing. When the products are used at-home as ingredients or frying agents, in preparing other foods, the amounts consumed are computed as percentages of the corresponding amounts of the end dishes in which they were used. These percentages are based on estimates obtained from standard recipes.

Intake Amounts. The quantity of the substance consumed by a given individual, from a single eating occasion of a given item of food on the list, is thus calculated by multiplying the average grams per eating occasion of that food, for an individual of that age and sex group, by the concentration of the substance in that food. It is displayed in Table 3, in the case of Aspartame.

Table 3. The Estimated Total Intake of Aspartame by a Given Person, in a Given Day, is the Sum Over All APM Containing Food Items Eaten that Day, by that Person, by the Multiplication of the Following Three Terms

Number of ***times*** an APM containing item of food was eaten on that day by that person	*	Average number of ***grams*** per eating occasion of that food for a person of that age and sex group	*	Number of milligrams of ***APM*** per gram of that food as eaten

These estimated intakes of the substance are then aggregated for all eatings of that food item by that same person, *separately for each day, and in total for all 14 days combined.* Corresponding sub-totals are accumulated for the intake of the substance from each food sub-category, and from the total diet, separately for each person, and for each day, treating each day for each person as an independent observation, to provide estimates of intake on a *Person-Day Basis.* Intakes are also accumulated for each person over all 14 days combined, to provide estimates on a *14-Day Average Daily Basis.*

Intake Study Reports. Frequency distributions of the average daily intake of the substance are then tabulated by age groups, in five percentile increments, for the eaters only, and separately for the total sample of eaters plus non-eaters, from the lowest to the highest, with finer breakouts for the heavier eaters of the substance above the 95th percentile, usually showing the 97.5th, the 99th, the 99.5th, and the 100th percentiles.

Separate reports show these intakes in milligrams (MG), vs in milligrams per kilogram (MPK) of the body weight of the eater. Reports are usually produced for the intake of the substance from individual food sub-categories; from cumulative food groups; and from all foods combined.

Multiple reports are frequently produced, separately on a Person-Day Basis vs. on a 14-Day Average-Daily Basis, as needed to differentiate analyses of the potential "acute" effects from the "chronic" effects of the consumption of the substance.

Corresponding reports are provided for the Sources of Mean Intake, which break out the contributions to the mean intake, by age groups, of each specific food included in the original list of concentrations of the substance.

Concentrations by Brand and Flavor. In addition, since the Menu Census Surveys identify the brands of each commercial product, reports can be produced tracking the "actual intake" of a food additive, once it is approved and introduced into the market, and as it expands its distribution over time; instead of the "prospective intake" of the substance, which is estimated when the petition is first prepared. This is the case, for example, with the Post Marketing Survey Phase II of Aspartame, in which MRCA has been tracking the intake of APM since April 1984, using its actual concentrations in each brand, as reported by their manufacturers.

Nutrition Surveys. The intake of any nutrient in the food is estimated using essentially a similar procedure, with the concentrations coming from a standard nutrient composition database. For this purpose, each food in the Menu Census Survey is assigned its corresponding code in the nutrient composition database, and the associated amount of the nutrient per 100 grams of the food is treated as the appropriate "concentration".

The reports for nutrition studies usually show the contributions of each food category and sub-category to the total daily intake of each vitamin or mineral, for individuals classified by age and sex groups. Additional reports distribute the intakes of nutrients by meal occasion or by some other food consumption patterns. These intakes are usually reported by age and sex groups, in absolute quantities, or as percentages of the Recommended Daily Allowances for the corresponding nutrient.

Special Diets. Since the Menu Census Study contains information on any special diets which a person may follow, reports are also produced comparing the intake of food additives, or of vitamins and minerals, by persons who are, for example, on a diet to reduce weight, to avoid cholesterol, or to control diabetes, etc.

Homemaker's Attitudes. Using the responses of the homemakers to the attitudes, awareness, and interests questionnaire, which has been administered to all Menu Census households since 1972, MRCA has been able to classify these homemakers by their concern about low-calories, nutrition, food additives, prepared foods, etc., and then correlate these classifications with the frequency of consumption of various food categories by individuals in the same households. Using the same classifications, it is possible to trend the nutrient intakes by the homemakers, and by other household members, over the past 18 years, a period during which interest in proper nutrition has substantially grown. Similar trends can be produced for selected food additives whose consumption may be correlated with specific homemaker's attitudes, such as towards fat or cholesterol, sugar, low-calorie sugar substitutes, caffeine, preservatives, and the like.

The Post Marketing Survey of Aspartame - An Intake Study Example

As stated earlier, the Post Marketing Survey (PMS) of Aspartame, PHASE I, was begun in January 1982. It was designed to track the percentage of children under 13 years old who consumed any food containing added Aspartame during an average 14-day period. PHASE I was to continue until the level of about 30% was reached, corresponding to the previously existing level of exposure by children to Saccharin containing products. PHASE II would then begin, in which the frequency distributions of the "actual" amount of Aspartame consumed through the diet will be tracked on a quarterly basis.

Phase I - Exposure by Children 0-12 Years Old. Figure 1 displays the growth in the exposure of children to Aspartame containing products from April 1982 through June 1984. For the first year, less than 2.1% of children 0-12 years old consumed any products containing Aspartame in an average two-week period. Beginning with April 1983, the exposure to Aspartame grew rapidly, and reached a level of about 24% by June 1984. In fact, the percentage of children 2-5 years old consuming added Aspartame reached 30% by June 1984, thus initiating the Post Market Survey Phase II.

Phase II - Intake by Children 6-12 Years Old. Figure 2 shows the 90th percentile of the 14 day average-daily intake of Aspartame, for eaters only, in milligrams per kilogram of body weight of the eater, for children 6-12 years old vs. the total sample. Note that this corresponds to the estimated "high" chronic level of intake of Aspartame, since it is based on the average intake in 14 days, including days in which Aspartame was not consumed, as well as those in which it was consumed by each individual.

As can be seen, the 90th percentile of intake for the 6-12 year old children has grown from a low of about 1.6 MPK in 1984 to 3.6 MPK in the first two quarters of 1986, with a single "extreme value" of 5.3 MPK in the first quarter of 1985. Note that such a variation in the estimates can be expected since it is based on a relatively small number of eaters, in this particular case on only 29 children in this quarter.

As is typical in these distributions, the intake in milligrams per kilogram of body weight by children is higher than that by adults, and thus the corresponding intake by the total sample is below that for children 6-12 years old. As can been seen in this figure, the 90th percentile of intake by the total sample has grown from 1.2 MPK in the first two quarters to about 2.0 MPK in the last three quarters on the figure.

An estimate of the "acute" intake of Aspartame for "eaters only" is shown in Figure 3. It is based on computing the estimated intake of Aspartame separately on each day,

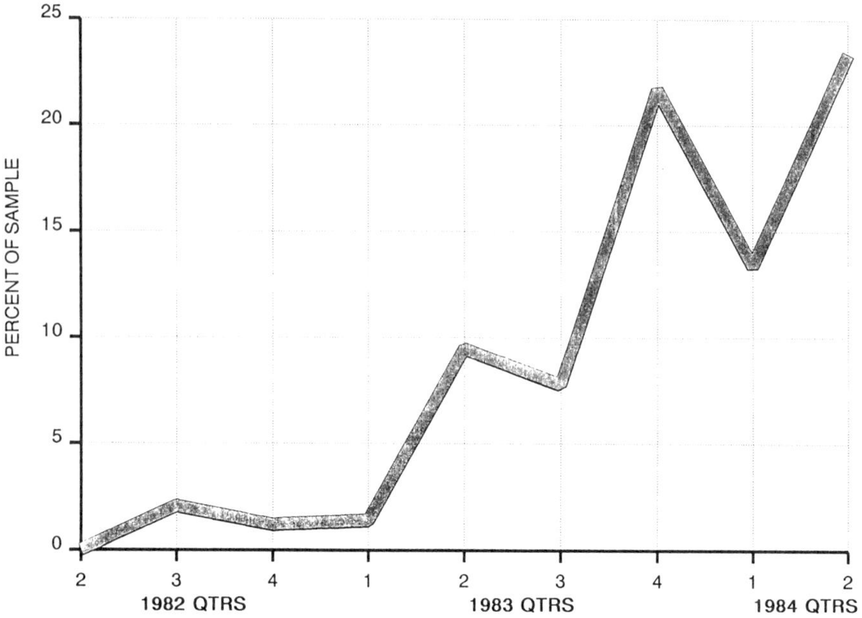

Figure 1. Growth in exposure of children 0–12 years old to aspartame in a 14-day period.

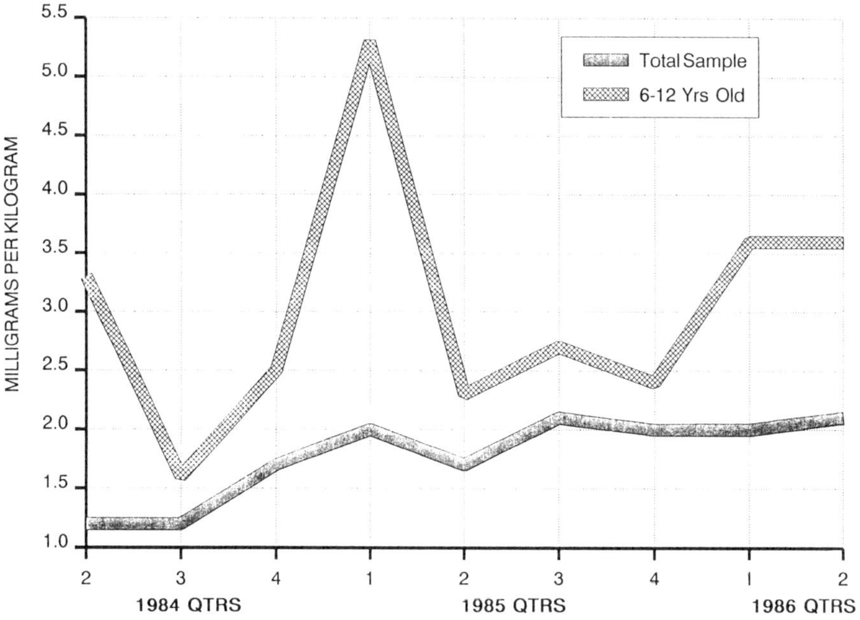

Figure 2. Aspartame intake by children 6–12 years old, 90th percentile of 14-day average daily intake.

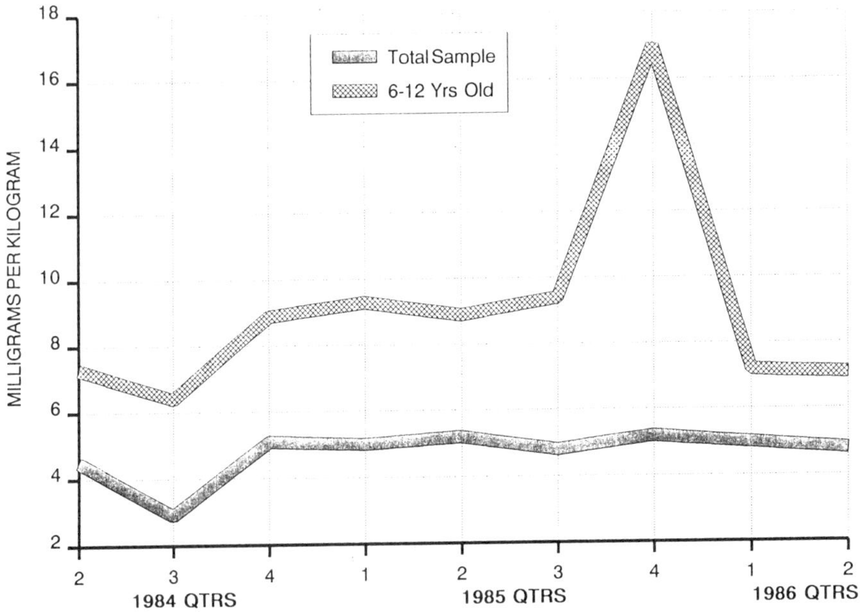

Figure 3. Acute levels of daily aspartame intake by children 6–12 years old.

for each individual, and excluding from the analysis all days in which no Aspartame containing products were consumed by the individuals. As can be seen from this figure, the 90th percentile of the intake of Aspartame per person-day, based on eaters only, in milligrams per kilogram of body weight, for children 6-12 years old, has increased from about 7 MPK during the third quarter of 1984, to about 9 MPK during the second and third quarters of 1985, jumped to 17 MPK in the fourth quarter, and then returned to about 7 MPK in the first two quarters of 1986.

As in the previous figure, the 90th percentile of intake per person-day for the total sample was well below that for the 6-12 year old children, and essentially remained constant over this entire period at about 5 MPK per day.

Note that all of these estimates, including the extreme values, are well below the FDA guideline for the Acceptable Daily Intake (ADI) of Aspartame on a "chronic" basis.

Phase III - Intake by Children 2-5 Years Old, by Reducers, and Diabetics. Figures 4 through 9 compare the "chronic" and the "acute" 90th percentile intakes, of eaters only, for children 2-5 years old, and for persons on a diabetic or on a reducing diet, vs. the corresponding intakes for the total sample.

Note that the intakes, in MPK, of children 2-5 years old are usually higher then those of children 6-12 years old. This, again, is due to their higher food intake relative to body weight.

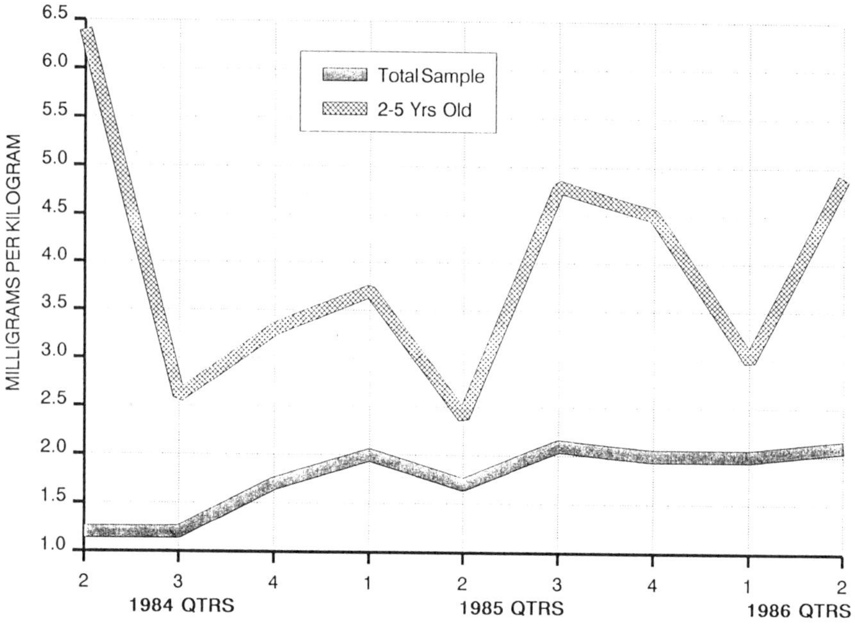

Figure 4. Aspartame intake by children 2–5 years old, 90th percentile of 14-day average daily intake.

Figure 5. Acute levels of daily aspartame intake by children 2–5 years old.

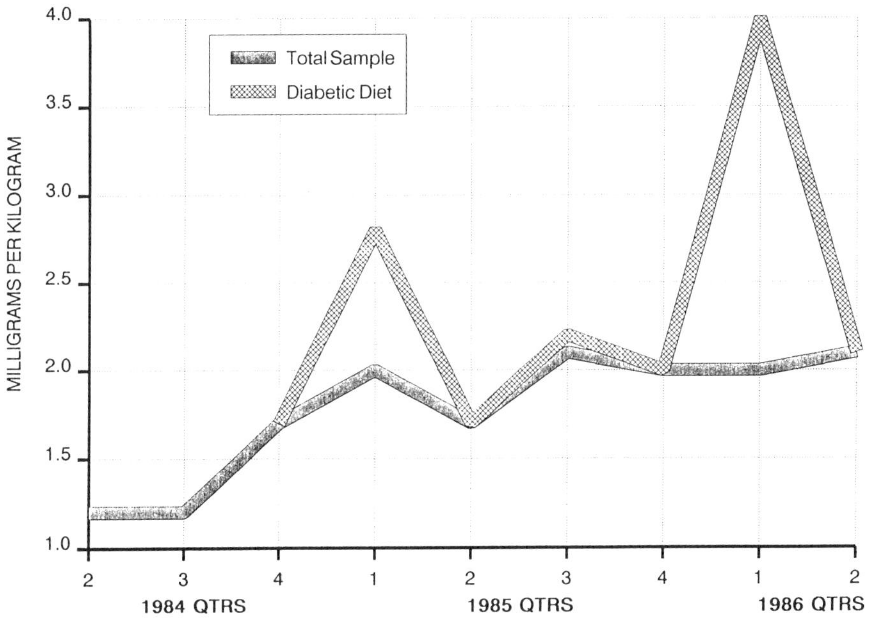

Figure 6. Aspartame intake by persons on a diabetic diet, 90th percentile of 14-day average daily intake.

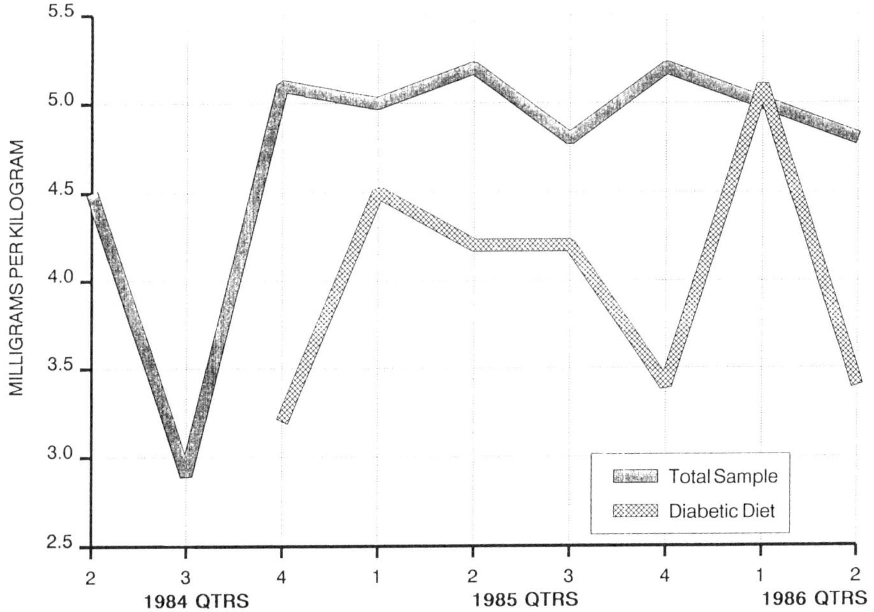

Figure 7. Acute levels of daily aspartame intake by persons on a diabetic diet.

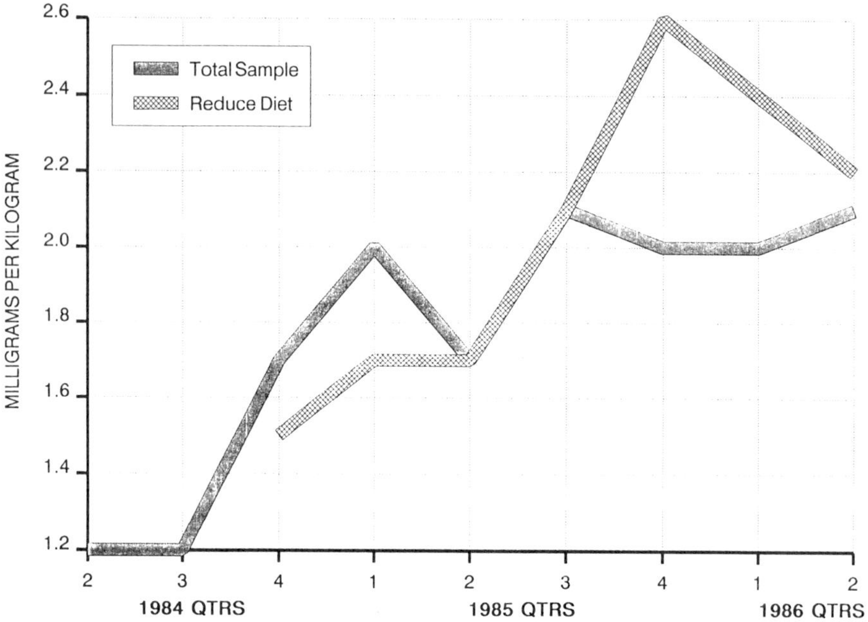

Figure 8. Aspartame intake by persons on a reducing diet, 90th percentile of 14-day average daily intake.

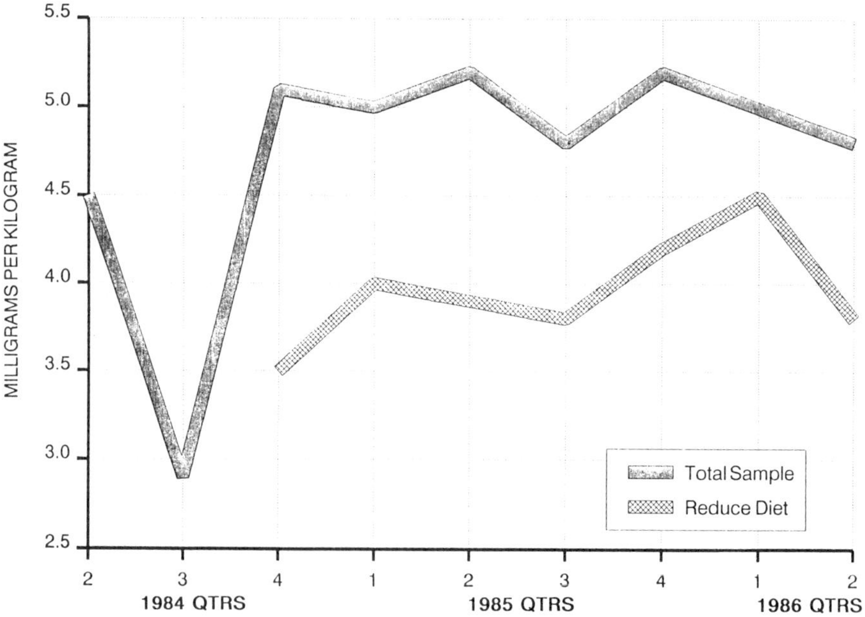

Figure 9. Acute levels of daily aspartame intake by persons on a reducing diet.

As may be expected, a substantially higher percentage of persons on a reducing or on a diabetic diet consume Aspartame containing products than those in the total sample. Nevertheless, their levels of intake of Aspartame are in general similar to those of the total sample.

Conclusion

The Menu Census Survey has proven over the years to be a very effective instrument for estimating the intake of food additives from the diet for both "chronic" as well as "acute" levels of exposure. The richness of the database has supported the extreme demands of estimating intake from foods which are "ready-to-eat", as well as from those which are used by the homemaker only as ingredients for preparing other dishes. The extended 14-day period of observation is indispensable in estimating the long-term average intake of nutrients from the diet; the "chronic" exposure to direct food additives; to food animal drug residues; as well as to contaminants in the food, such as naturally occurring lead, or that which migrates into the food from sodered cans.

Furthermore, once a substance has been approved by FDA, the same Menu Census database has frequently been used to set priorities for the introduction of the product into different food categories, to define the segments of the prospective market, to estimate trends, to explore for new uses in additional food items, and for various other market research and marketing purposes.

RECEIVED August 15, 1991

Chapter 21

Dietary Exposure Assessment in the Analysis of Risk from Pesticides in Foods

Michele Leparulo-Loftus, Barbara J. Petersen, Christine F. Chaisson, and J. Robert Tomerlin

Technical Assessment Systems, Inc., 1000 Potomac Street, NW, Washington, DC 20007

Pesticides are sometimes called "beneficial poisons." The judicious use of these chemicals has improved our standard of living and controlled some of our worst public health threats. Their benefits do not alter the fact, though, that pesticides are poisons.

Pesticide residues occur in our foods as a result of pesticide treatment of crops in the field, treatment to seeds, and post-harvest treatment of crops and processed foods. Secondary residues occur in meat, milk and eggs from the pesticide residues in livestock feeds. Primary residues in meat, milk and eggs occur as a result of direct treatment of livestock.

Some pesticides are "acutely toxic," i.e., the adverse effect occurs as a result of one or a few days of exposure. These effects may be headache, eye and skin irritation, breathing distress, severe nervous system interference, or birth defects.

Some pesticides may not cause acute effects, even in rather large doses, but they can cause serious health problems as a result of persistent exposure. They may cause cancer or liver and kidney disorders after a long duration of low-level exposure. Such adverse effects are referred to as "chronic effects."

In all these cases, the critical parameters which determine whether or not a pesticide will cause any adverse effect are (a) the intrinsic biochemical activity of the chemical; (b) the level of exposure to the chemical; and (c) the duration of exposure to the chemical. No matter what the intrinsic toxicological properties of the chemical, the level and duration of exposure are critical elements in assessing the potential risk. In this paper, the exposure component in the risk assessment of pesticides in our foods is discussed.

The exposure level of a chemical in a food is determined from the magnitude of the chemical in the food item and the amount of consumption of that food item.

$$\text{Exposure} = \text{Residue Level X Consumption}$$

An exposure assessment strives to obtain the most accurate estimate of the magnitude of the residue in the food and the consumption of that food. Exactly what type of number is used for the estimate of the magnitude of the residue and the

0097–6156/92/0484–0214$06.00/0

consumption in the exposure equation is driven by the nature of the risk. The question must first be asked, "What is the adverse health effect?" "Is the adverse effect due to a long-term exposure (a chronic effect) or to one or a few critical exposures (an acute effect)?" Estimation of the first element in the exposure equation, the magnitude of the residue, is discussed in the first section of this chapter. Estimation of the second element, consumption, is discussed in the second section. Combining the two elements to obtain the estimate of the exposure is discussed in the third section of this chapter.

Estimates of the Magnitude of the Residue

Exposure assessments based on tolerances — the theoretical maximum residue contribution (TMRC)—dramatically overestimate potential exposure from pesticides in foods. In the past, even though the exposure was dramatically overestimated, exposure assessments were routinely carried out using tolerances or the TMRC. Either more realistic residue data were not available, or when such data were available, they were not incorporated into the risk assessments conducted by regulators. That is no longer the case. Now the U.S. Environmental Protection Agency (EPA) not only uses such results in risk assessments, but also requires that marketplace monitoring be conducted by pesticide registrants.

Nature of the Residue. Before determining the residue level in food from use of a pesticide, the question must first be asked: "What is the residue?" This question must be answered before the magnitude of the residue can be determined. To obtain a registration allowing use of a pesticide on a crop, EPA requires crop metabolism studies to be conducted (*1*).

The nature of the residue in foods is determined by plant and animal metabolism studies. In plant metabolism studies, the pesticide is labelled with a radioactive atom, usually ^{14}C, and is applied to the crop in a manner simulating actual use. To avoid loss of radioactivity, the study is usually conducted in the greenhouse. At plant maturity, or sometimes prior to plant maturity, the crop is harvested and the radioactivity in the plant parts used as food or feed are extracted and separated into various fractions. The chemicals comprising the radioactivity are identified to the extent possible (*2*).

If residues of a pesticide result in the feed or if the pesticide is directly applied to livestock, animal metabolism studies are required by EPA to delineate the residue in meat, milk and eggs. Typically, lactating goats are used to represent ruminants and laying hens are used to represent poultry. If the metabolism profiles in ruminants and poultry are different, studies in swine may also be required (*1*).

Depending on the route of exposure, the animals are dosed either orally or dermally with the radiolabelled pesticide. Milk and eggs are collected and the animals are sacrificed within 24 hours of the last dose. The level of radioactivity in the edible tissues (muscle, fat, kidney and liver), milk and eggs is measured. If the level is significant (> 0.01 ppm), EPA requires the radioactivity to be characterized (*3*).

The metabolism studies should determine the total terminal residue. The total terminal residue is the sum of the parent compound and its degradation products, metabolites (free or bound), and impurities. Not all components of the total terminal residue are necessarily regulated. Only those components which are considered to be "toxicologically significant" are regulated.

Once the total terminal residue in the plant and animal tissues, milk and eggs are determined, the "total toxic residue" or "toxicologically significant residues" is defined by EPA. The term total toxic residue is used by EPA to describe the sum of the parent pesticide and "toxicologically significant" metabolites, degradation products and impurities (*2*). The tolerance expression is defined by the total toxic residue.

Residue Analytical Methods. The residue analytical method should determine the total toxic residue. The sensitivity necessary for the analytical method depends on the purpose of the analysis. If the purpose of the analysis is to enforce a tolerance, the method usually does not need to be as sensitive as one to conduct an exposure assessment; for enforcement, the limit of quantitation of the analytical method only needs to be as low as the tolerance level.

Residue Estimates-Tolerances. When a pesticide is registered for use on a particular crop by U.S. EPA, a tolerance is set on the basis of the results of controlled field studies (*1*). In controlled field studies, the crop is treated with the formulated pesticide product according to the maximum use allowed on the proposed label and typical cultural practices. The field trials represent a wide range of climatic and geographic conditions. The residue data from the study result in a distribution of residues with a median in the low residue range. The outliers in the high residue range are given considerable weight when setting tolerances. The tolerance is the maximum concentration of total toxic residue (the active ingredient and toxicologically significant metabolites, degradation products and impurities) of a pesticide which is legally allowed on a particular food. If a metabolite is more potent than the parent compound, a separate lower tolerance may be set for the more potent metabolite. A typical example of the distribution of residue data from a crop field study and its comparison to the tolerance is shown in Figure 1.

If residues result in the feed or if dermal treatments are registered for use, tolerances for meat, milk and eggs are established on the basis of animal feeding studies (*1*). In these studies, cattle and poultry are dosed at three different levels with the pesticide and metabolites for at least 30 days and are sacrificed within 24 hours of the last dose. The residue data from the feeding studies are adjusted to reflect a dose based on tolerance-level residues in the feeds.

The purpose of the tolerance is to enforce proper use of the pesticide. If the tolerance is exceeded, the presumption is that the use directions on the label were not followed. The U.S Food and Drug Administration (FDA) is charged with enforcing the tolerances. FDA accomplishes this by taking samples of crops "at the farmgate" and analyzing the crop to determine whether the crop has any illegal residues of pesticides. Illegal residues are residues in which the tolerance for the particular crop has been exceeded or has not been established. If illegal residues are found, the crop is labelled "adulterated" and not allowed to move in commerce.

Since the purpose of a tolerance is to prevent misuse, it greatly overestimates exposure. Although tolerances overestimate exposure, EPA often uses tolerances to calculate the Theoretical Maximum Residue Contribution (TMRC), an estimate of exposure to pesticides that assumes that tolerance level residues are in or on all foods for which a tolerance is proposed or established.

Residue Estimates — Anticipated Residues. A more refined approach to estimating the residue level is the "anticipated residue" concept (*4*). The anticipated residue is a more realistic residue estimate than the tolerance. The anticipated residue can be determined from controlled studies or from monitoring data.

Anticipated Residues from Controlled Studies. The types of controlled residue studies which can be used to determine the anticipated residue are crop field trials, processing studies, livestock feeding studies, food preparation studies and storage degradation studies. Except for very old chemicals, residue data from the first three types of controlled studies are usually available because such data are required by EPA to set a tolerance. The controlled studies used to establish the tolerance are reevaluated to obtain an anticipated residue for a particular food. In addition, market share data can be incorporated into the analysis for further refinement of the residue estimate.

As discussed above, crop field trials designed to set tolerance levels reflect the most extreme case. The trials reflect the maximum number of treatments and the maximum application rate. In addition, early maturing crop varieties are better represented in crop field studies as compared to actual cultural practice. Thus, a large portion of the residue data reflect short times between the last treatment and harvest allowed by the pesticide label.

Even though residue data generated to set tolerances provide an exaggerated residue estimate, these residue data can be evaluated to determine a more realistic residue estimate than the tolerance. Of course, residue data from field trials reflecting typical use patterns can be generated. Although EPA does not require such data to establish tolerances, if the data are submitted with documentation that the data reflect typical cultural practice, EPA will use these data to estimate the anticipated residue. From controlled crop field studies, a mean or an upper ninety-fifth percentile can be calculated from the residue data. Typically, the mean value is used for risk assessment of a chronic effect and the upper ninety-fifth percentile is used for risk assessment of an acute effect.

Examining the effects of processing is the next step towards obtaining a more realistic estimate of the residue in foods as eaten. The level of residue in processed foods may differ considerably, higher or lower, from the raw agricultural commodity (r.a.c.). Processing studies are often available because EPA requires certain processing studies to be conducted for many crops (*1*). In these studies, the treated crop is processed into various byproducts and the residue in the byproduct is determined. If the residue in the byproduct is greater than that in the raw agricultural commodity (r.a.c.), a separate tolerance for the byproduct is established. If the residue in the byproduct does not exceed that in the r.a.c., a separate tolerance is not established. Although EPA requires processing studies to determine the concentration of residue, these data can be used to determine the reduction in residue, as well. An example of reduction in residue from processing is shown in Figure 2.

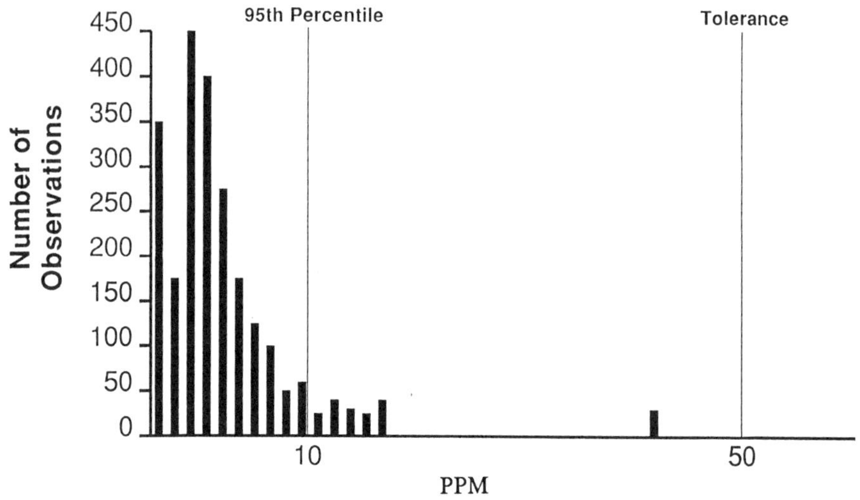

Figure 1. Residue data for chemical "Y" in sampled potatoes 1979-1984.

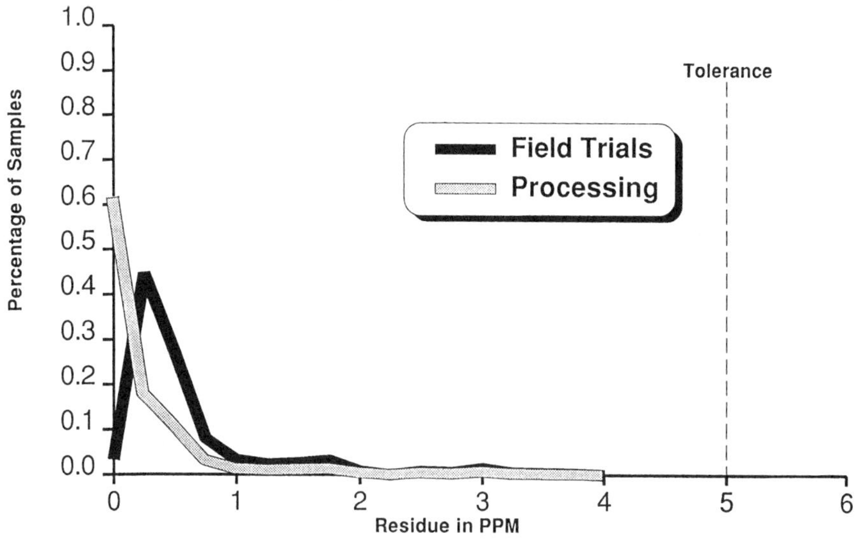

Figure 2. Comparison of residue data of raw agricultural commodity from crop field trials to residue data of processed by-products from processing study.

Food preparation and storage degradation will also have an effect on the level of residue in foods as eaten. Studies investigating the effect of food preparation and storage on the magnitude of the residue can also be conducted. Typically, food preparation and storage result in reduction of residue. An example of storage degradation which may occur in the chains of commerce is the residue data for acephate in head lettuce from field to supermarket, presented in Table 1.

Table 1. Summary of Acephate Residues on Crisp Head Lettuce

Location	*Residues[1] (Acephate & Methamidophos, ppm)*	*% Reduction of Residues*
Field	0.32	—
Field to Cooler (Head & Cap Leaf)	0.05	84
Cooler to Distributor (Head & Cap Leaf)	0.06	81
Distributor to Supermarket	0.04	87.5
Supermarket to Shelf	0.03	91

[1]One ground application (0.56 lb AI/A) plus one aerial application (1 lb AI/A) harvested 21 days after last application.

A reduction in residue during food preparation and storage is not always the case, though. For example, when Alar-treated produce is cooked, the metabolite 1,1-(unsymmetrical dimethylhydrazine), UDMH, increases, as compared to that in the raw agricultural commodity. UDMH, a moiety much more potent than the parent compound (daminozide), increases during cooking via degradation of the parent compound.

When market share data are available, the residue estimates can be further refined to reflect the acreage of the particular crop which is treated. For example, if usage data indicate that 20% of a food is treated with a given pesticide, the assumption is that over the long term, 20% of that food eaten by the typical person contains pesticide residues. One drawback, though, in refining the chronic exposure assessment for usage data is that pesticide usage changes from year to year, depending on the population of the pests and the weather conditions. However, if enough data are available, a running average may be used to compensate for annual variation. In spite of the necessity to periodically update usage data, it is more reasonable to use the available percent crop treated data than to assume that 100% of the crop is treated.

The anticipated residue in meat, milk and eggs can be determined from livestock feeding studies by comparing the livestock dietary burden to the results of the feeding study. Livestock feeding studies are required by EPA when pesticide residues may occur in the feed or when direct treatment to livestock is proposed. To obtain the anticipated residue in meat, milk and eggs for a chronic toxic effect, the livestock dietary burden is often determined from the mean residue levels of the pesticide in the feeds, corrected for the percent crop treated. For an acute effect, the livestock dietary burden typically reflects the upper ninety-fifth percentile of the pesticides in the feeds.

Anticipated Residue from Monitoring Data. The anticipated risk can be determined by residue monitoring data. Residue data from monitoring is distinguished from residue data from controlled field trials in that the samples are obtained from a "real life" situation using "real" cultural practices; i.e., the treatment rate, number of treatments and schedule of treatments for monitoring samples were chosen to obtain protection from a specific pest or to regulate the growth of the plant. In controlled field studies, the treatment regimen and cultural practices are not chosen to protect the crop but to provide data on the magnitude of the residue, usually under the most extreme conditions, in order to set tolerances. Figure 3 compares the anticipated residue from controlled studies, uncorrected for percent crop treated, and the tolerance with monitoring residue data for captan on grapes.

Many types of monitoring data are gathered for a variety of purposes. The treatment history of the sample may or may not be known. Monitoring programs include surveillance by regulatory agencies, residue monitoring by private industry for quality control/quality assurance, monitoring by special interest groups, and statistically representative food surveys. There are many focal points for food monitoring: at the farmgate, the food processing plant, the wholesaler, the port of entry and the supermarket.

For surveillance, the FDA samples at the farmgate and the port of entry. CDFA (California Department of Food and Agriculture) typically samples at the wholesaler and farmgate, but also collects samples at the processing plant. The objective of monitoring for surveillance enforcement activity is to prevent misuse of pesticides by catching violators. The monitoring data from surveillance activity can present a residue picture which is exaggerated as compared to overall food supply. The data may be quite adequate for the intended purpose, but usually cannot be used for quantitative risk assessment.

Private industry may conduct residue monitoring or screens for their own quality assurance. Likewise, "independent surveys" may be conducted by special interest groups. The monitoring may be focused (treatment history is known) or general (treatment history is not known). The studies may or may not be statistically valid. The data may be quite adequate for a specific commercial brand or region of the country, but are not necessarily representative of the whole food supply. Sometimes, questionable analytical methodologies are used which erroneously detect the chemical or, conversely, are too insensitive to measure residues, if present. Such independent monitoring studies should be carefully evaluated before using the data to estimate the anticipated residue.

Statistically valid market basket surveys or national food surveys are becoming more prevalent. National food surveys are difficult to design and expensive to conduct, but, when carefully designed, present the most accurate picture of residues in food as eaten by the public.

To obtain the most realistic snapshot of the magnitude of the residue as consumed, the sampling location usually must be as close to the consumer as possible. In most

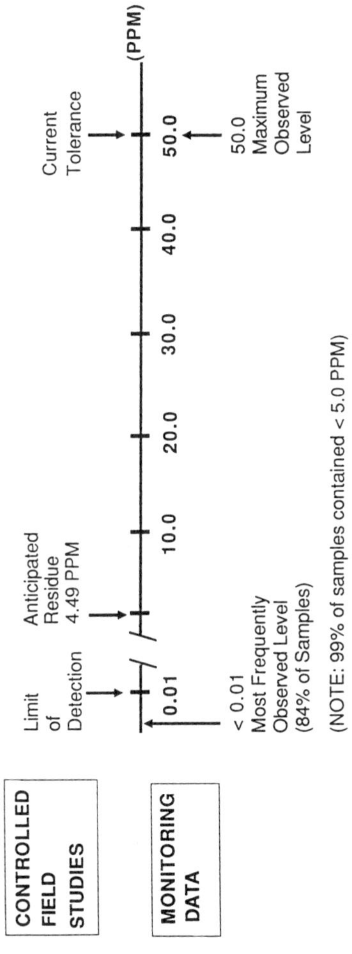

Figure 3. Residue data for captan on grapes.

cases, the grocery store is the sampling site that is closest to the consumer. Samples collected at the grocery store take into account the effect of commercial washing and storage of foods on the magnitude of the residue. However, if residue levels are not expected to change, sampling may be more efficient at earlier stages in the food production process.

The food survey should be tailored to specific issues and commodities. The population at risk should be considered when selecting foods to be monitored. All available residue data should be used to focus the survey on the foods constituting the most exposure to the relevant population subgroup. Foods that will have the greatest impact on the overall exposure assessment should be selected. This will depend on the consumption profile of the population at risk, as well as on the differences expected between the magnitude of the residue determined from controlled studies versus monitoring at the grocery store. Later in this chapter, in the section on Estimating Dietary Exposure, examples are provided which show how consumption can affect the exposure assessment.

The food survey should be designed to be statistically representative for criteria such as geography, urbanization, and store size. The best available information on the universe of sampling locations should be used for selecting the samples. Another part of the survey design is to determine the number of sampling locations. The number is determined by the desired level of precision. The best available residue data should be used for this purpose, in combination with information on market share and dynamics. A pilot survey will provide the best guidance in designing the study.

Another question to ask is what part of the sample to analyze. For enforcement purposes, FDA analyzes the whole fruit because the tolerance is set based on the residue in the whole fruit. In many cases, though, the whole fruit is not consumed. For example, only the inner flesh of melons is consumed. A decision whether to analyze the whole fruit or inner flesh should be made. If the pesticide in question is systemic (throughout the whole fruit), it will not make much difference whether the whole fruit or the inner flesh only is analyzed. However, if the residue is a surface residue, there will be a large difference between the results of the analysis of the whole fruit versus the inner flesh only. For a case such as a surface residue on melons, the residue estimate for the food as consumed would be exaggerated if the whole fruit were analyzed.

Consumption Estimates

In response to the need to more realistically estimate exposure to pesticide residues in food, EPA developed the Tolerance Assessment System to estimate exposure to pesticides for tolerance petitions, registration standards, and special reviews (5). Since the Tolerance Assessment System could do much more than assess tolerances, it was renamed the Dietary Risk Evaluation System (DRES). DRES combines the U.S. Department of Agriculture's (USDA) food consumption data, EPA's tolerance listings, EPA summary toxicology summary information, and statutory requirements into a mainframe-based system for analyzing food consumption data and assessing dietary intake. Technical Assessment Systems, Inc. has created the EXPOSURE Series, a microcomputer version of DRES (*6*).

The dietary estimates used in DRES and in the EXPOSURE Series were derived from the 1977-78 nationwide survey of individual food consumption conducted by the USDA (7). In this survey, three-day dietary records of 30,770 individuals were collected representing the four seasons. The records were collected via an interview in which the amount of each food item consumed and the weight of the individual were specified. Each person's food record in the USDA survey was accompanied by demographic and socioeconomic information about that person, and this information was incorporated into DRES and the EXPOSURE Series.

That the USDA food survey recorded information on the demographic and socioeconomic background of the individuals is important. Different age groups and ethnic groups may consume vastly different amounts of any given commodity. For example, children consume much more milk than adults. Asians consume much more rice than the rest of the population. Even if two people consume the same amount of a certain commodity (e.g., apples), they may consume it in different forms. One group may eat a great deal of raw apples (e.g., adults) and another may eat a great deal of apple sauce (e.g., infants). The geographic location of a person also will have an impact on the relative amounts of foods consumed. As compared to the entire U.S. population, Southerners, for instance, consume more greens while Californians consume more alfalfa sprouts.

Tolerances are established by EPA in or on foods as they enter commerce; i.e., in or on the raw agricultural commodity (apples, corn, milk), and only on a few processed foods. The USDA data was reported in terms of the consumption of food as eaten (e.g., apple pie). In DRES and the EXPOSURE Series, each of the 3734 individual food items reported in the USDA survey was broken down into its constituents (e.g., apples, wheat, etc.) to be compatible with the EPA tolerance-setting system (*8*). Standard recipes were devised according to the percentage of the raw agricultural commodity in the dish. The "EPA foods and feeds" are described by a crop to food map shown in Table 2.

Table 2. Example From the Crop-To-Food Map

Crop	*RACs*	*Food*
Apples		
	Food	
	Apples	Apples, fresh
		Apples, dried
		Apples, juice
	Feed	
	Apple, pomace	Meat, milk
	dried	poultry, eggs
	apples, pomace	Meat, milk
	wet & dry	poultry, eggs

The food consumption data from the 1987-88 USDA are now available (*9*). Unfortunately, the 1987–1988 survey is smaller and has been criticized for its low response rate (*10*). Technical Assessment Systems has incorporated the 1987-88 consumption data into the EXPOSURE Series, and EPA is considering various options. However, updating the database is important because dietary patterns change. For example, since 1978, consumption of apple juice by children has increased significantly, while consumption of red meat by the population as a whole has decreased.

Estimating Dietary Exposure

EXPOSURE 1. EXPOSURE 1 in the EXPOSURE Series is a dietary analytical system used to assess chronic exposure. As discussed before, the function of chronic exposure analysis is to estimate exposure over an extended period of time, presumably a lifetime. The annualized mean daily consumption for the U.S. population and 22 subgroups is used in this analysis.

The exposure from a particular food is calculated by multiplying the annualized mean daily consumption by the residue level in the food. The total exposure is calculated by summing the exposures for the individual foods. The exposure is expressed in terms of mg chemical/kg body weight/day and as a percentage of the Reference Dose (RfD) of a given chemical for the U.S. population and 22 subgroups. The Reference Dose is the No Observed Effect Level (NOEL) from a chronic study divided by a safety factor, usually 100 for a chronic study.

The best estimate for the residue level in the exposure equation should be used, and will depend on the available residue data. For older chemicals that have not been reregistered, the paucity of available data sometimes precludes the use of the anticipated residue for the residue estimate, and tolerances must be used. For the typical pesticide, though, residue data from controlled studies are available, and anticipated residues can be estimated. The best estimate for the anticipated residue of a food in chronic dietary exposure assessment is usually the mean residue obtained from controlled study residue data corrected for market share data (percent crop treated). If statistically valid monitoring data are available, the best estimate for the magnitude of the residue often is the mean residue from monitoring data.

EXPOSURE 4. EXPOSURE 4 in the EXPOSURE Series is a dietary analytical system used to assess the distribution of acute exposure. A chemical which is acutely toxic is one in which the adverse effect occurs after a very short duration, possibly after one critical exposure. For this reason, in contrast to EXPOSURE 1, the daily consumption data in EXPOSURE 4 are not annualized. The EXPOSURE 4 consumption database consists of the individual daily food consumption data. When combined with the magnitude of the residue in foods, EXPOSURE 4 characterizes the distribution of dietary exposure among individuals in the population. The results can be summarized as percentiles of the population with exposure above specified levels, as a ratio of the No Observed Effect Level to the calculated exposure estimate (also known as the margin of safety), or in a graphic format. Analyses can be tailored to a selection of population characteristics (e.g., age, sex, seasons, region of the country, ethnic group).

EXPOSURE 4 characterizes the distribution of exposure in the following way. Assume that residues of a Wonderchem occur on apples at 5 ppm and potatoes at 1 ppm. EXPOSURE 4 loads the consumption data for person-day #1, multiplies his consumption of apples by 5 ppm and his consumption of potatoes by 1 ppm, then sums the two estimates to produce an estimate of the daily intake of the Wonderchem by person-day #1. Person-day #1 is individual #1 in the USDA Nationwide Food Consumption Survey on the first day that consumption data is recorded for individual #1. This analysis is repeated for each of the person-days in the USDA Survey results. Individuals consuming neither apple nor potatoes drop out of the analysis. Examples of the distribution of the total dietary exposure for "Wonderchem" are shown in Figure 4a for the U.S. population and in Figure 4b for non-nursing infants.

Once again, the best estimate for the magnitude of the residue should be used in the dietary exposure assessment. The anticipated residue from controlled residue studies or statistically valid monitoring data, versus the tolerance, should be used as the best estimate when such data are available. However, the anticipated residue for acute effects is usually a different number, as compared to the anticipated residue for chronic effects. For acute effects, the anticipated residue for a food is usually the ninety-fifth percentile of controlled study residue data or, when available, the ninety-fifth percentile of statistically valid monitoring residue data (compare with mean, corrected for market share data, used for chronic effects).

As discussed in the first section, the residue profile for a particular food can be described by a distribution of residue values in a population of samples of that food. A refinement of EXPOSURE 4 can be performed where the distribution of the residue can be used as the anticipated residue. In this computation of the exposure, the distribution of the consumption per unit body weight is multiplied by the distribution of residues per unit of commodity.

Dietary Exposure Assessment and National Food Surveys. National food surveys are very expensive. For that reason, resources should be focused on the foods contributing the most to the dietary exposure of the populations at risk. When planning a national food survey, one should first determine what population is at risk. The available residue data should be evaluated to obtain the best estimate of the magnitude of the residue in the exposed foods. Then, an exposure assessment for the populations at risk should be conducted using the best estimate of the residue from the available residue data. The foods which contribute the most to the exposure of the populations at risk are the foods which should be monitored in the national food survey. Figures 5, 6 and 7 illustrate the importance of the population subgroup and the particular food on the exposure assessment.

The total dietary exposure of a pesticide in all treated foods is plotted in Figures 5, 6, and 7. The graphs demonstrate that refinement of the residue estimate for some foods will have a greater effect on the exposure assessment as compared to other foods. The graphs also show that the exposure assessment is dependent on the population subgroup.

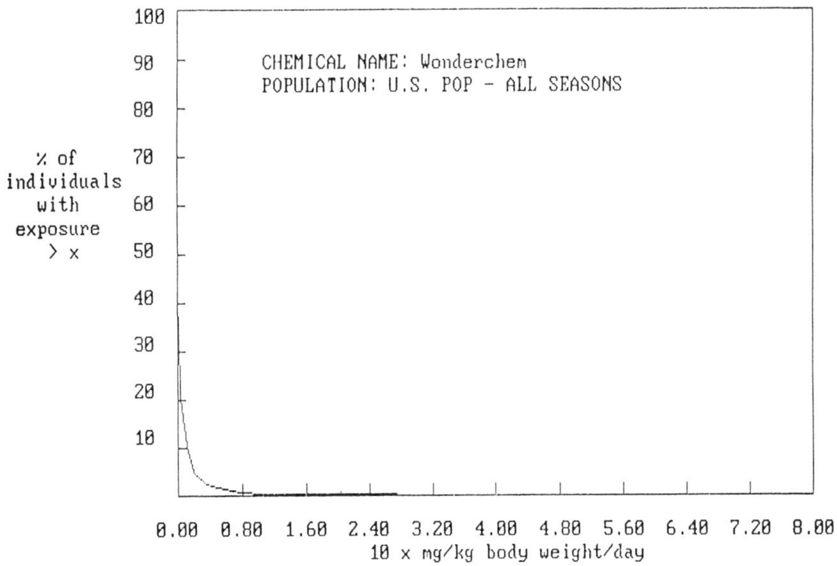

Figure 4a. Distribution of exposure from Wonderchem in foods for U.S. population.

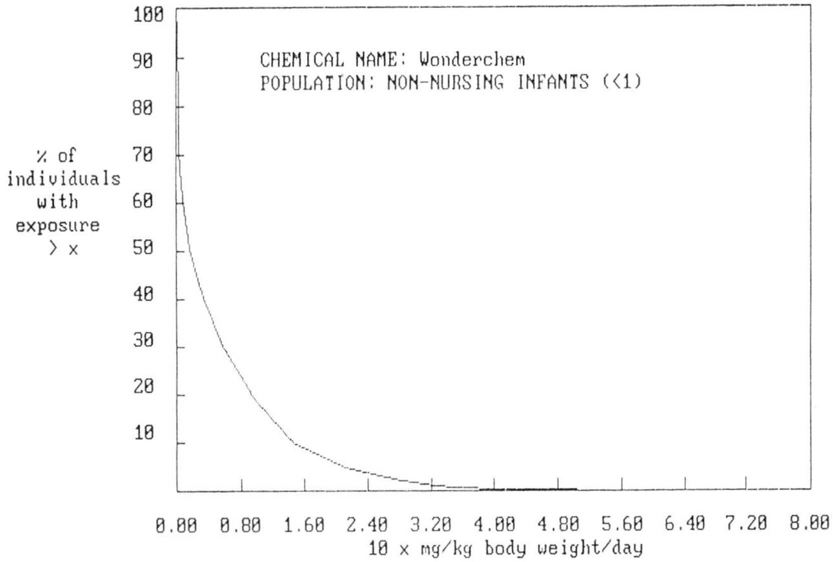

Figure 4b. Distribution of exposure from Wonderchem in foods for non-nursing infants.

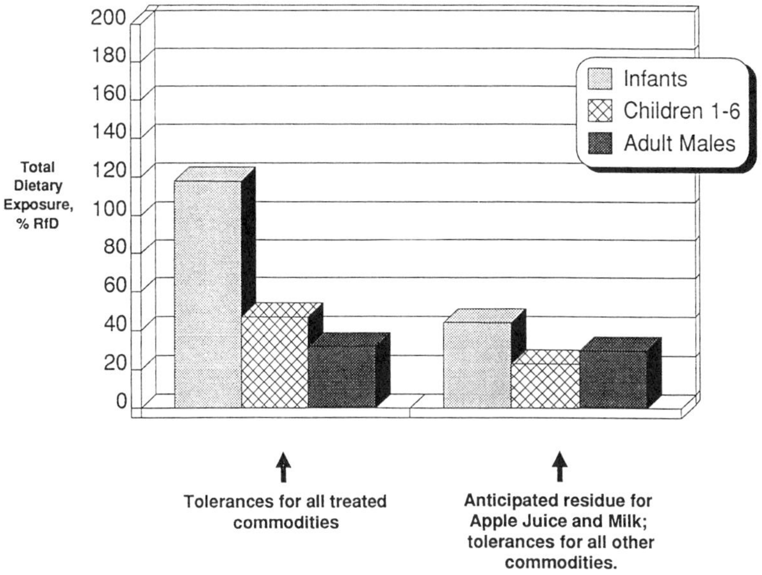

Figure 5. The effect of substituting apple juice and milk tolerances with anticipated residue data on the total dietary exposure using tolerance level residues for all treated commodities.

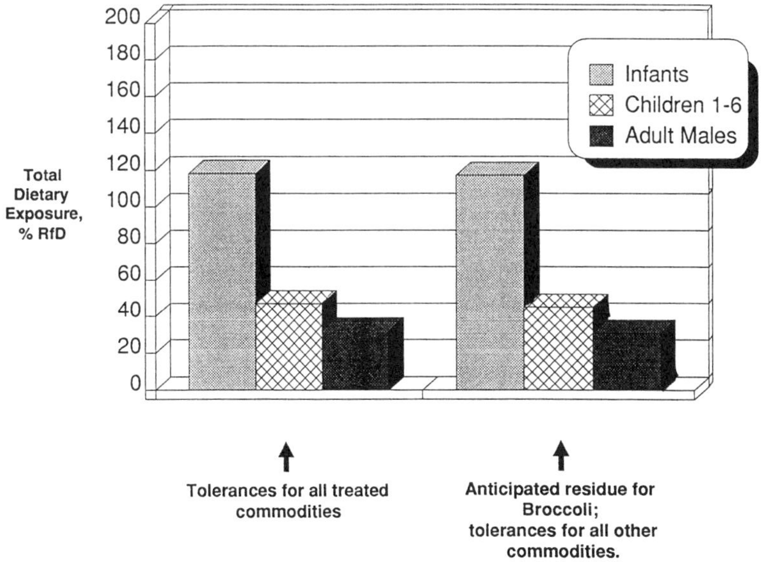

Figure 6. The effect of substituting broccoli tolerances with anticipated residue data on the total dietary exposure using tolerance level residues for all treated commodities.

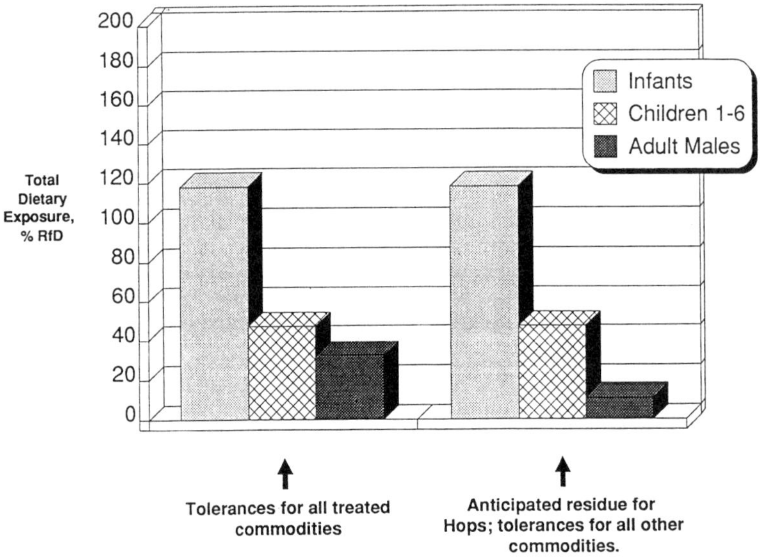

Figure 7. The effect of substituting hops tolerances with anticipated residue data on the total dietary exposure using tolerance level residues for all treated commodities.

Figure 5 compares the dietary exposure for infants, children under 6 years, and adult men using tolerances for the residue estimate for the various treated foods to the dietary exposure of these same population subgroups using the anticipated residue for apple juice and milk and tolerances for the remaining foods. The estimate of the dietary exposure of all three population subgroups declines by the substitution of the tolerance for apple juice and milk with their anticipated residues. The degree of reduction is greatest for infants because their consumption of apple juice and milk is highest as compared to the other subgroups. It must be emphasized that exposure has not changed, only the accuracy with which we estimate exposure has changed. Figure 6 presents the same sort of refinement of the residue estimate for broccoli. Substituting the anticipated residue for broccoli for the tolerance does not change the dietary exposure assessment for any of the three subgroups. In Figure 7, the refinement of the residue estimate is made for hops. Only the dietary exposure assessment for adult men is affected, demonstrating the importance of determining the population subgroup at risk when selecting the foods to be monitored.

Literature Cited

1. Schmitt, R.D. *Subdivision O, Residue Chemistry Guideline*; U.S. Environmental Protection Agency: Washington, DC; 63 pp.
2. Kovacs, Jr., M.F. *Metabolism (Quantitative Nature of the Residue): Plants, Hazard Evaluation Division Standard Evaluation Procedure*; U.S. Environmental Protection Agency: Washington, DC, 1988; 18 pp.
3. Schmitt, R.D. *Guidance on When and How to Conduct Livestock Metabolism Studies*; U.S. Environmental Protection Agency: Washington, DC, Internal Memorandum dated January 25, 1989.
4. Clayton, C.A.; Petersen, B.J.; White, S.B. *Issues Concerning the Development of Statistics for Characterizing "Anticipated Actual Residues"*; Research Triangle Institut, Research Triangle Park, NC, Unnumbered Report; 44 pp.
5. Saunders, D.S.; Petersen, B.J. *An Introduction to the Tolerance Assessment System*; U.S. Environmental Protection Agency: Washington, DC, 1987; 35 pp.
6. TAS EXPOSURE® Series; Technical Assessment Systems, Inc.: Washington, DC.
7. Nationwide Food Consumption Survey, 1977-78; U.S. Department of Agriculture: Washington, DC.
8. Alexander, B.V.; Clayton, C.A. *Documentation of the Food Consumption Files Used in the Tolerance Assessment System*; Research Triangle Institute: Research Triangle Park, NC, 1986; 49 pp.
9. *Nationwide Food Consumption Survey, 1987-88*; U.S. Department of Agriculture: Washington, DC.
10. *Pesticides—Food Consumption Data of Little Value to Estimate Some Exposures*; U.S. General Accounting Office: Washington, DC, 1991, 20 pp.

RECEIVED November 14, 1991

Assessing Microbial Safety in Food

Chapter 22

Current Concerns in Food Safety

R. V. Lechowich

National Center for Food Safety and Technology, Bedford Park, IL 60501

Consumers are concerned about the safety of their food supply. Their major concerns are identified as food additives, pesticide residues, food processing aids, and preservatives. This is in direct opposition to the relative priority placed upon food safety concerns as evaluated by food microbiologists and the U. S. Food and Drug Administration who rank microbiological hazards as *the* major food safety issue. For at least the past fifteen years a regulatory priority assessment has ranked food safety issues in descending priority as shown in Table I.

Table 1. Relative Risks of Food Safety Issues

1.	Microbiological Hazards
2.	Nutritional
3.	Natural Toxicants
4.	Environmental Contaminants
5.	Pesticide Residues
6.	Food Additives

While microbiological problems can and do appear in our food supply, these problems arise from infrequent breakdowns in the food production, processing, storage, distribution and final handling of our foods. These situations are the exception rather than the rule when we consider the more than 250 billion meals that are prepared and eaten in the U. S. each year.

Summaries of annual foodborne illness outbreaks reported in the U.S. have a lag time of several years and may not reflect the latest outbreaks. In 1988 there were more than 107,000 cases of reportable foodborne diseases according to the Centers for Disease Control (*1*) (Table 2.).

Many public health authorities feel that the number of cases of foodborne illness are substantially under-reported and the Centers for Disease Control has estimated that there are possibly more than six million cases of foodborne illness involving 9,100 deaths each year in the U. S. (2). The ratio of reported cases to estimated cases of illness has been reported by several authorities including Archer and Kvenberg (*3*) to range from 1:10 to 1:25. Todd (*4*) has stated that the number of cases of foodborne disease

0097–6156/92/0484–0232$06.00/0

Table 2. Cases of Reportable Foodborne Illness in the U. S., 1988

Illness	*No. of Cases*
Salmonellosis	49,000
Shigellosis	30,000
Hepatitis A	28,500
Trichinosis	45
Botulism	28

in the U. S. could be as high as five million per year with an attendant annual cost of $1 to $10 billion associated with medical costs, loss of productivity, and product loss.

Food poisoning microorganisms can originate from animal or plant products due to the wide distribution of microorganisms in nature or because of agricultural practices. Everyone in the food chain, from the farmer or rancher to the consumer, has responsibilities in maintaining the safety of our foods. Many factors influence the relative safety of our foods; the type of food itself, how and where it is grown, harvested, processed, packaged, distributed, and stored. The most critical step in prevention of foodborne illness is the ultimate preparation step as carried out in the home, restaurant, hospital, or food service establishment.

The major factor contributing to all foodborne illness is mishandling of food in final preparation steps which results in microbiological problems. The media and educational institutions should emphasize consumer education concerning safe food handling practices to prevent these outbreaks rather than "sensationalizing" relatively minor hazards like small amounts of pesticide residues in apples.

An extensive review of foodborne illness reported in the U. S. from 1977 to 1984 was published by Bryan (5) (Table 3). He tabulated the major food categories and frequency of involvement in more than 1500 food poisoning outbreaks that were reported during this eight-year period.

Table 3. Foods Implicated in U. S. Foodborne Illness Outbreaks 1977-84

Food Group	*Frequency (%)*
Seafoods	24.8
Meat	23.2
Poultry	9.8
Salads	8.8
Vegetables	4.9
Chinese Foods	4.9
Mexican Foods	4.9
Milk and Dairy Products	4.2
Baked Goods	3.3
Beverages	2.3
Other Foods	<5.0

SOURCE: Reproduced with permission from ref. 5. Copyright 1988 *Journal of Food Protection.*

Animal protein products were by far the most commonly implicated foods in foodborne outbreaks accounting for 57.8% of the outbreaks.

Hooper (*6*) tabulated the factors that contributed to foodborne outbreaks (Table 4) over a six-year period.

Until the 1970's, the major food poisoning microorganisms were *Staphylococcus aureus, Salmonella*, and *Clostridium botulinum* with fewer problems caused by *Shigella* and *Bacillus cereus*.

Clostridium perfringens was recognized in the late 1970's as a cause of food poisoning and was joined by several of the *Vibrio* species (*V. parahemolyticus, V. vulnificus* and *V. mimicus*) and by *Listeria monocytogenes*, enteropathogenic *Escherichia coli* 0157:H7, *Campylobacter jejuni*, and *Yersinia enterocolitica* as more recently recognized causative agents of food poisoning. In about 15 years, we have gone from dealing with about five food poisoning microorganisms to having to deal with 13. *Aeromonas hydrophila* is also a microorganism that has been reported to cause food poisoning in certain areas of the world.

The microorganisms associated most generally with foodborne illness include *Salmonella, S. aureus*, and *C. perfringens*. The percent of total reported confirmed foodborne outbreaks caused by specific microorganisms reported by the Centers for Disease Control over the six-year period of 1977 to 1982 are shown in Table 5. The three top microorganisms accounted for 74% of the bacterial outbreaks.

Campylobacter, Yersinia, and *Listeria* have been only recently recognized as microorganisms which can cause food poisonings in humans. The data in Table 5 do not currently reflect their present frequency of involvement in foodborne illness. Additionally, *Listeria* is quite widely distributed and *Listeria* and *Yersinia* can survive and grow at common refrigerator temperatures (about 38°F, 3.3°C) which increases their risk potential.

Salmonella

Salmonella is the generic name of a group of about 2,000 related bacteria. These bacteria are reported to have caused 37% of confirmed foodborne outbreaks during 1977-82. Foods of animal origin are the main sources in the U. S. with beef, turkey, and homemade ice cream the most frequently reported. Fresh tomatoes were recently implicated as a source of *Salmonella javania*. Other outbreaks include a Cheddar cheese outbreak in Colorado in 1976 that involved between 28,000 to 30,000 cases, a number of cases from 1976 to 1984 from consumption of improperly pre-cooked roast beef and another Cheddar cheese outbreak in Canada in 1984. A specific *S. typhimurium* strain was reported to be involved in the largest Salmonellosis outbreak in U. S. history. This outbreak involved 16,000 cases of residents of six states who had consumed two different brands of 2% lowfat milk processed at one dairy in Chicago (*7*). Post-process contamination of the pasteurized milk was suspected as the cause.

Salmonella enteriditis foodborne outbreaks that cause severe illness have been increasing in the U. S. In 1989, there were 49 outbreaks with 1,628 cases and 13 deaths. From 1985 through 1988, there were 140 outbreaks with 4,976 illnesses (896 hospitalized) and 30 deaths. Grade A shell eggs were the implicated vehicle in 65 (73%) of the 89 outbreaks involving contaminated food. As a result of concern about

Table 4. Factors Which Contribute to Foodborne Outbreaks CDC Five-Year Summary 1980 - 1985

Factors	*% of Outbreaks*
Improper cooling	46
Time lapse between preparing and serving	21
Infected persons touching food	20
Inadequate processing, cooking	16
Improper hot storage	16
Inadequate reheating	12
Contaminated raw food	11
Cross-contamination	7
Improper cleaning	7
Use of leftovers	4

Table 5. Confirmed U. S. Foodborne Outbreaks Caused by Specific Microorganisms, 1977-1982

Bacteria	*Number of Outbreaks*	*% of Total*
Salmonella	290	37
S. aureus	181	23
C. perfringens	110	14
C. botulinum	85	11
Shigella	40	5
B. cereus	31	4
C. jejuni	17	2
Vibrio parahemolyticus	15	2
V. cholera	6	0.8
E. coli	4	0.6
Yersinia	4	0.6
*Listeria**	0	0

**Listeria* not reported as foodborne during this period.

Salmonella enteriditis, health authorities have advised consumers, especially the ill and elderly, to avoid consuming raw or partially cooked eggs or dishes containing them as ingredients.

Salmonella are heat sensitive and are destroyed by ordinary heating and pasteurizing procedures. Most outbreaks are the result of mishandling, inadequate cooking, improper processing, improper cooling, and cross contamination from raw to cooked product. Both *Salmonella* and *Listeria* are infective types of food poisoning microorganisms and *any* cells detected in a food render them "adulterated."

S. aureus

S. aureus food poisoning has been estimated to be the second most common cause of foodborne illness in the U. S. (*8*). The microorganism is commonly found in or on the nose, throat, hair, and skin of 50% of healthy people. Discharge from infections, sneezes, and coughs of food workers can be sources of additional contamination. Any food which requires post-heating worker handling can become contaminated.

Cooked protein foods (such as meats, fish, or poultry), dairy products, salads (ham, chicken, and potato), custards, and cream-filled bakery items best support the growth of staphylococci. Keeping foods below 40°F (4°C) or above 140°F (60°C) will sufficiently limit growth to prevent food poisoning problems.

C. perfringens

C. perfringens is widely distributed in nature, and has often caused foodborne illness due to preparing, holding, and serving large quantities of food as in institutional food service operations. Improperly cooked, cooled, held, or stored foods of animal origin and rehydrated dry mixes that were improperly stored are frequently involved.

C. botulinum

C. botulinum causes the very serious disease of botulism and has been largely associated with improperly processed home-canned foods. Between 1899 and 1976, home-processed foods accounted for 72% of the outbreaks while commercially-processed foods accounted for 8.6% of the outbreaks.

Several botulism outbreaks occurred due to foods improperly handled in food service establishments. Potato salad prepared from leftover foil-baked potatoes, salsa sauce, sauteed onions, cold smoked uneviscerated fish (kapchunka), and chopped garlic were the foods involved. An improperly handled commercial chopped garlic/olive oil product was also involved (*9*).

Because of the widespread distribution of this organism and its substantial resistance to heat, food industry scientists are concerned about the potential risks of botulism (as well as listeriosis) from the many extended shelf life, minimally processed, refrigerated food products that are available (*10*).

Emerging Pathogens

The terms "new" or "emerging" pathogens are currently used to describe a group of microorganisms that were identified by scientists years ago but only recently shown to be causes of foodborne illness. New methods of food analysis developed in the last five years uncovered the presence of these "emerging" pathogens in food products. One particularly disturbing characteristic of the "emerging" pathogens is their ability to grow to significant populations at normal refrigerator temperatures. We all have experienced the growth of spoilage bacteria at refrigerated temperatures as we discard leftover food items kept too long in the refrigerator. But, evidence of the growth of disease-producing bacteria at refrigerator temperatures is a new and serious concern for food safety specialists.

Listeria

Listeria was first recognized as a human and animal pathogen in the 1920's. It is widely distributed in soils, water, animal species (more than 40), including man, birds, fish and shellfish, and vegetables. This wide distribution, coupled with its ability to survive for long periods under adverse conditions (such as when dry or in high salt concentration), its low temperature growth ability, and serious pathogenicity for immunocompromised individuals, renders it a foodborne pathogen of great concern (Table 6).

Table 6. Listeriosis

Symptoms in Normal Population	*Symptoms in At-Risk* Populations*
Fever	Serious infection
"Flu-like" symptoms	Inflammation of the brain and surrounding tissues
	Abortion
	Death of newborn
	Overall mortality, 25%

*At risk: Infants, pregnant women, elderly, patients on chemotherapy, organ transplant patients, AIDS patients, and other immune-compromised persons.

Documented outbreaks implicating *Listeria monocytogenes* (Table 7) occurred in Nova Scotia in 1981 with 39 individuals ill and an overall fatality rate of 41%. The food implicated was coleslaw made from farm contaminated cabbage (*11*) Pasteurized milk was responsible for a listeriosis outbreak in Boston in late 1983 involving 49 individuals (42 immunocompromised and 7 infants). The mortality rate was 29% of those ill from drinking a specific brand of whole or 2% fat milk. No deficiencies in pasteurization were found but the herds supplying this dairy were 12% positive for *Listeria*. The most serious listeriosis outbreak occurred in 1985 when 142 cases were found in people consuming fresh cheese (Queso Fresco) manufactured by one plant

Table 7. Foods Implicated in Listeriosis Outbreaks

Food	*Cases*	*Fatality Rate (%)*	*Place, Date*
Coleslaw	39	41	Nova Scotia, 1981
Pasteurized Milk	49	29	Boston, 1983
Fresh Hispanic-type Cheese	142	30	Southern Calif., 1985
Turkey Frankfurter	1	-	Texas, 1989
Cheeses recalled but no reported or documented illnesses:			
Liederkranz Cheese	-	-	Ohio, 1985
French Brie and Camembert	-	-	At import sites, 1986

in Southern California. More than half of the patients were mother-infant pairs, with an overall mortality rate of 30%. The microorganism was isolated from patients, cheese, and almost every environmental sample from the manufacturing plant. The most probable cause of listeriosis was a mixture of raw milk added to the pasteurized milk during cheese manufacture (*12*).

Campylobacter

Campylobacter was first recognized about 10 years ago as an animal pathogen causing abortion in sheep. Subsequently, it has become recognized a a major cause of human food poisoning (*13*) Foods involved have included raw milk, pasteurized milk, poultry, raw beef, clams, cake (contaminated by a food handler), municipal water systems, and undercooked chicken (Table 8).

The microorganism is not likely to be found in processed foods heated to pasteurizing temperatures, and is most often found on and transmitted by raw refrigerated products.

Table 8. Foods Implicated in Campylobacteriosis Outbreaks

- Raw Milk
- Pasteurized Milk
- Undercooked Chicken
- Raw Beef
- Clams
- Cake
- Untreated Municipal Water Supplies
- Fresh Retail Mushrooms

NOTE: More than 6,000 documented cases since 1979.

Yersinia

Yersiniosis is an infrequent cause of a severe foodborne illness that causes intense abdominal pain that resembles appendicitis, especially in children. Yersiniosis may rarely produce a subsequent arthritis, systemic infection, or meningitis in serious cases. Yersiniosis outbreaks from 1976 to 1982 resulted in 222 illnesses from chocolate milk, 239 from reconstituted dry milk and chow mein, and 87 from tofu packed in contaminated spring water.

Yersinia is found in foods due to poor sanitary practices. Like *Listeria* and *Aeromonas*, it is able to grow at refrigerator temperatures to significant populations at 44.5°F (7°C) within 10 days (*14*).

Vibrios

Consumption of seafood contaminated with four species of *Vibrio* including *V. cholera*, *V. parahemolyticus*, *V. vulnificus*, and *V. mimicus* have been responsible for about 3% of foodborne outbreaks. All of the disease producing vibrios occur naturally in the marine environment and are naturally occurring contaminants of seafood. They are also linked to potential pathogenicity of aquaculture products (*15*). Proper control of vibrios is achieved by refrigeration to prevent growth and by proper cooking and prevention of recontamination of cooked seafoods (*16*).

Enteropathogenic *E. coli*

The first reported major outbreak of enteropathogenic *Escherichia coli* 0157:H7 occurred in 1971 from mold-ripened French cheese. Since 1982 several outbreaks due to this microorganism occurred from consumption of contaminated hamburger from a fast-food company, causing 26 cases in Oregon and 21 cases in Michigan. French Brie cheese caused illnesses in five states in 1983, and 53 illnesses and 12 deaths were reported in 1985 in residents of an Ontario nursing home after they ate contaminated sandwiches. Another Ontario outbreak occurred in 1986 that involved 43 of 62 children who consumed raw milk during a visit to a dairy farm. Serious gastrointestinal illness also occurred in 40 school children in Minnesota from eating hamburgers in 1988. One form of illness produces bloody diarrhea and a serious urinary tract infection. This latter infection is found almost exclusively in children and can be a leading cause of acute kidney failure.

Consumer Trends Concerning Foods

Consumers currently favor foods that claim to be preservative and additive free, minimally processed, natural, and of low-salt content. Many of the traditional processes that render foods safe microbiologically are rejected by consumers as unnatural and undesirable. Radiation is a food preservation method that has been subjected to decades of research and found safe. It is approved for use in a number of countries, including the U. S., for treatment of chicken. Consumer activist groups are adamantly opposed to irradiation of foods, which could tremendously increase the microbiological safety, as well as the shelf life, of many foods.

Studies show that consumers have inadequate knowledge about basic food microbiology. Only one-third of consumers knew that raw meat and poultry contain living microorganisms and these same consumers also incorrectly identified food processing plants as the most likely place for food safety problems to occur. They chose the home and the farm as the least likely sources of food safety problems, when in fact, the kitchen is the most likely source of food abuse (*17*).

Control Measures

The food processing industry uses a number of methods to control the safety and stability of foods. Control or destruction of microbial cells and spores (the heat resistant form) is achieved by heat sterilization, pasteurization, cooking, refrigeration, freezing, moisture control, acidification, fermentation, vacuum or modified atmosphere packaging, or use of preservatives. Extensive use of these quality control procedures has maintained the low-risk status of our food supply.

However, the new generation of minimally processed, fresher, preservative and additive-free foods, an example of which are the "new generation refrigerated foods," have less resistance to spoilage and pose greater risks of food poisoning if abused. While these food products have not been implicated in increased food poisoning, the potential for problems is great.

Food processors are dealing with this situation by developing hazard analysis systems for critical control points (HACCP) in their food production plants. HACCP was pioneered about 20 years ago by Pillsbury and there has been considerable food industry interest in this procedure. HACCP procedures consider the potential risk of raw materials or ingredients to every handling and heating step in the process. Documentation of what is occurring at each critical step in the process allows the processor to determine the exact number of critical control points and their effect on eventual product safety. Most recently, HACCP procedure systems are being developed by the U. S. Food and Drug Administration jointly with the National Marine Fisheries Service for HACCP programs for processing of seafood products and by the U.S.D.A. for meat processing systems.

The use of HACCP procedures will permit the processing and distribution of perishable products while maintaining the excellent public health safety of such products.

Food Industry Safety Concerns

The food industry and the FDA recognize the safety issues (Table 1) addressed in the priority assessment previously described. They are dealing with these issues on an individual company basis as well as through research conducted at academic-industry research centers.

I would like to describe the food safety research priorities established by a new and unique consortium - the National Center for Food Safety and Technology (NCFST).

The National Center is a consortium of the Illinois Institute of Technology (IIT), the IIT Research Institute, the University of Illinois at Urbana-Champaign, the food science research laboratory of the U.S.Food and Drug Administration (the former Cincinnati lab), and 30 food-related companies which are food processers or suppliers to the food processing industry. The NCFST was formed to exchange scientific information leading to a better understanding of the science and engineering behind food safety decisions and to conduct the research upon which future food safety decisions will be based.

The NCFST is concerned with the effects of advanced food processing and packaging technologies upon food safety. Safety areas include formation of potentially detrimental compounds, detoxification of naturally occurring toxicants, nutrient losses, and adequacy of processing methods. Research will also be conducted on new foods and food ingredients produced through biotechnology. Research will be conducted to define hazard analysis of critical control points (HACCP) procedures for products and processing, packaging, and distribution systems.

The ten founding member companies that determine the overall policy and operation of the NCFST, together with the additional seven affiliate member companies who make up the Technical Advisory Committee that determine research project areas and project priorities have determined that the NCFST should begin its food safety research by focusing upon six initial project areas. These research areas have been determined to be in order of descending priority (Table 9):

Table 9. Six Highest Priority Research Topics in Four Major Areas of NCFST

1. Recycling of food-grade plastics
2. Development of a Moffett Food Safety Index for shelf-stable foods
3. Package/Seal Integrity
4. Sensor/Biosensor Technology
5. Food Process Automation
6. Modified Atmosphere Packaging

The NCFST's Technical Advisory Committee (TAC) has recently received project proposals from both the academic and FDA arms of the NCFST, and the TAC is currently reviewing, evaluating, and determining the research direction that will take place at the NCFST. This is an exciting concept in food safety research, and this cooperative program should produce research results that will considerably improve U. S. food safety.

Literature Cited

1. *Summary of Notifiable Diseases, U.S. 1988*, Centers for Disease Control, *MMWR*, **1989**, *37*; 51.
2. Cohen, M. L. In *Hearing before the Committee on Agriculture, Nutrition, and Forestry*, U. S. Senate, 100th Congress, First Session, June 4, 1987, U. S. Government Printing Office, Washington, D. C., 1988, p 28.
3. Archer, D. L.; Kvenberg, J. E. *J. Food Protect.* **1985**, *48*, 887-889.
4. Todd, E. C. D. *Economic Loss Resulting from Microbial Contamination of Food.* In Proceedings of the Second National Conference for Food Protection. U. S. Food and Drug Adminisration, Contract No. 223-84-2087, Washington, D. C., 1984.

5. Bryan, F. L. *J. Food Protect.* **1988**, *51*, 498-508.
6. Hooper, A. J. *Dairy, Food, and Environ. Sanit.* 1989, *9*, 549-551.
7. Cohen, M. L.; Tauxe, R. V. *Science* **1986**, *234*, 964.
8. Bennett, R. W. In *Microbiological Safety of Foods in Feeding Systems*, Am. Board Military Supplies, Report No. 125, National Academy of Sciences, National Acad. Press, Washington, D. C., 1982.
9. Anonymous. *Food Chemical News* **1989**, *32*, 22.
10. Lechowich, R. V. *Food Technol.* **1988**, *42*, 84-86.
11. Schlech, W. F.; Lavioue, P. M.; Bortolussi, R. A.; Allen, A. C.; Haldane, E. V.; Wart, A. G.; Hightower, A. W.; Johnson, S. E.; King, S. H.; Nicholls, E. S.; Bromme, C. F. *New Eng. J. Med.* **1983**, *308*, 203-204.
12. Lovett, J.; Twedt, R. M. *Food Technol.* **1988**, *40*, 188-191.
13. Foster, E. M. *Food Technol.* **1986**, *40*, 20.
14. Doyle, M. P. *Food Technol.* **1988**, *40*, 187-188.
15. Ward, D. R. *Food Technol.* **1989**, *43*, 82-86.
16. Madden, J. M. *Food Technol.* **1988**, *40*, 191-192.
17. Penner, K. P.; Kramer, C. *A Survey of Consumers in Eight Kansas Counties.* Kansas State University, Cooperative Extension Service, *MF-774*, 1985.

RECEIVED September 4, 1991

Chapter 23

High-Technology Approaches to Microbial Safety in Foods with Extended Shelf Life

Myron Solberg

Center for Advanced Food Technology, New Jersey Agricultural Experiment Station, Rutgers, The State University, New Brunswick, NJ 08903

The target for food processors and consumers that remains off in the distance is the shelf stable "chef-like" and "fresh-like" food which is ready to heat and eat. The interim goal is the extended shelf-life refrigerated product which, after reheating, will have quality attributes equivalent to food freshly prepared by a master chef. The first steps toward the identified objectives are manifest in supermarket and food service operations today. The supermarkets of England and Finland have far more space dedicated to refrigerated modified atmosphere packaged precooked entrees than to frozen foods. In the USA, the refrigerated shelf space is increasing as precooked items, in many cases prepared within the store but in some situations brought into the store, continue to fill the demand for quick to prepare and "fresh" which is important to the two income with or without children families. The food service sector is now introducing the *sous-vide* entree. This precooked vacuum packaged refrigerated entree has had a few years of success in France and has now entered into the U.S. menu. At least one manufacturer has elected to distribute the product frozen in the U.S.

The prerequisite for any food item is safety. Chemical and microbial considerations cannot be subject to compromise. Safety has traditionally been assured by sacrificing aesthetic quality. Military commanders have traditionally preferred a battle ready soldier griping about the overcooked food to one who is incapacitated, albeit temporarily, due to microbially induced food-borne illness. A similar approach is the "12D" thermal process concept which yields overcooked but safe shelf stable canned, jarred or pouched food.

Freezing of foods may also be looked upon as overprocessing. Concern for safety takes precedence over quality, which is often seriously affected by the thawing process.

Transfer of the "12D" thermal process concept to sterilization of food by ionizing irradiation is another overprocessing treatment which limits applications due to quality and has perpetrated a host of health related concerns. Physical approaches

0097–6156/92/0484–0243$06.00/0

have been used to improve heat transfer through conversion of the package from a round "tin" can to a flat tray or pouch. The increased surface area tends to even out the process and reduce the overprocessing. The product quality is improved but remains overprocessed in the name of microbial safety.

There is a striving toward the aseptic processing of particulate containing food systems. This process permits efficient heat transfer in small diameter or thickness heat exchangers followed by aseptic transfer of the commercially sterile food into presterilized containers. This system has the potential to reduce the overprocessing to a greater extent than the reconfigured packages described previously, but the overriding microbial safety need again demands overprocessing.

Still another approach is the seldom used direct steam injection under pressure aseptic systems which overcome the problems of particulate flow systems since the product is in a confined space, eliminating the variable flow rate and therefore heat transfer concerns. In these systems, vacuum cooling prior to sealing the container removed the steam condensate which was added to the container during the direct steam injection heating. This seemingly most effective treatment system still produces highly overprocessed food relative to the amount of cooking which would take place in home, restaurant or food service establishment preparation. The overprocessing is demanded in the name of microbial safety.

Attempts to reach the interim goal of extended shelf-life refrigerated products have relied upon minimal processing approaches. Thermal treatment is almost always the basis underlying these products. Some use of the hurdle concept (*1*) approach has been implemented. Lowered pH, incorporation of water immobilizing ingredients, and antimicrobial spices and condiments contribute to the progress toward the interim goal.

All of the approaches mentioned thus far, may be considered as "sledgehammer" in nature. They consist of processes or systems which are interactive with every molecule of the food system and therefore result in considerable change from the fresh unprocessed food. The degree of change is often far greater than that which would be effected by home or food service establishment food preparation.

There is a need to define the problem before solutions may be proposed. The microbial problems are similar for both the ultimate goal products which are shelf stable, "chef-like" and "fresh-like" and the interim goal products which are refrigerated, extended shelf-life, "chef-like" and "fresh-like."

The primary target organism in all case is *Clostridium botulinum*. The objective is to prevent outgrowth and toxin production. It is now clear that the non-proteolytic *C. botulinum* types B and E may grow and produce toxin at temperatures just slightly above 3°C and that time in the growth supporting temperature range tends to be cumulative (*2*). It is also clear that the potential for growth in an organic substrate of this strictly anaerobic organism in the presence of oxygen is possible due to the existence of anaerobic microenvironments within the system (*3*).

Among the secondary target organisms are two pathogens capable of growth at refrigerator temperatures. These are *Yersinia enterocolitica* and *Listeria monocytogenes*. Among the room temperature growing mesophilic pathogens are the familiar *Salmonella* species, *Staphylococcus aureus*, *Clostridium perfringens*, and the less well known *Bacillus cereus* and *Campylobater jejeuni*. The discouraging truth is that there are pathogens of concern in both the refrigerated and the shelf stable products. There are also spoilage causing microorganisms which cause quality losses and need to be controlled if success is to be achieved. The quality losses of concern include texture, color, flavor and nutritional value.

The quality losses described are attributable to chemical changes within the food system. The rates at which the chemical reactions occur can be increased by the presence of active enzymes. Intrinsic properties of the food such as acidity, moisture, salt and free radical moderators as well as extrinsic factors in the food environment, such as temperature, gaseous environment, light, and humidity; can regulate the chemical reactions.

The problem may be reduced to preventing microbial growth, enzymatic activity and chemical reaction, including changes of physical state within a food system after it has reached an optimum state of quality which may be described as "chef-like" and "fresh-like." The key question relative to microbial safety is, "why do we need to overprocess the food?" The answer is that since we have no way of knowing how many of what type of microorganisms are present, we are unable to design a system with a precise effect. The treatments used have excessive safety factors built into them so as to err on the side of caution. The result is highly overprocessed food.

There are two approaches which appear obvious and probably many more which are obscure. The first approach is to devise, design and develop a non-destructive, non-invasive system for on-line identification and enumeration of microorganisms. Such information could be fed to a computer which would determine the minimum safe treatment time and choice of end products for which such treatment could be used with the expectation of quality required. The computer integrated manufacturing system would determine which of the suitable products was needed for inventory and would drive the production toward the product.

The second approach is "magic bullet" based. The objective is to attack the microorganisms without affecting the food. Energy, specifically targeted to a relatively unique and critical site of an enzyme key to all microbial life, would be a possibility. A substance which could be added or produced from a precursor in the product which would specifically interact with microbial membranes to prevent transport of key components through the membranes thus terminating cell viability (*3*) is another possibility. Other similar cell mobility disrupting scenarios can be imagined.

Let us examine the present state of the art with respect to detecting microorganisms so as to estimate the time frame for real time identification and enumeration. There is no need to discuss the traditional selective media approaches which require one or more days. The furthest advanced methods emanating from the new biotechnology are those which depend upon monoclonal or polyclonal antibodies or upon DNA hybridization. These identification devices are linked to detectors which may be enzymes, fluors, or radioisotopes. The challenge to which these systems have been directed is specific organism detection by genus or specie. The developed techniques do this task well but are lacking in sensitivity. Thus, there is a need for large concentrations of the specific microorganism to be present for detection to be effected.

The time to detection is short, generally a few hours at most; the time to reach a detectable level is as long as several days if there are only a few organisms present initially.

If the antibody or hybridization techniques are to be useful in defining process requirements, they will need to be made very sensitive, quantitative and capable of broad as well as specific interaction with microorganisms. One scenario might be a cocktail of antibodies with differing detectors which would simultaneously indicate the presence of a variety of microorganisms. There is a potential biotechnology approach using both the antibody and hybridization techniques. Antibodies could be used to "fish" out specific cells from a production slip stream. The DNA of the cells could be amplified quickly using a polymerase chain reaction (PCR) (*4*). An enzyme linked DNA probe could then be used for identification. Such a system would identify a microorganism in 3 or 4 hours from the capture of a single cell. Problems that remain include quantification and the inability to determine whether the cell originally recovered was dead or alive at the time of capture. An approach to quantification may be possible by fixing the number of available capture sites, controlling the DNA amplification rigorously and measuring relative enzyme catalyzed response quantitatively. Every living cell and only living cells contain adenosine triphosphate (ATP) which when released from cells can be reacted with the enzyme luciferase to produce luminescence. The relatively new technique of photon counting imaging in combination with an optical microscope is able to directly visualize bioluminescence in a single cell (*5*). If such a system could be integrated with the antibody capture:PCR DNA amplification system previously described it would indicate whether the captured cell was dead or alive at the time that its DNA was released for initiation of the PCR process.

Another biotechnology based microorganism identification and enumeration scheme utilizes bacteriophages which are bacteria specific virus like systems. The *lux* gene which causes bioluminescence can be inserted into the genome of a bacteriophage (*6*). When this phage infects a bacteria it becomes amplified and utilizes the bacterial ATP to make the bacteria luminescent. This event occurs within 30-50 minutes and the light emission intensity is directly related to the number of bacterial cells available for infection. Standard bioluminescence measuring technology is capable of detecting emissions representing a few hundred cells (*5*) and improved reagent systems can detect ATP from as few as 10 cells (*7*). The possibility of combining this phage system to the previously described photon counting imaging microscope system could increase the sensitivity dramatically.

Although all of the described systems are existent, there is considerable research and development needed before any one of them will be ready for commercial application within a broad range of food products. It does seem reasonable to believe that the 1990's decade will see some of these advanced methods in use.

Predictive mathematical models of microbial growth and survival in foods could permit quality, shelf life and stability judgments to be made (*8*). Such predictions are critical as a processor moves toward perceived freshness via minimal processing or to reduced salt or preservative formulations to make foods which will be perceived as more healthful. This approach could be applied on an interim basis through two systems. The first would be dependent upon real time detection and enumeration of microorganisms followed by predictive modelling of product life. A second approach, one which is closest to application, would be to assure conditions within the product through ingredient and environment control which would insure, beyond any doubt, the prevention of growth or toxin production by the more difficult to destroy organisms. These would include the sporeforming bacilli and clostridia. Processing could then be done on a traditional probability basis to reduce the more susceptible organisms to an essentially zero level without significant loss in "chef-like" quality.

Predictive mathematical modelling is dependent upon experimental definition of combined effects of various factors affecting microbial growth and survival. Traditionally such data collection has been very difficult and time consuming involving standard microbial methods of incubation and plate counting or turbidity measurements when clear broths are used. Recent development of automated systems such as the Lab Systems Bioscreen machine, which can examine up to 200 cultures simultaneously using optical density for measurement, has simplified the model medium part of the data collection. The establishment of a more useful data base appears feasible. The combination of the data base with high speed computerized curve fitting (*9*) will permit some application of predictive modelling in the near future.

Next, let us consider the state of know-how relative to "magic bullet" approaches to selective inhibition of microorganisms and chemical reactions.

Availability of high-power pulsed lasers with output frequencies in the ultraviolet region as well as broadly tunable infrared color center lasers provide a range of high intensity stable energy sources which may interact uniquely with proteins and thus control enzyme activity and microbial viability (*10*). Although still far from economically feasible, even tunable monowavelength X-ray lasers are available and offer opportunities to exert closer control over irradiation processes by conferring some specificity of interaction (*11*). These outgrowths of the "Star Wars" program yield an opportunity for specific targeting of finely tuned energy which may result in excitation of molecular regions which could inactivate enzymes and microorganisms. The end result is not completely dependent upon the total energy input but may be in part dependent upon the vibrational mode established by the laser. The approach is to utilize ultra short pulses of high energy so that molecules are made to vibrate with ensuing bond disruptions prior to the energy being dissipated as heat (*12*). Successes to date are limited but the concept is intriguing.

The availability of high pressure hydrostatic based process equipment with working volumes as large as 1 liter, for use in the cold isostatic pressing of ceramic powders, opens possibilities in food processing (*13*). High pressure unfolds proteins and disrupts membrane structure. These two events, occurring simultaneously result in inactivation of enzymes and loss of viability in microorganisms. The question remaining is in specificity. Is the pressure required for enzyme inactivation one at which structural protein will be left intact? Is the pressure which disrupts the microbial membrane one which will also disrupt plant or animal cells? The Japanese formed a Research and Development Association for High Pressure Technology in the Food Industry within the past year (*13*). Answers to many questions should emanate from this four year program which is funded with $1.0 million in its first year.

An approach of high interest is through molecular biology and biotechnology. The ability to transfer genes into plants and animals and have them express anti-microbial substances which become part of a food system is an exciting prospect. The elements of such a scenario could include a substance like chitosan, a deacetylation product of chitin. Chitosan is a polycation which probably affects the permeability of cell membranes. Other substances which would fit the scenario are the bacteriocins. These are substances produced by microorganisms which specifically inhibit other microorganisms. Some bacteriocins are active against *C. botulinum*, the primary target microorganism previously mentioned. Other bacteriocins are active against *L. monocytogenes*, a low temperature growing pathogen which was previously cited among the secondary target microorganisms. It is possible that a unique bacteriocin exists which would be active against both of these bacteria and possibly others as well.

The present day approach is to add the antimicrobial substance or to have the producing bacteria grow in the food system to produce it. The goal is to identify the genetic components responsible for production of the antimicrobial, whether it be bacteriocin or chitosan or some other substance. Insert those components into the animal or plant so that the substance is produced at the proper time and in the proper amount. The result would be the elimination of the primary and secondary threats thus permitting minimal processing and extended storage without microbiological concerns for safety.

The potential for utilizing some of the presented approaches in combination could yield the desired results. The combined effects of "magic bullet" control and on-line real time detection would provide a potentially ideal system leading to a fulfillment of our desires. These desires include not only wholesomeness, healthfulness and freshness but also life-extension or adding years to our existence, satisfaction and pleasure or enjoyment in our eating. The additional factors of timeliness representing quick to prepare, availability all through the year and variety to satisfy our whims at any moment round out the list of desires represented by the terms "chef-like" and "fresh-like."

Acknowledgment

This is publication No. D10535-1-91 of the New Jersey Agricultural Experiment Station supported by State Funds and the Center for Advanced Food Technology (CAFT). The Center for Advanced Food Technology is a New Jersey Commission on Science and Technology Center.

Literature Cited

1. Scott, V.N. *J. Food Protec.* **1989**, *52*, 431-5.
2. Post, L.S.; Lee, D.A.; Solberg, M.; Furgang, D.; Specchio, J.; Graham, C. *J. Food Sci.* **1985**, *50*, 990-6.
3. Knorr, D.; Popper, L.; Kuhne, K.; Boguslawki, S. Institute of Food Technologists Annual Meeting Abstracts. 1990, p. 241.
4. Van Brunt, J. *Biotechnol.* **1990**, *8*, 291-4.
5. Stewart, G.S.A.B. *Lett. Appl. Microbiol.* **1990**, *10*, 1-8.

6. Ulitzur, S.; Kuhn, J. In *Bioluminescence and chemiluminescence New Perspectives*; Schlomerich, J.; Ed.; Wiley: New York, 1987, pp. 463-73.
7. Simpson, W.J.; Fernandez, J.L.; Hammond, J.R.M.; Senior, P.S.; McCarthy, B.J.; Jago, P.H.; Sidorowicz, S.; Jassim, S.A.A.; Donyers, S.P. *Lett. Appl. Microbiol.* **1990**, *11*, 208-10.
8. Gould, G. *Food Sci. Technol. Today.* **1989**, **3**(2), 89-92.
9. Buchanan, R.L.; Phillips, J.G. *J. Food Protec.* **1990**, *53*, 370-6.
10. Grygon, C.A.; Perno, J.R. ; Fodor, S.P.A.; Spiro, T.G. *Biotechniques.* **1988**, *6*(1), 50-5.
11. Kim, K-J.; Sessler, A. *Science.* **1990**, *250*, 88-93.
12. Rotman, D. *Industrial Chemist.* **1987**, *Aug.*, 32-5.
13. Farr, D. *Trends in Food Sci. and Technol.* **1990**, *1*, 14-6.

RECEIVED August 15, 1991

Chapter 24

Predictive Microbiology

Mathematical Modeling of Microbial Growth in Foods

Robert L. Buchanan

Microbial Food Safety Research Unit, Eastern Regional Research Center, Agricultural Research Service, U.S. Department of Agriculture, 600 East Mermaid Lane, Philadelphia, PA 19118

One of the basic precepts of modern food microbiology is that the growth of microorganisms is a function of the food as an environment. The species most capable of dealing with the environment that a particular food represents will thrive and predominate. Each environment can be considered the integration of a finite number of factors that influence a microorganism's physiological responses. Theoretically, the large number of factors that influence the growth of bacteria in foods could be quantified so that specific information on the growth characteristics of individual foodborne microorganisms would be available for each food. However, consideration of the thousands of different foods eaten worldwide and the high level of biological variation within single foods quickly leads to the realization that such a goal is virtually impossible. Luckily, in most foods the number of factors that are the primary determinants of growth for foodborne microorganisms is limited. If microorganisms' responses to these variables could be derived, their behavior in foods could be estimated. This is the underlying goal of predictive microbiology, a rapidly growing subdiscipline of food microbiology. This includes a primary objective of overcoming the need for an infinite amount of data by determining quantitative relationships between microbial growth or survival and identified primary determinants. The general approach involves the acquisition of data derived under controlled conditions and the use of that information to establish mathematical relationships that can depict the effects and interactions of the variables. The mathematical models derived can then be used to predict how microorganisms are likely to behave in a range of foods based on physical measurements of the primary determinants.

Historical

Most successful research on modeling the effects of multiple variables on the growth or survival of foodborne microorganisms has been achieved during the past decade, particularly the development of models related to the growth of pathogenic bacteria. There are a number of reasons for this recent burst of activity, not the least of which is the ready availability of increasingly sophisticated personal computers. However, attempts to understand and mathematically describe the interactions of various factors have been made throughout the history of microbiology. Much of the early work with microbiological modeling had an emphasis that was not pertinent to conditions associated with foods. There has been extensive modeling of conditions that occur during various industrial fermentations, including the development of a body of equations, such as those introduced by Monad (*1*), that describe the impact of variables on yield. Fermentation models seldom considered many of the variables of concern with foodborne microorganisms. Further, these models assume a nutrient-limited system that has already reached stationary phase or a steady state, a condition not generally pertinent to the growth of bacteria in food matrices.

The acquisition of data for elucidating the interactions of multiple variables associated with food systems has been underway for several decades, particularly in relation to the determining how the activity of antimicrobials is affected by other parameters. Research characterizing the effectiveness of nitrite in model and cured meat systems was an area of early emphasis due to interest in controlling nitrosamine formation without loss of antibotulinal activity. Nitrite's antimicrobial activity is dependent on its interaction with temperature, pH, water activity, oxygen availability, iron content, etc., and accordingly required the consideration of multiple variables (*2-7*). Studies of this type provided an understanding of the relative importance of multiple variables, and demonstrated the desirability of modeling techniques. For example, in one of the early applications of response surface analysis techniques to food microbiology, Schroder and Busta (*8*) demonstrated that only four of sixteen ingredients in a soy-based ground meat analog had a significant impact on the growth of *Clostridium perfringens*. However, little of this earlier research extended beyond limited research applications due to a lack of sufficient databases or effective modeling techniques. Farber (*9*) has provided an excellent review of the various modeling approaches that were investigated during that period.

The various models that have been developed to describe the growth of foodborne bacteria can be subdivided into two major approaches: probability-based models and kinetics-based models. The choice of approach is largely dependent on the type of bacterium being considered and the impact of growth on the safety of the product. Probability-based models have been usually employed with endospore forming bacteria, particularly *Clostridium botulinum*, where any growth is considered hazardous. Kinetics-based models have been employed more often with non-endospore forming pathogens, particularly those where the microorganism is not considered hazardous until there has been some degree of growth.

Probability-Based Models

Much of the work on probability-based models is similar to that pioneered by Hauschild (*10*) who attempted to estimate the probability that a single spore of *C. botulinum* would germinate and produce toxin in a food. This approach takes into account the strong effect that cultural conditions have on the germination of bacterial spores. For example, Montville (*11*) reported that almost all *C. botulinum* spores germinated in a medium with 0% added NaCl and a pH of 7.0, whereas only 1 in

100,000 spores germinated when the salt level was 2% and the pH was 5.5. If the number of spores in a product is low and conditions for germination are non-optimal, the probability that a population of spores includes one that is capable of initiating growth has a large impact on any model for predicting product safety. Other investigators (*12-18*) have systematically estimated the effects and interactions of various variables on the probability of germination, outgrowth, and toxigenesis of *C. botulinum*. Various forms of regression analysis have been used to model the individual contributions of the variables, providing a series of mathematical expressions that could be used to predict the bacterium behavior in foods. For example, Genigeorgis et al. (*17*) modeled the effects of temperature, inoculum size, and % brine on the lag time to toxigenesis (which includes time for sufficient growth to yield toxin formation) for non-proteolytic *C. botulinum* types B and E in cooked turkey, deriving the relationship

$$\text{Log}_{10}\text{LP} = 0.625 + 6.710(1/T) + 0.0005(I{*}T) - 0.033(T) + 0.102(B) - 0.102(I)$$

where, LP = Lag to toxigenesis; T = temperature; I = inoculum size; and B = % Brine.

This model achieved an acceptable degree of agreement between predicted and observed values (Table 1), though the authors concluded that a larger database was necessary for enhanced confidence levels.

The limiting factor for probability based models has been their adaptation for use by non-research personnel. One of the key questions is what is a realistic probability of failure that one should be willing to accept, particularly in relation to potentially fatal intoxications such as those that could occur with *C. botulinum*. Other issues include the level of spores that one could anticipate in products, and translation of the probabilities into values that can be used to set the safe shelflife of a product. This latter question is increasingly being addressed using an integration of probability- and kinetics-based models similar to that employed by Genigeorgis et al. (*17*) which addressed both the probability of germination and the time to achieve sufficient growth to yield toxin formation.

Kinetics-Based Models

The second major class of models depict the effects of cultural parameters on the growth kinetics of a microorganism, particularly its lag and exponential growth phases. The complexity of the models have varied with the complexity of target food system. Although a variety of factors can influence the growth kinetics of foodborne pathogens, in many instances growth is overwhelmingly dependent on a single prime determinant. For example, the primary determinant of microbial growth in a highly homogenous food such as fluid milk is temperature.

Various models have been developed to depict the effect of incubation temperature on exponential growth rates and/or lag phase durations including the "square root" (*19, 20*), "linear Arrhenius" (*21, 22*), and "non-linear Arrhenius" (*23-25*) models. The "square root" model has been studied extensively, particularly for refrigerated foods.

Table 1. Comparison of representative predicted versus observed "lag to toxigenesis" for cooked turkey inoculated with spores of Clostridium botulinum

Temp (°C)	Inoculum (Log cfu/g)	% Brine	Lag to Toxigenesis (days)	
			Observed	Predicted
30	3	0	0.5	0.3
	1	0	0.5	0.5
	2	1.5	0.5	0.6
20	0	0	2.5	2
	3	1.5	1.8	1.3
	1	2.2	2.5	2.5
16	4	0	1	1.2
	2	1.5	2	2.8
	0	2.2	7	5.4
12	2	0	5	4
	0	1.5	9	9
	4	2.2	7	4
8	3	2.2	16	12
	2	0	8	10
	1	1.5	14	17
4	3	1.5	110	101
	0	0	>180	149
	4	2.2	120	95

SOURCE: Based on the probability models of Genigeorgis et al., (*16*).

For the temperature range below a microorganism's optimum, the relationship is

$$(r)^{0.5} = b(T - T_O)$$

where r = growth rate constant, b = slope of the regression line, T = incubation temperature in °K, and T_O = notational minimal growth temperature in °K. The latter term is derived by extrapolating the regression line to zero (*19*). The function is very easy to use once the linear relationship between growth rate and the square root temperature function has been established. Above an organism's optimum growth temperature, its rate of growth declines, and a more complex equation is required (*20*). This technique has been used successfully to describe the relationship between storage temperature and the microbiological safety or quality of various refrigerated foods (*26-30*), particularly dairy products. Using a large database depicting *Lactobacillus plantarum* growth in a microbiological medium over a wide range of temperatures, Zwietering et al. (*31*) assessed various models for describing the effect of temperature on microbial growth. They concluded that two modifications of the Ratkowsky equations were most effective for modeling growth rates and maximum population densities, whereas a hyperbolic function was more effective for lag phase duration. An integrated combination of the three equations was used in conjunction with the Gompertz function to develop a model for predicting the organism's growth curve over its entire temperature range.

Several investigations have extended this approach to develop models describing the combined effects of temperature and water activity (*32*). A modification of the square root function was used to model the effect of cooling schedules on the potential growth of *C. perfringens* in a meat product (*33*).

While the above models have been effective in relatively simple food systems, attempts to model more complex systems that are dependent on the interaction of multiple variables have generally used a polynomial or response surface analysis approach (*34-38*). These approaches employed non-linear regression techniques to generate "best-fit," multidimensional response surface equations that describe the effects and interactions of the variables. This approach to kinetics modeling has been greatly enhanced by coupling it to model equations, such as the logistics and Gompertz functions (*35, 39*), that can be used to depict growth curves mathematically. Used in conjunction with curve fitting computer routines, these sigmoidal functions allow the growth curve to be described mathematically as a series of coefficients. For example, the Gompertz function describes a growth curve as four values

$$L(t) = A + Ce^{-e^{(-B(t-M))}}$$

where, L(t) = Log count of bacteria at time (in hours) t; A = Asymptotic log count of bacteria as time decreases indefinitely(i.e., initial level of bacteria); C = Asymptotic amount of growth that occurs as t increases indefinitely (i.e., number of log cycles of growth); M = Time at which the absolute growth rate is maximal; and B = Relative growth rate at M.

The Gompertz function has been the one most extensively used due to the combination of its relative simplicity and overall effectiveness (*39*). Once sufficient databases have generated, the coefficients (or suitable transformations of the coefficients) of the sigmoidal functions are fitted against the independent variables using either quadratic (*36*, *38*) or cubic polynomial models (*37*) (Table 2). When effective models are developed, the predicted values of the sigmoidal functions can be used to calculate parameters such as predicted generation times, lag phase durations, or time to reach a designated population density. Fits between predicted and observed values have been satisfactory, providing reasonable estimates of an organism's growth kinetics. For example, a comparison of representative data from Gibson et al. (*36*), who modeled the effects of temperature, pH and NaCl content on the growth of *Salmonella* (Table 3), indicates that the model provides reasonable predictions of the microorganism's growth rate over a wide range of variable combinations. A similar effectiveness was reported by Buchanan and Phillips (*37*) who modeled the effect of temperature, pH, sodium chloride content, sodium nitrite concentration, and oxygen availability on the growth kinetics of *Listeria monocytogenes.*

Most response surface models that have been released have been based on experimental data generated in microbiological media (*36-38*), wherein variables could be controlled rigorously. Although specific databases could be generated for individual commodities, the media-derived, Gompertz-based response surface models provide reasonable "first estimates" of the behavior of foodborne pathogens in a variety of food systems. This is demonstrated in Table 4 which compares predicted values for *L. monocytogenes* (*37*) against reported values for different commodities. The ability to use media-derived models to predict behavior in foods is an important advantage considering the experimental effort required to acquire sufficient data to generate accurate models when dealing with three or more variables.

Once developed, a key to the successful use of multi-variable models is reducing the calculations to a "user-friendly" form. The USDA/ARS Microbial Food Safety Research Unit (*40*) recently developed application computer software to demonstrate the potential usefulness of predictive microbiological approaches. The program, which automates the use of response surface models for *Salmonella* spp. (*36*), *L. monocytogenes* (*37*), *Staphylococcus aureus* (Smith et al., in preparation), *Shigella flexneri* (Zaika et al., in preparation), *Aeromonas hydrophila* (*38*), and *Escherichia coli* O157:H7 (Buchanan et al., in preparation), has been distributed widely to food microbiology laboratories in industry, government, and academia. Similar applications software are being developed by researchers in Europe. The development of computer programs of this type must be an integral part of predictive microbiology.

Concluding Remarks

There is a great deal of excitement among researchers in predictive microbiology as new techniques and findings appear almost weekly and as international teams of scientists begin to share their knowledge and databases. It seems reasonable to predict that the next five years will see the introduction of increasingly more comprehensive

Table 2. Cubic models for the effects and interactions of temperature (T)(5 - 37 °C), initial pH (P)(4.5 - 7.5), sodium chloride content (S)(5 - 50 g/l), and sodium nitrite concentration (N)(0 - 1000 mg/l) on the aerobic and anaerobic growth of *Listeria monocytogenes* Scott A, using Ln(M) and Ln(B) transformations (*37*)

Aerobic

Ln(M) =

$37.657 + 0.0135T - 13.7331P + 0.4013S + 0.0713N +$
$0.00372T^2 + 1.9759P^2 -0.000667S^2 - 0.000007051N^2 -$
$0.083TP + 0.000842TS - 0.000214TN - 0.1155PS -$
$0.0167PN - 0.000125SN + 0.0000292T^3 - 0.0935P^3$
$0.00000328S^3 + 0.000286TPS + 0.0000315TPN +$
$0.00000014TSN + 0.0000175PSN - 0.000384T^2P -$
$0.00000855T^2S - 0.00000043T^2N + 0.00731TP^2 -$
$0.0000441TS^2 + 0.00672P^2S + 0.000968P^2N +$
$0.000294PS^2 + 0.00000062PN^2 - 0.00000016S^2N$

Degrees of freedom = 308
R2 = 0.967

Ln(B) =

$-47.709 + 0.1631T + 18.6861P - 0.3609S + 0.01N -$
$0.00161T^2 - 2.7074P^2 + 0.00623S^2 - 0.0000863N^2 +$
$0.0242TPS - 0.000906TS + 0.000594TN + 0.0671PS -$
$0.00715PN + 0.000337SN - 0.0000648T^3 + 0.1276P^3 -$
$0.000029S^3 - 0.000551TPS - 0.0000733TPN -$
$0.00000033TSN - 0.0000431PSN + 0.000189T^2P +$
$0.0000549T^2S - 0.00000047T^2N - 0.00222TP^2 +$
$0.0000459TS^2 - 0.00000002TN^2 - 0.0007781P^2S +$
$0.000777P^2N - 0.000872PS^2 + 0.0000112PN^2 -$
$0.00000038S^2N$

Degrees of freedom = 308
R2 = 0.942

Anaerobic

Ln(M) =

$89.9195 - 0.5378T - 38.8065P + 1.735S + 0.2175N +$
$0.00284T^2 + 5.9583P^2 + 0.00962S^2 - 0.000186N^2 +$
$0.1063TP - 0.00159TS + 0.000397TN - 0.567PS -$
$0.0574PN + 0.0000813SN - 0.0000321T^3 - 0.3024P^3 -$
$0.000107S^3 - 0.0000148TPS - 0.0000468TPN -$
$0.00000118TSN - 0.0000143PSN + 0.000397T^2P -$
$0.0000126T^2S - 0.00000184T^2N - 0.00964TP^2 +$
$0.0000487TS2 + 0.0000001TN2 + 0.0436P2S +$
$0.0038P^2N - 0.000461PS^2 + 0.0000247PN^2 +$
$0.00000123S^2N - 0.00000001SN^2$

Degrees of freedom = 211
R2 = 0.974

Ln(B) =

$78.2567 + 0.7928T + 34.3598P - 0.913S - 0.4437N +$
$0.00218T^2 - 5.3119P^2 - 0.00394S^2 + 0.000233N^2 -$
$0.2134TP - 0.00174TS -0.00094TN + 0.3002PS +$
$0.1272PN - 0.00015SN + 0.0000274T^3 + 0.2693P^3 +$
$0.0000493S^3 + 0.000442TPS + 0.0000985TPN -$
$0.00000047TSN + 0.0000304PSN - 0.00104T^2P +$
$0.00000175T^2S + 0.00000584T^2N + 0.0194TP^2 -$
$0.0000318TS^2 - 0.00000011TN^2 + 0.0238P^2S -$
$0.00911P^2N + 0.000215PS^2 - 0.0000298PN^2 -$
$0.00000068S^2N - 0.00000003SN^2$

Degrees of freedom = 211
R2 = 0.944

Table 3. Comparison of representative predicted versus observed times to achieve a 1000-fold increase in the numbers of salmonellae in tryptone soya broth

			Time (hr)	
Temp (°C)	*% NaCl*	*Initial pH*	*Observed*	*Predicted*
10	0.82	6.22	176	180
	4.56	6.02	394	372
15	1.33	6.13	41	45
	3.75	5.95	85	70
20	0.77	6.50	14	17
	4.50	5.90	36	38
25	1.32	6.20	11	9
	4.06	6.02	16	17
30	1.32	6.20	10	7
	4.5	5.99	14	17

SOURCE: Based on the response surface models of Gibson et al. (*36*).

Table 4. Comparison of selected reported growth kinetics values for Listeria monocytogenes versus those predicted by the models of Buchanan and Phillips (37)

				Generation Time (h)		*Lag Phase (h)*		
Food[a]	*(°C)*	*pH*	*% NaCl*	*Reported*	*Predicted*	*Reported*	*Predicted*	*Reference*
1. Clarified cabbage juice	30	6.1	0.5	1.6-1.8	0.4	10	3	41
2. Clarified cabbage	30	6.1	2.0	2.2-2.3	0.5	—[d]	3	41
3. Whole milk	10	(6.7)[b]	(0.5)	10	4.4	24	25	42
4. Chocolate milk	13	(6.7)	(2.5)[c]	3.9-4.7	2.1	—	12	43
5. 2% Milk	13	(6.7)	(0.5)	4.4-4.5	2.6	12	16	43
6. Uncultured whey	6	6.2	(0.5)	14.8-21.1	8.9	72	50	44
7. Cultured whey	6	6.8	(0.5)	16.3-17.4	9.4	72	48	44
8. Ice cream mix	9.5	6.4	(4.5)[c]	8.7-13.3	7.5	22-95	33	Smith & Holsinger[f]
9. Whole milk	9.5	6.7	(0.5)	5.2-9.0	4.8	7-12	27	Smith & Holsinger[f]
10. Tryptose broth	4	5.6	0.5	27.1	16.3	144	91	45
11. Tryptose broth	21	5.0	0.5	8.0	1.8	12	16	45
12. Tryptose broth	35	5.0	0.5	1.1	1.1	6	9	45
13. Tryptose broth	35	5.6	0.5	0.8	0.5	3	3	45
14. Tryptic soy broth	4	7.0	0.5	33.5	14.7	—	68	46
15. Tryptic soy broth	13	7.0	0.5	4.8	2.8	—	17	46
16. Tryptic soy broth	35	7.0	0.5	0.7	0.5	—	1	46
17. Tryptic soy broth	30	4.7	0.5	6.2	1.7	—	19	46
18. Tryptose phosphate broth	4	5.6	0.5	26.4	16.2	144	91	47
19. Tryptose phosphate broth	4	5.0	0.5	NG[e]	29.9	NG	144	47
20. Tryptose phosphate broth	13	5.0	0.5	8	5.4	12	38	47
21. Tryptose phosphate broth	21	5.6	0.5	1.8	0.9	6	8	47
22. Tryptose phosphate broth	35	5.6	0.5	0.8	0.5	3	3	47

[a]All food samples were assumed to be aerobic.
[b]Parentheses indicate that value not given and had to be assumed.
[c]An elevated value was assumed to estimate effect of ingredients other than NaCl that would affect the water activity of the food system. The ice cream mix samples had measured a_W (10C) values of 0.930-0.956.
[d]Value not reported.
[e]No growth.
[f]Unpublished data

computer-based models and expert systems applicable for a range of food products. These techniques should be a boon to food microbiologists, allowing them to quickly explore the microbiological impact of varying conditions within a food. This new area of research will undoubtedly provide a powerful set of new tools that will allow us to get one step closer to the long term goal of being able to design microbiological quality and safety into a product, instead of attempting to infer these attributes after the fact using end product testing.

Literature Cited

1. J. Monod *Ann. Rev. Microbiol.* **1949**, *3*, 371-394.
2. A. C. Baird-Parker and B. Freame *J. Appl. Bacteriol.* **1967**, *30*, 420-429.
3. A. S. Emodi and R. V. Lechowich *J. Food Sci.* **1969**, *34*, 78-81.
4. R. L. Buchanan and M. Solberg *J. Food Sci.* **1972**, *37*, 81-85.
5. T. A. Roberts and M. Ingram *J. Food Technol.* **1973**, *8*, 467-475.
6. T. A. Roberts, C. R. Britton, and N. N. Shroff In *Food Microbiology and Technology*; B. Jarvis, J. H. B. Christian, and H. D. Michener, Eds.; Parma Italy: Medicina Viva, 1979, pp.57-71. .
7. T. J. Montville *Appl. Environ. Microbiol.* **1984a**, *48*, 311-316.
8. D. J. Schroder and F. F. Busta. *J. Milk Food Technol.* **1973**, *36*, 189-193.
9. J. M. Farber. In *Foodborne Microorganisms and their Toxins: Developing Methodology*; M. D. Pierson and N. J. Stern, Eds.; Marcel Dekker: NY., 1986, pp. 57-90.
10. A. H. W. Hauschild *Food Technol.* **1982**, *36*(12), 95-104.
11. T. J. Montville *Appl. Environ. Microbiol.* **1984b**, *47*, 28-30.
12. S. Lindroth and C. Genigeorgis *Int. J. Food Microbiol.* **1986**, *3*, 167-181.
13. M. J. Jensen, C. Genigeorgis, and S. Lindroth. *J. Food Safety* **1987**, *8*, 109-126.
14. G. Garcia and C. Genigeorgis *J. Food Protection* **1987**, *50*, 390-397.
15. G. Garcia, C. Genigoeorgis, and S.Lindroth *J. Food Prot.* **1987**, *50*, 330-336
16. B. M. Lund, A. F. Graham, S. M. George, and D. Brown *J. Appl. Bacteriol.* **1990**, *69*, 481-492.
17. C. Genigeorgis, J. Meng and D. A. Baker *J. Food Sci.* **1991**, *56*, 373-379
18. D. A. Baker and C. Genigoergis *J. Food Protection* **1990**, *53*, 131-140.
19. D. A. Ratkowsky, J. Olley, T. A. McMeekin, and A. Ball *J. Bacteriol.* **1982**, *149*, 1-5.
20. D. A. Ratkowsky, R. K. Lowry, T. A. McMeekin, A. N. Stokes, and R. E. Chandler *J. Bacteriol.* **1983**, *154*, 1222-1226.
21. K. R. Davey *J. Appl. Bacteriol.* **1989**, *67*, 483-488.
22. K. R. Davey *J. Appl. Bacteriol.* **1991**, *70*, 253-257.
23. J. M. Broughall, P. A. Anslow, and D. C. Kilsby *J. Appl. Bacteriol.* **1983**, *55*, 101-110.
24. J. M. Broughall and C. Brown *Food Microbiol.* **1984**, *1*, 12-22.
25. C. Adair, D. C. Kilsby, and P. T. Whitall *Food Microbiol.* **1989**, *6*, 7-18.
26. M. W. Griffiths and J. D. Phillips *J. Appl. Bacteriol.* **1988**, *65*, 269-278.

27. R. E. Chandler and T. A. McMeekin *Australian J. Dairy Technol.* **1985**, *40*(1), 10-13.
28. R. E. Chandler and T. A. McMeekin *Australian J. Dairy Technol.* **1985**, *40*(1), 37-41.
29. G. S. Pooni and G. C. Mead *Food Microbiol.* **1984**, *1*, 67-78.
30. J. D. Phillips and M. W. Griffiths *Food Microbiol.* **1987**, *4*, 173-185.
31. M. H. Zwietering, J. T. de Koos, B. E. Hasenack, J. C. de Wit, and K. van 't Riet *Appl. Environ. Microbiol.* **1991**, *57*, 1094-1101.
32. T. A. McMeekin, R. E. Chandler, P. E. Doe, C. D. Garland, J. Olley, S. Putros, and D. A. Ratkowsky *J. Appl. Bacteriol.* **1987**, *62*, 543-550.
33. L. C. Blankenship, S. E. Craven, R. G. Leffler, and C. Custer *Appl. Environ. Microbiol.* **1988**, *54*, 1104-1108.
34. D. W. Thayer, W. S. Muller, R. L. Buchanan, and J. G. Phillips *Appl. Environ. Microbiol.* **1987**, *53*, 1311-1315.
35. A. M. Gibson, N. Bratchell, and T. A. Roberts *J. Appl. Bacteriol.* **1987**, *62*, 479-490.
36. A. M. Gibson, N. Bratchell, and T. A. Roberts *Internat. J. Food Microbiol.* **1988**, *6*, 155-178.
37. R. L. Buchanan and J. G. Phillips *J. Food Protection.* **1990**, *53*, 370-376.
38. S. A. Palumbo, A. C. Williams, R. L. Buchanan, and J. G. Phillips *J. Food Protection* **1991**, *54*, 429-435.
39. M. H. Zwietering, I. Jongenburger, F. M. Rombouts, and K. van 't Riet *Appl. Environ. Microbiol.* **1990**, *56*, 1875-1881.
40. R. L. Buchanan *J. Food Safety* **1991**, *11*, 123-134.
41. D. E. Conner, R. E. Brackett, and L. R. Beuchat *Appl. Environ. Microbiol.* **1986**, *52*, 59-63.
42. D. L. Marshall and R. H. Schmidt *J. Food Protection* **1988**, *51*, 277-282.
43. E. M. Rosenow and E. H. Marth *J. Food Protection* **1987**, *50*, 726-729.
44. E. T. Ryser and E. H. Marth *Can. J. Microbiol.* **1988**, *34*, 730-734.
45. M. A. El-Shenawy and E. H. Marth *J. Food Protection* **1988**, *51*, 842-847.
46. L. Petran and E. A. Zottola *J. Food Sci.* **1989**, *54*, 458-460.
47. M. A. El-Shenawy and E. H. Marth *Internat. J. Food Microbiol.* **1989**, *8*, 85-94.

RECEIVED September 4, 1991

Chapter 25

Mycotoxins in Foods and Their Safety Ramifications

Garnett E. Wood and Albert E. Pohland

Division of Contaminants Chemistry, U.S. Food and Drug Administration, 200 C Street, SW, Washington, DC 20204

Mycotoxins are toxic metabolites produced by certain fungi in/on foods and feeds. These toxins have been associated with various diseases (mycotoxicoses) in livestock, domestic animals and humans throughout the world. The occurrence of mycotoxins is influenced by certain environmental factors; hence the extent of contamination will vary with geographic location, agricultural and agronomic practices and the susceptibility of commodities to fungal invasion during preharvest, storage and/or processing periods. Mycotoxins differ widely in their chemical and toxicological properties. The aflatoxins have received greater attention than any of the other mycotoxins because of their demonstrated potent carcinogenic effects in susceptible laboratory animals and their acute toxicological effects in humans. Many countries have attempted to limit exposure to aflatoxins and other selected mycotoxins by imposing regulatory limits on commodities in commercial channels. Recent FDA monitoring data generated on mycotoxins will be discussed.

Mycotoxins are toxic metabolites produced by certain fungi that may grow on foods and feeds under favorable conditions of temperature and humidity. These metabolites can exhibit acute, subchronic and chronic toxicological manifestations in humans and susceptible animals. Some mycotoxins have proven to be teratogenic, mutagenic and/or carcinogenic in certain susceptible animal species and have been associated with various diseases (mycotoxicoses) in domestic animals, livestock and humans in many parts of the world. The occurrence of mycotoxins in foods and feeds is unavoidable and influenced by certain environmental factors; hence the extent of mycotoxin contamination is unpredictable and may vary with geographic location, agricultural and agronomic practices and the susceptibility of commodities to fungal invasion during preharvest, storage and/or processing periods. Contamination of milk, meat and eggs can result from the consumption of mycotoxin-contaminated feed by farm animals. It has been estimated by the United Nations' Food and Agriculture Organization

(FAO) that at least 25% of the world's food crops are affected by mycotoxin contamination. Within the United States (U.S.) the incidence of mycotoxin contamination of a particular food crop has been noted to vary not only from region to region but also from year to year.

Mycotoxins can enter the food chain by one of two major routes — (1) direct contamination resulting from the growth of toxigenic fungi on the food item itself and (2) indirect contamination resulting from the use of a food component contaminated with mycotoxins. In regard to humans, direct exposure is more likely to be a problem in tropical areas and certain underdeveloped countries where the consumption of moldy food may be unavoidable because of shortages of good quality food and/or proper storage and processing facilities. In developed countries, foodstuff is usually discarded or fed to animals if it appears to be visibly moldy; hence, indirect exposure, resulting from prepared or processed foods or from the consumption of animal products, becomes more significant.

There are many reports in the literature with respect to the occurrence of mycotoxins in foods and feed. Unfortunately many of the reports dealing with the incidence and levels of contamination leave much to be desired. A major concern is the accuracy of the data. In most developed countries the levels of mycotoxins in human foods are quite low, frequently lying in the ng/g (ppb) and sub-ng/g range. It is a tenet of analytical chemistry that the data variation (coefficient of variation or CV) becomes greater as the analyte concentration decreases. As an example, the interlaboratory CVs for analytes at the µg/g [parts per million (ppm)] level are usually about 16%, whereas at the ng/g [parts per billion (ppb)] level, the CVs are invariably in the range of 45-50% (*1*). There are at least 3 types of errors associated with obtaining an accurate estimate of the true concentration of a component in a given lot of foodstuff, they are: (1) sampling, (2) sample preparation and (3) analysis (*2*). Of these, the largest relative errors encountered are associated with sampling, followed by the analytical procedure (*3*). It is indeed difficult to obtain an analytical-sized sample from large nonhomogeneously contaminated lots of any particular food or feed that accurately represents the concentration of the lot. Specific sampling plans have been developed and rigorously tested for only a few commodities, such as corn, peanuts and tree nuts; sampling plans for some other commodities have been modeled after these. In a study designed to determine the magnitude of the relative errors associated with an analytical result, it was calculated that the sampling variability, using the sampling plans currently in use for aflatoxins, was 55% for peanuts and 100% for cottonseed for a lot containing 20 ng aflatoxins/g; these variations were obtained by using well-developed sampling plans (*3,4*). In addition to this sampling variability, consideration must also be given to subsampling variability and the analytical variability. An overview of sampling and sample preparation for the identification and quantitation of natural toxicants in foods and feeds was published recently (*5*).

Errors may be associated with some of the analytical data presented in the scientific literature on mycotoxins. Of the hundreds of analytical methods published for mycotoxins, only a few have been subjected to a formal interlaboratory (collaborative) study. Survey data obtained by use of analytical procedures that have not been evaluated by an interlaboratory study are often reported. In some cases, even the limit of determination of the method is not given; hence the reader cannot determine the meaning of negative samples or samples labeled "trace." Even when a well-studied method is used, one is frequently troubled by the lack of a suitable confirmation of identity of the analyte. This is particularly true of most of the data that are generated by the increasingly popular immunoassay techniques. Various organizations have tried to improve data quality through check sample programs in which a sample of known contamination level is sent to each laboratory for analysis by the method routinely used in that laboratory. The results obtained enable the laboratory director to judge the capability of the analysts in his laboratory and to identify problems which may be otherwise unnoticed. Examples of such check sample programs are the International Aflatoxin Check Sample Program administered each year by the International Agency for Research on Cancer (IARC), and the Smalley Check Sample Series sponsored by the American Oil Chemists' Society (AOCS). In both series the interlaboratory coefficient of variations (CVs) at the 20 ng/g level were >45%. In the 1989 IARC survey, the CVs for total aflatoxin in corn were about 55%; most interestingly, with aflatoxin M in milk at the 0.3 ng/ml level, 17% of the participants found no aflatoxin, while 14% of the results had to be excluded as statistical outliers (Friesen, M., IARC, personal communication). Therefore, although there is no doubt about the worldwide occurrence of mycotoxins in foods, there is considerable doubt about the accuracy of the levels reported for various mycotoxins in foods.

The mycotoxins found to occur significantly in naturally contaminated foods and feeds include the aflatoxins, ochratoxin A, some trichothecenes, zearalenone, citrinin, patulin, sterigmatocystin, penicillic acid, cyclopiazonic acid and the fumonisins (Table 1). The crops that may be affected by these mycotoxins include corn, peanuts, cottonseed, tree nuts, cereal grains (wheat, barley, rice, oats) dried beans and apples.

Background exposure data, along with toxicological evaluation are essential in order to establish the need for formal regulatory control programs. Past concerns by the Food and Drug Administration (FDA) of mycotoxins such as patulin, zearalenone, deoxynivalenol, ochratoxin A and penicillic acid have resulted in surveys of susceptible commodities for these mycotoxins. From the results of these surveys and the available toxicological data, no formal regulatory programs were warranted. However, because of the random, unpredictable contamination of food by mycotoxins, the control of these toxins is expected to be a difficult task; therefore it is not logical to envision a food supply that can be guaranteed to be "mycotoxin-free." Continuous efforts are being made by the FDA to minimize the levels to which consumers may be exposed to mycotoxins.

For most mycotoxins, there is only limited information available regarding their natural occurrence, stability in foods and feeds, toxicity and carcinogenicity to humans. Our knowledge of mycotoxins is biased by the emphasis placed on carcinogenic mycotoxins, in particular the aflatoxins. A review of the mycotoxin literature reveals that information on aflatoxin contamination far exceeds that for all other mycotoxins. The aflatoxins have received greater attention by scientists because of their demonstrated potent carcinogenic effect in susceptible laboratory animals and their acute toxicological effects in humans. In view of this, the rest of this presentation will focus on measures that have been taken to ensure that our food supply is relatively safe and free from aflatoxin contamination; hopefully this will stimulate similar measures to be taken regarding other mycotoxins as more information about them becomes available.

The worldwide occurrence of the aflatoxins (B_1, B_2, G_1, and G_2) is well documented in the literature (*6-10*), with the major contamination occurring in areas of high moisture and temperature. The aflatoxins are the only mycotoxins currently being regulated in the U.S. From a regulatory viewpoint, the aflatoxins are considered to be added poisonous and unavoidable contaminants because they cannot be completely prevented or eliminated from food or feeds by current good agronomic and manufacturing practices. Many countries have attempted to limit exposure to aflatoxins and other selected mycotoxins by imposing legal restrictions or regulatory limits on food and feeds in domestic and commercial import channels (*11*). The legal basis for regulating poisonous substances in food in the U.S. is the Federal Food, Drug, and Cosmetic Act which prohibits the entry of adulterated food into interstate commerce. A food is considered adulterated if it contains "any poisonous or deleterious substance which may render it injurious to health" [sec 402(a)(1)]. The FDA has enforced limits on the aflatoxin content of foods and feeds involved in interstate commerce since 1965. The industries are routinely monitored by FDA through compliance programs to ensure adherence to the limits (action levels) that have been established for aflatoxins in various commodities. The monitoring effort includes both a formal Compliance Program and exploratory surveillance action. This presentation combines data generated by 3 compliance programs (Aflatoxins in Domestic Foods, Import Foods and Animal Feeds) for the fiscal years 1987, 1988 and 1989.

The strategies used by FDA for implementing compliance programs have been published elsewhere (*12*) and are only briefly summarized here. The three objectives of the compliance programs are:

(1) collect and analyze samples of various foods and feeds to determine the occurrence and levels of aflatoxins;
(2) remove from interstate commerce those foods and feeds which contain aflatoxins at levels judged to be of regulatory significance; and
(3) determine awareness of potential problems and control measures employed by distributors, manufacturers and/or processors.

Each FDA District involved is provided with a list of commodities known to be susceptible to aflatoxin contamination, a sampling plan (including product sample size) used by FDA in the regulatory control of mycotoxins in foods and feeds (2) and a quota of samples to be collected. Sampling of corn, corn-based products, milk, peanuts, peanut products and animal feeds is stressed. Previous incidence data obtained from surveys of small food grains such as soybeans, barley, oats, rye and rice indicated that these grains were not a significant source of aflatoxin exposure unless they were abused in storage or after preparation (*13*). All samples are analyzed for aflatoxins by the official collaboratively studied methods specific for each product (14) at the FDA Mycotoxin Analytical Laboratory in New Orleans, LA. The limit of determination using the AOAC procedures, is 1 ng/g for aflatoxins in grains, nuts and their products and 0.05 ng/ml for aflatoxin M in milk.

Regulatory actions are directed in accordance with existing compliance policy guides. The guides specify that legal actions are to be recommended when the level of total aflatoxins (B_1, B_2, G_1, and G_2) in all foods and feeds for dairy cattle or immature animals exceeds 20 ng/g (300 ng/g for cottonseed meal used as a feed ingredient for beef cattle, swine and poultry) and other identifying or restricting criteria are met. The action level for aflatoxin M in milk is 0.5 ng/ml. The following action levels apply to corn designated for specific animal species: 100 ng/g for breeding cattle, breeding swine and mature poultry; 200 ng/g for finishing swine (> 100 lb) and 300 ng/g for finishing beef cattle (Table 2). FDA announced in 1988 (Fed. Regist. 53:5043, 1988) that its current action levels are not binding on the courts, the public (including food processors), or the agency. The current action levels do, however, represent the best guidance available on chemical and other contaminant levels that FDA would consider to be of regulatory interest; the agency intends to initiate notice and comment rule-making proceedings to amend certain of its regulations in the very near future.

Results and Discussion

The monitoring data obtained over the years in general show that in the U.S. aflatoxin is a frequent and major contaminant of corn, peanuts and cottonseed and an occasional contaminant of almonds, pecans, pistachio nuts and walnuts. Initially it was feared that aflatoxin contamination of animal feeds might result in significant human exposure to aflatoxin and/or its metabolites in meat, milk and eggs, although transmission studies had indicated a very low potential for contamination of the edible tissues in this way. When the corn crops in 1977 and 1980 were heavily contaminated by aflatoxin because of adverse weather conditions, a study of swine, beef and chicken livers from slaughter houses and egg-cracking plants was conducted, using methodology capable of determining B_1 and M_1 at levels as low as 0.05 ng/g; no aflatoxin was detected in 1453 samples that were almost equally divided among the four sample types (*10*). In the follow-up survey of 1980 of swine (251) and turkey (114) livers, again no aflatoxins were found. The only case in the U.S. of aflatoxin in eggs was documented in an early FDA survey (1977) of eggs collected from cracking plants in the southern U.S.; only one of 112 samples contained aflatoxin (0.06 ng/g) (*15*). On the other hand, aflatoxin M_1 has been found on occasion in milk, particularly in the Southeast where aflatoxin is a frequent contaminant of corn, and in the Southwest where cottonseed used in feed is frequently contaminated. In these areas, strong state supervision and control have effectively limited human exposure to aflatoxin M_1.

Monitoring data obtained for the years 1987-1989 are tabulated in Tables 3 to 11. These data highlight the unpredictable occurrence of aflatoxins in foods and the relatively low continuous levels present in certain commodities. These data are biased in the sense that the samples collected under compliance programs target those areas

Table 1. Mycotoxins identified as natural contaminants

Mycotoxin	*Food or feed contaminated*
Aflatoxins	Corn, cottonseed, peanuts, milk, tree nuts
Citrinin	Barley, oats, rice
Cyclopiazonic acid	Corn, peanuts, cheese
Deoxynivalenol	Corn, barley, rye, wheat
Ergot alkaloids	Wheat, rye, oats
Fumonisins	Corn
Nivalenol	Corn, barley, wheat
Ochratoxin A	Corn, barley, wheat, oats, green coffee beans
Patulin	Apples, pears
Penicillic acid	Corn, dried beans
Sterigmatocystin	Cheese, green coffee beans
T-2 toxin	Corn, barley, sorghum
Zearalenone	Corn, wheat

Table 2. FDA action levels for aflatoxins

Commodity	*Level (ng/g)*
All products, except milk, designated for humans	20
Milk	0.5
Corn for immature animals and dairy cattle	20
Corn for breeding beef cattle, swine and mature poultry	100
Corn for finishing swine	200
Corn for finishing beef cattle	300
Cottonseed meal (as a feed ingredient)	300
All feedstuff other than corn	20

NOTE: Compliance Policy Guides 7120.26, 7106.10, 7126.33.

Table 3. Peanut products examined for aflatoxin and levels

			Determinable Aflatoxins	
Peanut product	*Year*	*Total*[a]	*Percent of products >1 ng/g*	*Percent of products >20 ng/g*
Peanut butter	1987	146	13.7	2.7
	1988	372	1.9	0.5
	1989	158	0.6	0.0
Shelled, roasted	1987	63	3.2	3.1
	1988	357	2.5	0.5
	1989	243	0.0	0.0
In-shell roasted	1987	15	0.0	0.0
	1988	80	1.2	0.0
	1989	52	0.0	0.0

[a]Total number of products examined.

Table 4. Domestic tree nut products examined for aflatoxin and levels

Tree nut Product	*Year*	*Total*[a]	*Determinable aflatoxins* *Percent of products >1 ng/g*	*Percent of products >20 ng/g*
Almond	1987	44	0.0	0.0
	1988	241	0.4	0.0
	1989	108	0.0	0.0
Pecan	1987	73	12.3	8.2
	1988	225	0.4	0.0
	1989	212	0.5	0.5
Pistachio	1987	27	22.0	3.7
	1988	56	10.7	5.3
	1989	104	1.9	0.9
Walnut	1987	51	2.0	0.0
	1988	249	0.4	0.0
	1989	228	4.4	2.6

[a] Total number of products examined.

Table 5. Aflatoxins in shelled corn designated for human consumption

Area of U.S.	*Year*	*Total*[a]	*Determinable aflatoxins* *Percent of samples >1 ng/g*	*Percent of samples >20 ng/g*
Southeast[b]	1987	105	30.4	20.9
Corn belt[c]		49	12.2	0.0
Virginia-Maryland		5	0.0	0.0
Arkansas-Texas-Oklahoma		49	14.2	2.0
Rest of U.S.		32	12.5	0.0
Southeast	1988	299	9.0	7.0
Corn belt		100	1.0	1.0
Virginia-Maryland		44	6.8	2.2
Arkansas-Texas-Oklahoma		115	9.5	5.2
Rest of U.S.		224	0.0	0.0
Southeast	1989	262	7.6	3.4
Corn belt		736	14.1	4.3
Virginia-Maryland		8	0.0	0.0
Arkansas-Texas-Oklahoma		252	19.8	12.3
Rest of U.S.		261	14.5	8.0

[a]Total number of samples examined. [b]AL, FL, GA, LA, MS, NC, SC, TN. [c]IA, IL, IN, KS, MI, MN, MO, NE, OH, SD, WI.

Table 6. Aflatoxins in shelled corn designated for animal feed

Area of U.S.	Year	Total[a]	Determinable aflatoxins: Percent of samples >1 ng/g	Percent of samples >20 ng/g
Southeast[b]	1987	17	47.0	35.2
Corn belt[c]		75	8.0	0.0
Virginia-Maryland		4	0.0	0.0
Arkansas-Texas-Oklahoma		13	30.7	7.6
Rest of U.S.		7	0.0	0.0
Southeast	1988	74	13.5	8.1
Corn belt		15	0.0	0.0
Virginia-Maryland		28	3.5	0.0
Arkansas-Texas-Oklahoma		78	29.0	25.6
Rest of U.S.		71	0.0	0.0
Southeast	1989	159	13.2	8.1
Corn belt		784	15.0	3.4
Virginia-Maryland		16	0.0	0.0
Arkansas-Texas-Oklahoma		25	40.0	32.0
Rest of U.S.		28	0.0	0.0

[a]Total number of samples examined. [b]AL, FL, GA, LA, MS, NC, SC, TN. [c]IA, IL, IN, KS, MI, MN, MO, NE, OH, SD, WI.

Table 7. Aflatoxins in milled corn products

Area of U.S.	Year	Total[a]	Determinable aflatoxins: Percent of samples >1 ng/g	Percent of samples >20 ng/g
Southeast[b]	1987	94	14.8	3.1
Corn belt[c]		45	4.4	0.0
Virginia-Maryland		4	0.0	0.0
Arkansas-Texas-Oklahoma		24	0.0	0.0
Rest of U.S.		52	0.0	0.0
Southeast	1988	206	3.3	1.4
Corn belt		180	0.0	0.0
Virginia-Maryland		40	2.5	0.0
Arkansas-Texas-Oklahoma		40	2.5	0.0
Rest of U.S.		288	0.0	0.0
Southeast	1989	155	1.0	0.6
Corn belt		112	3.5	0.0
Virginia-Maryland		16	0.0	0.0
Arkansas-Texas-Oklahoma		64	1.5	0.0
Rest of U.S.		435	9.4	0.0

[a]Total number of samples examined. [b]AL, FL, GA, LA, MS, NC, SC, TN. [c]IA, IL, IN, KS, MI, MN, MO, NE, OH, SD, WI.

Table 8. Survey for aflatoxins in sweet corn in the U.S. (1988)

	Canned, whole kernel	*Canned, cream style*	*Frozen, whole kernel*	*Number of samples analyzed*	*Determinable aflatoxins ng/g*
Arizona	6	0	0	6	0
Arkansas	6	2	0	8	0
Delaware	2	0	3	5	0
Idaho	3	1	0	4	0
Illinois	8	1	1	10	0
Maryland	2	0	1	3	0
Minnesota	6	6	0	12	0
New York	4	2	4	10	0
Oklahoma	1	0	0	1	0
Oregon	3	3	2	8	0
Pennsylvania	0	0	2	2	0
Utah	1	0	0	1	0
Washington	1	1	7	9	0
Wisconsin	6	2	0	8	0

Table 9. Cottonseed and cottonseed meal lots examined for aflatoxins

	Year	*Total*[a]	*Percent of lots >1 ng/g*	*Percent of lots >20 ng/g*	*Percent of lots >300 ng/g*[b]
Cottonseed	1987	0	0.0	0.0	0.0
Cottonseed meal					
AZ, CA		3	0.0	0.0	0.0
Rest of U.S.		7	57.1	43.0	0.0
Cottonseed	1988				
AZ, CA		56	7.1	0.0	0.0
Rest of U.S.		20	5.0	0.0	0.0
Cottonseed meal					
AZ, CA		16	25.0	6.2	0.0
Rest of U.S.		9	22.0	0.0	0.0
Cottonseed	1989				
AZ, CA		8	12.5	12.5	0.0
Rest of U.S.		29	10.3	10.3	3.4
Cottonseed meal					
AZ, CA		26	31.0	7.6	7.6
Rest of U.S.		1	0.0	0.0	0.0

[a]Total number of lots examined. [b]Action level for cottonseed meal as a feed ingredient.

Table 10. Fluid milk and milk products examined for aflatoxin

		Determinable aflatoxin	
		Percent of Products	*Percent of products*
Year	*Total*[a]	*>0.05 ng/ml*	*>0.5 ng/ml*
1987	67	0.0	0.0
1988	155	9.0	0.0
1989	632	43.5	1.1

[a]Total number of products examined.

Table 11. Miscellaneous imported food products examined for aflatoxins

		Determinable aflatoxins	
		Percent of products	*Percent of products*
Product	*Total*[a]	*>1ng/g*	*>20ng/g*
Brazil nut	95	16.8	3.1
Corn meal/flour	54	22.2	1.8
Crackers, nut	59	11.8	5.0
Figs	92	2.1	2.1
Melon seed	88	5.6	1.1
Nutmeg	33	18.1	12.1
Peanut butter	29	13.7	3.4
Peanut candy	129	25.0	7.7
Peanuts,shelled	67	10.4	5.9
Pecans	56	7.1	3.5
Pistachio	53	7.5	3.7
Pumpkin seed	117	9.4	5.1
Sesame seed	264	1.1	0.0

[a]Total number of products examined

and commodities where one is most likely to find contamination. Even with the bias limitation and the relatively small numbers of samples involved, it is still possible to find some useful trends in the data obtained.

Peanut Products. All raw shelled peanuts in commercial channels in this country are marketed under a USDA/industry agreement that requires analysis and certification of each lot for aflatoxin content by the U.S. Department of Agriculture (USDA). This agreement forms the basis for a Memorandum of Understanding between USDA and FDA (Compliance Policy Guides 7155a.11, 7155a.13 and 7155a.14). The testing for aflatoxins in roasted shelled and in-shell peanuts, and in processed peanut products for consumer use is the responsibilty of FDA. The contamination of peanuts by aflatoxins has been noted in all peanut-growing areas of the U.S. In a study designed to highlight the uneven distribution of a few highly contaminated peanut kernels among a large number of uncontaminated ones in a given sample or subsample, aflatoxin concentration ranged from a trace to 1,100,000 ng aflatoxin B_1/g in individual kernels (*16*). The levels of aflatoxin contaminations in peanut butter and shelled roasted peanuts were high in 1987 but decreased in succeeding years (Table 3). It has been generally observed over the years that when aflatoxin contamination occurred in peanut products, higher levels were usually noted in peanut butter and roasted shelled peanuts than in roasted in-shell (ball park) peanuts. Apparently the roasting process, coupled with the grade and cultivar used, keeps these peanuts (roasted in-shell) relatively free of aflatoxins from year to year. None of the peanut products analyzed in 1989 contained total aflatoxins above the action level of 20 ng/g.

Tree Nuts (Domestic). Measurable levels of aflatoxins were found in almonds, pecans, pistachio nuts and walnuts (Table 4). No aflatoxins were noted in any of the 60 cashew, filbert and macadamia nut samples examined. Generally the levels of contamination were lower in almonds and walnuts than in pecans and pistachio nuts. Aflatoxin contamination in almonds has been attributed to kernel damage incurred before harvest. From a survey of California almond and walnut crops (*17*) it was established that the probability of aflatoxin contamination in almonds is one kernel in 26,500 unsorted in-shell nuts from the field; in walnuts it is one in 28,250 nuts. Current techniques of removing visibly damaged nuts immediately after harvest, followed by cool, dry storage conditions, seem to be effective in controlling the extent of aflatoxin contamination of almonds and walnuts in commercial channels. The major cause of contamination of pecans has not been determined; however, data obtained from a 6-year survey of late-harvested pecans suggest that the incidence of contamination was greatly influenced by the prevailing orchard temperatures during the latter part of the harvest season (*18*). Domestic pistachio nuts have been commercially available since 1980; aflatoxin contamination has been observed consistently during the past 4 years.

Corn. The grain most susceptible to aflatoxin contamination in the U.S. is corn. In most of the southeastern states (Alabama, Florida, Georgia, Louisiana, Mississippi,

North Carolina, South Carolina and Tennessee), aflatoxin contamination of corn is a major concern each year. The incidence of aflatoxin contamination of corn grown in other areas of the country may vary from year to year depending primarily on the weather conditions prevailing during the preharvesting and harvesting periods. The bulk of the corn found in commercial channels in the U.S. is grown in the midwest (Iowa, Illinois, Indiana, Kansas, Michigan, Minnesota, Missouri, Nebraska, Ohio, South Dakota and Wisconsin). Data available from earlier surveys of aflatoxin in corn indicated that this area was virtually aflatoxin-free (*10*). Unfortunately, in 1983 and more recently in 1988, this area along with other areas of the country experienced atypical weather conditions (severe drought and late rainfall) during the latter part of the growing season. Because of these conditions, it was expected that the harvested corn would contain higher than usual levels of aflatoxin contamination. Since flexibility is one of the features of the FDA Compliance Programs, it was not a difficult task for FDA to shift its monitoring efforts and resources so that a larger than planned number of samples could be collected and analyzed for aflatoxin contamination. The data in Tables 5 and 6 show that the number of corn samples collected and analyzed from the midwest or corn belt states during 1989 significantly exceeded the number of samples processed in previous years. The bulk of the corn from the 1988 crop began showing up in commercial channels after the beginning of the 1989 fiscal year (October 1). The data in Table 5 show a significant increase during 1989 in the percentage of corn lots designated for human consumption that contained aflatoxins at levels greater than 20 ng/g as compared with the years 1987 and 1988. The states of Virginia and Maryland were exceptions; however, because of the earlier harvesting season in those states, some samples were analyzed during the latter part of the fiscal year 1988. The data in Table 6 reflect a similar increase in levels and incidences of contamination in corn designated for animal feed from the various areas. In the case of milled corn products analyzed (Table 7), significant levels were observed only in samples collected from the Southeast, which was not unexpected. Studies have shown that two general methods for processing corn (dry and wet milling) significantly reduce the level of aflatoxins that may be present in those fractions that are used for human consumption. For example, corn starch derived from the wet milling process was found to contain only 1% of the aflatoxin level present in the raw corn; grits, low fat meal and flour derived from the dry milling process contained only 6-10% of the original aflatoxin level (*19,20*). Heat processing and cooking are effective in reducing the aflatoxin content in foods (*21*). None of the 410 manufactured corn-based products (e.g., corn chips, hush puppies, mixes, tortilla, breakfast cereals ready-to-eat and popcorn), collected from all over the country during the 3-year period, contained aflatoxin levels in excess of 20 ng/g.

Aflatoxin contamination is not believed to be a problem in sweet corn, that is "eating ears" or whole kernel corn. Studies have shown that corn is most vulnerable to invasion by aflatoxin- producing fungi during the late milk and early dough stages of development. Sweet corn is usually harvested at the early milk stage of development and processed immediately; therefore, the possibility for invasion of sweet corn by the aflatoxin-producing fungi is very remote. To confirm that aflatoxin contamination was not a problem in the 1988 sweet corn crop, a limited survey was conducted.

Samples of sweet corn (canned whole kernel, canned cream style and frozen whole kernel) were collected in 14 states from packing plants or primary storage warehouses. The results of this survey are shown in Table 8. None of the 87 samples collected and analyzed contained observable levels of aflatoxins. The results obtained in this limited survey, when considered along with the negative results obtained from a larger survey (263 samples) conducted in 1976 and 1977 (*22*), suggest that concerns about the occurrence of aflatoxins in processed sweet corn are not warranted.

Cottonseed and Cottonseed Meal. Cottonseed and cottonseed meal are used as ingredients in animal feed; therefore, contaminated cottonseed may pose a possible hazard to humans who ingest meat or milk from such animals. High levels of aflatoxins in the feed of dairy cattle can result in aflatoxin M in milk. The incidence and levels of contamination found from 1987 to 1989 are shown in Table 9. With the exception of the data for 1987, the higher percentage of cottonseed and cottonseed meal contamination were noted in the Southwest (Arizona-California).

Milk. Aflatoxin M_1 is a metabolite of aflatoxin B, formed in the milk of mammals that have ingested aflatoxin B in their feed. The ratio of conversion from aflatoxin B in naturally contaminated feed to aflatoxin M in milk by dairy cattle has been reported by two independent laboratories to be 66:1 and 75:1 (*23, 24*). Therefore, dairy cattle fed rations containing <20 ng/g aflatoxin B should not have aflatoxin M_1 concentrations greater than the action level of 0.5 ng/ml in their milk. The data in Table 10, show that in 1989 only 1.1% of the samples analyzed contained aflatoxin M above the action level.

Miscellaneous Import Food Products. In addition to the domestic monitoring program for aflatoxins, import foods that are susceptible to aflatoxin contamination are also monitored. The foods monitored included nut and nut products, edible seeds, spices and animal feeds. Products found to contain aflatoxin concentrations greater than the action level are denied entry into the country. The data in Table 11 show some of the import products examined between 1987 and 1989.

Conclusions

Mycotoxins are considered unavoidable contaminants in foods because agronomic technology has not yet advanced to the stage where preharvest infection of susceptible crops can be eliminated. The control of mycotoxins in foods and feeds is a constantly evolving process. In highly developed countries, acute toxic effects of mycotoxins are rarely observed because advances in processing technology and quality control programs prevent heavily molded food from entering the human food supply. From a public health viewpoint, there is concern regarding the health effects resulting from long-term exposure to low levels of toxins in foods. The possibility that aflatoxins specifically may be human hepatocarcinogens is still a matter of contention; however, efforts are being made to restrict their presence in foods to the lowest level practically attainable.

One way the safety of our food supply can be improved is by the use of effective monitoring techniques and the establishment of quality control safeguards for food processing operations. FDA's surveillance and regulatory programs are designed to keep aflatoxins in foods and feeds at the lowest levels consistent with maintaining an adequate food supply at a reasonable cost. By enforcing the Federal Food, Drug, and Cosmetic Act, FDA can remove from interstate commerce any food or feed found to be adulterated according to the provisions of the Act. The data obtained from monitoring a major commodity, like corn, over a 3-year period show that the guidelines and policies of the compliance programs are flexible enough to cope with emergency situations that may be encountered within a given year. The emergency brought on by the 1988 corn crop resulted in increased sample collection and analyses, as well as the enactment of new tiered action levels, so that large amounts of contaminated corn could be used effectively as feed for certain animal species. FDA's efforts to ensure the safety and quality of foods and feeds are complemented by control programs carried out by the USDA, state departments of agriculture, and various trade associations. Based upon the levels of aflatoxins found in foods, human exposure to the aflatoxins has been estimated, human risk assessments have been made and government controls set in place in at least 50 countries. Five of these countries claim to have set regulatory levels on the basis of a risk analysis of some type. The wide range of regulatory limits put into effect in the various countries (0 to 50 ng/g for peanuts as an example) reflects the difficulty in coming to a conclusion about the risk resulting from low level exposure to carcinogens in foods; not only do the analytical data upon which the exposure is based have wide confidence limits, but the toxicological effects of low level exposure are open to considerable debate.

Current technology cannot prevent mycotoxin contamination of field crops before harvest. Research under way in several laboratories is aimed at controlling preharvest contamination of peanuts and corn specifically through genetic manipulations and the use of various chemicals. In spite of the random, unpredictable pattern of contamination of foods by mycotoxins, continuous progress is being made to minimize the levels to which humans may be exposed since it is impossible to guarantee a "mycotoxin-free" food supply. Enforcement of current regulatory programs will continue at the current levels until breakthroughs in research and/or control procedures allow further reduction in the aflatoxin content of foods and feeds, or until the risk to man is better established.

Literature Cited

1. Horowitz, W; Kamps, L. R.; Boyer, K. W. *J. Assoc. Off. Anal. Chem.* **1980**, *63*, 1344-54.
2. Campbell, A. D.; Whitaker, T. B.; Pohland, A. E.; Dickens, J. W.; Park, D. L. *Pure Appl. Chem.* **1986**, *58*, 305-14.
3. Whitaker, T. B. *Pure Appl. Chem.* **1977**, *49*, 1709-17.
4. Whitaker, T. B.; Whitten, M. E.; Monroe, R. J. *J. Am. Oil Chem. Soc.* **1976**, *53*, 502-5.
5. Park, D. L.; Pohland, A. E. *J. Assoc. Off. Anal. Chem.* **1989**, *72*, 399-404

6. Pohland, A. E.; Wood, G. E. *Proc. Mycotoxins, Cancer and Health Conference*, Pennington Biomedical Research Center, Baton Rouge, LA **1990** (In Press).
7. Pohland, A. E., Wood, G. E. In *Mycotoxins in Foods*; Krogh, P., Ed.; Academic: New York, 1987; pp 35-64.
8. *Mycotoxins: Economic and Health Risks*, Council for Agricultural Science and Technology, Task Force Report No. 116, Ames, Iowa, 1989.
9. Jelinek, C. F.; Pohland, A. E.; Wood, G. E. *J. Assoc. Off. Anal. Chem.* **1989**, *72*, 223-30.
10. Stoloff, L. In *Carcinogens and Mutagens in the Environment*; Stich, H. F., Ed.; CRC: Boca Raton, FL, 1982; Vol. 1 pp 97-120.
11. van Egmond, H. P. *Food Addit. Contam.* **1989**, *6*, 139-88.
12. Wood, G. E. *J. Assoc. Off. Anal. Chem.* **1989**, *72*, 543-8.
13. Stoloff, L. In *Mycotoxins in Human and Animal Health*; Rodricks, J. V.; Hesseltine, C. W.; Mehlman, M. A., Eds.; Pathotox: Park Forest South, Il, 1977; pp 7-28.
14. *Official Methods of Analysis*, Assoc. Off. Anal. Chem. 15th ed. Arlington, VA **1990**; secs 26.001-26.150.
15. Stoloff, L.; Trucksess, M. W. *J. Assoc. Off. Anal. Chem.* **1977**, *61*, 995-6.
16. Cucullu, A. F.; Lee, L. S.; Mayne, R. Y.; Goldblatt, L. A. *J. Am. Oil Chem. Soc.* **1966**, *43*, 89-92.
17. Fuller, G.; Spooncer, W. W.; King, A. D.; Schade, J.; Mackey, B. *J. Am. Oil Chem. Soc.* **1977**, *54*, 231A-4A.
18. Wells, J. M.; Payne, J. A.; Stoloff, L. *Plant Dis.* **1983**, *76*, 751-3.
19. Yahl, K. R.; Watson, S. A.; Smith, R. J.; Barabolok, R. *Cereal Chem.* **1971**, *48*, 385-91.
20. Brekke, O. L.; Peplinski, A. J.; Nelson, E. N.; Griffin, E. L. *Cereal Chem.* **1975**, *52*, 205-11.
21. Scott, P. M. *J. Food Prot.* **1984**, *41*, 489-99.
22. Stoloff, L.; Francis, O. J. *J. Assoc. Off. Anal. Chem.* **1980**, *63*, 180-1.
23. Frobish, R. A.; Bradley, B. D.; Wagner, D. D.; Long-Bradley, P. E.; Hairston, H. *J. Food Prot.* **1986**, *49*, 781-5.
24. Price, R. L.; Paulson, J. H.; Lough, O. G.; Gingg, C.; Kurtz, A. G. *J. Food Prot.* **1985**, *48*, 11-15.

RECEIVED August 15, 1991

IMPACT OF DIET

Chapter 26

Diet–Health Relationship

Paul A. Lachance

Graduate Program in Food Science, Rutgers, The State University, New Brunswick, NJ 08903–0231

Diet

Diet can be defined as either "Food and drink regularly consumed in the habitual course of living" or "A prescribed allowance or regimen of food and drink with reference to a particular state of health". Given the diverse functions of food, the challenge is to reconcile the pleasure of food and drink with the promotion of optimal health.

When food and drink are reduced to chemical terms, the pleasure is essentially absent. Since the number of chemicals which constitute food and drink is substantial and a minority fraction is essential for physiological health, the equivocal chemicals which impart sensory appeal must be distinguished from the chemicals that should be avoided. In all instances, (essential, equivocal, to be avoided) there is a determination of relative risk which must be made.

Health

Health is defined as "a continued state of soundness and vigor of body and mind". We have no direct measures of health and so we redefine health "as the absence of disease" but the disease state must be pathological to be recognized and indexed. Indicator conditions such as hypertension; or risk factors such as obesity; or risky practices such as smoking; "at-risk" environments such as considerable exposure to elevated ozone and other pollutant levels are invariably "silent" and idiopathic. Practices (dietary, exercise, ecological) which promote health are not a major domain of medicine, not only because the financial rewards to medicine are considerably inferior but because the connection to specialized diagnosis related (DRG's) disease categories are not specific and thus there are few legitimate categories for third party reimbursement.

Diet-Health Association

The association of diet with health (or disease) has been "in the news" since the 1969 White House Conference on Food, Nutrition and Health. As a result of the conference,

0097–6156/92/0484–0278$06.00/0

nutrition labeling was instituted and is now in the final process of a major revision (*1*). Labeling now will be mandated for most foods rather than triggered only by advertising claims and otherwise being voluntary. The "Dietary Goals" which emanated in the dying gasps of the Senate Select Committee on Nutrition and Human Needs (*2*) eventually led to the current joint USDA/DHHS Dietary Guidelines. The documentation for the health benefits of such guidelines are to be found in the over 5,000 scientific references cited in the 1989 Diet and Health report (*3*) of the National Academy of Sciences. These diet-health association pronouncements are also in part attributable to the yeoman lobbying and news media activity of the Center for Science and the Public Interest and related "Nader type" activists, as well as the peripatetic activities of the civil servants of the government agencies prodding and being prodded by zealous congressional legislative staffers. Both the major professional nutrition societies and the food industry provided debate, reaction and stabilizing inertia. The consumer "at best" has accomodated.

Consumer Beliefs and Practices

A Gallup survey of consumers conducted in December 1989 (commissioned by the International Food Information Council and the Am. Dietetic Association) revealed (with ±4% accuracy) that 95% of Americans "believe balance, variety and moderation are the keys to healthy eating" and further 83% recognized that what they eat may affect their future health"; but 67% mistakenly choose food based on "good food"/ "bad food" perceptions. In other words, Americans are more apt to opt for quick fixes and the latest health fads. For example, 52% of these (over 18) adult respondents reported increasing their consumption of oat bran but only 8% reported eating more vegetables and only 6% reported eating more fruit or fruit juices. Again "tunnel vision" nutrition is strong and well. Consumers believe one food or ingredient will prevent or maybe cure a disease! It rather fits the fact that Americans are the highest per capita users of over-the-counter and prescription drugs. We prefer and expect "quick fixes". Moreover, the diet-health connection translated into good food/bad food perception leads Americans to state they don't find eating a pleasure because 56% worry about fat and cholesterol and 35% are unsure of the difference between food cholesterol and blood cholesterol! Fifty percent say they gain weight if they eat what they like and 35% believe high fat foods cannot be part of a healthy diet even if balanced with low fat foods. In a nation that has serious levels of functional illiteracy, maybe we cannot expect better performance when prerequisites to performance (e.g. poverty, literacy) are limiting.

But Americans are healthier today than ever before in our history. We live longer. Since a peak in the '50'S, we now have a lower rate of coronary heart disease deaths and strokes. We have fewer dental cavities and other positive indicators of a lower morbidity yet we have only recently recognized the need to avoid sexual practices that promote AIDS; the need to avoid smoking that promotes both lung cancer and coronary heart disease etc. There are certain types of information the average consumer fears because they do not understand and certain mismanaged public incidents have served to rationalize an attitude that is unscientific.

A two day symposium at these ACS meetings is focusing on the realities of chemophobia (*4*). Abelson (*5*) states that "for most of the public, the word "chemical" elicits antipathy and fear". No doubt lack of and/or the poor quality of science education at all levels of our education system, even at the University level, can be implicated; however, there is no generic American consumer. If there were, the number of items in the supermarket would not have risen from 12,000 to 25,000 in the last 10 years in order to meet the demands of an increasingly segmented consumer market place.

The need for market segmentation began when "Rosie" became the riveter during World War II and the double income family life style was born. It really did not flourish until more recently when marriage at a later age and planning on fewer children became the norm. By the year 2,000, more people in the USA will be over 50 than under age 18. One reason is a significant increase in life span.

In theory, the "balance" of the diet was better in 1900 but life span was 50 years or so. The discovery of vitamins, minerals and antibiotics and other miracle drugs has shifted the causes of infirmities and death from infections and deficiency diseases to chronic diseases and decreased resistance to acute diseases. The decreased resistance is associated with the decreased integrity of aging organ systems.

The household is smaller. Both the delay in marriage, and the increase in older Americans has produced more one and two person households (*6*).

An unknown countervening force could be the yet to be fully established reality that the fastest growing segment of our population is Asian Americans followed by the already larger Hispanic-American population. A major variable contrasting these population segments is the initial educational level and goals of these two emerging population segments, both of which also have quite different food heritages and preferences. The realities of the foregoing facts are simply not challenging the scientific and medical communities.

Disease-Health Practices Associations

The report of the Surgeon General (*7*) and the Diet & Health report of the National Academy of Sciences (*3*) has gelled recognition by both the public health community and the scientific community of the medical establishment that many of today's major causes of mortality and morbidity are associated with clinical indicators (e.g.. obesity, blood pressure and blood lipid profiles) that are amenable to routine public health screening; and that certain modifiable practices of individuals (smoking, dietary choices) have significant effects in or on the pathogenesis of the diseases associated with the leading causes of mortality and morbidity. A listing of the chapter headings of the Diet and Health report (*see* Box) efficiently serves to identify the associations now recognized.

Diet and Health: Implications for Reducing Chronic Disease Risk

Part III: Impact of Dietary Patterns on Chronic Diseases

While few scientists now disagree with the associations that have emerged, considerable debate ensues relevant to the practical aspects of the recommendations made in the Diet and Health report. The Surgeon General's report is less controversial because it makes less judgemental conclusions. There is considerable agreement in both reports on the importance of the role of obesity, hypertension and smoking as clinical indicators but less agreement (and more complex scientific interactions and unknowns) on the role of type and quantities of dietary polyunsaturated fatty acids;the type and quantity of micronutrient intakes above levels needed to thwart deficiency diseases;the predictive value of total or HDL or lipoprotein (a) levels in clinical screenings etc. The Surgeon General's report (7) mentions the issues of microbiological food safety. Neither report considers the issues of chemical food safety. The fact is that the lack of an extensive body of human pathology attributable to naturally occurring food chemicals or food additives does not signal a lack of concern or knowledge or both. The beneficial and deleterious linkage between the chemistry of food and the pathogenesis of diseases are only now being explored.

A considered approach until controversial recommendations can be clarified with more quantitative scientific insights is to apply the concept of "limiting" and accordingly to determine the most valuable strategy to pursue. I am of the opinion that the control of obesity should have the greatest priority relative to its strong association with atherosclerosis and thus coronary heart disease and stroke, and that the limiting and most valuable strategy would be to control saturated fat intake and increase exercise. Relevant to the second greatest cause of mortality, namely cancer, the limiting and most valuable strategy is to curtail smoking and sources of carcinogens (including food). The most valuable strategy available is to enhance the intake of dark green and yellow vegetables for factors such as beta carotene, ascorbic acid and other antioxidant food sources which are highly associated with lower risks of several major types of cancer. The necessary dietary strategy for both a decrease in saturated fat (animal fats)and an increase in plant foods has existed as a dietary guideline for more than two generations and can be readily communicated to and practiced by the consumer (Figure 1). It should supercede the current nebulous dietary guideline of "eat a variety of foods". The benefits of this simple "peace symbol" food array guideline is the concomitant assurance of low fat, low saturated fat, high fiber, high carotenoid and a high nutrient density dietary without the need to account for individual nutrient or chemical components.

Obesity

The most prevalent indicator of chronic diseases today is obesity and the incidence increases (Figure 2). It is associated with diabetes, especially type II (adult onset), hypertension and heart disease. Obesity complicates physiological conditions such as pregnancy and increases the risk of several other common medical conditions such as cholecystitis and appendicitis, etc.

The daily per capita grams of saturated fat in the food supply has not dramatically changed since about 1920 but the daily per capita grams of both monounsaturated and polyunsaturated fat has increased with PUFA essentially doubling since 1935 (Figure 3). In the relatively same time period, per capita energy intake has dropped and the percent calories from fat has also dropped from 42% to 37% of calories. The food categories from which dietary energy has been derived has shifted substantially. Cereal grain products were a much more important contributor to the diet prior to WWII. With affluence came an increased consumption of higher quality (and more expensive) sources of protein. Balance in food group proportions began as a dietary goal circa post WWII, and these guideline proportions (the basic four) have remained the practical definition of balance, yet national surveys have repeatedly and consistently demonstrated that Americans fail to meet these proportions (Figure 4). Although we "eat a variety of foods", we consume an expensive profile with emphasis on high quality protein entrees and inadequate quantities of cereal grain products and fruits and vegetables.

The increase in the quality of foods per se coupled with the discovery, manufacture and utilization of micronutrient fortificants has permitted the lowering of energy intakes coupled to a less energy demanding environment, and the incidents of frank nutrient deficiency diseases so evident in the first half of the century are exceedingly rare. We now have an increasingly sedentary society that has shifted its caloric intake downward but not sufficiently to thwart obesity and other diet associated risk factors in the chronic disease causes of morbidity and mortality.

Smoking

Thirty percent of persons (in 1985) aged 18 and older were smokers, and the prevelance was equal for men and women under 30 years of age. Smoking is a risk factor for cancer and cardiovascular disease. An important but not frequently reported fact is that the food consumption patterns and dietary intakes of smokers and nonsmokers differ. Smoking is a marker for a poor diet.

Whereas, a similar percentage (38%) of smokers and nonsmokers report eating snacks on a daily basis, 38% of smokers habitually skip breakfast in contrast to 18% of nonsmokers. In a study of career age women in 1985-86 women smokers consumed less fruits and vegetables and more coffee and alcoholic beverages than nonsmokers (*8*). NHANES II results showed that median vitamin C intakes were lower in current

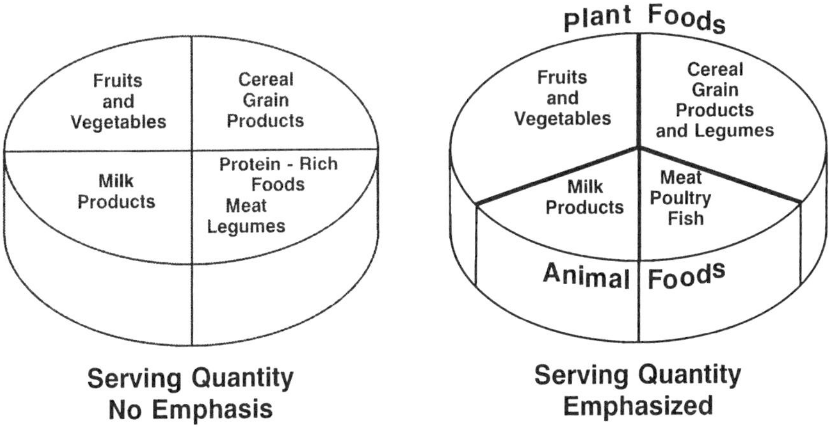

Figure 1. The basic four food groups required, remembering number of servings recommended per group, namely 4:4:2:2. A graphic representation immediately illustrates the proportions of the plate or day that should be allocated to each food group to promote balance. (Reproduced with permission from ref. 23. Copyright 1981 *Food Technology.*)

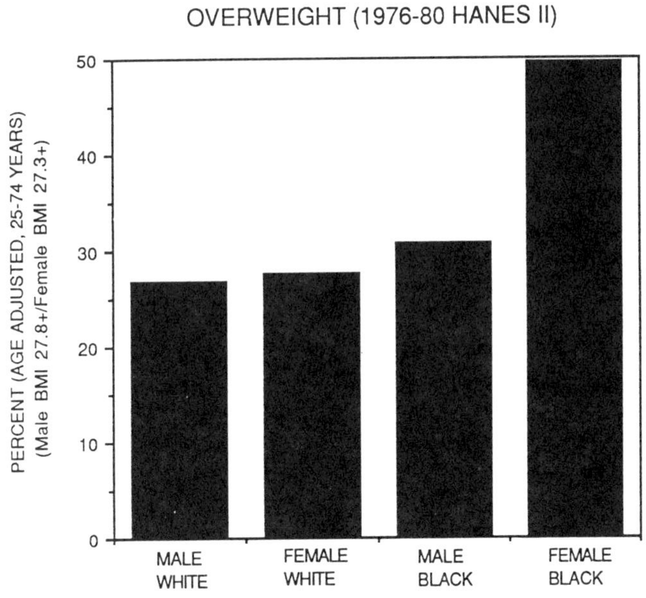

Figure 2. The prevalence of obesity defined as a BMI > 27.8 in male and >27.3 in female participants in the 1976-80 HANES II survey exceeds 25 percent in whites, 30 percent in black males and approaches 50 percent in black females. Adapted from ref. # 8.

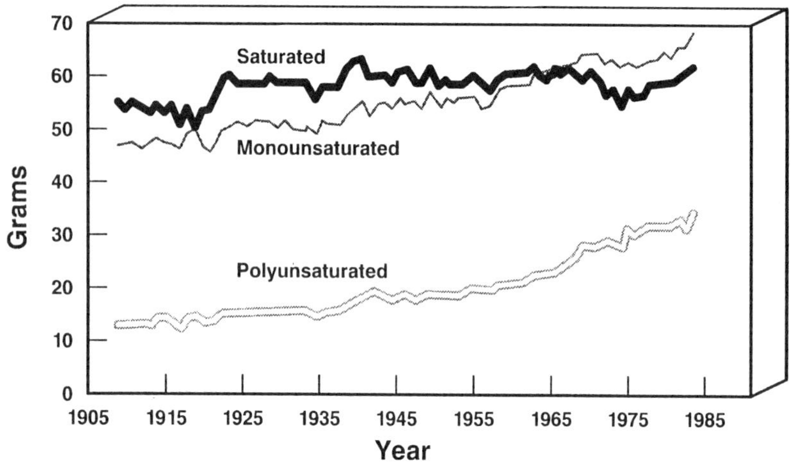

Figure 3. Per capita amounts of saturated, monounsaturated, and polyunsaturated fats in the U.S. food supply: U.S. Food Supply Series, 1909-85.

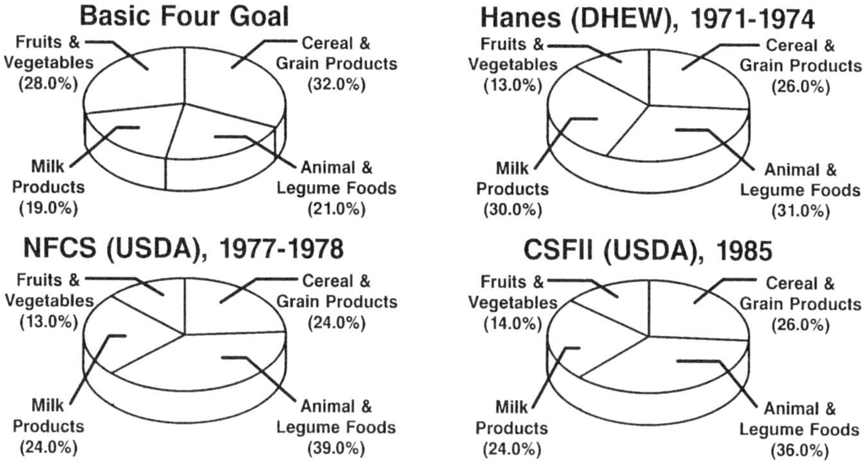

Figure 4. The Basic Four recommendation approximates the recommended 4:4:2:2 servings per day of key food groups. What Americans have been eating is reflected in the 1971-74 HANES data, the 1977-78 NFCS data, and the 1985 CSFII data. Rather than consuming 2 servings of plant derived food for each serving of animal derived food, as recommended by the Basic Four food guide concept, the consumer is doing the opposite.

smokers than in nonsmokers (*9*). Thiamin and fiber intakes per 1,000 kilocalories in career age women were lower in smokers.

Low birth weight (<2,500 grams) has become a bench mark (although not the most limiting) of a healthy outcome and is strongly associated with maternal nutrition. However, smokers have a substantially greater probability of giving birth to a low birth weight infant and whereas the probability does not increase with maternal age in nonsmokers, it nearly doubles in smokers (Figure 5). This phenomena is somewhat compensated by an increase in the mother's pregravid weight which means that smoking effects energy metabolism and nutrition (*8*).

A compeling body of prospective and retrospective human studies, in addition to hundreds of animal studies, associate a decrease in the incidence of several cancers with increased dietary intakes of dark green and yellow vegetables, often inaccurately labeled as high vitamin A food intakes. One can calculate (*10*) that the recommended dietaries of both the USDA, for dietary guideline purposes, and the NCI, for cancer prevention purposes would provide over 5 mg of daily dietary beta carotene, yet the average adult intake is in the vicinity of 1.5 mg (a value that the 1980 RDA's would have considered a respectable contribution to total vitamin A intake). The epidemiological study of Shekelle et al (*11*) on the incidence of lung carcinoma in smokers and nonsmokers reveals a very significant lower incidence in individuals (even smokers) who consume 5 or more mg of beta carotene equivalent daily.

Hypertension

The occurence of hypertension increases with age and is higher for Black Americans (Figure 6). It is a major risk factor for both heart disease and stroke.

Some individuals maintain normal blood pressure over a wide range of sodium or chloride intake, whereas others appear to be "salt sensitive" and display a decreased blood pressure in response to a decrease in sodium/chloride intake. Not all individuals are equally susceptable to the effect of sodium, but there is a lack of a practical biological marker for individual sodium sensitivity. The safe and adequate level of intake for adults is 1.1 to 3.3 grams per day, but the current intake ranges from 4 to 6 grams per day. One third of the intake is in the control of the consumer in the form of added "table" salt. Another third of the intake of sodium is associated with processed foods, however, the intake of certain foods which utilize sodium chloride as a preservative or processing aid can be avoided by the consumer - namely fermented foods such as soy sauce, most cheeses, and cured meats such as salami, pepperoni, hot dogs, ham and bacon.

Rates of both hypertension and diabetes are nearly tripled in persons 20 percent or more overweight. The rates for hypertension (except in white males) have been dropping. Whereas, both the consumer and manufacturers have decreased their use of sodium, the concomitant increases in obesity may be preventing further major decreases in hypertension.

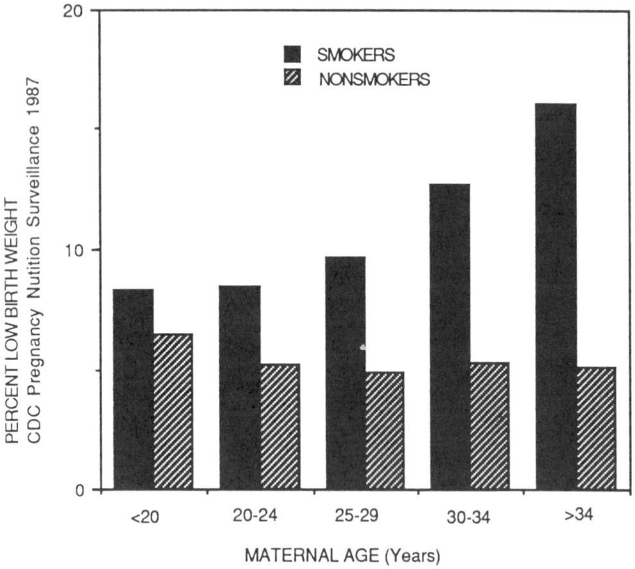

Figure 5. Low birth weight defined as < 2500 grams at birth has a prevalence greater in smokers than nonsmokers in all maternal age groups. Adapted from ref. # 8.

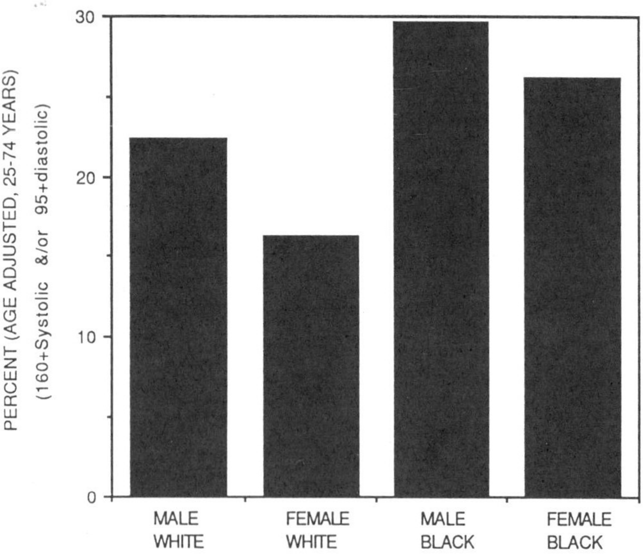

Figure 6. Hypertension defined as a systolic pressure greater than 160 or a diastolic pressure greater than 95 was most prevalent in black males (29%) and least prevalent in white females (16%) in the 1976-80 HANES II study. Adapted from ref. # 8.

RDA's (RDI's) and Dietary Intakes

The 1989 edition of the Recommended Dietary Allowances has emerged. The FDA is proposing to promulgate Reference Daily Intakes to replace the USRDA for a major nutrition labeling revision (*1*). A reality of both RDA's and RDI's and similar nutrient-by-nutrient scientific acumen reflections is their complexity and lack of practical relevance to the consumer for arriving at a balanced and healthy diet.

The dietary intakes of career age (19-50) women in the United States reveals a number of interesting insights. The mean caloric intake is at 1,517 kcal and ranges from 783 kcal to 2,431 kcal. On any given day fifty percent of these women are consuming less than the mean (Figure 7). The implications of low energy intakes upon nutrient status are potentially serious. The following are examples. Fully 70% of women (18-50) are consuming less than the RDA for Vitamin E (Figure 8), Vitamin B6 (Figure 9) and calcium (Figure 10).

In the case of folacin, the RDA (1989) has been lowered to 180 micrograms (from 400). It is noteworthy that a dietary based on the dietary USDA/NCI guidelines would afford 350-375 mg/day - the amount recommended to thwart neurological deformities (neural tube defects) during early pregnancy (*12*). Yet, (Figure 11) 45% of women ingest less than the RDA for folacin and practically all the women are in double jeopardy relevant to the role of folacin in fetal development and the recommendation that the intake at that time should be 400 micrograms (*13*). If one discounted the six percent of folacin intakes or B6 intakes that can be attributed to food fortification (*14*) one realizes that nutrition recommendations for optimal intakes would be enhanced if based upon dietary (food proportion) standards rather than numerical theoretical recommendations based on evidence for minimal needs to thwart deficiency states plus added margins.

Oxidant Stress

Oxidant stress, from both endogenous and exogenous origin, leads to degenerative processes and thus contributes to the development and exacerbation of cancer and chronic diseases. Increased exposure to such oxidants is an integral part of aging.

Antioxidants and free radical scavengers (*15*, *16*), both those recognized as nutrients and those which are not, have to be considered critical components to optimal nutrition. A dietary cannot be solely based upon an RDA/RDI etc philosophy. Nor should nutrition be divorced from food safety. Concerns with naturally occurring toxins, direct and indirect food additives, oxidation of lipids (*17*) and other chemical changes with food processing (*18*) and food preparation need to be counterbalanced with a better understanding of the chemoprevention role many natural chemicals exert. These properties often are labeled nutrient, or biochemical or pharmaceutical depending upon the bias of the investigator. Again, the consumer would benefit from a practical dietary food proportion guideline that does not require technical training.

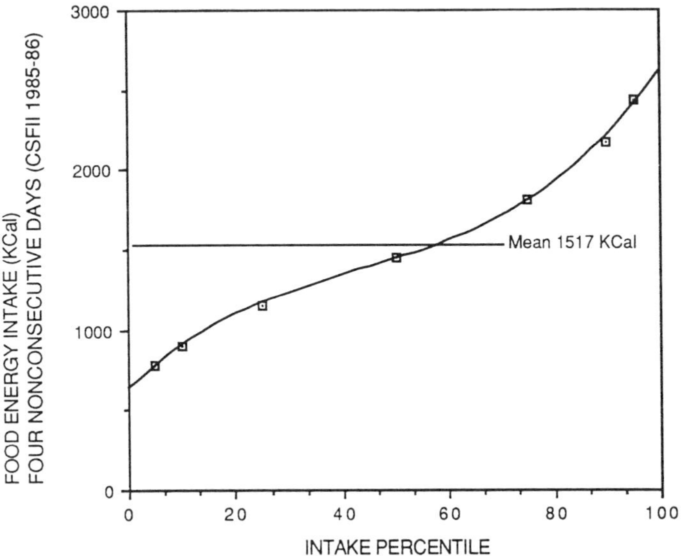

Figure 7. The mean energy intake of women 19-50 years of age was 1517 Kcal. More than 55% of these women consumed less than this mean in four non-consecutive days of the USDA Continuing Survey of Food Intakes by Individuals in 1986. Adapted from ref. # 8.

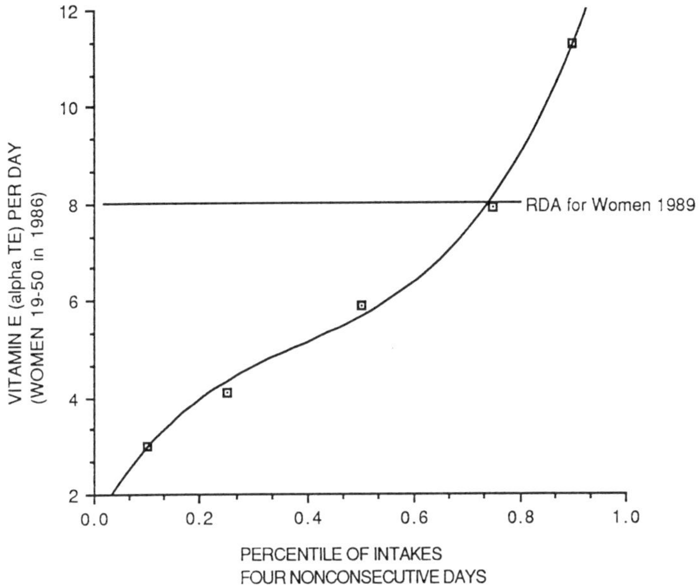

Figure 8. Seventy percent of women 19-50 years of age failed to obtain the RDA for vitamin E on four non-consecutive days of the USDA Continuing Survey of Food Intakes by Individuals in 1986. Adapted from ref. # 24.

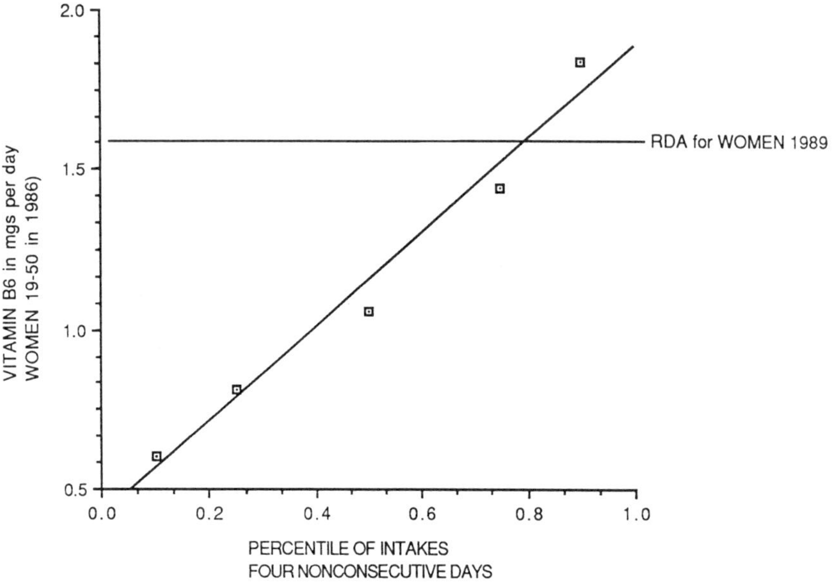

Figure 9. Seventy-five percent of women 19-50 years of age failed to obtain the RDA for vitamin B_6 on four non-consecutive days of the USDA Continuing Survey of Food Intakes by Individuals in 1986. Adapted from ref. # 24.

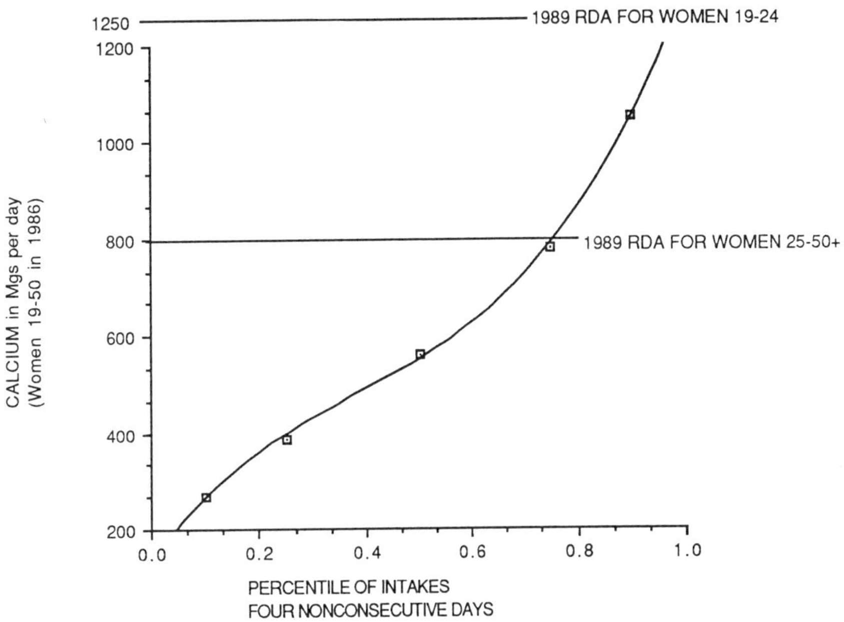

Figure 10. Ninety-five percent of women 19-24 and seventy-three percent of women 25-50 years of age failed to obtain the RDA for calcium on four non-consecutive days of the USDA Continuing Survey of Food Intakes by Individuals in 1986. Adapted from ref. # 24.

Table 1. Percentage contribution of selected food groups to total intake of selected food components for women aged 19-50 years, 4 nonconsecutive days

Food Group	*Total Fat*		*Saturated Fatty Acids*	
	As Reported	*Mixtures Separated*	*As Reported*	*Mixtures Separated*
Meat, poultry, Fish	31	26	30	27
Milk and milk products	15	19	25	33
Eggs	4	4	3	3
Legumes, nuts, seeds	4	4	2	2
Grain products	22	9	20	6
Vegetables	8	6	7	4
Fats and oils	14	30	10	21
Sugars, sweets	1	1	2	2

SOURCE: Reference 20.

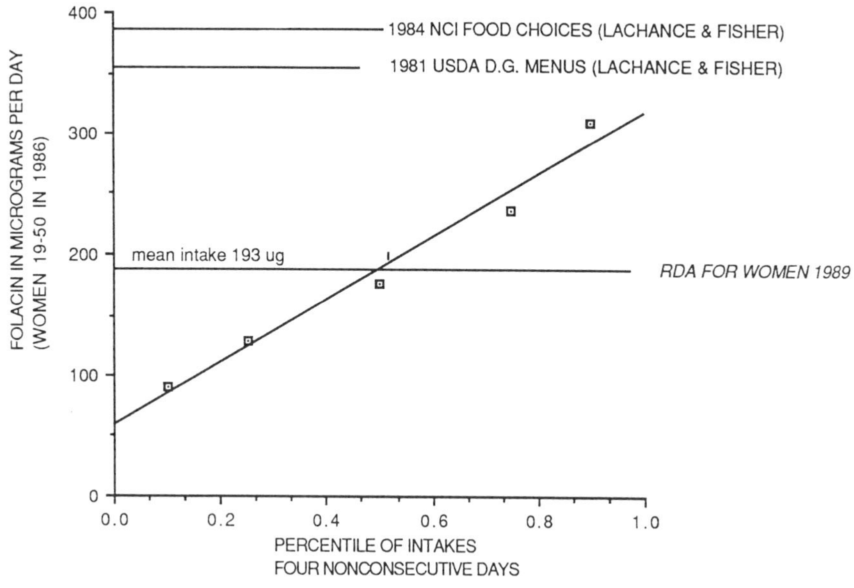

Figure 11. The mean folacin intake of women 19–50 of 193 micrograms per day approximated the 1989 RDA of 200 micrograms in only 55% of women in the USDA Continuing Survey of Food Intakes by Individuals of four non-consecutive days in 1986. None of the women received the folacin levels which would be delivered by the recommended dietaries of the USDA/DHHS Dietary Guidelines or of the National Cancer Institute as calculated by Lachance and Fisher. (Adapted from ref. 19.)

Disease Risk and Eating Patterns

Nearly seventy two percent of deaths (1987) in the USA can be attributed to eight causes with a dietary association namely; heart disease, cancers, strokes, alcohol related accidents and suicides, diabetes, chronic liver disease (and cirrohsis) and atherosclerosis. The major risk factors for these diseases are obesity, high blood pressure, and pollutants especially cigarette smoking. The most prevelant morbidity conditions are osteoporosis and diverticular disease. Both of the latter have dietary associations, namely calcium metabolism and intestinal physiology as affected by dietary fiber respectively.

It should also be self evident that placing emphasis on a disease (tunnel vision medicine) has no advantage over placing emphasis on an associated dietary parameter (tunnel vision nutrition). It is therefore far wiser to advocate dietary recommendations that provide for thwarting as many dietary associated diseases as possible. In order to avoid the politics of agricultural commodity group pressures, the dietary guideline "eat a variety of food" is practically worthless. It can be agreed that few Americans desire to live on one ration. "Away from home" eating practices as well as the increasing choices in the retail marketplace are realistic indicators that Americans demand and eat a variety of foods. What we need to do is define the proportions of the variety needed. It is interesting to note that appendices of the Surgeon General's report listed (a) the historical events of nutritional sciences (1500 BC to 1950) and (b) the Nutrition Policy Initiatives (1862-1988) without mention to the USDA dietary food groups beyond their first issuance in 1917! Americans are <u>not</u> consuming the proper proportion of foods! Every recent national survey has observed this fact (*14*). The "first" dietary guideline needs to explicitly state that Americans should eat 2/3 of their daily food as "plant" foods and only 1/3 as "animal" food. In essence we must reverse the proportions currently consumed. A concomitant guideline -"avoid too much saturated fat"... needs to explicitly state that one serving per day of meat, poultry or dairy product is fully adequate to meet high quality protein needs. This practice will automatically curtail saturated fat intakes. Very low fat dairy products are important sources of key nutrients such as calcium. Consumers have decreased their intake of full fat milk but have increased their intake of high fat cheeses.

"Dietary...saturated fat raises blood cholesterol" (*7*). The major dietary sources of saturated fat in the American diet (of women 19-50) are meat, poultry, fish (27%) dairy products (33%) and fats and oils (21%). Please note that eggs contribute only 3% of saturated fat a quantity analagous to the contribution of vegetables to saturated fat intake in the dietary (Table 1).

Dietaries that follow food group proportions (Table 2) have been issued by both USDA and NCI. These dietaries provide for (1) low calorie combinations that satiate, (2) low saturated fat intakes, (3) high dietary fiber (complex carbohydrates), (4) high dark green/yellow carotenoid and antioxidant foods, and (5) high nutrient density.

Table 2. Health Factors of USDA/NCI Recommended Dietaries

Dietary Health Factor	Recommended Diet Values		
	USDA[a]	HHS(NCI)[b]	RDA(1989) Adult
Calories	1695	1604	>1520
Protein, gm	84 ±8	84 ±5	50-63
Total fat, gm	59 ±6	52 ±6	Not specified
Percent calories from fat	31%	30%	30%
Polyunsaturated fat, gm	15 ±4	12 ±4	Not specified
Saturated fat, gm	19 ±4	17 ±4	<10% of calories
P/S ratio	0.8	0.8	Not specified
Cholesterol, mg	238 ±97	188 ±33	300
Total carbohydrate, gm	216 ±15	212 ±12	>200
Fiber, gm	19.5 ±2.4	22.2±2.2	Not specified
Total vitamin A activity, mg	9689	1118 3	800-1000 RE
Preformed vitamin A, iu	919	1018	Not specified
Provit A (carotine), mg	5.2	6.0	Not specified
Percent Provit A (carotine)	90.5	90.0	Not specified
Vitamin E, total	27	23	8-10
Vitamin C, mg	225	217	60
Thiamin (B_1), mg	1.7	1.6	1.1-1.5
Riboflavin (B_2), mg	1.9	1.8	1.3-1.7
Niacin (B_3), mg	24	24	15-19
Vitamin B_6, mg	1.	1.3	1.6-2.0
Vitamin B_{12}, mg	3.2	2.9	2.0
Folic acid, mg	353	381	180-200
Calcium, mg	1004	1017	800
Phosphorus, mg	1371	1420	800
Sodium, mg	1887	1955	>500
Potassium, mg	3464	3480	>2000
Magnesium, mg	362	388	280-350
Iron, mg	14	14	10-15
Zinc, mg	13	13	12-15

[a]SOURCE: Reference 21.
[b]SOURCE: Reference 22.

Summary

To thwart disease and enhance the quality of life, Americans do not need (1) *tunnel vision medicine* — Medicine is stuck in the financial benefit of specialization or (2) *tunnel vision nutrition* — Nutrition begins at the lips and the chemical we call nutrients are invisible because we eat food, not nutrients.

Americans do need: (1) simple yet operable dietary (food type) guidelines which promote plant foods over animal foods and (2) simple health indices relevant to risk factors and thus how to best avoid and/or control risk factors and optimize health.

Acknowledgment

A contribution of the NJ Agricultural Experiment Station 10405-90.

Literature Cited

1. FDA *Federal Register* **1990**, *55*(139), 29476-29533.
2. U.S. Senate, Select Committe on Nutrition and Human Needs *Dietary Goals*, Second Edition; U.S. Government Printing Office: Washington DC., 1977, 83 pp.
3. Committee on Diet and Health, Food and Nutrition Board, NRC *Diet and Health:Implications for Reducing Chronic Disease Risk*; National Academy Press: Washington, DC, 1989, 749 pp.
4. American Chemical Society 1990 Symposium on Chemophobia-Realities and Prescriptions.
5. Abelson, Philip H *Science* **1990**, *249*, 225.
6. U.S. Bureau of the Census *Current Population Reports Series P-20* 1985, #424.
7. Surgeon General Report on Nutrition and Health, DHHS (PHS) Publication V #88-50211; U.S. Government Printing Office: Washington, DC., 1988.
8. DHHS/USDA Nutrition Monitoring in the United States - An Update Report on Nutrition Monitoring. DHHS Publication No. (PHS) 89-1255 Washington, DC, 1989.
9. Woteki, C., Johnson, and R. Murphy In *What is America Eating?*; National Academy Press: Washington, DC, 1986.
10. Lachance, P.A. *Clinical Nutrition* **1988**, *7*, 118-122.
11. Shekelle, R.B., Lepper, M., Liu, S., et al *Lancet* **1981**, *2*, 1185-1190.
12. Milunsky, A., Hershel, J., Jick, S.S., Bruell, C.L., MacLaughlin, D.S., Rothman, K.J., and W. Willett *JAMA 262* **1989**, *20*, 2847-52.
13. Institute of Medicine *Nutrition during pregnancy.* National Academy Press: Washington, DC, 1990.
14. Lachance, P.A. *Food Technol.* **1989**, *43(4),* 144-150.
15. Balz, F., England, L., and B.N. Ames *Proc. Natl. Acad. Sci. USA* **1989,** *86*, 6377-6381.

16. Esterbauer, H., Striegl, G., Puhl, H., Oberreither, S., Rotheneder, M., El-Saadani, M., Jurgens, G. *Annals of the New York Acadamy of Sciences* **1989**, *570*, 254-267.
17. Addis, P.B. and Park, S-W. In *Food Toxicology - a perspective on the relative risk*; Taylor, S.L. and R.A. Scanlan, Eds.; Marcel Dekker, Inc.: New York, 1989, pp 297-330.
18. Richardson, T. and Finley, J.W. Eds. *Chemical Changes in Food During Processing;* Avi. Publishing Co.: Westport, CT., 1985.
19. Lachance, P.A. and Fisher, M.C. *Calculated nutritive value of USDA Dietary Guideline menus and NCI Food Choices* Unpublished, 1988.
20. Krebs-Smith, Cronin, and Haytowitz *Food Sources of Energy Yielding Nutrients Cholesterol and Fiber in Diets of Women*; Prepared for the Expert Panel on Nutrition. (Tables 5-3 in Reference 8).
21. US Department of Agriculture. *Ideas for better eating —Menus and recipes to make use of dietary guidelines*; US Government Printing Office: Washington, DC., 1981.
22. US Department of Health and Human Services, National Cancer Institute, National Institutes of Health. *Diet, nutrition and cancer prevention; A guide to food choices* (NIH publication No. 85-2711). Washington, DC: US Government Printing Office, 1984.
23. Lachance, P.A. *Food Technology* **1981**, *35*(3), 58.
24. US Department of Agriculture, Nationwide Food Consumption Survey of Food Intakes by Individuals. Report No. 86-3. Women 19-50 years and their children 1-5 years, 4 days. 1986 (issued Sept. 1988) USDA, Nutrition Monitoring Division, Hyattsville, Maryland.

RECEIVED August 15, 1991

Chapter 27

Diet and Carcinogenesis

John A. Milner

Department of Nutrition, The Pennsylvania State University, University Park, PA 16803

Etiological relationships between dietary practices and the risk of cancer are supported by data obtained from a variety of sources. Extensive laboratory, epidemiological and clinical data support modification of the typical American diet as a means of reducing cancer risk. Experimentally, the dietary intake of several macro- and micro-constituents significantly alters cancer incidence and severity. Specific dietary constituents can alter the formation and bioactivation of carcinogens, modify the promotion of neoplastic cells, lead to variation in rates of tumor growth and modulate immunocompetence. In humans, variations in the relationship between dietary practices and cancer risk may relate to the numerous and complex interactions known to exist among dietary constituents and to the specific cancer examined. Research aimed at establishing the specific role and interactions of nutrients, and non-nutrients, in the carcinogenic process will assist in the identification of critical times for intervention and will lead to sound and accurate dietary advice that is tailored to individual needs. New or novel foods containing non-traditional quantities of individual nutrients and non-nutrients offer exciting opportunities for producers to develop additional food choices for the consumer.

Diet and Cancer Risk

Cancer is no longer viewed as an inevitable consequence of aging. Changes in death rates over a relatively short period of time point to significant environmental factors, rather than to genetic predisposition, as determinants of cancer risk. Environmental factors correlate with approximately 90% of all cancer cases. Within the environmental factors, dietary practice is one which likely modifies cancer risk. Wynder and Gori (*1*) implicated food factors in 60% of cancers in women and in more than 40% in men. Geographic correlations of per capita intake of fat and selenium are examples of observed relationships between dietary components and cancer risk (*1-5*). While these

0097–6156/92/0484–0297$06.00/0

and other geographical correlations of dietary intakes and cancer risk are insufficient to establish a causal relationship, they are nonetheless useful in the generation of testable hypotheses. Variability observed in detecting the significance of dietary practices on cancer risk is not surprising and is explicable, due to a number of environmental and genetic factors. Dietary relationships with cancer are not solely dependent on the presence or absence of carcinogens in the food supply, but likely depend on the impact of nutrients and non-nutrients on the complex and poorly understood cancer process. Given the multitude of complex interactions that are possible among dietary nutrients and non-nutrients, it is not surprising that inconsistencies exist in the relationship between dietary practices and cancer risk. Likewise, since all metabolic and phenotypic characteristics are determined by one's hereditary material, it is inconceivable that a simple change in diet will substantially alter the cancer susceptibility of all individuals. Additional information is desperately needed with regards to the interrelationships between genetics and diet. Such information should assist in tailoring dietary recommendations.

Some of the strongest evidence supporting the role of nutrition in the development of cancer comes from studies of migrant populations (*6-8*). Generally, the cancer patterns of migrants shift from that of their native countries to that observed among residents of the host country. While these shifts could reflect environmental changes or lifestyles, it is generally concluded that pollution and food contaminations do not explain the observed differences in cancer incidence. Additional evidence that diet, rather than other environmental or genetic factors, is involved in the development of cancer risk comes from studies of homogeneous populations living and working in the same environment as their cohorts; though exposed to the same environmental pollutants, all have markedly different rates of cancer (*9-12*).

While it has long been recognized that dietary habits can alter the course of tumor formation and development, only in recent years have investigators actively examined specific dietary constituents for their impact on the cancer process. Table 1 indicates some of the dietary factors known to alter significantly carcinogenesis in experimental animals. During recent years considerable attention has been given to the influence of dietary intakes of macro-constituents as factors in cancer development. While these dietary constituents are likely suspects, much less attention has been given to nutrients and non-nutrients within the diet that are consumed in much smaller quantities. Although limitations exist in knowledge concerning the precise role of individual food constituents in carcinogenesis, the issue was addressed by in the recently released Surgeon General's Report Nu*trition and Health* and the National Academy of Science's report on *Diet and Health.* (*11*, *12*). Continued examination of dietary constituents and their interactions will likely result in more specific dietary recommendations for the general population and indicate critical times where nutritional intervention may be most beneficial.

This review will briefly summarize the impact of several major and minor dietary constituents on cancer risk. This review is limited to (1) those nutrients and non-

Table 1. Incomplete List of Dietary Factors Known to Alter Experimental Carcinogenesis

General	*Amino Acids*	*Lipids*	*Vitamins*	*Minerals*
Calories	Methionine	Linoleic	Riboflavin	Calcium
Lipids	Cysteine	ω-3	Folic acid	Zinc
Carbohydrates	Tryptophan	Stearic	B_{12}	Copper
Proteins	Arginine		Vitamin A, C	Iron
Fiber			Vitamin D, E	Selenium
Alcohol			Choline	Iodine
			Niacin	
			Pyridoxine	

nutrients about which sufficient information is available to verify an effect upon the process of carcinogenesis; and (2) generally to those dietary constituents in which possible mechanisms can be explored. The complex interrelationships existing between nutrients and non-nutrients in practical diets make it extremely difficult to generalize about the impact of specific dietary constituents on cancer risk. It must be emphasized that purified constituents are not consumed and that single nutrients must be considered as part of a complex diet. The continued examination of diet as a factor in cancer risk will hopefully result in the ability to make dietary recommendations tailored to an individual's dietary habits and needs.

Nutritionists have long recognized that the quantity of a dietary constituent does not dictate relative importance in either disease prevention or treatment. Thus the consumption of several micronutrients and non-nutrients may be as significant on cancer incidence as the intake of the macro-constituents. Examination of the interrelationships among the various constituents of the diet should likewise assist in explaining some of the observed variation in cancer risk.

The Carcinogenesis Process

Cancers represent unique cell populations that have acquired the ability to multiply and spread without normal restraints. The two predominant theories of the cancer process are: (a) the initiation-promotion hypothesis (*9*) and (b) the electrophilic hypothesis (*13*, *14*). The former proposes that carcinogenesis occurs in three, or more, discrete and distinct phases: (a) initiation; (b) promotion and (c) progression. The initiation phase represents an early, rapid and largely irreversible change in the hereditary material, culminating in a permanently altered cell. The promotion phase of carcinogenesis constitutes a more gradual process(es), during which an initiated cell is converted or transformed into a tumor cell. The spread of transformed cells, via the blood or lymph, to distant sites constitutes the progression phase of carcinogenesis. The electrophilic hypothesis of carcinogenesis concludes that metabolic activation of most carcinogens occurs through reactive, electrophilic intermediates, which, by altering critical cell nucleophilic targets such as DNA, are directly responsible for the carcinogenic activity of a compound. Evidence for the involvement of free radicals in carcinogenesis is also available (*15*). The immune system is generally thought to influence the host's resistance to cancer, both prophylactically, by destroying neoplastic cells, and protectively, by retarding the growth of established tumors.

The process of carcinogenesis may be separated both mechanistically and temporally. Normally during the initiation phase some critical factor(s) within the cell is(are) rapidly modified. While cells can frequently repair some defects, others are unable to remove the damage and, thus, either die or become transformed. For nutrients to impact this phase, they must modify the biological behavior of the cell prior to or concurrent with the time of modification. In contrast with the initiation phase's short duration, the promotion phase is a lengthier process which is considered to last decades in human beings. In this extremely complex phase, a variety of cellular changes occur. As discussed below, several nutrients can influence the early and late stages of promotion. Occasionally, early pathological changes appear to be reversed by some dietary constituents, such as ascorbic acid and the retinoids. However, it is unknown if these changes represent a complete reversibility to normalcy. To date little information is available on the effect of specific nutrients on the progression of malignant cells to cells with increasing invasive or metastatic properties.

It is impossible to extrapolate data obtained with animal models to human risk, conclusively, due to problems in evaluating the influence of dosages, metabolism, physiology, etc. Nevertheless, there is overwhelming evidence that human cancers develop at the cellular level, with considerable similarity to those observed in experimental animals. These similarities became evident when agents known to cause cancers in humans were found to produce tumors in animals (*3*, *16*). Similarities in the cancer process are also made evident by the presence of similar classes of oncogenes in tumors of rodents and human beings (*17*). Increased cell division induced by either external or internal stimuli appears to be a common denominator in the pathogenesis of cancer.

Carcinogenic Exposure and Metabolic Activation

Human beings are inevitably exposed to compounds which are not essential for life or not "natural," from the standpoint of evolution. These compounds may be acutely toxic or potentially toxic, following activation; or they may exhibit long-term effects, such as cancer development. The ability of the diet to modify the formation of potentially carcinogenic agents is indicated by studies demonstrating that vitamin C can reduce the formation of N-nitrosamines (*18*). Most N-nitrosamines are recognized as being capable of inducing tumors (*19*). While dietary ascorbic acid may significantly depress the formation of this class of carcinogen (*20*), other constituents may promote cancer development. Some phenols, such as ones present in vegetables, or thiocyanate and iodide, may enhance cancer development by stimulating nitrosation and, consequently, the formation of carcinogenic nitrosoamines (*21*).

Largely due to the pioneering work of Miller and Miller (*22*), substantial evidence exists that drug-metabolizing enzymes play an essential role in the bioactivation of the carcinogens to biologically reactive carcinogenic intermediates. Dietary nutrients may modify cancer development by modifying the formation of specific carcinogens, by altering the metabolic activation of carcinogens, or, possibly, by changing the occurrence of cocarcinogens or anticarcinogens. The ability of selective nutrients, such as vitamin A and selenium, to modify the metabolic activation of carcinogens is recognized and is discussed below. Similarly, several non-nutrients may shift the metabolism of carcinogens. The metabolism of benzo(a)pyrene by non-nutrients, such as butylated hydroxytoluene, has been found to lead to a reduction in cancer in experimental animals (*23,24*). Several minor constituents of commonly consumed plant foods, such as flavones, dithiothiones, thioethers, isothiocyanates, phenols and indoles, have also been reported to have the ability to alter carcinogens' metabolic activation and, consequently, tumor risk (*24-27*). The inverse association between the risk of cancer of the gastrointestinal tract and the consumption of selected vegetables, particularly the cruciferae, suggests a possible role for some of these non-nutrient dietary constituents in the bioactivation of carcinogens and, ultimately, in the risk of human beings to cancer.

Mechanism of Action of Selective Nutrients and Non-nutrients

Caloric Intake. Caloric restriction has long been recognized as a method of inhibiting or significantly delaying the formation and/or development of tumors (*28-34*). Caloric restriction inhibits both chemically-induced and spontaneous cancer, including cancers of the skin, breast, lung, liver, colon, pancreas, muscle, lymphatic system, and endocrine systems. Generally, the degree of caloric restriction determines the magnitude of the depression of tumorigenesis. Tannenbaum proposed that a sigmoid-like response best characterized the relationship between caloric restriction and tumorigenesis (*31*). In those pioneering studies, the latency period for tumor appearance was frequently delayed by rather modest caloric restriction. However, overall tumor numbers were not altered until restrictions became rather severe. Although the impact of caloric restriction depends on the type of tumor examined, a 50% decrease in food intake generally led to a depression in tumorigenesis and, frequently, to an increased longevity.

Considerable evidence indicates the ability of caloric restriction to alter the promotion, rather than the initiation, phase of carcinogenesis (*34*). Although the mechanism by which caloric restriction inhibits tumorigenesis is unknown, a direct effect is unlikely. Changes in tumor development may result from marked physiological changes that occur as a result of inadequate caloric intake. Alterations in hormonal regulation and/or depressed tissue mitotic activity may account for the ability of caloric restriction to depress tumorigenesis (*35-37*).

The ability of ad *libitum* feeding to increase the incidence of chemically or virally induced tumors has been ascribed to excess calories. It is clear that all animals, including human beings, do not consume the same quantity of calories when free choice of food is provided. Studies by Clinton and Visek (*38*) demonstrated that tumor frequency in *ad libitum* fed rats, previously treated with a mammary carcinogen, was dependent upon total caloric intake, regardless of the diet's actual composition. Furthermore, in animal studies tumor development does not necessarily correlate with changes in body composition (*37*). Thus, considerable evidence suggests that subtle changes in energy metabolism may have a significant impact on tumorigenesis. In human beings the lowest overall cancer mortality is generally observed in individuals whose body weights ranged from approximately 10% below to 20% above the average for their age and height. While not particularly novel, it remains advisable to eat a variety of foods in moderation.

Dietary Fat. Epidemiological studies reveal a significant, positive association between dietary fat intake and mortality from cancer at several sites (*10*, *39-42*). However, liver and stomach cancers do not correlate with dietary lipid intake (*43*). The site of action of dietary lipids on the cancer process likely depends, at least in part, on the quantity and types of fatty acids esterified to glycerol. Laboratory investigations typically reveal that increased lipid intake is associated with an increased tumor frequency, regardless of the form consumed. Nevertheless, polyunsaturated fats tend to enhance tumor yields to a greater degree than does saturated fat (*44*). Interestingly, epidemiological data frequently reveal a strong positive relationship between cancer susceptibility and saturated fat intake, whereas they demonstrate little or no relationship between cancer susceptibility and unsaturated fat intake. This discrepancy between epidemiological and laboratory investigations may relate to the minimal quantity of unsaturated fatty acids required to enhance tumorigenesis. Furthermore, total caloric intake may be more important as a risk factor than the source of the calories. Recent findings with ω-3 fatty acids raise new issues about the overgeneralization of all unsaturated fatty acids as potentially detrimental (*45*, *46*).

Consumption of a high fat diet was recognized more than 50 years ago as a promoter of skin tumors in mice, at a greater frequency than that occurring in animals fed a low fat diet. Since that time numerous studies have documented the synergistic ability of dietary fat to promote tumorigenesis, following treatment with a wide variety of carcinogens (*44*). Although the initiation phase of carcinogenesis has not been examined thoroughly, limited evidence suggests that this phase is altered by the quantity and type of lipid consumed. While the exact mechanism is unknown, changes in carcinogen metabolism or membrane fluidity are logical sites of modification induced by dietary lipids. Studies that have examined high fat diets while controlling caloric intake have generally detected an alteration in the promotional effect of lipid on tumorigenesis (*44*). Alterations in membrane fluidity, hormonal milieu, immunocompetence, biologically active intermediates or formation of fatty acid metabolism by-products may account for lipid-induced carcinogenesis (*47-49*).

Additional evidence for the ability of lipids to alter the promotion phase of carcinogenesis comes from studies of caloric restriction. While caloric restriction appears to inhibit all types of tumors, the inhibitory effect of restricting dietary fat appears to be more selective. Consumption of high fat diets has been reported to have little influence on the incidence of lung tumors or leukemia (*28*).

The influence of dietary cholesterol on cancer incidence is difficult, if not impossible, to determine. While some studies suggest low plasma concentrations of cholesterol are a risk factor for cancer, it must be emphasized that plasma cholesterol reflects many factors, and not merely dietary intake. Considerably more information is needed before firm conclusions can be made about the relationship between cancer and either blood or dietary cholesterol levels.

In view of the vast number of experiments that have examined the relationship between dietary fat intake and cancer risk, it is unfortunate that more specific recommendations cannot be made to the consumer. Reducing the contribution of fat to total caloric intake from the present day 38% to 30% or less appears to be prudent until additional information becomes available. A balanced intake of saturated, monounsaturated and polyunsaturated fatty acids also appears prudent until more precise information can be gained that assesses the long term benefits/risks of altering the proportion of dietary fatty acids.

Protein. The impact of dietary protein or specific amino acids on the cancer process has received far less attention than has been given to several other dietary constituents. The interpretation of epidemiological studies suggesting that the intake of high protein foods positively correlates with cancer are generally complicated by the simultaneous presence of excess fat. The information that is available on the influence of protein and amino acids on cancer risk comes primarily from laboratory investigations. In one of the more detailed studies, increasing the dietary protein concentration from 9% to 45% was found to have minimal effects on cancer, as indicated by the development of mammary, skin or skeletal muscle tumors (*50*). However, Silverstone and Tannenbaum (*51*) observed that a low protein diet (9%) suppressed the development of spontaneous hepatomas. Studies that have examined the impact of dietary protein on chemically induced tumors have generally revealed that tumor formation and growth are depressed when diets contain limiting quantities of protein or essential amino acids. Changes in drug-metabolizing enzymes are proposed to account for the observed inhibitor effect of inadequate protein or amino acid intake on tumor induction (*52*).

The impact of dietary protein likely depends upon its content of individual amino acids. While deficiency of an essential amino acid typically retards tumor growth, the specificity of this effect is questionable. The impact of supplemental amino acids on cancer risk has not been as thoroughly evaluated. Nevertheless, some amino acids have been shown to modify the incidence of chemically induced tumors or to modify the growth of transplantable tumors (*53, 54*). A significant body of literature suggests that arginine supplementation reduces tumorigenesis in experimental animals (*54*). The mechanism by which arginine alters tumor growth is unknown, but it may relate to hormonal or immunological changes (*54, 55*).

Carbohydrates. The association between dietary fiber and cancer has received widespread attention in recent years. A variety of approaches has been used to address this relationship. Mixed results about the association between cereal consumption and several types of cancer have been reported, although the vast majority of studies indicated the protective effect of foods containing fiber (*56*). Likewise, animal studies are equally inconsistent with regards to the impact of dietary fiber on colon cancer risk. At least part of the variation is explained by qualitative and quantitative differences in the fiber sources, animal strains examined, and duration of the experiment (*57*). Fiber's possible effect may be to reduce transit time in the bowel, thereby reducing the time the bowel is in contact with potential carcinogens; to alter the intestinal microflora; to bind potentially carcinogenic agents; or to dilute toxic compounds, by virtue of its hydrophilic property (*56*).

Vitamin A. Interest in retinol has stemmed from several epidemiologic investigations that have observed an inverse relationship between vitamin A intake and cancer risk. It has been recognized, for many years, that vitamin A deficiency increases the susceptibility of experimental animals to some chemically-induced tumors; more recently, it has been recognized as increasing susceptibility to tumor incidence associated with viral induction (*57-61*). Carcinogen treatment in many cases markedly reduces liver stores of vitamin A. Marginal deficiencies may therefore become evident following carcinogen exposure and thereby facilitate cellular transformation. Interestingly, several studies have reported an inverse association between hepatic and plasma vitamin A concentrations with vitamin C levels. The cause of this increase in vitamin C is unknown but may relate the compensatory mechanisms needed to resist changes in metabolism, as indicated below.

Retinoids is a term coined to refer to both natural and synthetic analogues of retinol. The vitamin A activity of the retinoids is known to depend on specific structural characteristics. Nevertheless, several retinoids are known to be effective in inhibiting chemical carcinogenesis in the skin, mammary gland, esophagus, respiratory tract, pancreas, and urinary bladder of experimental animals, in some cases with compounds that do not possess vitamin A activity (*59-61*). Retinoids appear to be most efficient when provided continuously (*62*). While retinoids are generally more effective when administered shortly after carcinogen treatment, delaying the treatment frequently results in at least partial protection (*63*). However, a critical point can be exceeded when retinoids are no longer effective. These data suggest that retinoids may function in inhibiting the early phase of the promotion stage of carcinogenesis (*63*). Among laboratories inconsistencies in data on the biological effects of retinoids may come as a result of investigators choosing, for examination, inappropriate retinoids, dosage administered, species tested or initiating carcinogen.

Dietary β-carotene may also be as important in cancer prevention as preformed retinol (*64-65*). At present it is not clear if vitamin A activity or the presence of constituent groups similar to those found in retinol, such as unsaturated double bonds or the aromatic characteristic, are the most important determinants of the retinoids' efficacy. Nevertheless, it is known that at least one retinoid without vitamin A activity is capable of reducing UV induced skin cancer (*66*). Interestingly, at least some retinoids enhance carcinogenesis during some experimental circumstances (*67*). At present it is difficult to generalize about the benefits of vitamin A and retinoid supplements.

The mechanism(s) by which retinoids inhibit experimental carcinogenesis remains largely unknown. While vitamin A deficiency enhances carcinogenesis and vitamin A excess inhibits it, there is little evidence that a continuum exists between the deficiency and excess states. Thus, the increased risks of animals to cancer during a deficiency state are likely mechanistically and physiologically different from the prophylactic and therapeutic effects of near-toxic dosages of retinoids. Several vitamin A compounds and analogs are known to inhibit the *in vitro* microsomal mixed function oxidases that metabolize carcinogenic compounds (*68*). Nevertheless, the greatest effect appears to involve the promotional phase of tumor growth (*63*,*69*,*70*).

The ability of vitamin A and the retinoids to regulate cellular differentiation is often cited as their mechanism of action in reducing cancer risk. The importance of this action is emphasized by the fact that most primary cancers in human beings arise in epithelial tissues dependent upon vitamin A for differentiation. However, vitamin A and/or the retinoids may also modify preneoplastic and/or neoplastic cells by blocking cellular division or enhancing cellular destruction. In some tissues, retinoids can inhibit differentiation, proliferation and DNA synthesis (*70*, *71*). No single mechanism appears to examine all of the actions of vitamin A and the retinoids. Of the various theories accounting for the effects of retinoids upon cancer promotion, the following have received the greatest attention: (A) retinoids participate in sugar transfer reactions by means of the intermediate retinyl phosphate mannose, which plays a critical role in controlling differentiation and carcinogenesis; (B) retinoids control gene expression by interacting with cyclic AMP-dependent and -independent protein kinase; and (C) retinoids control gene expression by mediating specific intracellular binding proteins in a manner analogous to that observed in steroid hormones. Data are available which both support and refute each of these hypotheses, as recently reviewed by Roberts and Sporn (*72*).

Vitamin A has long been known to be associated with immunocompetence, as it alters both humoral and cell-mediated immunity (*73*). In cell-mediated immunity, retinoids not only influence the total T-cell population but also stimulate cellular lysis and growth inhibition. During the early processing of the antigen by macrophages during humoral immunity, retinoids may directly interact with the B-cell or activate the B-cell by way of an increased number of T-helper cells (*73*).

Significant process has been made in the understanding of the role of vitamin A/ retinoids in the cancer process. Future research will hopefully identify under what circumstances and by which mechanism(s) these anticarcinogenic agents function.

Vitamin C. Vitamin C has been proposed as a protective against cancer for over 50 years. In 1936 Eufinger and Gaehtgens (*74*) proposed that vitamin C influenced the pattern of pathologically-modified white blood cells and described the successful treatment of a patient with myeloid leukemia. Since then, vitamin C has been implicated in human cancer prevention largely upon the basis of several epidemiologic studies showing that consumption of foods containing high concentrations of this vitamin is associated with a reduction in cancer incidence, particularly cancer incidence in the stomach and esophagus (*75*,*76*). Results from several clinical studies have provided inconclusive evidence about the specific role of vitamin C in human carcinogenesis.

Experimental data from animals and cell culture systems suggest several mechanisms by which vitamin C might function in cancer prevention. An important mechanism by which vitamin C may interfere with cancer development is to inhibit N-nitrosamine formation. N-nitrosamines' carcinogenicity has been recognized for years. Vitamin C effectively competes for nitrite, thereby inhibiting its reaction with amines or amides to form carcinogenic nitroso compounds (*77,78*). Reed and co-workers (*79*) demonstrated that vitamin C treatment reduced N-nitroso compound formation in a selected group of human subjects at risk for gastric cancer. Supplementation of the diet with ascorbic acid and a-tocopherol was found to reduce significantly the mutagenic compounds excreted in human feces, suggesting that antioxidants in the diet may lower the body's exposure to endogenously formed mutagens (*80*).

Glatthaar, et al. (*76*) have reviewed the mechanisms by which vitamin C may be involved in cancer development. The potential for bioactivation of carcinogens by microsomal hydroxylation and demethylation systems, and the associated electron transport protein components, such as cytochrome P-450, have been observed in animal studies to decrease during conditions of ascorbic acid depletion (*81*). Overall, the protection against carcinogen-induced neoplasms that is provided by supplemental ascorbic acid is not impressive, except in experiments where nitroso-compound formation is involved.

The ability of vitamin C to modify tumor cells has been documented through studies examining synergism with various drugs, inhibition of the action of cytotoxic drugs and interference with tumor cell metabolism (*76,82,83*). Ascorbate may react with free copper ions, thereby leading to enhanced peroxide formation. While such metabolic effects may account for the inhibition of some tumor cells, the effectiveness of vitamin C may depend upon cellular catalase and peroxidase activities (*76*). Under some conditions it is possible to cause a reversion of chemically transformed cells to apparently normal morphological phenotypes, simply by adding ascorbic acid (*82*). Recent studies suggest that vitamin C enhances the immune function, thereby reducing the risk of certain types of cancer (*83*). As a biologic redox reagent, vitamin C can interfere with oxidative processes during the functional stimulation of polymorphonuclear leukocytes and macrophages involving, among other functions, chemotaxis and phagocytosis (*83*).

Vitamin D and Calcium. Recently 1,25 dihydroxyvitamin D3 has been shown to inhibit the proliferation of some neoplastic cells, including the differentiation of murine and human myeloid leukemia cells in vitro (*84-86*). Suppression of growth of these malignant cells appears to depend upon a receptor-mediated process. Such results point to a previously unsuspected involvement of vitamin D in cell proliferation and differentiation and suggest that analogs of the vitamin D may possibly act as therapeutic agents in the treatment of malignancy. The mechanism of vitamin D's action in the induction of differentiation of the neoplastic cells remains largely unknown. Ornithine decarboxylase induction by vitamin D in intestinal cells indicates an involvement of the polyamine biosynthesis pathways (*87*). However, Wood et al. (*88*) suggest that vitamin D is effective in inhibiting papillomas induced by methylcholanthrene, either before or after treatment with phorbol esters, suggesting that some other mechanism is involved in the action of this vitamin.

The function of vitamin D in differentiation may relate to its action on intracellular calcium metabolism. Several studies provide indirect evidence of calcium's possible involvement in carcinogenesis, based on calcium's effects on the activity and pharmokinetics of carcinogens and promoters and on the ability of carcinogen or tumor proliferation to induce disturbances in calcium homeostasis (*89,90*). Whatever the mechanism, calcium seems to have an active role in the promotion of, rather than the initiation of, carcinogenesis. The association of calcium-activated oxygen release and the expression of tumor promoters is of considerable interest (*91*). Normal epithelial cells cease to proliferate or differentiate in calcium-deprived media; however, under the influence of carcinogens or promoters, cells may become resistant to the regulatory calcium signals and proliferate.

Vitamin E. Vitamin E (α-tocopherol) is the major radical trap in lipid membranes. Few studies have adequately evaluated the role of the lipid phase antioxidants in experimental carcinogenesis. However, their potential role is supported by the hypothesis that cellular damage produced by active oxygen contributes to the promotional phase of carcinogenesis (*15*) and that antioxidants such as a-tocopherol can, at times, protect against this damage. It has been shown that consumption of a diet devoid of vitamin E for an extended period alters the mixed function oxidase system required for carcinogen activation (*92*). Additionally, it has been demonstrated that a-tocopherol inhibits endogenous nitrosation reactions leading to the formation of carcinogenic nitrosamines, implying that if nitrosation is a determinant of human cancer, then a-tocopherol should be protective. Support for this hypothesis comes from the observation that administering vitamin E and ascorbic acid to volunteers consuming a Western diet causes a dramatic reduction in fecal mutagenicity (*80*). The anticarcinogenic effects of vitamin E have largely been observed with extremely high and nonphysiological concentrations of this vitamin (*93, 94*). While the observed effects may relate to reduced lipid peroxidation, other studies suggest that vitamin E may also be involved in cellular mitosis and DNA production (*95*). Data by Ip and Horvath (*96*) suggest that vitamin E may potentiate the inhibitory effects of selenium on the promotion or proliferative phases of carcinogenesis.

Choline and Other B Vitamins. Choline deficiency produces pathologic lesions in virtually every organ. A dietary deficiency of this vitamin is known to enhance the initiating potency of several carcinogens, possibly by modifying the promotion phase of carcinogenesis (*97-99*). Interestingly, choline deficiency may directly impact tumor formation (*97*). Chronic feeding of a deficient diet may increase hepatic tumors by allowing for the expression of naturally occurring preneoplastic lesions. Choline deficiency is known to enhance liver cell proliferation, reduce the supply of methyl groups and cause hypomethylation of DNA (*97,100*). Enzymatic methylation of DNA is recognized as a mechanism for genetic control. Hypomethylation resulting from choline deficiency may result in oncogene activation (*101*). Continued research of this nutrient is likely to supply valuable information on the mechanism of carcinogenesis.

Several B vitamin deficiencies are known to reduce the growth rate of tumor cells and to interfere with the normal functioning of the organism. Since these vitamins are essential components of any adequate diet and are necessary for the continued maintenance of cellular integrity and metabolic function, such results are not surprising. Inadequate information is available to evaluate broadly the impact of B vitamin intake on the cancer process. B vitamins, in general, may function in carcinogenesis by modulating cellular processes, including growth and immunosurveillance. The ability of a deficiency of either vitamin B_{12} or folic acid to enhance the carcinogenicity of several chemicals suggests that B vitamins may be anticarcinogenic (*102-104*). A deficiency of either vitamin B_{12} or folic acid may result in a hypomethylation, as is the case during choline deficiency, and, subsequently, in the activation of oncogenes. Failure of the immune system to defend the host may also account for increased tumor development that is observed in cases of deficiencies of these vitamins (*105*). Folic acid and choline metabolism are also interrelated. Perturbations in one lead to alterations in the other. Continued examinations of these vitamins and their interrelationships should add valuable information to the understanding of the cancer process.

Iron, Zinc and Copper. Iron deficiency has been correlated with cancers of the upper alimentary tract. Iron deficiency in experimental animals is associated with a fatty liver and with a decreased delay in the time of onset of liver tumors (*106*) Nevertheless, insufficient information is available to evaluate adequately the importance of dietary iron intakes in cancer risk.

Copper is an essential nutrient implicated as a positive factor in cancer susceptibility. Although some clinical and epidemiological studies have suggested a direct relationship betweem plasma copper and cancer risks, little evidence is available relating dietary intakes of copper to cancer development. Experimentally, large dosages of copper tend to protect against chemically induced tumors (*107*). The influence of normal intakes of copper does not appear to have been examined adequately.

Zinc, an essential constituent of numerous enzymes, is known to function in cell replication and tissue repair. The impact of excess zinc on chemically induced tumors has been mixed, some studies showing increased tumor growth, while others report decreased tumor development (*108, 109*). Zinc deficiency is also known to modify the growth of neoplasms (*110*). Whether these effects on the growth of established tumors are a result of alterations in other nutrients, changes in the immune system or a specific effect on cell proliferation, has not been completely resolved. Several studies reveal important interactions between zinc and vitamin A. Zinc deficiency influences retinaldehyde reductase activity, resulting in a change in the oxidation of retinaldehyde to retinoic acid. Indirectly, zinc may also affect vitamin A homeostasis through a variety of zinc-dependent enzymes.

Selenium. The ability of various forms of selenium to inhibit experimentally induced tumors is well documented (*111-114*). Selenium supplementation at concentrations beyond those necessary to optimize gluthathione peroxidase activity is frequently effective in inhibiting the formation of chemically induced tumors in the gastrointestinal tract, liver, breast, skin and pancreas. Several studies also document the ability of selenium, as sodium selenite, to inhibit virally induced and transplantable tumors. Although selenomethionine and several other organic selenium compounds present in foods have not been extensively examined, they appear to be less effective than selenite supplementation (*115*). The ability of this trace element to exhibit such dramatic effects across such a variety of experimental conditions suggests a generalized mechanism, rather than a tissue- or cell-specific reaction (*111, 112*). As observed with vitamin A, the continuous intake of selenium appears to be necessary for maximum cancer (*116*). As with retinoids, selenium supplementation during some experimental conditions may actually enhance tumor development (*117*). However, there is overwhelming evidence supporting the anticarcinogenic role of this trace element.

The mechanisms of selenium inhibition of carcinogenesis appear complex and are poorly understood. Studies have shown that selenium inhibits the initiation phase of carcinogenesis (*118*). In studies with 7,12-dimethylbenz(a)anthracene, at least part of the protection appears to relate to the inhibition of an enzyme(s) responsible for the formation of *anti*-dihydrodiol epoxide adducts (*119*). Other studies have clearly demonstrated the effects of this trace element when provided following carcinogen administration. (*75, 76, 88, 120*). These data suggest that at least part of the inhibitory effects relate to a decrease in the promotion phase of carcinogenesis. The intracellular form of selenium resulting in cancer inhibition remains unknown but may relate to a metabolite formed during detoxification. Selenodiglutathione, a compound formed during selenium detoxification, has been shown to be far more effective in inhibiting carcinogen binding to DNA and tumor cell growth than equivalent quantities of selenium supplied as sodium selenite (*111, 119*).

Several studies reveal the interactive nature of selenium with other dietary constituents. Two of the most promising candidates for chemoprevention, vitamin A and selenium, have been shown to act additively in the inhibition of chemically induced mammary cancer (*121*). Likewise, vitamin E provides a more favorable environment by protecting against oxidative stress, thereby potentiating the action of selenium (*122*). Other nutrients may have the opposite effect on selenium. For example, it has been reported that Vitamin C reduces the chemopreventive action of selenite (*122*). This inhibitory effect of vitamin C likely relates to changes in the valence state of selenium.

Other agents. Considerable evidence indicates that the liberation or generation of activated oxygen species (hydrogen peroxide, superoxide anions, hydroxyl radicals, etc) within the cell is highly damaging and may directly or indirectly contribute to carcinogenesis (*123*). As indicated above, several dietary antioxidants, including β-carotene, ascorbic acid or α-tocopherol, can reduce cancer incidence in experimental animals. Other agents, such as synthetic antioxidants (e.g. BHA and BHT), also have been reported to inhibit tumor formation significantly. Although it is difficult to draw firm conclusions, since numerous inconsistencies exist in laboratory and epidemiological data about antioxidants *per se,* there is a strong rationale for a belief in their potential protective effects. Various investigators have suggested that phenolic antioxidants (BHT, BHA) possibly inhibit carcinogenesis by their ability to alter metabolic activation and not by their antioxidant properties (*27*). β-carotene, as indicated previously, could likewise owe its anticarcinogenic effects to properties other than its ability to inhibit free radical generation. A number of metal ions, including nickel, cadmium and cobalt, are known to be carcinogenic in animals (*124*). Inflammatory responses resulting from exposure to these metals may involve the formation of active oxygen species.

An increased risk for developing esophageal cancer has been associated with the consumption of alcoholic beverages (*125*) and of salt-cured, salt-pickled and moldy foods (*126*). The molds which can contaminate foods produce several toxins which promote the formation of N-nitrosamine compounds (*127*).

Several studies have examined the influence of a number of dietary non-nutrients, particularly those in vegetables and fruits, for their influence on the cancer process. Many of these chemicals have been shown to protect against the induction of a number of neoplasms. Table 2 list some of the compounds found to have an anticarcinogenic property. These compounds inhibit carcinogenesis by blocking the bioactivation of carcinogens, inducting enzymes that lead to detoxification, or by binding (trapping) of the parent carcinogen. Some non-nutrients occurring in foods appear to able to block the promotion phase of carcinogenesis. The desire to magnify the beneficial effects of the nutrients and non-nutrients against several disorders, including cancer, will likely foster the development of new foods or blends of foods. The complex interactions of nutrients and non-nutrients may prevent the development of a "magic bullet" food, that is capable of substantially reducing cancer risk. Nevertheless, this complexity should not serve as the basis for delaying the search for new or novel foods. At the very least, the development of new or novel foods will offer the consumer greater flexibility in choosing an acceptable diet. Until these new or novel foods are developed, it is prudent to incorporate generous amounts of fruits and vegetables into ones diet.

Table 2. Incomplete List of Non-Nutrients Found in Foods Known to Alter Experimental Carcinogenesis

Coumarins	Ellagic Acid
Isothiocyanates	Dithiolethiones
Phenols	Flavones
Terpenes	Indoles
Organosulfides	Glucarates
Conjugated fatty acids	Tannins

Summary and Conclusions

Cancer remains a major threat to many Americans. Epidemiological and laboratory findings provide rather convincing evidence for a relationship between dietary habits and the incidence of cancer. Several nutrients appear to be capable of modifying the carcinogenic process at specific sites, including carcinogen formation and metabolism; initiation, promotion and tumor progression; cellular and host defenses; cellular differentiation; and tumor growth. Unquestionably, dietary habits are not the sole determinant of cancer; but they may represent a significant factor in which control is possible. Adjustment of dietary practices to conform to generalized dietary goals may not be necessary, or even appropriate, for all segments of the population. Sophisticated techniques and procedures are desperately needed for adequate assessment of the potential merit of nutritional intervention for each individual in relationship to his/her cancer risk. A thorough appreciation or understanding of how dietary components contribute to or modify the cancer process will require the continuation of carefully controlled, probing investigations. Hopefully, future research will lead to the recognition of the critical sites in which nutrition intervention would be appropriate and will allow for sound and realistic recommendations for dietary practices.

Literature Cited

1. Wynder, E.L. and Gori, G.B J. Natl. Cancer Inst. **1977**, *58*, 825.
2. Armstrong, B. and Doll, R. Int. J. Cancer 197**5**, *15*, 617.
3. Doll, R. and Peto, R. *J. Natl. Cancer Inst.* **1981**, 66-1192.
4. Carroll, K. K. Cancer Res. 1975, *35*, 3374.
5. Hunter and Willett In *Nutrition and Cancer Prevention, Investigating the Role of Micronutrients*; Moon, T. E. and Micozzi, M. S., Eds.; Marcell Dekker, Inc.: New York, 1990, p. 83.
6. Haenszel, W. and Kurihara, M. J. Natl. Cancer Inst. 1968, *40*, 43.
7. Gori, G. B. Cancer 1979, *43*, 2151.
8. Phillips, R.L. Cancer Res. 1975, *35*, 3513.
9. Enstrom, J.E. *Cancer* 1978, *42*, 1943.
10. Kolonel, L. N., Nomura, A. M., Hirohata, T., Hankin, J. H. and Hinds, M. W. Am. J. Clin. Nutr. 1981, *34*, 2478.
11. Cancer In The Surgeon General's Report on Nutrition and Health. U. S. Dept. Health and Human Services, Public Health Service, DHHS (PHS) Publication No. 88-50210: Washington, D.C., 1989, p. 177.
12. Cancer In Die*t and Health: Implications for Reducing Chronic Disease Risk. Committee on Diet and Health.* Food and Nutrition Board, Commission on Life Sciences, National Research Council, National Academy Press: Washington, D.C., 1989, pp 593.
13. Boutwell, R.K. Crit. Rev. Toxicol. **1974,** *2*, 419.
14. Miller, E.C. Cancer R*es*. 1978, *38*, 1479.
15. Mason, R.P. In *Free Radical in Biology*; W.A. Pryor, Ed.; Academic Press: New York, 1982, Vol. 5, p. 161.
16. Krontiris, T.G. N. Engl. J. Med. 19**83**, *309*, 404.
17. Weinberg, R.A. Science **1985,** *230*, 770.
18. Mirvish, S.S., L. Wallcave, M. Eagen, and P. Shubik. Science 1972, *177*, 65.
19. Schmaehl, D. and Habs, M. Oncology 1980, *37*, 237.
20. Ranieri, R. and Weisburger, J. H. Ann. NY Acad. Sci. 1975, *258*, 181.
21. Mirvish, S.S., Cardesa, A., Wallcave, L. and Shubik, P. J. Natl. Cancer Inst. 1975, *55*, 633.

22. Miller, E.C. and Miller, J.A. *J. Natl. Cancer Inst.* **1966,** *48*, 1425.
24. Wilson, A.G.E., Kung, H.C., Boroujerdi, M., and Anderson, M.W. *Cancer Res.* 1981, *41*, 3453.
25. Wattenberg, L.W. *Cancer Res.* 1985, *45*, 1.
26. Slaga, T.J. and Digiovanni, J. In *Chemical Carcinogens*; C.E. Searle, Ed; American Chemical Society: Washington, DC, 1984, Vol. 2, p. 1279.
27. Wattenberg, L.W. *Cancer Res.* 1983, *43*, 2448s.
28. Rous, P. *J. Exp. Med.* 1914, *20*, 433.
29. Tannenbaum, A. *Cancer Res.* 1942, *2*, 460.
30. Bischoff, M., Long, L. and Maxwell, L. C. *Am. J. Cancer* 1935, *24*, 549.
31. Tannenbaum, A. In Approaches to *Tumor Chemotherapy*; Moulton, F. R., Ed.; American Association for the Advancement of Science: Washington, D.C., 1947.
32. Birt, D. F. Am. J. Clin. Nutr. 1987, *45*, 203.
33. Pollard, M., Luckert, P. H., and Snyder, D. *Cancer* 1989, *64*, 686.
34. Birt, D. F., Pelling, J. C., White, L. T., Dimitroff, K. and Barnett, T. *Cancer Res.* 1991, *51*, 1851.
35. Lok, E., Scott, F. W., Mongeau, R., Nera, E. A., Malcolm, S., and Clayson, D. B. *Cancer Lett* 1990, *51*, 67.
36. Klurfeld, D. M., Welch, C. B., Davis, M. J. and Kritchevsky, D. *J. Nutr.* 1989, *119*, 286.
37. Ruggeri, B. A., Klurfeld, D. M., Kritchevsky, D. and Furlanetto, R. W. Cancer Res. YEAR?, 49, 4130.
38. Clinton, S. K., Imrey, P. B., Alster, L. M. et al. *J. Nutr.* 1984, *114*, 1213.
39. Carroll, K. K., Gammal, E. B. and Plunkett, E. R. *Can. Med. Assoc. J.* **1968**, *98*, 590.
40. Lea, A. J. *Lancet* 1966, *2*, 332.
41. Kolonel., L. N., Hankin, J. H., Nomura, Lee, J., Chu, S. Y., Normura, A. M. Y. and Hinds, M. W. *Br. J. Cancer* 1981, *44*, 332.
42. Blair, A. and Fraumeni, J. F. Jr. *J. Natl. Cancer Inst.* 1978, *61*, 1379.
43. Carroll, K. K. and Khor, H. T. *Prog. Biochem. Pharmacol.* 1975, *10*, 308.
44. Birt, D. F. In Essential Nutrients in Carcinogenesis; Poirier, L. A., Newberne, P. M. and Pariza, M. W., Eds.; Plenum Press: New York, 1986, p. 69.
45. Braden, L. M. and Carroll, K. *Lipids* 1984, *21*, 285.
46. Reddy, B. S. and Sugie, S. *Cancer Res.* 1988, *48*, 6642.
47. Welsch, C. W. *Amer. J. Clin. Nutr.* 1987, *45*, 192.
48. Goldin, B. R. In *Dietary Fat and Cancer*; Ip, C., Birt, D. F, Rogers, A. E. and Mettlin, C., Eds.; New York, 1986, p 655.
49. Delicoinstantinos, G. *Anticancer Res.* 1987, *7*, 1011.
50. Tannenbaum, A. and Silverstone, H. *Cancer Res.* 1949, *9*, 162.
51. Silverstone, H. and Tannenbaum, A. *Cancer Res.* 1951, *11*, 442.
52. Clinton, S. K., Truex, C. R. and Visek, W. J. *J. Nutr.* 1979, *109*, 55.
53. Dunning, W. F., Curtis, M. R. and Maun, M. E. *Cancer Res.* **1950**, *10*, 454.
54. Barbul, A. *J. Parenteral Enteral Nutr.* 1986, 10, 227.

55. Barbul, A. Wasserkrug, H. L., Sisto, D. A., Seifter, E., Rettura, G., Levenson, S. M. and Efron, G. *J. Parenteral Enteral Nutr.* 1980, *4*, 446.
56. Klurfeld, D. M. and Kritchevsky, D. In *Essential Nutrients in Carcinogenesis*; Poirier, L. A., Newberne, P. M. and Pariza, M. W.,Eds.; Plenum Press: New York, 1986, p. 119.
57. Pilch, S. M., Ed. P*hysiological effects and health consequences of dietary fiber.* Federation of American Societies for Experimental Biology: Bethesda, MD., 1987.
57. Mori, S. *Johns Hopkins Hosp Bull* 1922, *33*, 357.
58. Sporn, M.B. and Roberts, A.B. *Cancer Res.* 1983, *43*, 3034.
59. Meyskens, F.L. and Prasad, K., Eds. *The Modulation and Mediation of Cancer by Retinoids*; S. Karger: Basel, 1983.
60. Moon, R.C. and Itri, L.M In The Retinoids; M.B. Sporn, A.B. Roberts, D.S. Goodman, Eds; Academic Press: New York, 1984, Vol. 2, p. 327.
61. Olson, J.A. In *Essential Nutrients in Carcinogenesis*; Poirier, L.A., Newberne, P.M. and Pariza, M.W., Eds.; Plenum Press: New York, 1986, p. 379.
62. Thompson, H.J., Becci, P.J., Brown, C.C., and Moon, R.C. *Cancer Res.* **1979,** *39*, 3977.
63. McCormick, D.L. and Moon, R.C. *Cancer Res.* 1982, *42*, 2639.
64. Santamaria, L., Bianchi, A., Annaboldi, A., Anderoni, L., and Bermond, P. *Experientia (Basel)* **1983,** *39*, 1043.
65. Peto, R., Doll, R., Buckley, J.D. and Sporn, M.B. *Nature* 1981, *290*, 201.
66. Matthews-Roth, M.M. *Oncology (Basel)* 1982, *39*, 33.
67. Birt, D., Davies, M.H., Pour, P.M. et al. *Carcinogenesis* 1984, *4*, 1215.
68. Hill, D.L. and Shih, T.W. *Cancer Res.* 1974, *34*, 564.
69. Verma, A.K. and Boutwell, R.K. *Cancer Res.* 1977, *37*, 2196.
70. Bertram, J.S. **1983,** *2*, 243.
71. Mehta, R.G. and Moon, R.C. *Cancer Res.* **1981,** *40*, 1109.
72. Roberts, A.B. and Sporn, M.B. In *Retinoids*; Sporn, M.B., Roberts, A.B. and Goodman, D.S., Eds.; Academic Press: New York, 1984, Vol 2, p. 210.
73. Dennert, G. In Retinoids; Sporn, M.B., Roberts, A.B. and Goodman, D.S., Eds.; Academic Press: New York, 1984, Vol. 2, p. 373.
74. Eufinger, H. and Gaehtgens, G. *Wochenschr.* **1936**, *15*, 150.
75. Weisburger, J.H., Harquardt, H., Mower, H. F., et al. *Prev. Med* **1980**, *9*, 352.
76. Glatthaar, B.E., Hornig, D.H. and Moser, U. In Ess*ential Nutrients in Carcinogenesis*; Poirier, L.A., Newberne, P.M. and Pariza, M.W., Eds.; Plenum Press: New York, 1986, p. 357.
77. Mirvish, S. S., Wallacave, L., Eagen, M. and Shubik, P. *Science* **1972**, *177*, 65.
78. Mirvish, S. S., Cardesa, A., Wallcave, L., and Shubik, P. *J. Natl. Cancer Inst.* **1975,** *55*, 633.
79. Reed, P. I., Summers, K., Smith, P. L. R. et al. *Gut* **1983**, *24*, 492.
80. Dion, P.W., Bright-See, E.B., Smith, C.C. et al. *Mutat. Res.* **1982**, *102*, 27.
81. Cameron, E., Pauling, E. and Leibovitz, B. *Cancer Res.* **1979**, *39*, 663.
82. Benedict, W.F., Wheatley, W.L. and Jones, P.A. *Cancer Res.* **1982**, *42*, 1041.

83. Prasad, K.N. *Life Sci.* 1980, *27*, 275.
83. Schmidt, K. and Moser, U. *Int. J. Vitam. Nutr. Res.* **1985**, *27*, 363.
84. Abe, E., Miyaura, H., Sakagai, M., Takeda, K., Konno, T., Yamasaki, S., Yoshiki, S. and Suda, T. *Proc. Natl. Acad. Sci. USA* **1981**, *78*, 4990.
85. Eisman, J. A., McIntyre, I., Martin, T., Moseley, J. M. *Lancet* **1979**, *2*, 1335.
86. Dokoh, S., Donaldson, C.A. and Haussler, M.R. *Cancer Res.* **1984**, *44*, 2103.
87. Shinki, T., Takahashi, N. and Miyaura, C. *Biochem. J.* **1981**, *195*, 685.
88. Wood, A.W., Chang, R.L. Huang, M. -T. Uskokovic, M., and Conney, A. H. *Biochem. Biophys. Res. Commun.* **1983**, *116*, 605.
89. MacManus, J.P. In Ions, Cell Proliferation and Cancer; Boynton, A.L., McKeehan, W.L. and Whitfield, J.F., Eds.; Academic Press: New York, 1982, 489.
90. Jaffe, L.F. In Ions, Cell Proliferation and Cancer; Boynton, A.L., McKeehan, W.L. and Whitfield, J.F., Eds.; Academic Press: New York, 1982, 295.
91. Green, T.R., Wu, D.E. and Wirtz, M.K. *Biochem. Biophys. Res. Commun.* **1983**, *122*, 734.
92. Gairola, C. and Chen, L.H. *Int. J. Vitam. Nutr. Res.* **1982**, *52*, 398.
93. Harber, S.L. and Wissler, R.W. *Proc. Soc. Exp. Biol. Med.* **1962**, *111*, 774
94. Harman, D. Clin. Res. **1969**, *17*, 125.
95. Konings, A.W.T. and Trieling, W.B. *Int. J. Radiat. Biol.* **1977**, *31*, 397.
96. Ip, C. and Horvath, P. *Proc. Am. Assoc. Cancer Res.* **1983**, *24*, 382.
97. Shinozuka, H., Katyal, S.L. and Perera, M.I.R. In Ess*ential Nutrients in Carcinogenesis*; Poirier, L.A., Newberne, P.M. and Pariza, M.W., Eds.; Plenum Press: New York, 1986, p. 253.
98. Poirier, L.A., Mikol, Y.B., Hoover, K., et al. *Proc. Am. Assoc. Cancer Res.* **1984**, *25*, 132.
99. Rogers, A.E. and Newberne, P.M. *Nutr. Cancer* **1982**, *2*, 104.
100. Ghoshal, A.K. and Farber, E. *Carcinogenesis* **1983**, *7*, 801.
101. Riggs, A.D. and Jones, P.A. *Adv. Cancer Res.* **1983**, *40*, 1.
102. Eto, I. and Krumdieck, C.L. In Essential Nutri*ents in Carcinogenesis*; Poirier, L.A., Newberne, P.M. and Pariza, M.W., Eds.; Plenum Press: New York, 1986, p. 313.
103. Bennett, M.A., Ramsey, J., and Donnelly, A.J. *Int. Z. Vitaminforsch.* **1956**, *26*, 417.
104. Horne, D. E., Cook, R. J, and Wagner, C. *J. Nutr.* **1989**, *119*, 618.
105. Nauss, K.M. and Newberne, P.M. *Adv. Exp. Med. Biol.* **1981**, *135*, 63.
106. Vitale, J.J., Broitman, S.A., Vavrousek-Jakuba, E., Roddy, P.W. and Gottlieb, L.S. *Adv. Exp. Med. Biol.* **1978**, *91*, 229.
107. Brada, Z. and Altman, N.H. *Adv. Exp. Med. Biol.* **1978**, *91*, 193.
108. Fong, L.Y.Y., Sivak, A. and Newberne, P.M. *J. Natl. Cancer Inst.* **1978**, *61*, 145.
109. Duncan, J.R., and Dreosti, I.E. *J. Natl. Cancer Inst.* **1975**, *55*, 195.
110. Fenton, M.R., Burke, J.P., Tursi, F.D. and Arena, F.P. *J. Natl. Cancer Inst.* **1980**, *65*, 1271.
111. Milner, J.A. In Essential Nutrients in Carcino*genesis*; Poirier, L.A., Newberne, P.M. and Pariza, M.W., Eds.; Plenum Press: New York, 1986, p. 449.

112. Shamberger, R.J. *J. Natl. Cancer Inst.* **1970**, *44*, 931.
113. Medina, D. In Diet, Nutrition *and Cancer: A critical Evaluation*; Reddy, B., and Cohen, L., Eds.; CRC Press: Boca Raton, Florida, 1986.
114. Whanger, P., Schmitz, J.A. and Exon, J.H. *Nutr. Cancer* **1982**, *3*, 240.
115. Thompson, H.J., Meeker, L.D. and Kokoska, S. *Cancer Res.* **1984**, *44*, 2803.
116. Ip, C. *Cancer Res.* **1981**, *41*, 4386.
117. Birt, D.F., Julius, A.D. and Pour, P.M. *Proc. Am. Assoc. Cancer Res.* **1984**, *25*, 133.
118. Liu, J., Gilbert, K., Parker, H., Haschek, W. and Milner, J. A. *Cancer Res.* **1991**, *51*, 4613.
119. Milner, J.A., Dipple, A. and Pigott, M.A. *Cancer Res.* **1985**, *45*, 6347.
120. Thompson, H.J. and Becci, P.J. *J. Natl. Cancer Inst.* **1980**, *65*, 1299.
121. Thompson, A.J., Meeker, L.D. and Becci, P. *Cancer Res.* **1981**, *41*, 1413.
122. Ip, C. In *Essential Nutrients in Carcinogenesis*; Poirier, L.A., Newberne, P.M. and Pariza, M.W., Eds.; Plenum Press: New York, 1986, p. 431.
123. Cerutti, P.A. 1985. *Science* **1985**, *227*, 375.
124. Sunderman, F. W. Jr. *Ann. Clin. Lab. Sci.* 1984, *14*, 93.
125. Walker, E. A., Castegnario, M., Garren, L., Toussaint, G. and Kowalski, B. *J. Natl. Cancer Inst.* **1979**, *63*, 947.
126. Tuyns, A. J., Pequignot, G. and Abbatucci, J. S. *Int. J. Cancer* **1979**, *23*, 443.
127. Li, M. H., Ji, C. and Cheng, S. J. *Nutr. Cancer* **1986**, *8*, 63.

RECEIVED October 13, 1991

Chapter 28

Food Allergies

Steve L. Taylor[1], Julie A. Nordlee[1], and Robert K. Bush[2,3]

[1]Department of Food Science and Technology, University of Nebraska, Lincoln, NE 68583–0919
[2]Department of Medicine, University of Wisconsin, Madison, WI 53792
[3]William S. Middleton VA Hospital, Madison, WI 53705

Food allergies afflict only certain individuals in the population, but the resultant adverse reactions can be quite serious on occasion. The foods eliciting these adverse reactions are safe to consume for the vast majority of consumers. Many different types of mechanisms are involved in these individualistic adverse reactions to foods (*1*,*2*). The various types of reactions are listed in Table 1 along with some of the common foods known to elicit these reactions.

True Food Allergies

True food allergies result from an abnormal immunologic response to some component of a food. Gell et al. (*3*) classified hypersensitivity into four distinct types designated as types I-IV on the basis of the type of immunologic mechanism involved in the reaction. Three of these types of true allergic reactions (types I, III, and IV) may occur with foods. The Type I reactions are mediated by immunoglobulin E (IgE). These reactions are often known as immediate hypersensitivity reactions because the onset of symptoms following the ingestion of the offending food is rapid. Type I reactions to foods are well known to occur and will be the primary focus of this review. Type III reactions are mediated by immune complexes. These reactions would typically occur 4-6 h after ingestion of the offending food. However, uncertainty exists regarding the importance of Type III reactions to foods (*4*), and they will not be discussed further. Type IV reactions, also sometimes referred to as delayed hypersensitivities, involve sensitized immune cells. These reactions would occur on a delayed basis (>6 h after ingestion of the food). Only a few examples such as contact dermatitis due to contact of the skin of a sensitive individual with a certain type of food are well established as Type IV reactions. Since these reactions tend to occur primarily in food handlers, they will not be discussed further. Celiac disease, an abnormal sensitivity to the cereal grains, wheat, rye, barley, and oats, may be a Type IV reaction (*5*) but considerable uncertainty exists regarding the mechanism of this illness at the present time.

0097–6156/92/0484–0316$06.00/0

Table 1. Individualistic Adverse Reactions to Foods

Reactions	*Implicated Foods*
True Food Allergies	
IgE-Mediated Reactions (Immediate Hypersensitivities)	Cows' milk, eggs, legumes, crustacea, fish, tree nuts, many others
Non-IgE-Mediated Reactions (Delayed Hypersensitivities)	Significance unknown
Other Food Sensitivities	
Anaphylactoid Reactions	Strawberries
Metabolic Food Disorders	
Lactose intolerance	Dairy products
Favism	Fava beans
Idiosyncratic Reactions	
Sulfite-induced asthma	Sulfites
Aspartame-induced urticaria	Aspartame
Celiac disease	Wheat, rye, barley, oats
Tartrazine-induced urticaria	Tartrazine
Many others	Many others

In Type I reactions, susceptible individuals produce an allergen-specific IgE in response to exposure to an allergen. In the case of food allergies, this exposure would occur primarily from the ingestion of foods containing allergenic proteins (*6,7*). The production of allergen-specific IgE distinguishes individuals with allergies from non-allergic individuals. The IgE antibodies bind to the surfaces of mast cells and basophils thereby sensitizing these cells.

Upon subsequent exposure to the allergen, the allergen cross-links IgE molecules on the surface of the sensitized cells. This event triggers the degranulation of the mast cells and basophils through a complex series of biochemical events. The mast cells and basophils possess numerous granules that contain the mediators of allergic disease. As many as 40 mediators of allergic reactions have been identified in these cells (*8*). These mediators are released into the bloodstream, interact with a wide variety of cellular and tissue receptors, and are responsible for the various symptoms associated with food allergies.

Table 2 provides a list of the symptoms most commonly associated with IgE-mediated food allergies. The cutaneous symptoms such as urticaria and atopic dermatitis are quite common. The gastrointestinal symptoms such as nausea, vomiting, and diarrhea are also frequently encountered but not particularly definitive. Respiratory symptoms are less commonly associated with food allergies than with other types of environmental allergens where the respiratory route is the major route of exposure. On rare occasions, very serious reactions can occur; several deaths have recently been attributed to allergic reactions to foods (*9-11*). Although many symptoms are associated with food allergies, most allergic individuals experience only a few of these symptoms.

The prevalence of IgE-mediated food allergies in the overall population is not known. However, this form of food allergy probably affects fewer than 1% of the overall population (*1,2*), although estimates of prevalence range from 0.3% to 7.5% of the population for all forms of food allergies (*12-14*). Certainly, infants and young children are more likely to suffer from IgE-mediated food allergies than adults. Careful clinical evaluations have demonstrated that adverse reactions to foods, many of which are likely to be IgE-mediated food allergies, occur in 4-6% of infants and 1-2% of young children (*15,16*). Public perceptions of the prevalence of food allergies are often much higher but many of these complaints involve adverse reactions that are unproven and/or likely non-allergic in nature (*17*). Some investigators have demonstrated that some reports of food allergies can be attributed instead to psychological and emotional ailments (*18-20*).

Infants and young children have a higher prevalence of IgE-mediated food allergies than adults because some infants will outgrow their sensitivities to foods (*21-25*). Infants who develop food allergies before their first birthdays are especially likely to outgrow this condition (*21*). Allergies to some foods such as cows' milk and eggs are more likely to be outgrown than allergies to certain other foods especially peanuts (*21-24*). Presumably, this loss of sensitivity is due to enhanced protein processing in the gastrointestinal tract, increased maturity of the gastrointestinal tract, and the development of blocking antibody responses especially secretory IgA responses in the gut lumen (*22*).

Table 2. Common Symptoms of IgE-Mediated Food Allergies

Gastrointestinal Symptoms
Nausea
Vomiting
Abdominal cramping
Diarrhea
Colic
Cutaneous Symptoms
Urticaria
Angioedema
Atopic Dermatitis
Respiratory Symptoms
Rhinitis
Laryngeal Edema
Bronchospasm
Asthma
Other Symptoms
Migraine Headache
Anaphylactic Shock

The diagnosis of the various immunological and non-immunological food sensitivities first requires the establishment of a causal basis for the suspect food in the adverse reaction. Sometimes, especially in the case of severe, IgE- mediated reactions, the diagnosis can be made accurately on the basis of the patient's history. However, historical accounts can be misleading, and over- reliance on this diagnostic procedure is not advocated. Challenge tests are often needed to establish the causal role of specific foods in adverse reactions.

Although several approaches are available in challenge testing, the double-blind challenge test (DBCT) is the most reliable type of challenge procedure (*26-29*). DBCTs should not be used in situations involving patients who provide histories of severe life-threatening reactions (*26*). Other challenge procedures involve single-blind or open challenges which have been used in some studies (*30,31*).

Once the cause-and-effect relationship with foods has been established, the role of IgE must be proven to establish the diagnosis of immediate hypersensitivity. Several procedures including the skin prick test and the radioallergosorbent test (RAST) are available to detect the existence of food allergen-specific IgE (*32,33*).

For all of the various types of true food allergies and other food sensitivities (Table 1), the principal method of treatment is the specific avoidance diet (*34*). The affected individuals must simply avoid the food(s) that provoke their adverse reactions. In the case of IgE-mediated allergic reactions, the tolerance for the offending food can be extremely low. The exquisite sensitivity of some patients with IgE-mediated food allergies causes them to react adversely to trace levels of the offending food that might contaminate other foods (*2,34,35*). Exposure to very small amounts of the offending food has provoked adverse reactions in several well established cases (*36,37*). The failure to clean processing equipment or serving utensils thoroughly are examples of the situations that can lead to the contamination of foods with potentially hazardous allergenic residues (*2,35*). The severity of the adverse reaction will usually depend on the degree of exposure to the offending food. Therefore, reactions to trace quantities of the offending foods are often relatively mild. However, inadvertant exposure to larger amounts of the offending food can elicit very severe reactions in some cases. Several deaths and extremely severe reactions have been associated with inadvertant exposures to foods which the patients realized would cause severe reactions (*9,37,38*). Clearly, patients with IgE-mediated food allergies especially those with severe sensitivities must carefully scrutinize food labels. However, these individuals can be plagued by the undeclared uses of certain ingredients in specific foods, the lack of labelling on some types of foods, the lack of labelling in restaurant and other foodservice settings, and the inadvertant contamination of one food with another.

The presence of the allergen in the food is obviously a prerequisite for provocation of an IgE-mediated allergic reaction. Some foods may not contain the allergen. For example, edible oils do not appear to contain the proteinaceous allergens that exist in the seeds from which the oil is extracted (*39-41*). Some food processes may drastically modify allergenic proteins to the point that they lose their allergenicity. Hydrolysates of some food proteins are apparently not allergenic (*42,43*) but the extent of hydrolysis is extremely important and partially hydrolyzed proteins should be viewed with skepticism. The allergenicity of specific foods can be compared to the native food product through use of the RAST inhibition test (*42*).

Allergic reactions can also occur due to the cross reactions that exist between closely related foods. Common examples include the cross reactions between goats' milk and cows' milk (*44*) and the various species of avian eggs (*45*). Some cross reactions occur much less frequently such as those between the various species of crustacea including shrimp, crab, and lobster (*46*) and the various species of legumes such as peanuts and soybeans (*47,48*).

Breast feeding has been advocated by many physicians as a prophylactic treatment to prevent the development of food allergies among infants but the results of controlled trials of breast feeding versus formula feeding have yielded inconsistent and controversial results (*49*). Despite the confusion, many physicians continue to advocate breast feeding for a period of 6-12 months especially for infants with a high risk for the development of food allergies such as those infants born into families with histories of allergic disease. Curiously, breast feeding probably also provides the best example of the exquisite sensitivity of allergic individuals. Numerous reports exist of infants who have developed food allergies during an extended period of exclusive breast feeding or upon the first known exposure to a specific food (*50-52*). Apparently, these infants were sensitized by exposure to the trace amounts of food proteins secreted in breast milk and emanating from the mothers' diet (*53*).

Because of the exquisite sensitivity of some individuals with IgE-mediated food allergies, it is desirable to have methods to detect trace quantities of food allergens in other foods or in breast milk. Typically, radioimmunoassays or RAST inhibition tests have been used for this purpose (*36,53,54*). These tests do not necessarily need to be specific for the allergenic protein(s). If one or a mixture of peanut proteins is detected for example, it is easy to assume that the allergenic protein(s) are likely to be present. In the case of celiac disease, several highly specific immunoassays based on monoclonal antibodies have been developed to detect traces of gliadin or gluten in other foods (*55,56*).

While the specific avoidance diet is clearly the most effective and widely used treatment for IgE-mediated food allergies and for many of the other types of food sensitivities, other treatments do exist. Elimination diets involve the simultaneous removal of a variety of possible allergenic foods (*57,58*). These diets are difficult to sustain over long periods of time but do find application in the initial treatment of conditions where the offending foods are difficult to identify. Attempts to use elimination diets outside of the controlled clinical setting are fraught with difficulty (*59*).

Pharmacological treatments are available to treat the symptoms of food allergies. Patients with histories of life-threatening reactions to foods are advised to carry epinephrine-filled syringes for emergency use (*60*), and epinephrine is the treatment of choice for severe reactions. Antihistamines are a common form of intervention in mild cases of allergic reactions (*61*), but cannot be relied upon to forestall severe reactions. Prophylactic pharmacologic treatments remain in the investigational stage of development. Although several agents, including sodium cromolyn, have been tested, the results are inconsistent (2) and their use cannot be recommended. Immunotherapy is a controversial prophylactic treatment for food allergies (*2,61*) which carries the risk of producing severe allergie reaction, and therefore, is contraindicated at the present time.

While a wide variety of foods can be implicated in IgE-mediated food allergies, some foods are implicated much more frequently than others. Since most known food allergens are proteins, any food containing protein could theoretically have the potential for sensitization and the subsequent provocation of allergic reactions. However, some foods with high levels of protein such as beef, pork, and chicken are rarely implicated in food allergies. The most common allergenic foods in infants in the U.S. are cows' milk, eggs, peanuts, and soybeans (*2,34*). The most common allergenic foods among adults in the U.S. are legumes (particularly peanuts and soybeans), crustacea, molluscs, fish, tree nuts, eggs, and wheat (*2,34*). The situation may vary in other countries in relation to the importance of specific foods in the diet. Soybeans are common food allergens in Japan, and codfish is a common allergenic food in Scandanavian countries (*34*).

Only a few of the allergenic food proteins have been isolated, identified, and characterized. Table 3 lists these known allergenic proteins from cows' milk, eggs, peanuts, soybeans, codfish, shrimp, green peas, rice, cottonseed, and tomato (*6,7,34*). In many cases, these allergens have been only crudely purified into fractions containing a number of proteins. It is widely accepted that many foods contain multiple allergens, although the reasons for this multiplicity are unclear. The characteristics of these known food allergens have been extensively reviewed elsewhere (*6,7,62*).

Allergenic food proteins tend to have several features in common (*6*). First and most obviously, the proteins must be capable of stimulating IgE production in susceptible individuals. Some proteins appear to be more immunogenic than others.

Table 3. Known Allergenic Food Proteins

Food	*Allergenic Proteins*
Cows' milk	Casein
	β-lactoglobulin
	α-lactalbumin
	others
Egg whites (Gallus domesticus)[a]	Ovomucoid (*Gal d* I)[a]
	Ovalbumin (*Gal d* II)
	Ovotransferrin (conalbumin) (*Gal d* III)
	Lysozyme
	Ovomucin
Egg yolks	Livetin
	Apovitellenin I
	Apovitellenin VI
	Phosvitin
Peanuts	Peanut I
	Lectin-reactive glycoprotein
	Arachin
	Conarachin
Soybeans	β-Conglycinin (7S fraction)
	Glycinin (11S fraction)
	2S Fraction
	Kunitz trypsin inhibitor
	Unidentified 20kD protein
Codfish (*Gadus callarias*)[a]	Allergen M (parvalbumin) (*Gad c* I)[a]
Shrimp	Antigen II
Green peas	Albumin fraction
Rice	Glutelin fraction
	Globulin fraction
Cottonseed	Glycoprotein fraction
Tomato	Several glycoproteins
Papain	Papain

Source: Adapted from Reference 34.
[a] Reference 82.

The reasons for these differences in immunogenicity are not entirely clear but the host obviously recognizes these highly immunogenic proteins as being more foreign (*6*). Complex proteins with numerous antigenic determinants are most likely to be highly immunogenic (*6*). The severity and persistence of peanut allergy could possibly indicate that the peanut allergens are particularly potent immunogens.

The molecular size and shape of the protein is important for several reasons. Histamine release from mast cells and basophils requires the allergen to bridge between IgE antibody molecules on the surface of the mast cell or basophil membrane. Thus, the allergens must have the proper geometry to allow such bridging. Most food allergens fall in the molecular weight range of 10,000 to 70,000 which may be ideal for bridging (*6,63*). Bridging also requires that the allergen possess two or more allergenic determinants or epitopes located at suitable distances equivalent to the spacing between the IgE molecules on the membrane surface (*6,63*). Bridging may be less dependent on molecular size than molecular shape. Also, there is no evidence to conclude that the two or more epitopes on an allergen have to be identical (*6,63*). The spacing between the IgE molecules on the membrane surface may be variable allowing for greater flexibility in allergen structure. The optimal size for food allergens is further dictated by two other requirements: the immunogenicity of the protein and its intestinal permeability (*6*). Smaller proteins would tend to be less immunogenic and a molecular weight of 10,000 probably represents the lower limit needed to stimulate an immunogenic response. Intestinal permeability may constrain the upper limits on the size of food allergens. Proteins in excess of a molecular weight of 70,000 are not likely to be efficiently absorbed in the intestinal tract. Some of the proteins listed in Table 3 are considerably larger than the upper molecular weight limit of 70,000 but these proteins are known to exist in polymeric form and have monomers within the predicted molecular weight range.

Allergenic food proteins must possess several other key attributes. First, these proteins must be reasonable stable to food processing treatments (*1,6*). Many of the known food allergens are comparatively heat-stable and acid-stable (*6,43*). Food allergens must also survive the proteolytic processes of digestion (*1,43*). Although these proteins can be partially hydrolyzed in the process of digestion, immunologically active fragments must survive digestion and reach the IgE-producing cells of the gut wall. In general, most food allergens are known to be comparatively more resistant to proteolysis than other proteins. This feature coupled with the acid stability insures that these proteins will reach the intestinal mucosa in immunogenic form.

The structure of most of the known food allergens is not well understood. Although the general structures of some of these proteins such as the milk proteins have received considerable study, the features which are responsible for their allergenic activity are unknown. The notable exception is codfish allergen M or by new nomenclature *Gad c* I which has been extensively purified and characterized (*64-66*). *Gad c* I is a sarcoplasmic protein from codfish muscle belonging to a group of calcium-binding proteins known as parvalbumins (*64*). *Gad c* I has a molecular weight of 12,328 and is comprised of 113 amino acid residues and one glucose molecule (*67*). *Gad c I* is arranged in three domains designated loops AB, CD, and EF (*68*). Loops CD and EF bind calcium so allergen M is a calmodulin (*68*). *Gad c* I contains several

IgE-binding sites as expected (*68*). Hexadecapeptides located in loop CD from residues 49 to 64 and in loop EF from residues 88 to 103 are able to bind specific IgE (*69,70*). These two binding sites are not identical (*69,70*). A third IgE-binding site located in loop AB from residues 13 to 32 may also exist (*68*) but further confirmation is needed. Thus, *Gad c* I appears to possess the multiple IgE-binding sites that would be necessary to bridge IgE molecules on the surfaces of mast cell and basophil membranes. *Gad c* I also survives partial proteolysis with trypsin, pepsin, subtilisin, and pronase, although extensive proteolysis will destroy its immunogenicity (*71*). The ability of small peptide fragments of *Gad c* I to bind IgE from cod-allergic patients (*69,70*) suggests that fairly extensive proteolysis would be required for a loss of immunologic activity. The allergenicity of *Gad c* I is resistant to denaturation with urea/β-mercaptoethanol or guanidinium HCl suggesting that conformational structure was not especially important to its immunogenicity (*72*). *Gad c* I is the only food allergen that is understood in such detail. Obviously, our understanding of IgE-mediated food allergies would increase substantially with greater knowledge of the chemistry of other food allergens.

Other Food Sensitivities

In addition to the true food allergies discussed above, many other individualistic adverse reactions to foods occur (*2,43*). These food sensitivities occur through many different non-immunological mechanisms. Like true food allergies, these other food sensitivities affect only certain individuals in the population. Unlike food allergies, abnormal immunological responses to food components are not involved in these illnesses. The three major classes of non-immunological food sensitivities are anaphylactoid reactions, metabolic food disorders, and idiosyncratic reactions (Table 1).

Anaphylactoid reactions involve the non-immunological release of mediators from mast cells and basophils. The mediators are identical to those implicated in the true food allergies, but IgE is not involved in mediating the release of the mediators in anaphylactoid reactions. The nature of the non-immunological release of mediators from mast cells and basophils in anaphylactoid reactions remains undefined. In fact, the evidence for the existence of anaphylactoid reactions is largely circumstantial. Presumably, some substance in the offending food destabilizes the mast cell membranes causing a spontaneous release of histamine and other mediators. However, none of these substances has ever been identified in foods. The most persuasive evidence for the existence of anaphylactoid reactions is the lack of evidence for immunological involvement in a few types of food sensitivity. The classic example is strawberry allergy (*1,43*). Strawberries are well known to cause urticaria in some individuals, but strawberries contain little protein and no strawberry allergen or strawberry-specific IgE has ever been proven to exist. The symptoms of strawberry sensitivity resemble some of the common symptoms observed in true food allergies so the in vivo release of histamine and other mediators through non-immunological mechanisms seems a plausible explanation.

Metabolic food disorders result from genetically based defects in the ability to metabolize a food component or that enhance the sensitivity to an ingested chemical through some alteration in the metabolic pattern (*2,34*). The best examples of metabolic food disorders are lactose intolerance and favism (Table 1). Lactose intolerance is based on an inherited deficiency in the activity of intestinal β-galactosidase (*73*). The result is an intolerance to lactose, the principal sugar in milk and other dairy products. Since the lactose cannot be hydrolyzed and absorbed in the small intestine, the lactose passes into the colon where bacteria metabolize the lactose to CO_2, H_2, and H_2O (*2*). This bacterial metabolism of lactose in the colon gives rise to the hallmark symptoms of lactose intolerance - abdominal cramping, flatulence, and frothy diarrhea (*2,34*). Lactose intolerance affects a substantial number of people on a worldwide basis, although certain ethnic groups such as Greeks, Arabs, Jews, black Americans, Hispanics, Japanese, and other Asians are affected in greater proportion than Caucasians (*73*). Favism is an intolerance to the ingestion of fava beans, a commonly consumed legume in some parts of the world (*2,34*). Favism affects individuals with an inherited deficiency of the enzyme, glucose-6- phosphate dehydrogenase in their red blood cells (*34*). Without this enzyme, the red blood cells are susceptible to oxidative damage by several naturally occurring oxidants, known as vicine and convicine, in fava beans (*34*). The resulting symptoms are those of hemolytic anemia, namely pallor, fatigue, dyspnea, nausea, abdominal and/or back pain, fever, and chills (*34*). The deficiency of glucose-6-phosphate dehydrogenase is the most common enzymatic defect in humans, and favism would be much more common if fava beans were more widely consumed (*34*). Metabolic food disorders are typically treated by avoidance of the offending food, although most of the affected individuals can tolerate at least small quantities of the offending food (*2*).

Many adverse reactions to foods that affect certain individuals in the population occur through unknown mechanisms (*2*). These types of food sensitivities are known collectively as food idiosyncrasies (*2,34*). Obviously, this category of food sensitivities could include a broad range of illnesses associated with a large number of foods or food components involving a myriad of mechanisms and displaying a wide variety of symptoms from the trivial to severe, life-threatening reactions. Only a few food idiosyncrasies have well established links to the ingestion of specific foods or food components. For the vast majority of food idiosyncrasies, the association with specific foods has not been firmly established and thus the role of foods in these illnesses remains unproven and controversial (*34*). A few food idiosyncrasies such as hyperkinesis from the ingestion of food coloring agents have been largely disproven (*74*), although many consumers persist in their belief of the existence of these illnesses.

The established food idiosyncrasies include sulfite-induced asthma (*75,76*), celiac disease (*5,77*), and aspartame-induced urticaria (*78*). While the mechanisms involved in these illnesses remain unknown, the existence of these sensitivities and the role of foods in the causation of these conditions are indisputable. Sulfite-induced asthma is triggered in sensitive individuals by the ingestion of sulfites which are widely used food additives and ingredients (*75,76*). Severe asthmatics are more likely to experience this sensitivity but it affects only about 1-2% of all asthmatics (*79*). Sulfite sensitivity can provoke serious, life-threatening responses in sensitive individuals

(*75,76*). However, most sulfite-sensitive asthmatics can tolerate small quantities of sulfites with no ill effects (*80*). Celiac disease, also known as celiac sprue or gluten-sensitive enteropathy, is characterized by malabsorption of nutrients from the intestine as a consequence of damage to the absorptive epithelial cells of the small intestine. The intestinal damage occurs in sensitive individuals following the ingestion of certain cereal grains - wheat, rye, barley, and oats (*77*). The protein fraction of these grains is involved in the response. Several mechanisms have been proposed including Type IV food allergy (*5*) and metabolic food disorder (*2,77*). However, the mechanism remains unknown despite considerable research effort. Celiac disease affects about 1 in every 3000 persons in the U.S. (*77*). The tolerance for cereal grains among individuals with celiac disease is not precisely known but is thought to be quite low (*2,81*). Aspartame-induced urticaria is a well established condition (*78*). However, this adverse reaction is rather mild and self-limited. The number of individuals affected by this sensitivity is unknown pending further investigations.

Conclusion

Food allergies and related diseases affect only a small number of individuals in the overall population. However, the symptoms can occasionally be quite severe, and even life-threatening. Also, the tolerance for the offending foods is extremely low in the cases of true food allergies and celiac disease. Many different mechanisms are involved in food allergies and the many related diseases. Only IgE-mediated food allergies have been well defined in terms of mechanism. Much further research is needed to elucidate the mechanisms and nature of the foodborne substances provoking these conditions.

Literature Cited

1. Taylor, S. L. *Food Technol.* **1985**, *39* (2), 98-105.
2. Taylor, S. L. In *Nutritional Toxicology*; J. N. Hathcock, Ed.; Academic Press: Orlando, 1987; Vol. II; pp. 173-198.
3. Gell, P. G. H.; Coombs, R. R. A.; Lachmann, D. J. *Clinical Aspects of Immunology*, 3rd ed.; Lippincott: Philadelphia, 1975.
4. Paganelli, R.; Matricardi, P. M.; Aiuti, F. A. *Clin. Rev. Allergy* **1984**, *2*, 69-78.
5. Strober, W. *J. Allergy Clin. Immunol.* **1986**, *78*, 202-211.
6. Taylor, S. L.; Lemanske, R. F., Jr.; Bush, R. K., Busse, W. W. *Ann. Allergy* **1987**, *59*, 93-99.
7. Taylor, S. L.; Lemanske, R. F., Jr.; Bush, R. K. *Comments Agric. Food Chem.* **1987**, *1*, 51-70.
8. Wasserman, S. I. *J. Allergy Clin. Immunol.* **1983**, *72*, 101-115.
9. Yunginger, J. W.; Sweeney, K. G.; Sturner, W. Q.; Giannandrea, L. A.; Teigland, J. D.; Bray, M.; Benson, P. A.; York, J. A.; Biedrzycki, L.; Squillace, D. L.; Helm, R. M. *JAMA* **1988**, *260*, 1450-1452.
10. Yunginger, J. W.; Squillace, D. L.; Jones, R. T.; Helm, R. K. *New Engl. Reg. Allergy Proc.* **1989**, *10*, 249-274.

11. Settipane, G. A. *New Engl. Reg. Allergy Proc.* **1989**, *10*, 271-274.
12. Bahna, S. L.; Gandhi, M. D. *Ann. Allergy* **1983**, *50*, 218-224.
13. Wood, C. B. S. *Acta Paediatr. Scand.* **1986**, *323*, 76-83.
14. Metcalfe, D. D. *J. Allergy Clin. Immunol.* **1984**, *73*, 749-762.
15. Sampson, H. A. *J. Allergy Clin. Immunol.* **1988**, *81*, 495-504.
16. Bock, S. A. *Pediatrics* **1987**, *79*, 683-688.
17. Sloan, A. E.; Powers, M. E. *J. Allergy Clin. Immunol.* **1986**, *78*, 127-139.
18. Pearson, D. J.; Rix, K. J. B.; Bentley, S. J. *Lancet* **1983**, *i*, 1259-1261.
19. King, D. S. *Nutr. Health* **1984**, *3*, 137-151.
20. Selner, J. C.; Staudenmayer, H. In *Psychobiological Aspects of Allergic Disease*; Young, S. H.; Rubin, J. M.; Daman, H. R., Eds.; Praeger Publ.: Westport, CT, 1986; pp. 102-106.
21. Bock, S. A. *J. Allergy Clin. Immunol.* **1982**, *69*, 173-177.
22. Dannaeus, A.; Inganas, M. *Clin. Allergy* **1981**, *11*, 533-539.
23. Sampson, H. A.; Scanlon, S. M. *J. Pediatr.* **1989**, *115*, 23-27.
24. Ford, R. P. K.; Taylor, B. *Arch. Dis. Child.* **1982**, *57*, 649-652.
25. Bock, S. A. *J. Pediatr.* **1985**, *107*, 676-680.
26. Bock, S. A.; Sampson, H. A.; Atkins, F. M.; Zeiger, R. S.; Lehrer, S.; Sachs, M.; Bush, R. K.; Metcalfe, D. D. *J. Allergy Clin. Immunol.* **1988**, *82*, 986-997.
27. May, C. D.; Bock, S. A. *Allergy* **1978**, *33*, 166-188.
28. Sampson, H. A. *J. Allergy Clin. Immunol.* **1983**, *71*, 473-480.
29. Atkins, F. M.; Steinberg, S. S.; Metcalfe, D. D. *J. Allergy Clin. Immunol.* **1985**, *75*, 348-355.
30. Goldman, A. S.; Anderson, D. W.; Sellars, W. A.; Saperstein, S.; Kniker, W. T.; Halpern, S. R. *Pediatrics* **1963**, *32*, 425-443.
31. Pearson, D. J. *Clin. Exp. Allergy* **1989**, *19*, 83-85.
32. Adolphson, C. R.; Yunginger, J. W.; Gleich, G. J. In *Manual of Clinical Immunology*, 2nd ed; Rose, N. R.; Friedman, H., Eds.; Am. Soc. Microbiol.,: Washington, D.C., 1980; pp. 778-788.
33. Bock, S. A.; Buckley, J.; Holst, A.; May, C. D. *Clin. Allergy* **1977**, *7*, 375- 383.
34. Taylor, S. L.; Nordlee, J. A.; Rupnow, J. H. In *Food Toxicology - A Perspective on the Relative Risks*; Taylor, S. L.; Scanlan, R. A., Eds.; Marcel Dekker: New York, 1989; pp. 255-295.
35. Taylor, S. L.; Bush, R. K.; Busse, W. W. *New Engl. Reg. Allergy Proc.* **1986**, *7*, 527-532.
36. Yunginger, J. W.; Gauerke, M. B.; Jones, R. T.; Dahlberg, M. J. E.; Ackerman, S. J. *J. Food Prot.* **1983**, *46*, 625-628.
37. Fries, J. H. *Ann. Allergy* **1982**, *48*, 220-226.
38. Evans, S.; Skea, D.; Dolovich, J. *Canadian Med. Assoc. J.* **1988**, *139*, 231- 232.
39. Taylor, S. L.; Busse, W. W.; Sachs, M.I.; Parker, J. L.; Yunginger, J. W. *J. Allergy Clin. Immunol.* **1981**, *68*, 373-375.
40. Bush, R. K.; Taylor, S. L.; Nordlee, J. A.; Busse, W. W. *J. Allergy Clin. Immunol.* **1985**, *76*, 242-245.
41. Halsey, A. B.; Martin, M. E.; Ruff, M. E.; Jacobs, F. O.; Jacobs, R. I. *J. Allergy Clin. Immunol.* **1986**, *78*, 408-410.

42. Nordlee, J. A.; Taylor, S. L.; Jones, R. T.; Yunginger, J. W. *J. Allergy Clin. Immunol.* **1981**, *68*, 376-382.
43. Taylor, S. L. *J. Food Prot.* **1986**, *49*, 239-250.
44. Juntunen, K.; Ali-Yrkko, S. *Kiel. Milchwirt. Forschungsberg.* **1983**, *35*, 439-440.
45. Langeland, T. *Allergy* **1983**, *39*, 399-412.
46. Waring, N. P.; Daul, C. B.; deShazo, R. D.; McCants, M. L.; Lehrer, S. B. *J. Allergy Clin. Immunol.* **1985**, *76*, 440-445.
47. Bernhisel-Broadbent, J.; Taylor, S.; Sampson, H. A. *J. Allergy Clin. Immunol.* **1989**, *84*, 701-709.
48. Herian, A. M.; Taylor, S. L.; Bush, R. K. *Int. Arch. Allergy Appl. Immunol.* **1990**, *92*, 193-198.
49. Burr, M. L. *Arch. Dis. Child.* **1983**, *58*, 561-565.
50. Gerrard, J. W.; Shenassa, M. *Ann. Allergy* **1983**, *61*, 300-302.
51. Van Asperen, P. P.; Kemp, A. S.; Mellis, C. M. *Arch. Dis. Child.* **1983**, *58*, 253-256.
52. Warner, J. O. *Clin. Allergy* **1980**, *10*, 133-136.
53. Machtinger, S.; Moss, R. *J. Allergy Clin. Immunol.* **1986**, *77*, 341-347.
54. Keating, M. U.; Jones, R. T.; Worley, N. J.; Shively, C. A.; Yunginger, J. W. *J. Allergy Clin. Immunol.* **1990**, *86*, 41-44.
55. Skerritt, J. H. *J. Sci. Food Agric.* **1985**, *36*, 987-994.
56. Ellis, H. J.; Freedman, A. R.; Ciclitira, P. J. *J. Immunol. Meth.* **1989**, *120*, 17-22.
57. Dockhorn, R. J.; Smith, T. C. *Ann. Allergy* **1981**, *47*, 264-266.
58. Gallant, S. P.; Franz, M. L.; Walker, P.; Wells, I. D.; Lundak, R. L. *Am. J. Clin. Nutr.* **1977**, *30*, 512-516.
59. Taylor, S. L. *Allergy* **1989**, *44* (Suppl. 9), 97-100.
60. Atkins, F. M. *Nutr. Rev.* **1983**, *41*, 229-234.
61. Anderson, J. B.; Lessof, M. H. *Proc. Nutr. Soc.* **1983**, *42*, 257-262.
62. Taylor, S. L.; Lemanske, R. F., Jr.; Bush, R. K.; Busse, W. W. In *Food Allergy*; Chandra, R. K., Ed.; Nutrition Research Education Foundation: St. John's, Newfoundland, 1987; pp. 21-44.
63. Aas, K. *Allergy* **1978**, *33*, 3-14.
64 Aas, K. In *Progress Allergology Applied Immunology*; Oehling, A.; Glazer, I.; Malthove, E.; Arbesman, C., Eds.; Pergammon Press: New York, 1980; pp. 339-344.
65. Aas, K.; Jebsen, J. W. *Int. Arch. Allergy* **1967**, *32*, 1-20.
66. Elsayed, S.; Aas, K. *Int. Arch. Allergy* **1971**, *40*, 428-438.
67. Elsayed, S.; Bennich, H. Scand. *J. Immunol.* **1975**, *4*, 203-208.
68. Elsayed, S.; Apold, J. *Allergy* **1983**, *38*, 449-459.
69. Elsayed, S.; Titlestad, K., Apold, J.; Aas, K. *Scand. J. Immunol.* **1980**, *12*, 171-175.
70. Elsayed, S.; Ragnarsson, U.; Apold, J.; Florvaag, E.; Vik, H. *Scand. J. Immunol.* **1981**, *14*, 207-211.
71. Aas, K.; Elsayed, S. *J. Allergy* **1969**, *44*, 333-343.
72. Elsayed, S.; Aas, K. *J. Allergy* **1971**, *47*, 283-291.

73. Kocian, J. *Int. J. Biochem.* **1988,** *20,* 1-5.
74. Stare, F. J.; Whelan, E. M.; Sheridan, M. *Pediatrics* **1980,** *66,* 521-525.
75. Taylor, S. L.; Higley, N. A.; Bush, R. K. *Adv. Food Res.* **1986,** *30,* 1-76.
76. Taylor, S. L.; Nordlee, J. A. *ISI Atlas Sci. Immunol.* **1988,** *1,* 254-258.
77. Kasarda, D. D. *Cereal Foods World* **1978,** *23,* 240-244.
78. Kulczycki, A., Jr. *Ann. Int. Med.* **1986,** *104,* 207-208.
79. Bush, R. K.; Taylor, S. L.; Holden, K.; Nordlee, J. A.; Busse, W. W. *Am. J. Med.* **1986,** *81,* 816-820.
80. Taylor, S. L.; Bush, R. K.; Selner, J. C.; Nordlee, J. A.; Wiener, M. C.; Holden, K.; Koepke, J. W.; Busse, W. W. *J. Allergy Clin. Immunol.* **1988,** *81,* 1159-1167.
81. Hartsook, E. I. *Cereal Foods World* **1984,** *29,* 157-158.
82. Marsh, D. G.; Goodfriend, L.; King, T. P.; Lowenstein, H.; Platts-Mills, T. A. E. *J. Allergy Clin. Immunol.* **1987,** *80,* 639-645.

RECEIVED August 15, 1991

Evaluation of Specific Foods

Chapter 29

Chemical Safety of Irradiated Foods

George G. Giddings

Consultant, 61 Beech Road, Randolph, NJ 07869

Food irradiation, carried out according to well-established principles and procedures developed worldwide over decades, yields safe, wholesome foods and feeds, and their irradiated raw materials and ingredients, with minimal overall energy deposition, or noticeable change in their characteristics! In fact, often the irradiated product can be chemically and/or microbiologically "safer" than the non-irradiated counterpart, a feature that is being taken advantage of in a growing list of countries. Such statements can now be made with utmost confidence thanks to the global efforts of a legion of researchers and evaluators, past and present, who devoted significant portions of their professional careers to objectively addressing the question of irradiated food safety/ wholesomeness, competently employing state-of-the-art approaches and techniques of scientific research and peer review. Ironically, if not surprisingly given the name of this process/treatment, as the "fruits" of this decades-long global effort were appearing in the 1980s in the form of positive World Health Organization and Codex Alimentarius actions at the international level plus national approvals and industrialization initiatives, all based on unshakeable facts and absence of reasonable doubt, an unlikely (professionally) cast of predisposed characters, attracted by such highly visible public events, set about making part-or-full-time "careers", or sidelines out of opposing food irradiation. They employ such time-worn tactics as false and misleading misinformation-cum-politicization campaigns coupled to standard-fund-raising techniques. Business must be good, for as we begin the decade of the '90s such disparate individuals and groups have coalesced into something of a coordinated international, largely single-issue activist network.

The three fundamental aspects of "whole" irradiated food safety/whole-someness are, of course, (a) toxicological safety, (b) microbiological safety, and, (c) nutritional adequacy. Proof beyond reasonable doubt of toxicological safety also applies to, for example, dry ingredients such as irradiated spices and seasonings, but not nutritional adequacy or microbiological safety since such are not significant sources of dietary nutrients, and irradiation can only improve their microbiological hygiene and safety which is the reason for irradiating them in the first place. In the context of toxicological safety of irradiated whole foods, food raw materials (e.g, irradiated grains) and

0097–6156/92/0484–0332$06.00/0

ingredients (e.g., irradiated spices and seasonings), earlier testing, going at least as far back as the 1920s, relied virtually entirely on biological tests, notably animal feeding studies. In fact, animal feeding of irradiated foods was one of the very first uses of the so-called rodent bioassay, way back in the 1920s (*1*), but still over a quarter century after the concept of radiation preservation of foods arose and was published upon in the late 1890s (*2*)! The application of analytical chemical methods to the question of irradiated food toxicity came along later, and represents probably the most conclusive proof yet of their toxicological safety. Similarly, biological tests including microbiological "feeding studies" contributed considerably to addressing the question of the impact of irradiation on macro- and micronutrients, and nutritional adequacy of irradiated diets and foods. But, here again, results of analytical chemical evaluations contributed as much or more to addressing this aspect as well (with the caveat that analytical chemical methods can measure what is present and in what quantities, but not bioavailability and efficiency of utilization of nutrients). Finally, to complete the picture it must be noted that analytical chemistry is the foundation for several dosimetric methods of measuring absorbed ionizing energy as well as methods of detecting whether-or-not a food has been irradiated.

The remainder of this chapter focuses primarily on the role of analytical chemistry coupled with theoretical and experimental radiochemistry (of water and other food constituents in this instance) in establishing the toxicological safety of irradiated foods and their irradiated raw materials, ingredients and packaging materials, in keeping with the symposium theme. In doing so the intent of this chapter is <u>not</u> to present a comprehensive survey of irradiated food chemistry studies conducted worldwide over recent and not so recent decades. That would require at least one weighty volume, and several recent volumes on food irradiation devote considerable space to this aspect alone (e.g., *3, 4, 5, 6, 7*). Rather, the following is an attempt to briefly document, employing selected key references, chemistry's special place in the establishment of irradiated food (and ingredient, etc.) toxicological safety/wholesomeness, and to trace the "evolution" of this development. In the context that a few selected extracts from the enormous volume of published irradiated food chemistry data couldn't begin to do justice to the sum total, the interested reader is invited to consult the several references at the end of this chapter that consolidate much of such data. One set of data that played a central role in addressing toxicological safety of food irradiation (*see* below) is presented in Table 1.

The Chemical Safety of Irradiated Foods

In the context of chemical/toxicological safety of irradiated foods, the fundamental questions to be answered were, of course , what takes place chemically as a food absorbs discreet quanta, or accelerated electrons, of ionizing energy in the millions of electron volts — each range (now set at no greater than 10 MeV) to total absorbed energies ("doses") approximately between 0.01 and 100 kiloGrays (depending upon the application), and, what are the stable end-products and their toxicological ramifications, if any? Especially early-on, answering such questions was hampered by a far less than complete knowledge of food composition per-se, especially at the

Table1. Compounds Identified in Beef

	Cobalt Irradiation				*LINAC Irradiation*				*Thermal Sterilization*				*Frozen Control*				*Raw*	
	Cooked		*Uncooked*		*Cooked*		*Uncooked*		*Cooked*		*Uncooked*		*Cooked*		*Uncooked*			
	PPB	*SD*	*PPB*	*SD*	*PPB*	*SD*	*PPB*	*SD*	*PPB*	*SD*	*PPB*	*SD*	*PPB*	*SD*	*PPB*	*SD*	*PPB*	*SD*
Alkanes																		
Ethane	0	-	172	20.2	0	-	179	19.0	0	-	0	-	0	-	0	-	0	-
Propane	60	19.7	164	16.9	65	20.3	173	19.7	0	-	0	-	0	-	0	-	0	-
Butane	125	39.7	208	23.1	127	42.0	221	31.0	0	-	0	-	0	-	0	-	0	-
Pentane	170	54.1	205	21.4	180	52.8	203	49.7	4	1.9	8	2.1	2	0.7	1	0.7	1	-
Hexane	207	69.8	209	23.2	248	59.9	217	24.9	67	24.5	125	32.1	7	1.9	6	1.5	1	-
Heptane	281	96.1	417	43.6	298	88.1	438	50.4	62	16.6	102	25.9	8	2.2	10	1.7	1	1.1
Octane	284	91.2	348	37.4	302	84.9	367	42.2	0	-	47	8.3	0	-	0	-	0	-
Nonane	125[a]	-	266[a]	-	146[a]	-	-	-	0	-	0	-	0	-	0	-	0	-
Decane	175[a]	-	362	278-436[b]	184[a]	-	-	-	0	-	0[a]	-	0	-	0	-	1	0.4
Undecane	217[a]	-	176	108-267[b]	203[a]	-	-	-	-	-	0[a]	-	-	-	0[a]	-	-	-
Dodecane	326[a]	-	207	98-317[b]	286[a]	-	-	-	-	-	0[a]	-	-	-	0[a]	-	-	-
Tridecane	-	-	321	293-362[b]	-	-	-	-	-	-	0[a]	-	-	-	0[a]	-	-	-
Tetradecane	-	-	313	231-392[b]	-	-	-	-	-	-	0[a]	-	-	-	0[a]	-	-	-
Pentadecane	-	-	696	617-796[b]	-	-	-	-	-	-	0[a]	-	-	-	0[a]	-	-	-
Hexadecane	-	-	221	122-286[b]	-	-	-	-	-	-	0[a]	-	-	-	0[a]	-	-	-
Heptadecane	-	-	394	376-493[b]	-	-	-	-	-	-	0[a]	-	-	-	0[a]	-	-	-
Alkenes																		
Ethene	0	-	28	3.3	0	-	28	3.2	0	-	0	-	0	-	0	-	0	-
Butene	12	5.2	32	3.5	13	4.5	33	3.6	0	-	0	-	0	-	0	-	0	-
Pentene	2	1.0	36	4.5	2	0.9	38	3.6	3	0.8	4	4.4	3	1.0	0	-	0	-
Hexene	2	1.1	34	4.6	2	0.9	35	4.9	22	4.7	31	8.5	1	-	1	-	4	0.7
Heptene	46	14.1	111	13.5	45	15.7	116	14.5	12	5.7	31	8.0	2	6.2	1	-	0	-
Octene	22	8.4	95	11.7	20	7.4	97	10.1	0	-	0	-	0	-	0	-	0	-
Nonene	33[a]	-	59[a]	-	48[a]	-	61	9.1	0	-	0	-	0	-	0	-	0	-
Decene	101[a]	-	126	70-161[b]	116[a]	-	-	-	0	-	0[a]	-	-	-	0[a]	-	-	-
Undecene	82[a]	-	78	54-103[b]	93[a]	-	-	-	0	-	0[a]	-	-	-	0[a]	-	0	-
Dodecene	171[a]	-	156	113-229[b]	162[a]	-	-	-	0	-	0[a]	-	-	-	0[a]	-	0	-
Tridecene	-	-	178	121-268[b]	-	-	-	-	0	-	0[a]	-	-	-	0[a]	-	0	-
Tetradecene	-	-	488	324-588[b]	-	-	-	-	0	-	0[a]	-	-	-	0[a]	-	0	-

Pentadecene	-	-	121	98-187[b]	-	-	-	-	0	-	0[a]	-	-	-	0[a]	-	0	-
Hexadecene	-	-	156	116-234[b]	-	-	-	-	0	-	0[a]	-	-	-	0[a]	-	0	-
Heptadecene	-	-	618	583-803[b]	-	-	-		0	-	0[a]	-	-	-	0[a]	-	0	-
Iso-Alkanes																		
2-Methylpropane	27	8.1	45	6.9	28	9.2	47	7.4	0	-	9	1.4	0	-	0	-	0	-
2-Methylbutane	4	1.5	19	2.4	5	1.4	22	3.4	0	-	143	46.0	0	-	0	-	1	-
2-Methylpentane	10	3.4	33	4.3	11	4.1	34	3.1	1	-	28	8.5	1	-	0	-	0	-
2-Methylheptane	11	4.1	29	39.5	12	4.0	24	3.2	27	10.1	47	10.8	4	1.2	4	1.0	0	-
Iso-Alkenes																		
2-Methylpropene	2	0.8	37	4.6	2	0.9	39	5.3	0	-	4	1.0	0	-	0	-	0	-
Alkynes																		
Decyne	25[a]	-	23	18[b]	23[a]	-	-	-	-	-	0[a]	-	-	-	0	-	-	-
Undecyne	9[a]	-	4	0-6[b]	11[a]	-	-	-	-	-	0[a]	-	-	-	0[a]	-	-	-
Dienes																		
Tetradecadiene	-	-	98	83-113[b]	-	-	-	-	-	-	0[a]	-	-	-	0	-	-	-
Pentadecadiene	-	-	73	68-79[b]	-	-	-	-	-	-	0[a]	-	-	-	0	-	-	-
Hexadecadiene	-	-	706	626-829[b]	-	-	-	-	-	-	0[a]	-	-	-	0	-	-	-
Heptadecadiene	-	-	16	12-21[b]	-	-	-	-	-	-	0[a]	-	-	-	0	-	-	-
Aromatic Hydrocarbons																		
Benzene	15	5.1	18	2.5	14	5.0	19	2.1	2	-	0[a]	-	3	-	0	-	0	-
Toluene	50	17.9	65	7.1	50	17.1	66	7.4	48	14.5	73	20.5	3	1.6	6	1.5	0	-
Xylene	4	1.9	1	0.2	3	1.2	1	0.5	7	1.4	7	4.7	1	-	1	-	0	-
Alcohols																		
Methanol	16	6.2	20	3.2	15	5.2	19	2.6	23	6.7	40	9.7	41	12.0	0	-	0	-
Ethanol	76	25.7	122	28.3	73	30.3	124	16.2	9	2.9	15	3.9	18	1.0	0	-	0	-
Ketones																		
Acetone	108	31.3	137	14.9	106	32.8	140	14.1	65	20.1	120	28.1	3	1.1	4	1.1	1	0.6
2-Butanone	71	24.3	88	10.0	72	24.7	90	9.3	5	1.6	10	2.5	5	3.3	0	-	0	-
Aldehydes																		
2-Methylpentanal	11	3.2	10	1.8	10	3.3	11	1.9	0	-	0	-	0	-	0	-	0	-

Table 1. *Continued on next page.*

Table1. *Continued*

	Cobalt Irradiation				*LINAC Irradiation*				*Thermal Sterilization*				*Frozen Control*				*Raw*	
	Cooked		*Uncooked*		*Cooked*		*Uncooked*		*Cooked*		*Uncooked*		*Cooked*		*Uncooked*			
	PPB	*SD*	*PPB*	*SD*	*PPB*	*SD*	*PPB*	*SD*	*PPB*	*SD*	*PPB*	*SD*	*PPB*	*SD*	*PPB*	*SD*	*PPB*	*SD*
Undecanal	-	-	76	52-89[b]	-	-	-	-	-	-	0[a]	-	-	-	0[a]	-	-	-
Dodecanal	-	-	63	53-78[b]	-	-	-	-	-	-	0[a]	-	-	-	0[a]	-	-	-
Tetradecanal	-	-	54	47-65[b]	-	-	-	-	-	-	0[a]	-	-	-	0[a]	-	-	-
Pentadecanal	-	-	46	41-52[b]	-	-	-	-	-	-	0[a]	-	-	-	0[a]	-	-	-
Hexadecanal	-	-	127	94-187[b]	-	-	-	-	-	-	0[a]	-	-	-	0[a]	-	-	-
Octadecanal	-	-	30	26-34[b]	-	-	-	-	-	-	0[a]	-	-	-	0[a]	-	-	-
Hexadecenal	-	-	33	27-41[b]	-	-	-	-	-	-	0[a]	-	-	-	0[a]	-	-	-
Octadecenal	-	-	398	371-432[b]	-	-	-	-	-	-	0[a]	-	-	-	0[a]	-	-	-
Sulfur Compounds																		
Carbonylsulfide	2	0.8	2	0.9	2	0.8	2	0.8	22	16.0	75	16.0	0	-	0	-	0	-
HydrogenSulfide	2	0.8	2	0.6	1	0.6	1	1.7	1	1.4	5	1.3	0	-	0	-	0	-
Ethylmercaptan	7	2.4	9	3.1	7	2.5	11	2.2	0	-	0	-	0	-	0	-	0	-
Dimethylsulfide	4	1.3	5	1.5	4	1.4	6	1.4	0	-	0	-	0	-	0	-	0	-
Dimethyldisulfide	3	1.0	4	0.8	3	1.2	4	0.8	7	2.9	13	3.3	1	-	1	-	0	-
Chloro Compounds																		
Tetrachloroethylene	9	3.5	8	1.5	9	3.1	8	1.8	9	3.2	11	2.8	1	3.0	12	1.8	4	0.8
Miscellaneous																		
Acetonitrile	3	1.0	1	-	3	0.9	1	-	21	12.7	57	14.2	6	9.1	3	0.7	0	-

SOURCE: Reproduced from ref. 19.
[a]Determined for only one sample.
[b]Range of 4 determinations.
0 = below detectable limits
- = not determined

microconstituent level, and, a paucity of the kinds of sophisticated instrumental methods of analysis and information processing that we take for granted today. This situation changed rapidly in the course of recent decades, and it is worth pointing out that irradiated food chemical analysis contributed importantly to our general knowledge of food composition, just as early irradiated food animal feeding studies did for our knowledge of test animal nutrient (especially vitamin) requirements.

A major factor leading to the application of chemical analysis to determine the toxicological safety of irradiated foods was the classification of food irradiation as a food additive by the U.S. Congress in enacting the 1958 Food Additives Amendment of the Federal Food, Drug and Cosmetic Act; soon followed by the Food and Drug Administration determination that the whole product as irradiated and consumed shall be considered the additive for animal feeding test purposes and not extracts/fractions thereof (*8*). Nearly twenty years later, radiation chemist and food irradiation chemistry authority Jack Schubert in effect upheld this determination when he argued against the use of concentrated extracts in animal feeding studies, stating "There really is no substitute for feeding the intact food..."(*9*). This is not to say that qualitative/quantitative chemical analyses of radiolytic products in the context of establishing irradiated food safety wasn't already being done, or would not otherwise have been done; rather, it is to say that this classification by the Congress and follow-up ruling by the FDA over thirty years ago, for all practical purposes rendered chemical evaluation essential given the inherent insensitivity of feeding whole irradiated foods to test animals.

It must be noted in this regard that anti-food irradiation special interest activists who demand the feeding of concentrated radiolytic products (e.g., x1000-or-greater their concentration in the irradiated food in question) are apparently unaware of, or ignore the points that (a) very early-on the FDA required that the whole foods as irradiated and consumed be the feeding test subject, (b) nevertheless, concentrated extracts of irradiated foods have on occasion been fed to test animals in the U.S. and else where with no untoward results, and requiring (c) to justify the subjecting of an identified, individual, known radiolytic product or mixture of same to animal feeding study, it/they would first need to be toxicologically suspect from the structure/reactivity standpoints as well as be being present at significant levels in the food(s). This writer contends that no such case exists following decades of searching, and challenges the activists to submit any they feel otherwise about to scientific peer review with accompanying detailed arguments! It is the strongly held view of this writer that the activists' call for animal feeding trials with concentrated radiolytic products is no more than a "red-herring" obstructionist ploy to begin with; one that appeals to like-minded lay people including predisposed politicians, but is easily "seen through" by competent, knowledgeable scientist professionals, as illustrated in Pamela Zurer's incisive "Food Irradiation: A technology at a turning point" article in the May 5, 1986, *Chemical & Engineering News*. No amount of testing of any kind will ever win over the determined, hard-core, self-serving anti-food irradiation activists in any case and such demands should be treated accordingly. Finally, (d) the more direct testing method of analytical chemical analysis of irradiated foods is typically done on concentrated extracts of the foods in question! That is, the evaluation of concentrated extracts of irradiated foods has not only been done, it has been extensively done via the direct analytical chemical approach, in addition to what has been done on same via animal feeding.

Interestingly, the theoretical and experimental chemistry efforts that contributed so importantly to establishing the toxicological safety of irradiated foods and ingredients were actually for-the-most-part undertaken with the aim of elucidating the underlying chemical nature of changes in eating quality attributes, especially flavor, detected and described primarily through sensory evaluation. In the early Post-World War II period, with the gradual resumption of global food irradiation R&D, irradiation-induced flavor changes became of particular interest to researchers. Such were logically attributed to irradiation-induced changes in "volatile" (a relative term, of course) flavor-note components. With the growing availability of gas chromatographs in particular, beginning in the 1950s, instrumental flavor analysis in general, including "irradiation flavor" of foods took off, and "the rest is history". In connection with the U.S. Army's food irradiation program, which at that time was based at the Quartermaster Food and Container Institute, Chicago, volatile flavor component analysis (of radiation-sterilized meats in particular) began in-earnest in the mid-to-late 1950s at the Army's Natick, Massachusetts Research, Development and Engineering Laboratories. This monumental effort reached its zenith in the 1960s-70s in the hands of Drs. Charles Merritt, Jr. and Pio Angelini and co-workers of Natick's Pioneering Research Division, during which time gas chromatography was coupled to mass spectrometry and computerized data processing to provide a formidable state-of-the-art combination. Progress was also facilitated by the relocation of the Army food irradiation program from Chicago to a new fully equipped facility at Natick in 1962-63. During the earlier part of this period, from the late '50s through much of the '60s a nearby parallel and cooperative effort took place at the Massachusetts Institute of Technology Department of Nutrition and Food Science under the direction of Dr. Emily Wick. At the other end of the State, the analysis of irradiated food flavor components was joined in the early 1960s by Drs. Wassef Nawar and Irving Fagerson of the University of Massachusetts Department of Food Science & Technology. It continues to this day under Professor Nawar with emphasis on detection of irradiated foods through chemical changes in the lipid fraction.

The earlier work of Wick et al, and of Angelini and Merritt et al, along with that of others was reported at the first of three American Chemical Society food irradiation symposia, at Atlantic City in September, 1965, as part of the 150th ACS National Meet Meeting (10). The second ACS food irradiation symposium, at the 161st National ACS meeting at Los Angeles in April, 1971, was, like the first but less-so, a mixed bag of food irradiation chemistry research reports along with more applied papers (*11*). In the meantime, the American Institute of Chemical Engineers had held a similar symposium as well at its 60th National Meeting in September, 1966 (*12*). The third and latest in the series of ACS symposia on the subject, held at the First Chemical Congress of the North American Continent at Mexico City in December, 1975, co-sponsored with the Chemical Institute of Canada and the Mexican Chemical Society, was by contrast virtually exclusively devoted to "hard" irradiated food-related chemistry studies (*13*). This was reflective of the fact that considerable food irradiation chemistry data, accumulated in several countries, was in hand by the mid-1970s. One of those countries, the Federal Republic of Germany and specifically its Institute for Radiation Technology at the Federal Research Center for Nutrition, Karlsruhe, was host to the all-important International Project in the Field of Food Irradiation, or the IFIP, the acronym it was commonly known by.

The IFIP, devoted exclusively to irradiated food safety/wholesomeness research, was created by agreement of nineteen initial supporting countries, signed in Paris in October 1970. Such an international project was deemed desirable, even essential for once-and-for-all establishing unequivocally whether-or-not foods irradiated according to well established principles and procedures to a achieve specific beneficial objectives were safe/wholesome beyond reasonable doubt following the USFDA's ominous Spring, 1968 rejection of a U.S. Army petition to gain FDA approval of radiation-sterilized ham. This highly arguable and questionable action (*see* Ref. 14 for a retrospective analysis) had the effect of putting irradiated food safety in serious doubt, and of casting regulatory approval and industrial implementation under a global moratorium which ended up lasting over a decade, during which the IFIP labored in cooperation with the World Health Organization, etc., to resolve the impasse. Nongovernmental sponsors included the International Atomic Energy Agency which provided the cost of the project leader, and the Organization for Economic Cooperation and Development (OECD) Nuclear Energy Agency which provided the secretariat. The German Federal Research Center for Nutrition and its Institute for Radiation Technology provided facilities at no cost to the Project.

The basic strategy was to centralize irradiated food wholesomeness-safety research with interested countries contributing financial and/or in-kind support rather than have individual countries each pursuing and duplicating such costly, specialized research. After an initial five year period, the IFIP agreement was extended in December, 1975 for an additional three years. In fact, the Project continued on into the early 1980s wrap-up period and contributed enormously to gaining a World Health Organization-Joint Expert Committee on (the safety-wholesomeness of) Food Irradiation (JECFI) "clean-bill-of-health" for all applications up to an overall average dose of 10 kiloGrays (1 megarad; the cut-off dose for the IFIP whereas the wholesomeness part of the on-going U.S. Army food irradiation program was addressing sterilization doses above this cut-off as a de-facto "division-of-labor").

The IFIP sponsored and funded investigations at selected specialized laboratories in Europe (including at the Karlsruhe Center), North America and elsewhere on contract, employing the entire battery of food/additive wholesomeness study methods including animal feeding tests. In the mid-1970s it commissioned a team of internationally recognized authorities on food irradiation chemistry to prepare up-to-date reviews of published literature on (a) chemical effects of ionizing energy absorption in general, (b) radiation chemistry of lipids, (c) proteins, (d) carbohydrates, and (e) vitamins. These reviews became the 1977 hardcover book cited as Reference 3. The then IFIP Director, Dr. Peter Elias of the U.K. states in his "general introduction" to this landmark volume that "the various reviews provide evidence of the great similarity in radiolytic products in related foods treated with radiation doses in the megarad range. The reactions of the protein, lipid and carbohydrate constituents of foods to radiation display a remarkable uniformity. Most of the radiolytic products identified in irradiated foods can also be found in non-irradiated foods; furthermore, many of the identified chemical entities have been generated in foods by other processing techniques. The concentrations of the most abundant of these radiolytic substances are confined to the part per million range with radiation doses normally employed in food processing and these concentrations fall off considerably as the dose ranges are reduced. The available data on the chemical structure of radiolytic products in food and the very low concentrations which have been detected suggest the general conclusion that the health hazard they might represent is negligible."

Clearly, by the mid-1970s knowledge as to what goes on chemically during and following food irradiation had reached a level at which it was able to contribute importantly to growing confidence in irradiated food safety/whole-someness. In 1975 Diehl and Scherz of the Karlsruhe Center proposed the estimation of radiolytic products as a basis for evaluating the wholesomeness of irradiated foods (15), arguing, through the use of their example estimation, that such estimations could (then) readily be done and should be employed along with results of animal feeding studies for legal-regulatory purposes. The following year, Taub, Angelini and Merritt of the U.S. Army Natick Laboratories addressed the question of the validity of extrapolating the conclusions reached on the wholesomeness of an irradiated food receiving high doses to the same food receiving a lower dose (*16*). Employing data on selected volatiles identified and quantified from codfish irradiated to doses between 1 and 30 kiloGrays, plus results of three animal feeding studies on cod and haddock receiving the same range of doses, they concluded that such extrapolation is indeed valid when feeding studies at higher dose(s) indicate no adverse effects and radiolytic product yields are linear with dose (i.e., the products are the same qualitatively at various doses; they merely differ quantitatively, being proportionately less at lower doses), as in their example.

The following year South African radiation chemist-entrepreneur Dr. Rocco Basson summed up then current thinking in a word when he coined the term "chemiclearance" in 1977 (*17*) and applied the concept to evaluating the toxicological safety of irradiated fruits (*18*). At its 1976 meeting, the FAO/IAEA/WHO Joint Expert Committee on (the safety-wholesomeness of) Food Irradiation (JECFI) accepted the approach of evaluating the safety of irradiated foods on the basis of available radiolytic product data as well as on the basis of biological tests including animal feeding studies. Also in 1976, the U.S. Army Medical Research and Development Command contracted the Life Sciences Research Office of the Federation of American Societies for Experimental Biology to convene a committee to conduct a peer review of the volatile radiolytic compounds from radiation sterilized beef identified and quantified in the work of Merritt and Angelini et. al. at the Army Natick Research Center as to their toxicological significance, if any. This review culminated in an August 1977 report (*19*) which states as its "bottom-line" conclusion that "on the basis of the available data (Table 1), the Committee concluded that there were no grounds to suspect that the radiolytic compounds evaluated in this report would constitute any hazard to the health of persons consuming reasonable quantities of beef irradiated in the described manner". Subsequently, the same Committee revisited a handful of selected volatile radiolysis products from the above review, including benzene, since they had become the subjects of closer public health scrutiny in unrelated contexts (e.g., chronic inhalation exposure to them as solvents). Upon doing so, the Committee reaffirmed its earlier conclusion of no cause for concern in a March, 1979 supplement to the above-mentioned report, also available for FASEB (*see* Reference 19). A second March, 1979 supplement reports the view of the Committee on the likelihood of radiolytic compounds of a less volatile nature having toxicological significance forming in radiation-sterilized beef. For this the Committee relied mainly on knowledge of protein radiochemistry in the context that the more volatile compounds largely come from the lipid fraction while the less volatile ones can be expected to arise from the protein fraction of beef macroconstituents. Having much less direct analytical data to consider in this instance the Committee's understandably more guarded conclusion was that "many of the radiolytic products in the concentrations estimated to be present appear to pose no hazards to consumers of beef irradiated in the described manner. No evaluation can be made of other compounds theoretically possible in small amounts, but which have not been demonstrated in irradiated beef or model systems. Because no analysis, however exhaustive, can exclude the possibility of the presence of such theoretical but undetected constituents, no unequivocal demonstration of safety seems possible from consideration of individual radiolytic products alone. It is desirable to couple chemical studies as described in this report with suitable animal feeding studies to provide complementary approaches to ensure the wholesomeness and safety of irradiated foods". This combination of direct chemical analysis together with biological tests is, of course, exactly what has been done and what has provided the basis for approvals and clearances at the national and international levels. It is worth adding in passing that the fundamental uncertainty regarding the composition of foods per-se and in this case irradiated foods down to the "last molecule" is exploited by the hard-core, hopelessly biased anti-food irradiation activists as their tactic of calling for testing of radiolytic products ad-infinitum. Being for-the-most-part anti-nuclear activists at the core, they hope that such obstructionist tactics will indeterminably confound industrial implementation and public acceptance of food irradiation, and thereby preempt what they regard as an important technological manifestation of the atomic-nuclear age that must be thwarted at all costs; a heavy indictment for a proven beneficial technology that currently sterilizes at least 50% of non-heat sterilizable medical products to prevent infections.

In 1978, the International Project in the Field of Food Irradiation (IFIP) established a Coordinated Program on the Radiation Chemistry of Food and Food Components (CORC) in recognition of the increasingly important contribution such information was contributing to answering the questions of irradiated food toxicological safety and nutritional wholesomeness, and, in response to calls for such information to assist in the evaluation of same at the international and national levels. Extensive, up-to-date literature reviews resulting from this program provided much of the basis for the late-1980 Joint Expert Committee on (the safety/wholesomeness of) Food Irradiation (JECFI) meeting's conclusion "that the irradiation of any food commodity up to an overall average dose of 10 kiloGrays presented no toxicological hazard and, hence, toxicological testing is no longer required". The literature reviews that served to support this landmark conclusion appear in up-dated form along with other chapters, including ones on microbiological and nutritional aspects, in Reference 4, published a few years later. The overall theme of this volume, as expressed by Dr. Rocco Basson of the Republic of South Africa who, again, coined the term "chemiclearance", and by Drs. Charles Merritt and Irwin Taub of the U.S. Army's Natick, Mass. R&D Center in their chapter is that, based upon the considerable detailed knowledge of food irradiation chemistry it can be stated with confidence that reaction pathways and end products of food irradiation are highly uniform or common and highly predictable. Also, it is possible to extrapolate bioassay (i.e., animal feeding, mutagenicity, etc.) test results from one food to another within the same class (eg., from one "muscle food" to another regarding changes in the protein or lipid; from one fruit to another in terms of changes in the carbohydrate fraction, and-so-on). Further, one can extrapolate results obtained under one set of conditions (dose, temperature, etc.) to another, and, from food simulating model systems to the actual foods. Perhaps most importantly, not only can radiolytic products of food macroconstituents, and a number of microconstituents such as vitamins, be predicted both qualitatively and quantitatively under a given set of conditions, but exhaustive searching was unable to turn up a single truly unique radiolytic product; or any radiolysis product having significant toxicological potential from the structure/reactivity and quantity standpoints.

In his recent book (*7*) Diehl sums it up with the statement that "the recognition of commonality and predictability of radiolytic changes in irradiated foods has finally shown the absurdity of demanding ever more animal feeding studies...". Referring to the large volume of global food irradiation chemistry studies, Diehl makes the point that "such studies have made it clear, for instance, that radiation does not cause formation of aromatic rings, condensation of aromatic rings and formation of heterocyclic compounds, reactions known to take place at higher cooking temperatures". This real food safety matter is the subject of a recent review (*20*).

All of the foregoing through the first half of 1980 served as de-facto background to the deliberations and conclusions of the U.S.F.D.A. "Irradiated Foods Committee" of several staff scientists representing different disciplines that was authorized by the Agency in Fall, 1979 to (a) "review then current agency food irradiation policy", (b) "to examine the foundation and soundness of that policy and its past implementation", and (c) "to establish those toxicological requirements appropriate for assessing the safety of irradiated food consistent with the level of human exposure, where the degree of testing is consistent with the potential risk as predicated on the level of human exposure", as stated in the Introduction to the Committee's final report dated July, 1980 (*21*). This first intra-agency review of its food irradiation policy since May, 1967 was brought about by such factors as the great amount of new data and information pertaining to irradiated food safety since then, and the growing national and international activity and interest in the process/treatment that developed during the 1970s and was clearly going to continue.

The Committee expressed the view in its final report that little of substance had changed since the May, 1967 review regarding microbiological safety, to which a single short paragraph was devoted "in passing" in the 1980 report (later, during the 1980s microbiological safety became the FDA's primary preoccupation in promulgating new approvals, particularly the May 1990 approval of poultry irradiation, and in not taking action on a petition to gain approval of fishery product radurization). One section of the 1980 final report was devoted to "nutritional adequacy" and the impact of irradiation on essential nutrients, but the major focus was on toxicological safety and the matter of food irradiation chemistry and stable radiolytic end products. In fact, it was the first major application of chemistry-based "quantitative risk assessment" for food safety evaluation. The chemistry upon which the quantitative risk assessment of radiolytic products was based was primarily the data on radiation-sterilized beef developed by Angelini and Merritt and co-workers at the U.S. Army's Natick, Mass. R&D Center's Pioneering Research Division and evaluated by the aforementioned peer review committee assembled by the Life Sciences Research Office of FASEB in the late 1970s (Table 1). Short of going into the details of the July, 1980 FDA Committee report and its more arguable assumptions and estimations regarding radiolytic products, suffice it to say that in a very real sense it marked a food safety-related regulatory turning point at least insofar as FDA handling of food irradiation is concerned, but in a broader sense as well. While the report did not call for the abandoning of animal feeding and short-term biological (i.e., mutagenicity) testing, it predicates same on "qualitative and quantitative estimates of URPs". In other words, in the context of regulatory approval of a given food irradiation application, still treated as a food additive in the U.S., any biological toxicity testing requirements, if any, would hinge on an estimate of the kinds and amounts of "unique radiolytic products" (URPs) generated, which is essentially radiation dose-dependent for a specific food or ingredient under a given set of environmental conditions during and following irradiation.

This has been probably the most controversial aspect of the report in the sense that, besides inadvertently serving to popularize and lend credence to the since much abused and exploited "URPs" term, it came at a time at which those engaged in the field of irradiated food chemistry were reaching a consensus that the process/treatment, as carried out according to established principles and procedures simply should not be expected to, and does not result in the generation of truly unique radiolytic products per se, much less any that pose even a hypothetical significant health risk in terms of structure, reactivity and quantity. As previously mentioned, the JECFI of the World Health Organization plus its concerned sister U.N. agencies expressed this very conclusion at its review meeting later in 1980 in stating that no further toxicological testing is necessary for any food up to an overall average dose of 10 kiloGrays (again, deferring to the U.S. Army's wholesomeness program regarding sterilizing doses). In fairness, however, in applying such quantitative risk assessment for probably the first time in a major way to so confounding and potentially controversial a matter as irradiated food safety, the FDA can be empathized with for taking a most ultracon-servative approach back in 1979-80. Then, the informed consensus about there being no URPs in any meaningful sense was more tenuous than it is today. Nor does it serve any useful purpose to critique a report on this aspect, which has been for all practical purposes superseded by subsequent events and information.

As a case in point, in recently approving poultry irradiation up to a dose three times the original one kiloGray cut off, below which toxicological testing would not be required on the bases of the URP estimates plus "projected levels of human exposure" and "state-of-the-art sensitive toxicological tests" (the three criteria deciding safety testing recommended in the July 1980 report), the FDA's chief preoccupation during its late-1980s deliberations over a permitted dose maximum for poultry was microbiological safety (specifically, the agency's long-standing preoccupation with a hypothetical type-E *Cl. botulinum* risk. A preoccupation that the international food irradiation microbiology "community" does not share and many feel should not be applied to poultry and red meats in any case).

In closing it is best to dwell on the fact that the FDA Committee's landmark July 1980 report fulfilled the promise of all of the irradiated food chemistry research and evaluation that preceded it in recommending a trend-setting analytical chemistry-based policy the correctness of which has been borne out by subsequent related information, and events of the 80s. Not incidentally, the report also acknowledged the validity of "generic" (chemi-)clearance.

Literature Cited

1. Narat, J.K. *Radiobiology* **1927**, *8*, p 41.
2. Minsch, F. *Munchen Med. Wochensch.* **1896**, *5*, p. 101 and *9*, p. 202.
3. Elias, P.S. and Cohen, A.J., eds.. *Radiation chemistry of major foods components: Its relevance to the assessment of the wholesomeness of irradiated foods.* Elsevier Scientific Publishing Co.: Elsevier/North Holland Biomedical Press B.V., Amsterdam, **1977**.
4. Elias, P.S. and Cohen, A.J., eds. *Recent advances in food irradiation.* Elsevier Biomedical Press, Amsterdam, **1983**.
5. Josephson, E.S. and Peterson, M.S., eds. *Preservation of food by ionizing radiation.* Vol. II, CRC Press, Inc. Boca Raton, Florida, **1983**.
6. Urbain, W.M. *Food irradiation.* Academic Press, Inc. Orlando, Florida, and London (U.K. edition), **1986**.
7. Diehl, J.F. *Safety of irradiated foods.* Marcel Dekker, Inc., New York, **1990**.
8. Lehman, A. The new food additives law. Proc. Contractors Sixth Annual Meeting - Quartermaster Corps. Radiation Preservation of Foods Project. Highland Park, Il. 27-29 January, 1959.
9. Schubert, J. Toxicological studies on irradiated food and food constituents. In *Food Preservation by Irradiation, II.* Proceedings of a Symposium. Wageningen, Netherlands, 11/1987). International Atomic Energy Agency STI/PUB/470, Vienna, **1978**.
10. *Radiation Preservation of Foods*; Advances in Chemistry Series 65; American Chemical Society: Washington, DC, 1967.
11. Josephson, E.S. and Merritt, C.E., Jr. *Radiation ion Research Reviews* **1972**, *3*(4).
12. Brownell, L.E., Josephson, E.S., Manowitz, B. and Harmer, D.E., eds. *Chemical and food applications of radiation. Nuclear engineering - part XIX.* Chemical Engineering Progress Symposium Series No. 83, Vol. 64 **1968**.

13. *J. Agric. and Food Chemistry* **1978**, *26* , 1-2.
14. Giddings, G.G. *Food Reviews International* **1986**, *2*(1), 109-137.
15. Diehl, J.F. and Scherz, H. *International J. Applied Radiation and Isotopes* **1975**, *26*, pp. 499-501.
16. Taub, I.A., Angelini, P., Merritt, C., Jr. *J. Food Science* **1976**, *41*, 942-44.
17. Basson, R.A. *Nuclear Active* **1977**, *17*, 3
18. Basson, R.A., Beyers, M., Thomas, A.C. *Food Chemistry* **1979**, *4*, 131 (a related paper by the same authors appeared in *Food Preservation by Irradiation II*. Proceedings of a symposium. Waginingen, Netherlands, 11/1977. International Atomic Energy Agency, STI/PUB/470, Vienna, **1978**).
19. FASEB-Life Sciences Research Office. **1977**. Evaluation of the health aspects of certain compounds found in irradiated beef. FASEB-LSRO 9650 Rockville Pike, Bethesda, Md 20014. 113pp. (Also available, two March, 1979 supplements under the same title.)
20. Chem, C.; Pearson, A. M. ; and Gray, J. I. *Advances in Food and Nutrition Res.* **1990**, *34*, 387-449.
21. USFDA. *Recommendations for Evaluating the Safety of Irradiated Foods*. Report of the Irradiated Foods Committee, prepared in July, **1980**.

RECEIVED August 15, 1991

Chapter 30

Safety Issues with Antioxidants in Foods

P. B. Addis and C. A. Hassel

Department of Food Science and Nutrition, University of Minnesota, St. Paul, MN 55108

During the past decade, two important areas of research have emerged which have intensified interest in the risks and benefits of antioxidants. Increased awareness of potential health benefits of antioxidants has occurred because of three lines of evidence: (1) dietary lipid oxidation products (LOPS) have been shown to have numerous deleterious effects *in vivo* and *in vitro*; (2) *in vivo* oxidative modification of low-density lipoprotein (LDL) may greatly amplify the atherogenicity of the LDL particle; and (3) antioxidants reduce exposure of humans to dietary LOPS and reduce the oxidative modification of LDL in vitro and may well do so *in vivo*. A number of other biological benefits have been reported.

In contrast to the foregoing benefits, another emerging research area has encompassed the possible detrimental effects of antioxidants. Although usually seen only at very high levels of intake, the study of putative risks associated with antioxidant consumption has developed into an interesting and somewhat controversial area of research and certainly one worthy of consideration.

The present status of knowledge of antioxidants does not permit an accurate quantification of a risk/benefit ratio but it is possible to estimate the magnitude of some of the most important parameters that will be used in the calculation. We feel confident that our conclusions concerning the overall efficaciousness of antioxidants, weighing risks against benefits (including potential health benefits), are reasonable and should be used in decisions regarding whether to continue to use certain antioxidants. In addition, our hope is that this review will help researchers to formulate future plans for antioxidant research.

It would be impossible in the space of this review to attempt a comprehensive review of the literature published on the subject. Rather, after a brief review of types, mechanisms of action and uses of antioxidants, we will focus primarily on toxicology of α-tocopherol and carcinogenicity of butylated hydroxyanisole (BHA). Next, a discussion will ensue of the benefits of antioxidants, especially possible health benefits of antioxidants with regard to coronary artery disease (CAD). A discussion of important emerging issues of food processing as they pertain to antioxidant usage

0097–6156/92/0484–0346$08.75/0

will follow. We will end by estimating risk/benefit of antioxidants and by making some recommendations for future research.

Antioxidants: Types and Mechanisms

Oxidation is one of the primary modes of deterioration of food. As food processing becomes more complex and food products are subjected to ever lengthening storage periods, the opportunity for oxidative damage greatly increases. Therefore, the need for diligence with regard to prevention of lipid oxidation has never been greater than at the present.

Lipid oxidation influences several important properties including sensory quality (flavor, odor, texture and color), nutritional value, functionality and toxicity (1). Although oxidation usually begins with the lipid moiety, eventually other components are affected including proteins, nutrients and pigments. Oxidation of lipids proceeds as a true chain reaction with a three phase path: initiation (Equation 1), propagation (Equations 2, 3 and 4) and termination (Equations 5 and 6).

$$L{:}H \longrightarrow L\cdot + H\cdot \qquad (1)$$
$$L\cdot + O_2 \longrightarrow LOO\cdot \qquad (2)$$
$$LOO\cdot + L^1{:}H \longrightarrow LOOH + L^1\cdot \qquad (3)$$
$$LOOH \longrightarrow LO\cdot + OH\cdot \qquad (4)$$
$$L^1\cdot + L^1\cdot \longrightarrow L^1{:}L^1 \qquad (5)$$
$$LOO\cdot + L^1\cdot \longrightarrow LOOL^1 \qquad (6)$$

The initiation step involves the creation of radicals by reaction of the lipid with singlet oxygen, high energy radiation or transition metals (*1-3*). Propagation involves transforming the carbon-centered radical (L·) to a more reactive peroxyradical, thereby setting into motion a chain reaction (Equations 2, 3 and 4). Termination involves formation of non-radical products (Equations 5 and 6).

Antioxidants are chemicals which are absolutely critical in some cases to the retardation of rancidity in foods. There are two major types of antioxidants: chelators and radical scavengers (chain interruptors). An excellent review of iron chelation by phytate has been published by Graf and Eaton (*4*). Removal of iron from solution by chelation and/or chelation by phytate, which by occupying six coordination sites on iron prevents production of ·OH, substantially reduce lipid oxidation (*4*). Chain interruptors react with radicals formed in oxidation as illustrated in the following reactions outlined by Kahl and Hildebrandt (*3*):

$$LOO\cdot + AH \longrightarrow LOOH + A\cdot \qquad (7)$$
$$LO\cdot + AH \longrightarrow LOH + A\cdot \qquad (8)$$
$$L\cdot + AH \longrightarrow LH + A\cdot \qquad (9)$$
$$OH\cdot + AH \longrightarrow H_2O + A\cdot \qquad (10)$$

An important characteristic of a good radical scavenger antioxidant is illustrated in the following (*3*):

$$A\cdot + LH \longrightarrow N.R. \qquad (11)$$

The phenoxy radical should not abstract a hydrogen from a fatty acid (LH) and thereby become part of the propagation process (*3*).

Antioxidants are a diverse group of compounds, varying from lipid-soluble to water-soluble, natural to synthetic, and phenolic/chain-breaking to chelation. Structures of representative antioxidants are shown in Figures 1-5. Chelators (Figure 1) function as scavengers of transition metal ions, long recognized as key catalysts in lipid oxidation. The ratio of chelator to transition metal is critical (*2*). A recent review by Miller et al. (*6*) outlines in great detail the pivotal role of transition metals and provides substantial evidence for transition metals in a requisit role for "autoxidation." It is highly recommended to the reader.

Chain-breaking antioxidants (Figures 2 and 3) are scavengers of free radicals. The toxicity of some of these will be discussed later. Probucol has the interesting and potentially very significant antioxidant property of inhibiting the oxidative modification of LDL to the highly atherogenic modified LDL (mLDL). Figure 4 illustrates a few of the lipid-soluble, natural chain-breaking antioxidants available, some of which (tocopherols, tocotrienols) are commercially available in synthetic forms. Flavanoids are able to scavenge superoxide anion and express radical scavenging antioxidant activity (*7*).

In recent years there has developed an increased awareness of the important detrimental role played by enzyme-catalyzed oxidation in biological tissue and in certain foods post-harvest (*2*). Figure 5 illustrates some enzyme inhibitors which may be useful in retarding oxidation. Oleic acid inhibition of lipoxygenase (*8*) may be the most practical in terms of prompt application because of the stability and lack of toxic effects of oleic acid and the importance of lipoxygenase.

The application of antioxidants can be extremely complicated and diligence is required to avoid pitfalls. For example, low levels of vitamin C express a prooxidant effect in some systems whereas high levels of vitamin E are prooxidative in others. Chelators and radical scavengers are generally synergistic. Recent research of Niki (*9*) has elaborated the intimate interrelationship between vitamin C and E as illustrated in Figure 6. Vitamin C acts to regenerate vitamin E from the vitamin E radical.

Antioxidants are widely employed by the food industry and numerous benefits to both processors and consumers result from antioxidant usage. However, there are a number of controversies surrounding the use of antioxidants. The levels of some antioxidants being used in foods are considered by some to be too high. BHA has been suggested by some researchers to be carcinogenic. BHA recently has been placed on the Proposition 65 list in California. Even vitamin E has been shown to be toxic at high levels. Therefore, a review of the safety/toxicology issues related to antioxidant use appears to be appropriate. Subsequently, we will review the potential health benefits of high levels of antioxidant intake and the use of high levels in food products to prevent accumulation of LOPS.

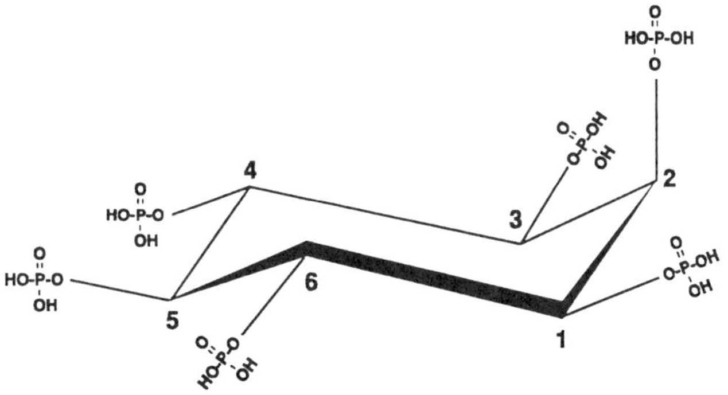

Phytic Acid
(Myo-inositol hexaphosphate)

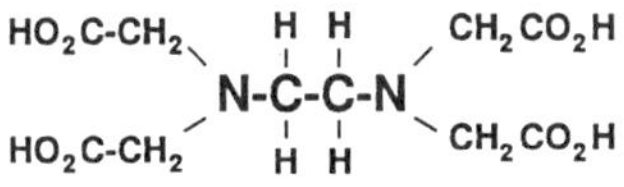

Ethylenediamine tetraacetic acid (EDTA)

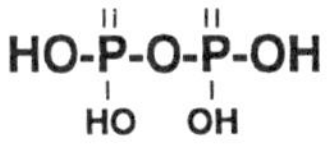

Pyrophosphate

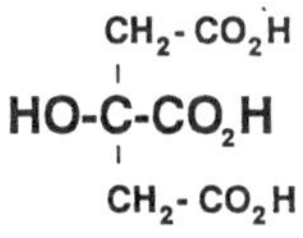

Citrate

Figure 1. Chelators known to possess biological or food antioxidant capability. Phytate derivatives possessing 3-5 phosphates are effective also at inhibiting "superoxide-driven generation of · OH" as discussed by Graf and Eaton (4). Pyrophosphate is but one of many condensed phosphates capable of iron-binding and antioxidant effects. All require some degree of deprotonation for antioxidant activity.

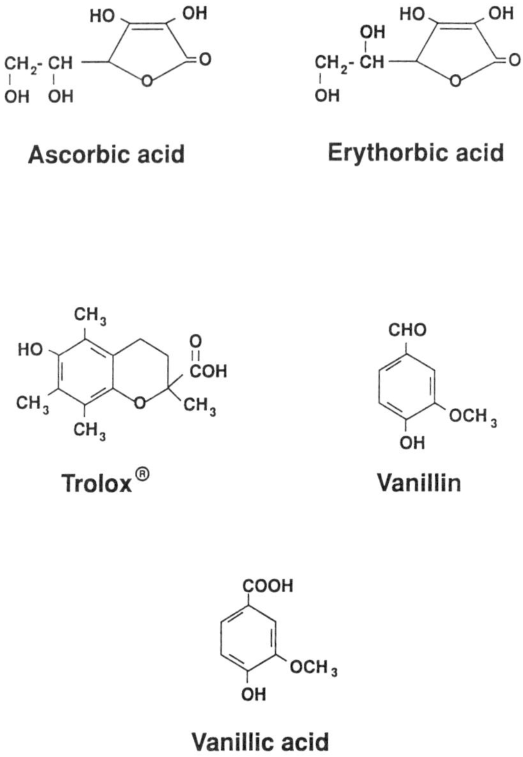

Figure 2. Water-soluble, chain-breaking antioxidants.

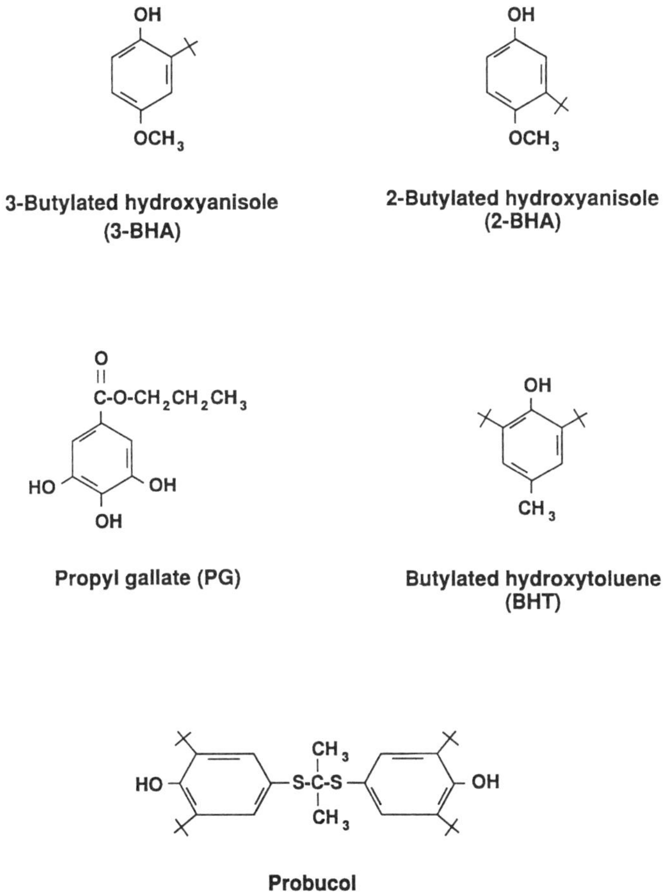

Figure 3. Lipid-soluble, synthetic, chain breaking antioxidants. Commercial BHA is a mixture of the 2- (4.5 %) and 3-BHA (95.5 %) isomers; 3-BHA predominants and is far the stronger antioxidant of the two and is more active at inducing forestomach carcinogenesis. Probucol, which possesses antioxidant activity, is a drug used in the treatment of coronary artery disease.

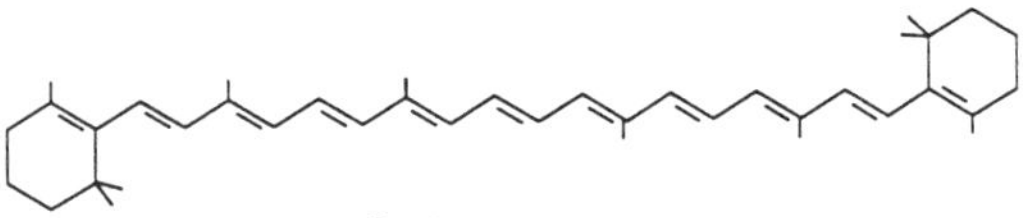

β-Carotene

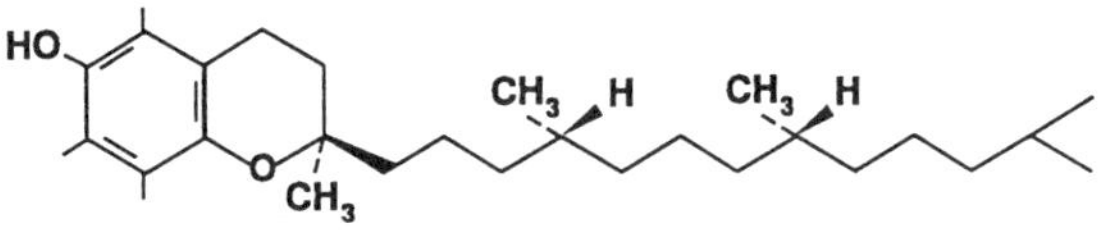

d-α-Tocopherol
(natural vitamin E)

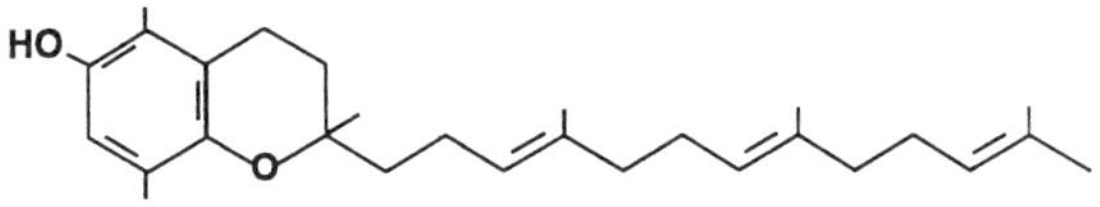

d-β-Tocotrienol

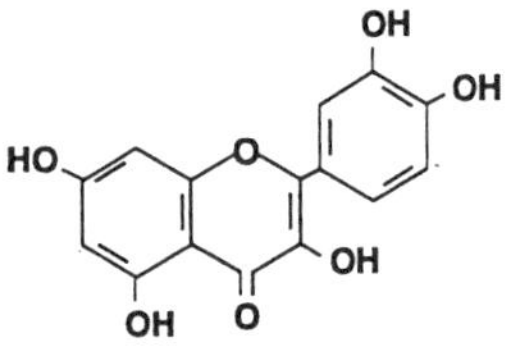

Quercetin

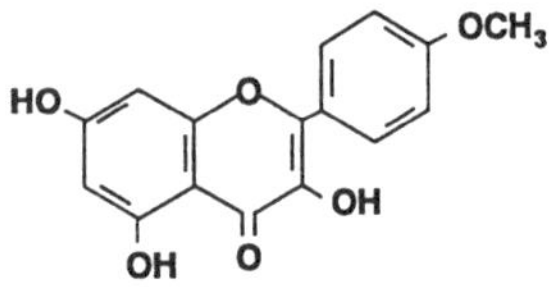

Acacetin

Figure 4. Lipid-soluble, natural chain-breaking antioxidants. Synthetic vitamin E also is available as a mixture of eight stereoisomers. Natural vitamin E and tocotrienol isomers include α, β, γ and δ-isomers.

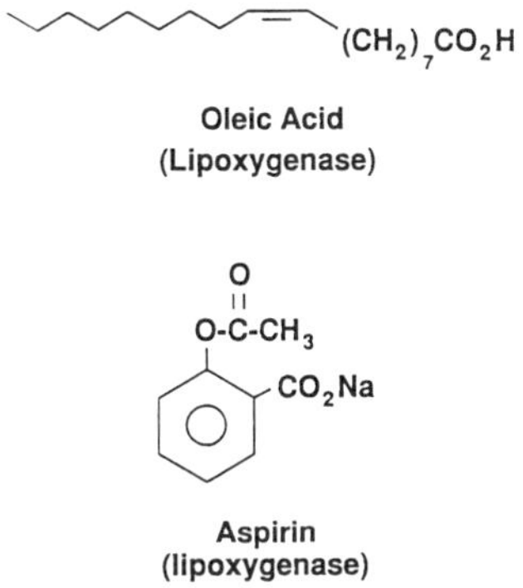

Figure 5. Inhibitors of enzyme-catalyzed lipid peroxidation along with the corresponding enzyme (listed in parentheses).

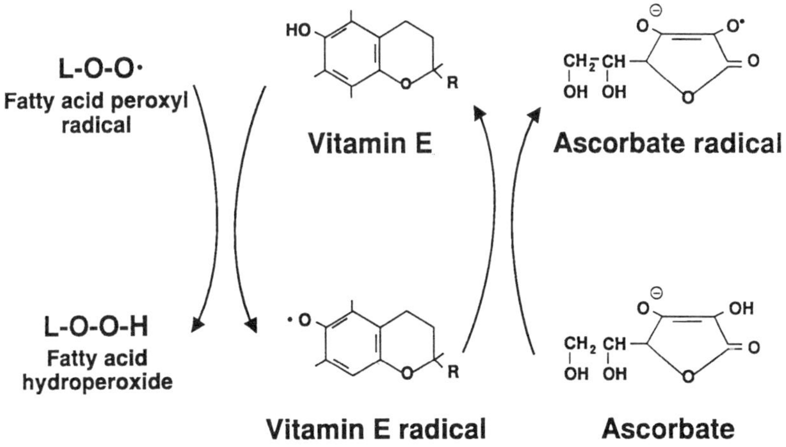

Figure 6. Vitamin E regeneration, after reaction with lipid peroxy radical, by vitamin C (as shown by Niki [9]).

Studies of Potential Health Risks of Antioxidants

Two main areas of research on health risks have involved various forms of (1) acute and chronic toxic effects and (2) carcinogenicity. For both types of studies, it has been necessary to use extremely high levels to achieve the sought after deleterious effect and as a result some of these studies are exceedingly controversial. The authors believe that it is interesting and significant that in some of the same studies which show "carcinogenic" effects of BHA the anticarcinogenic properties of BHA have also been demonstrated! We will first discuss toxicity of vitamins E and C and the food additive antioxidants (BHA, BHT), although some comments about mutagenicity of antioxidants will be made where appropriate. Next, we will focus on the carcinogenic effects of synthetic phenolic antioxidants.

Table 1. Dose Equivalents Between Rat and 100-kg Man for Toxicity Study of d-α-Tocopheryl Acetate

Groups	*Dose equivalents, added vitamin E*	
	Rat, mg/kg/day[a]	*Man, g/100 kg/day*[b]
Untreated controls	0	0
Corn oil controls	0	0
d-α-tocopheryl acetate-1	125	12.5
d-α-tocopheryl acetate-2	500	50.0
d-α-tocopheryl acetate-3	2000	200.0

SOURCE: Reference 10.
[a]Amount fed by gavage daily during 13-week study; 10 rats per group.
[b]Calculated equivalent intake for 100 kg man.

Toxicity. Abdo et al. (*10*) conducted a 13-week study in rats of d-α-tocopheryl acetate given by gavage. Daily doses of 0, 125, 500 or 2000 mg/kg body weight were given. Corn oil was used as the vitamin E vehicle at 3.5 ml/kg daily. In Table 1, we have outlined the design of this experiment and extrapolated the levels of tocopheryl acetate intake in the rat to that of a 100-kg man. No effect on body weight or food consumption was noted (*10*). The 2000-mg dose increased the liver-to-body weight ratio of females, the prothrombin time and the activated partial thromboplastin time, reticulocytosis, and decreased hematocrit values and hemoglobin concentration in males. Deaths occurred in 7/10 male rats at the 2000 mg level. Internal hemorrhage was the cause of death. Vitamin E at all levels studied caused interstitial inflammation and adenomatous hyperplasia of the lung. The authors concluded that the "findings indicate that vitamin E administration in excessive amounts is potentially toxic."
It is the opinion of the authors and several other scientists that the levels used in the study by Abdo et al (*10*) were simply too unrealistically high to raise any bona fide concern about toxicity of vitamin E. Using the extrapolated 200-g per day (per 100-

kg man) figure, it can be calculated that in 13 weeks the total consumption of vitamin E acetate would exceed 18 kg! We believe that, quite to the contrary to the conclusions of Abdo et al. (*10*), it is remarkable that 3/10 males and 10/10 females survived for 13 weeks at the highest intake level and conclude that either the rat is resistant to toxic effects of vitamin E or the vitamin is exceedingly safe!

A comprehensive review of the safety of tocopherols as food additives has been published by Tomassi and Silamo (*11*) who concluded that "tocopherols are safe food additives." This conclusion was based on consideration of acute, subchronic and chronic toxicity data, reproduction and teratogenesis data and results obtained on humans. Studies have, however, shown a number of interactions with other fat-soluble vitamins (A, D, K). The toxic effects of excessive intake of vitamin A can be either intensified or reduced by tocopherol, depending upon the specific effect. Vitamin E (10,000 I.U. per kg feed) can interfere with vitamin D utilization.

The hemorrhage seen by Abdo et al. (*10*) is at least partly due to interference by vitamin E with vitamin K (*11*). High levels of E could also influence the "redox tone" in the organism which is important in platelet aggregation. Therefore, the fact that hemorrhaging results from high level tocopherol intake is not surprising.

A very comprehensive review of vitamin E, including dietary sources, intestinal absorption, transport and tissue metabolism has been published by Parker (*12*). Absorption of vitamin E occurs via lymphatic chylomicrons and tocopherols rapidly exchange among other lipoprotein particles so that, with respect to tocopherol concentration in plasma very-low-density lipoprotein (VLDL) > LDL > high-density lipoprotein (HDL).

The fact that LDL is an important transport mechanism for vitamin E is significant because, as will be discussed later, oxidative modification of LDL is opposed by most common antioxidants.

Vitamin C is another vitamin which, like E, is megadosed by many people, perscribed for certain medical problems, used as a food additive and is fairly abundant in many foods. An evaluation of high level intake of vitamin C in humans has been published by Rivers (*13*). The issues reviewed included oxalate (and kidney stone) formation, uric acid excretion, vitamin B_{12} status, iron overload, ascorbic acid dependancy induced scurvy and mutagenicity. In every case, the proposed deleterious side-effects of high-level intake of vitamin C is dismissed by Rivers (*13*) based on his evaluation of the scientific literature (74 references). In particular, the data indicating mutagenic effects of vitamin C were unexpected because of the well-established antimutagenic properties of ascorbic acid. Rivers (*13*) concludes "there is no evidence that high intakes of ascorbic acid will be mutagenic in man." Nevertheless, concerns about potential mutagenic effects of ascorbate persist. Metal-ion catalyzed oxidation of ascorbate will produce ascorbate radicals and ·OH. Such concerns prompted Champagne et al. (*14*) to study ascorbate free radicals in infant formulas as affected by pH and phytate content. Alkaline pH tended to increase levels of ascorbate radical, possibly an important finding pertaining to infants and the elderly, two groups characterized by inability to maintain gastric acidity. Phytate was effective at reducing radical levels by up to 50% in model systems but not in soy-based and phytate-spiked milk-based formulas. The authors stated a concern for persons consuming high levels of ascorbic acid who also have reduced stomach acid production (*14*). Much more research needs to be done before these issues are resolved.

Carcinogenicity. Solid data showing bona fide carcinogenic effects of vitamins C and E are rare. Synthetic antioxidants such as BHT and BHA are known to have some toxicities at high levels but the primary concerns are carcinogenicity and these will be our primary focus. Although some of the research to be reviewed in detail includes both carcinogenic and toxicological data, because of space limitations we will focus primarily on BHA putative carcinogenicity. Several toxicological studies are recommended to the reader, including Van Esch (*15*) on *tert*-butylhydroquinone (TBHQ),

Table 2. Dose-response study of BHA in rats: forestomach neoplastic changes

Dietary BHA			*Rats, forestomach changes, %*		
%	*mg/kg/day*	*Ni/Nf*[a]	*Hyperplasia*[b]	*Papilloma*[c]	*Squamus-cell carcinoma*[d]
0	0	50/37	0	0	0
0.125	55	50/41	2	0	0
0.25	110	50/40	14	0	0
0.5	230	50/42	32	0	0
1	428	50/38	88	20	0
2	1323	50/42	100	100	22

SOURCE: Modified from Ito et al. (*20*) on changes in forestomachs of F344 rats fed control or BHA-diets for 104 weeks.
[a]Ni/Nf = number of rats starting and completing study.
[b]Epithelial cell hyperplasia.
[c]A benign tumor.
[d]A malignancy.

van der Heijden (*16*) on propyl gallates (PG), Wurtzen and Olsen (*17*) on chronic BHT effects in rats (in which BHT treated rats outlived control rats!), Tobe et al. (*18*) on BHA in dogs, and Wurtzen and Olsen (*19*) on BHA in pigs.

Studies which have suggested a carcinogenic effect of BHA have been of pivotal importance in the debate surrounding this important antioxidant additive.

An important contribution to BHA carcinogenicity knowledge is the work of Ito and coworkers on mice, rats and hamsters (*20*). BHA fed at either 1 or 2% of diet induced hyperplasia, pappilloma and squamous-cell carcinoma in the forestomachs of rats and hampsters at prevalences greater than the control group. Similar differences were noted in mice for hyperplasia and papilloma but not for squamous-cell carcinoma. Ito et al. (*20*) also conducted a dose-response study of BHA in rats (Table 2). The results indicated that 2% BHA (but not 1% BHA) induced an increased prevalence of squamous-cell carcinoma, the only true malignancy noted. Papillomas, a benign tumor, developed in 20 and 100% of the rats at 1% and 2%, respectively. A clear dose-response effect was seen for BHA on the incidence of epithelial hyperplasia, in some respects an indicator of epithelial irritation.

Other findings by Ito et al. (*20*) included the identification of 3-*tert*-BHA as the active component of crude BHA with respect to carcinogenicity, additive effects of several antioxidants when combined individually with BHA for forestomach carcinogenicity and both enhancement and inhibition of carcinogenesis induced by chemical carcinogens by BHA, depending upon the tissue and specific carcinogen applied. Ito et al. (*20*) failed, however, to report any carcinogenic effects of any antioxidant tested for glandular stomach carcinogenesis. Compounds tested on the glandular stomach were BHA, BHT, sodium ascorbate, sodium erythorbate, and α-tocopherol.

The study by Ito et al. (*20*) is somewhat representative of rat forestomach carcinogenicity studies conducted previously by the same laboratory and by others. Because it is obvious that high levels of BHA intake are carcinogenic in the rat forestomach, several issues need to be discussed relative to the potential carcinogenic effects of BHA in humans. First, the relevance of results obtained on a unique organ, the forestomach, needs to be considered. Most studies have shown that the BHA-cancer linkage can only clearly be established in the forestomach, an organ found almost exclusively in rodents. Rats, hampsters and mice infrequently show BHA-carcinogenicity in tissues other than the forestomach and only at very high levels. Second, the types of lesions quantified by Ito et al. (*20*) need clarification. Epithelial hyperplasia is a state of increased cell division, likely induced by the irritation effect of high levels of BHA exposure. A high rate of cell division renders tissues more susceptible to carcinogens but does not, in itself, constitute the development of cancer. Papillomas are benign tumors, an indication of neoplastic changes but not a malignant cancer. Of the three parameters studied, only the squamal-cell carcinoma constitutes an irreversible malignancy. It is not the purpose of our discussion to attempt to diminish the importance of the findings of Ito et al. (*20*); clearly, the results indicate the need for prudence with regard to using BHA and also the need for further study. We wish only to interpret for the reader clearly the nature of the results.

Also relevant to this discussion is the matter of BHA-intake in rats vs. the quantity consumed by humans. The results of Kirkpatrick and Lauer (*21*) indicate a mean intake of 0.13 mg/kg/ day of BHA in Canada. Therefore, levels used by Ito et al. (*20*) were about 10,000-fold higher than human intake. This figure is 100-fold greater than the usual 100-fold safety factor used for interspecies comparisons of food additive toxicities.

A more recent study by Nera et al. (*22*) emphasized the importance of cellular proliferation due to BHA-induced epithelial irritation as a key aspect of BHA-carcinogenicity. Tumor induction was found to be highly dependent upon cellular proliferation in 2% BHA-fed rats; proliferation in turn depends upon the 2% dietary BHA. A key finding by Nera et al. (*22*) was that BHA-induced carcinogenesis was reversible in rats fed 2% BHA for up to 12 months. It appeared that 2% BHA is the maximum tolerated dose (MTD) for rats as a 20% decrease in body weight was noted at this level of BHA intake. Squamal cell carcinoma was seen in 2 of 37 rats after 24 months. Papillomas were commonly seen at 2% BHA. BHA also induced increased rate of cellular proliferation in bladder tissue, an effect which was reversible upon removal of BHA from the diet. The authors (*22*) concluded that BHA is not genotoxic, not an initiator of cells, but is likely an epigenetic carcinogen (tumor promoter).

Carcinogenesis irritation in the rat forestomach is likely due to trace contaminants in the feed or refluxing of glandular stomach contents containing N-nitroso or other initiating compounds.

With regard to the relevance of rat-BHA data to humans, Nera et al. (*22*) concluded:

"BHA is a forestomach carcinogen that does not appear to be effective in species such as humans that do not possess this organ." Regarding proliferation: "The induction of this proliferation is thresholded at a level very considerable above that to which humans are normally exposed."

The overall conclusion (*20*): "It appears very unlikely therefore that BHA will be effective as a carcinogen in humans exposed to much lower and non-toxic levels of this antioxidant."

The controversial nature of carcinogenic effects of antioxidants stems in part from an incomplete appreciation of the potential health benefits of antioxidants. Serious assessment of the efficacy of a food additive must include both risks and benefits. In the case of antioxidant additives, a somewhat unusual situation exists because in addition to the well-known technological benefits of antioxidants (*1*), evidence is mounting that antioxidant usage may have substantial positive contributions to human health. Consideration of these potential benefits is worthy of careful consideration.

Health Benefits of Antioxidants

Antioxidants are not unique among food additives in conferring a health benefit. Additives which inhibit microorganisms help to prevent food borne disease. Nitrite, for example, inhibits the outgrowth of spores of *Clostridium botulinum*, thereby preventing botulism. What is unusual, if not unique, about the antioxidants is that they possess properties which, after digestion-absorption-assimilation, are beneficial to the organism. In other words, the benefits realized are in addition to the benefits conferred by inhibiting chemical deterioration reactions in food. The benefits directly enjoyed by the organism of antioxidant intake include nutrient protection in tissue, amplification of immune response, ameleoration of the detrimental effects of some drugs and some toxins and retardation of deleterious lipid peroxidative reactions in the organism. Antioxidants are also crucial to the protection of foods against oxidative deterioration, thereby reducing the exposure of consumers to toxic food-borne LOPS.

Of all of the foregoing benefits to be discussed, the possible beneficial effects of antioxidants on CAD are the most exciting for several reasons. The number of lives affected by CAD exceeds all other diseases combined. The advances made in the past decade of research on CAD, as related to both dietary and *in vivo* LOPS, have been exceptional. Therefore, our primary focus will be to explore the possibility that antioxidants might be able to slow the progression of atherosclerosis and reduce the prevalence of CAD. As will be seen, lipid oxidation in the atherosclerotic arterial wall has become a focal point of the most advanced and promising research being done on CAD at the current time. Therefore, antioxidants are playing an important role in helping scientists understand atherosclerosis at its most fundamental level and may help provide a therapy or "cure" for the disease as well.

Antioxidants and Coronary Artery Disease. Before dealing with specific aspects of the oxidative modification of LDL, it is desirable to develop an appreciation for the importance of CAD and the classical hypothesis on diet, lipoproteins and atherosclerosis.

1. Prevalence and Epidemiology of Coronary Artery Disease. CAD continues as the leading cause of death among Americans. Over one million heart attacks occur each year (two-thirds of them in men) and more than 500,000 people die as a result (*23*). It was estimated that 65 million Americans have some form of heart or blood vessel disease and that the cost of CAD in 1985 exceeded $ 50 billion in direct health care expenditures and lost productivity. Despite these alarming figures, an encouraging reduction in CAD mortality has occurred; between 1964 and 1985 a 42% reduction in CAD mortality accompanied medical advances in health care and lifestyle changes.

Epidemiological studies suggest quite strongly that there is no single cause of CAD, rather it is a disease of multifactorial etiology. In fact, the process of atherosclerosis leading to CAD may not be a single disease but a group of related diseases (*24*). Atherosclerosis is characterized by the presence of numerous macrophage-derived "foam cells" that develop on the inner portion of the artery wall, restricting blood flow (*25*). Occlusion of the coronary arteries leads to myocardial infarction, which accounts for about 60% of atherosclerosis-related deaths. Epidemiological studies have exposed statistical associations known as risk factors that often accompany atherosclerosis and CAD. A variety of risk factors for CAD have been uncovered, some amenable to lifestyle changes and others impervious to intervention. Family history of premature heart disease is a major risk factor, signifying the importance of heredity in CAD. Other unmodifiable risk factors include increasing age and male sex. The modifiable risk factors include smoking, hypertension, hypercholesterolemia, glucose intolerance, obesity and physical inactivity. These risk factors tend to interact synergistically; any combination of risk factors are known to impose greater risk than the sum of their independent effects. On the basis of available data, the modifiable risk factors together account for about 65% of the variance in CAD incidence (*26*). Within the U.S. population, roughly 35% of CAD incidence can be attributed to unknown or unmodifiable risk factors.

Current CAD prevention strategies focus attention on reducing the prevalence of major modifiable risk factors within the population (*23*). This approach has led to improved prospects for therapy and prevention. Yet the actual events of atherosclerosis leading to CAD remain poorly understood. Often, risk factors are taken to be the causes of atherosclerosis or CAD. Although risk factors may be causative in certain circumstances, such as familial hypercholesterolemia (*25*), they more often appear as contributors to the overall disease process. Knowledge of underlying causes is still fragmentary, and the actual causes probably vary considerably among individuals (*24*). Therefore, one of the foremost challenges now confronting cardiovascular research is to move beyond the risk factor concept to a comprehensive understanding of the actual events and mechanisms responsible for the initiation and progression of atherosclerosis (*27*). A fuller basic understanding of atherogenic processes will almost certainly identify additional factors that may be amenable to therapeutic intervention. We will see that new lines of research implicate oxidative processes as playing a central role in atherogenesis and explore the potential value of antioxidants as therapeutic agents.

2. Plasma Cholesterol, Lipoproteins and Atherosclerosis. Many lines of research suggest that elevated plasma cholesterol levels are consequential in promoting the formation of atherosclerotic lesions. Atherosclerosis of clinical significance develops when influx and deposition of cholesterol exceeds efflux of cholesterol from the artery wall (*28*). The cholesterol that accumulates within the artery wall is derived from plasma lipoproteins, spherical particles which function to transport non-polar lipids through the bloodstream. Because the plasma lipoproteins play very distinct roles in the metabolism of cholesterol, it is important to identify those lipoproteins responsible for the delivery of cholesterol to the site of atherosclerosis.

The plasma lipoproteins may be separated into five major classes (Table 3) according to their size, density and net surface charge: chylomicrons, very low density lipoproteins, intermediate density lipoproteins, low density lipoproteins, and high density lipoproteins. Each class is heterogeneous in size and composition, and several subclasses have been identified.

Chylomicrons are the largest and least dense of the lipoproteins, containing as much as 90% triglyceride in their hydrophobic core. Chylomicrons are synthesized in the intestine and are the vehicle by which dietary cholesterol and triglycerides are transported from the site of absorption to various tissues throughout the body.

Very low density lipoproteins (VLDL) are synthesized in the liver and function as carriers of endogenously synthesized triglycerides and cholesterol for distribution to peripheral tissues. VLDL and chylomicrons are generally referred to as triglyceride-rich lipoproteins.

Intermediate density lipoproteins (IDL) are actually VLDL "remnant" particles, generated through the progressive catabolism of VLDL core triglycerides (*29*). IDL particles are therefore smaller and more dense than VLDL. They appear in postprandial plasma and are rapidly cleared from the bloodstream.

Low density lipoproteins are the major carriers of cholesterol and cholesterol esters in human plasma. They are derived from VLDL through a catabolic cascade that first generates IDL. The plasma concentration of LDL is positively correlated with the pathogenesis of CAD (*30*).

High density lipoproteins (HDL) are the smallest and most dense of the plasma lipoproteins.

The synthesis of HDL appears to occur in the liver and intestine, as well as within plasma during the lipolytic processing of the triglyceride-rich lipoproteins (*31*). HDL are thought to function in the transport of excess cholesterol from peripheral tissues to the liver for excretion (*32*).

Epidemiological studies indicate that as LDL cholesterol levels increase, the relative risk of CAD also increases (*26*). Other lines of evidence have established that elevated LDL levels are associated with accelerated atherogenesis (*28,30*). By contrast, as high-density lipoprotein (HDL) cholesterol levels increase, the relative risk of CAD decreases (*33*). Thus, elevated LDL levels promote development of atherosclerosis whereas high HDL levels tend to have a protective effect. In view of these general trends, it is interesting to note that numerous exceptions do occur (*26*); if one

Table 3. Physical and Chemical Properties of the Major Plasma Lipoproteins

				Chemical Composition, % of dry mass				
Lipoprotein[a]	*Density (g/ml)*	*Molecular Weight*	*Diameter (nm)*	*Total Protein*	*Phospholipids*	*Free Cholesterol*	*Cholesteryl Esters*	*Triglycerides*
Chylomicrons	0.93	400,000,000	80-500	2	7	2	3	86
VLDL	0.93 -1.006	10-80,000,000	30-80	8	18	7	12	55
IDL	1.006-1.019	5-10,000,000	25-35	19	19	9	29	23
LDL	1.019-1.063	2,300,000	18-25	22	22	8	42	6
HDL[b]	1.063-1.210							
HDL_1[c]	1.04 -1.09	1,300,000	10-14	27	32	7	29	5
HDL_2	1.063-1.125	360,000	9-12	40	33	5	17	5
HDL_3	1.125-1.210	175,000	6-9	55	25	4	13	3

[a]Data represent human plasma lipoproteins except where noted.

[b]HDL were originally designated according to density 1.063-1.210; metabolically distinct HDL subclasses are presently recognized.

[c]Data represent rat plasma HDL_1' although this lipoprotein has also been identified in human plasma.

compares the plasma cholesterol levels of individuals with and without CAD, a significant degree of overlap is found. Even siblings very closely matched for elevated LDL cholesterol levels can have clinical CAD at vastly different ages (*34*).

These conflicting observations call attention to a conspicuous gap in our understanding of precisely how the LDL cholesterol transport process is linked to the events of atherogenesis leading to CAD. Given the importance of plasma LDL concentrations in the development of CAD, it is clear that unknown factors in addition to LDL cholesterol influence the rate at which atherogenic processes may occur. Quite recently, new understandings have begun to evolve from research focusing on the metabolism of lipoproteins at the actual sites of atherosclerosis within the artery wall. These newer lines of research may have profound implications for the role of antioxidants in the atherosclerotic process.

3. Oxidative Modification of LDL. The earliest recognizable lesion in atherosclerosis is the fatty streak, characterized by an accumulation cholesterol esters within foam cells. Studies have firmly established that circulating monocytes can adhere to the arterial endothelium, penetrate into the intima, take up residence as macrophages and accumulate cholesterol esters from LDL, thus becoming foam cells (*35,36*). However, these cells do not accumulate cholesterol even when exposed to excessive amounts of "native" LDL under in vitro conditions. This puzzling observation led to the notion that LDL must be in some way modified from its native form to an atherogenic form (mLDL) that results in cholesterol ester accumulation by macrophages.

A search for a modified forms of LDL led to the discovery that chemical acetylation of LDL modified it in a way that could stimulate cholesterol ester accumulation in macrophages and foam cell production (*37*). The reason is that chemical acetylation alters the uptake of LDL by redirecting the particle away from normally operating LDL receptors and toward receptors collectively referred to as "scavenger receptors". Cell surface LDL receptors actually serve to protect cells from over-accumulation of cholesterol because these receptors are under precise feedback control. As the levels of free cholesterol within the cell rise, synthesis of new LDL receptors is suppressed, thereby limiting the flow of external cholesterol into the cell, even in the face of very high LDL concentrations (*25,30*). Unlike the LDL receptor, the scavenger receptors recognize only modified forms of LDL; they are not under feedback regulation and will continue to take in large amounts of acetylated LDL fast enough to convert resident macrophages to cholesterol ester-enriched foam cells. Other chemical modifications, including malonaldehyde conjugation (*38*) similarly modify native LDL such that it is redirected to the acetyl LDL receptor. Although there is no evidence that such chemical modifications occur to any extent in vivo, these experiments raised the possibility that some other unidentified process could elicit a similar metabolic transformation at the site of atherogenesis.

A major advance occurred when investigators recognized that endothelial cells and smooth muscle cells were capable of inducing modifications of LDL that led to accelerated uptake in macrophages, again by way of LDL receptor independent mechanisms (*39-41*). Subsequent studies have confirmed these findings and established that the essential element in this cell-induced modification is the oxidation of the LDL particle itself (*40,42,43*). This oxidative process involves free radical peroxidation of LDL lipids (*42,43*). Additional studies have demonstrated that oxidative modification by cultured cells results in a number of compositional and structural changes within LDL. These include: 1) peroxidation of long-chain polyunsaturated fatty acids (*21*); 2) hydrolysis of lecithin to form lysolecithin and liberation of free fatty acids (*41*); 3) conversion of cholesterol into various oxysterol derivatives (*44*); 4) fragmentation of the protein moiety of LDL (apolipoprotein B-100) into polypeptides (*45*); and 5) conjugation of short-chain lipid oxidation products to polypeptide fragments (*46*). Incubation of LDL with transition metal ions such as copper yields an oxidized LDL with the same properties as the cell-induced oxidation as described above (*27,41,46*); in fact, many investigators now employ copper induced autoxidation of LDL as an study model.

Interestingly, none of these modifications occur if antioxidants such as BHT or α-tocopherol are present in the medium (*41,42*). When LDL is subjected to oxidative conditions, there appears to be a lag phase before any signs of LDL oxidation are observed (*47*). The lag phase may represent the time required to deplete the endogenous antioxidants within LDL, such as α-tocopherol and various carotenoids (*47*). Time course studies show that α-tocopherol is depleted preferentially over carotene, suggesting that vitamin E acts as a first line of defense in LDL oxidation (*27*).

The mechanisms responsible for uptake of oxidatively modified LDL by macrophages is currently the focus of much attention. In a manner analogous to the acetylated LDL described above, oxidatively modified LDL is directed away from normally operating LDL receptors and toward scavenger receptors. Under oxidative conditions, it has been established that derivatization of LDL lysine residues by short chain lipid oxidation products is responsible for directing the oxidized LDL particle away from the LDL receptor (*34*). Even mild oxidation (derivatization of 5-10% of available lysine residues) appears sufficient to negate interaction with the LDL receptor (*48*). Interestingly, this same mild degree of LDL oxidation is not sufficient to stimulate uptake by scavenger receptors and cholesterol accumulation by macrophages is low (*48*). Apparently, more complete oxidation of LDL is necessary to reach a threshold of recognition by scavenger receptors (*27*). Redirection of the oxidized LDL particle toward scavenger receptors appears to be solely the result of derivatization of apolipoprotein B by lipid peroxidation products and not a function of other oxidative events occurring within the particle (*49*). Finally, recent evidence suggests that while oxidized LDL is more potent than native LDL in stimulating cholesterol accumulation in cultured macrophages, it is less potent than equivalent amounts of acetyl LDL (*27,50*). The explanation may involve oxysterol derivatives which are produced from cholesterol during oxidation of LDL (*50,51*). The decreased cholesterol uptake by macrophages may be entirely accounted for by the reduced content of cholesterol within oxidized LDL due to conversion to oxysterols (*27*); alternatively, the oxysterols so produced may have a net inhibitory effect on intracellular cholesterol esterification of incoming cholesterol (*50*).

The events of oxidative modification of LDL have been recently outlined in full detail (*see* References 27, 34, and 52 for review) and will be presented briefly below. At the site of atherogenesis, cell lipids are oxidized, either by cellular lipoxygenases or the cyclooxygenase system. The cellular oxidizing potential could oxidize LDL lipids directly or oxidized cell lipids could be transferred into the LDL. In either case, once the LDL contains fatty acid lipoperoxides, a rapid propagation would lead to large numbers of free radicals and extensive fragmentation of fatty acid chains. This is followed by an increase in phospholipase A_2 activity; the subsequent release of fatty acids further propagates the oxidative chain reaction. Accompanying these changes is a drastic degradation of apolipoprotein B-100 into a heterogeneous mixture of polypeptides containing fragments of covalently bound oxidized fatty acids. Finally, LDL cholesterol is oxidized, producing a variety of oxysterol products.

Considerable experimental evidence implicates endothelial injury as an important factor in atherogenesis (*53*). Oxidized LDL is highly cytotoxic whereas native LDL is not (*42*). Most of this cytotoxic activity is attributable to the lipid components of LDL, including 2-alkenals, lipid hydroperoxides and hydroxy fatty acids (*54*). The cytotoxicity of oxidized LDL may lead to denudation of the endothelium, further accelerating development of atherogenesis.

Recent studies have shown that oxidatively modified LDL, but not native LDL, is a potent chemoattractant for circulating human monocytes (*55*). The chemotactic activity appears to reside within the lipid component, much being attributed to the lysolecithin generated within oxidized LDL (*56*). The chemoattractant activity of oxidized LDL may play a role in recruiting circulating monocytes to the site of atherogenesis (*55*), where they adhere to the arterial endothelium, penetrate into the intima, take up residence as macrophages and subsequently accumulate cholesterol and become foam cells. At the same time, oxidized LDL appears to be a potent inhibitor of motility of resident macrophages (*55*), in effect trapping these cells within the intimal layer of the endothelium and providing additional opportunity for cholesterol accumulation.

The information presented thus far can be summarized by describing four mechanisms by which oxidized LDL may contribute to atherogenesis (*34*). Native LDL circulating within the bloodstream have access to subendothelial tissue spaces beneath the artery wall; an equilibrium exists between circulating LDL and tissue space LDL. Once within the tissue space, LDL are subject to oxidative modification. Factors governing the rate of LDL oxidation are unknown at this time. Recent evidence indicates that at certain areas highly succeptable to lesion formation, the residence time of LDL within the tissue space is increased (*57*); in theory, the increased residence time allows greater opportunity for oxidative processes to occur. Oxidized LDL may express a chemotactic factor (lysolecithin) that aids in the recruitment of circulating monocytes to the subendothelial space. Within the arterial wall, the monocyte undergoes phenotypic modification, and its return to the plasma as a tissue macrophage is inhibited by oxidized LDL. Since the macrophage itself can modify LDL, the rate at which the oxidized LDL is produced accelerates rapidly as the number of subendothelial macrophages increases. Resident macrophages take up oxidized LDL via the scavenger receptor mechanism, and become foam cells. In addition, the cytotoxic effects of LDL could lead to a loss of endothelial integrity, further propagating the sequence of atherogenic events. Thus, it is possible to visualize a distinct series of events that lead to the development of fatty streak lesions.

Taken in their entirety, these experimental observations provide strong evidence that oxidative modification of the LDL particle structure plays a major role in its ability to induce atherosclerosis. Furthermore, they provide an entirely new array of mechanisms for possible intervention to postpone atherosclerotic processes. The proposed scheme outlined above suggests that the extent to which LDL undergo oxidative modification may be directly tied to the rate of lesion formation.

4. Does Oxidation of LDL Occur *in Vivo*?. The hypothesized mechanisms described above have been developed almost exclusively on the basis of *in vitro*, cell culture studies. One may legitimately argue that these observations are not meaningful unless they are substantiated by direct *in vivo* evidence. Therefore, it is essential to address several new lines of evidence suggesting that oxidation of LDL does indeed occur *in vivo*.

A recent study by Carew et al. (*58*) addresses the potential antiatherogenic role of antioxidants in experimental rabbits and is of particular interest. These investigators chose to study a specific strain called the Watanabe heritable hyperlipidemic (WHHL) rabbit for use in this study. The WHHL rabbit lacks the LDL receptor, and even without cholesterol feeding, has high plasma cholesterol levels (often greater than 500 mg/dl) and develops spontaneous atherosclerosis. They wished to test the antiatherogenic effects probucol, a drug which closely resembles 2 BHT molecules joined together (Figure 3). The drug is quite lipophilic and is transported in plasma almost exclusively within the hydrophobic core of lipoproteins, predominantly LDL (*59*). Probucol is a potent antioxidant and possesses cholesterol-lowering properties as well (*57*). In order to test the antioxidant hypothesis and not cholesterol lowering effects, the investigators had to match probucol treated animals with controls of very similar plasma cholesterol and LDL concentrations. To achieve this, control animals were treated with carefully titrated amounts of lovastatin (a potent hypocholesterolemic drug with no antioxidant properties) so that during the 7-month protocol, there were negligible differences in plasma lipids. The results showed a 50% reduction in atherosclerotic lesion development of probucol-treated animals relative to controls. Further, the investigators found that in lesions of probucol-treated animals, LDL was degraded much more slowly than in controls; by contrast, in nonlesion areas of artery, LDL degradation was similar in both groups of rabbits (*57*). These findings suggest that probucol acted as an antioxidant to inhibit oxidative modification of LDL and thereby inhibit lesion progression.

Other studies provide additional lines of evidence that oxidative modification of LDL occurs *in vivo*. Autoantibodies that react to aldehyde conjugates of LDL (formed during oxidative modification) are present in the sera of humans and rabbits (*38*). Aortic lesions in WHHL rabbits were immunostained using antibodies specific to protein-lipid conjugates that are present in oxidatively modified LDL (*61*). In another set of experiments, LDL was gently extracted from atherosclerotic tissue under strict antioxidant conditions (*60*). When compared to plasma LDL isolated from the same person or rabbit, LDL isolated from lesions had more extensive breakdown of apolipoprotein B, Western blotted with malonaldehyde using specific antibody and 4-hydroxynonenal specific antibodies. These results suggest that lipid peroxidation reactions occur in atherosclerotic lesions *in vivo* and are responsible for generation of stable lysine adducts on polypeptides.

Although these results do not prove a pathogenic role for oxidatively modified LDL, they do provide evidence for its occurrence *in vivo*. Additional studies are necessary to elucidate the mechanisms of oxidative modification of LDL *in vivo*. A role for antioxidant therapy in slowing atherosclerotic progression is intriguing; however, other antioxidants should be tested to confirm this theory.

5. The Role of Antioxidant Protection in CAD: Epidemiological Evidence. As stated earlier, population data suggest that about 35% of CAD occurrence cannot be accounted for by means of conventional risk factors. Given the multifactorial nature of the disease, it is not surprising that despite the current devotion to dietary lipids as causal factors of CAD, low intake of antioxidants has gained attention as a plausible hypothesis to help explain CAD incidence. To address the question of whether a low dietary intake of antioxidants is linked to high CAD mortality, several epidemiological approaches have been employed. The most common method to estimate antioxidant status is plasma measurement of several vitamins including vitamin E, vitamin C, and the provitamin beta-carotene. These measurements are often conducted in large-scale cross-sectional surveys in an effort to correlate status of one or more antioxidants with CAD incidence in population segments differing in their succeptability to the disease. Although interpretation of data of this kind is fraught with difficulty, the available evidence will be reviewed for each of the antioxidant nutrients described above.

a) Vitamin E. Vitamin E is a term often used interchangeably with α-tocopherol but actually refers to the collective antioxidant contribution of β-, δ-, and γ-tocopherols in addition. However, α-tocopherol is the most important chain-breaking antioxidant in humans and is preferentially depleted during copper-induced peroxidation of LDL particles (*47*), and is the major antioxidant present in LDL (*62*). In light of the experimental evidence already presented, one might reason that in the search for epidemiological support for the antioxidant hypothesis, vitamin E status would thus represent a logical starting point.

A recent cross-cultural survey compared vitamin E plasma levels of males aged 40-49 years from regions varying substantially in CAD mortality rates (from a low of 107 annual deaths due to CAD per 100,000 population in southern Italy to a high of 469 deaths per 100,000 in North Karelia, Finland) (*63*). Plasma vitamin E concentration was significantly correlated with plasma lipids, including cholesterol, triglycerides and the sum of cholesterol and triglycerides. In an attempt to circumvent the confounding effects of plasma lipids, the plasma vitamin E levels were mathematically standardized to a plasma cholesterol level of 220 mg/dl, and a triglyceride level of 110 mg/dl. So adjusted, plasma ∝-tocopherol levels among all 12 populations studied maintained a strong inverse relation with CAD mortality ($r^2 = 0.49$). In a partial regression analysis of eight populations which did not differ in plasma cholesterol levels ($p > 0.05$), absolute plasma tocopherol concentrations related inversely to CAD mortality ($r^2 = 0.55$). These results support the concept of a protective role for vitamin E activity in CAD.

A second type of epidemiological approach is the case-control study, in which plasma vitamin E levels of individuals free of clinical CAD are compared to the vitamin E levels of individuals with symptomology for CAD. A preliminary report of a case-control study of over 550 Scottish men, aged 35-54, recently appeared in which the antioxidant hypothesis was examined (*64*). Vitamin E levels of 125 patients with angina (positive for chest pain as determined by World Health Organization questionnaire) were not significantly different from the 430 controls. However, when the results were expressed as plasma vitamin E/cholesterol molar ratio, patients with CAD had significantly lower vitamin E status. The authors conclude that a low plasma vitamin E content predisposes to angina (*64*).

Several clinical trials studying the effects of vitamin E on CAD symptomology have been conducted. Clinical trials involve administration of set amounts of vitamin E or placebo to experimental and control groups, respectively, for a pre-determined period of time to determine treatment effect on end-point measurements, such as myocardial infarcts, coronary bypass surgery, or angina. A recent review of vitamin E clinical trials (*65*) indicates only two of six trials of vitamin E and angina were randomized and double-blind. Rinzler (*66*) gave 200 mg α-tocopherol acetate for two weeks and thereafter 300 mg daily or placebo to 20 patients with chronic chest pain for an average of 16 weeks. No outcome measurement differences were reported (*66*). In a more recent trial, Anderson gave 3200 IU vitamin E or placebo daily to 20 patients with stable angina for 9 weeks (*67*). Again, nonsignificant differences in outcome measures were reported. A major shortcoming of such studies is the extremely short time-span employed by researchers in patients with advanced atherosclerosis. Financial and ethical considerations often limit the scope of well-intentioned clinical trials; both trials tested short-term efficacy of vitamin E therapy on advanced angina. These studies do not directly address the issue of primary prevention of CAD through long-term intake of moderate amounts of vitamin E. In order to address this question, a larger, longer, well-designed double blind trial of inevitably high expense is necessary to confirm or reject the results of existing studies. Although epidemiological data provide marginal support for such an endeavor, the experimental data discussed earlier may provide impetus in this direction.

b) Vitamin C. Cigarette smokers have lower vitamin C concentrations than nonsmokers and this could result from increased oxidant stress due to free radical damage associated with smoking (*68*). The case control study of Reimersma et al. (*64*) revealed a significantly lower plasma vitamin C concentration in patients with angina compared to controls. Cross-sectional studies of Gey and Puska (*63*) suggest that lipid standardized vitamin E levels parallel the absolute vitamin C level; this correlation appears particularly strong in populations with low vitamin C medians. In addition, vitamin C concentrations are lower in aortic tissue from patients with occlusive CAD (*69*). However, it is difficult to separate a potential independent effect of vitamin C from the confounding effects of smoking and vitamin E concentration; these observations may not necessarily provide direct support for the antioxidant hypothesis.

c) Beta-carotene. Beta-carotene is a provitamin, the precursor to vitamin A. Beta-carotene has both singlet oxygen quenching and antioxidant properties and its antioxidant capacity may exceed that of vitamin E under certain conditions (*70*). Unlike the parallel relationship between vitamins E and C, no such intercorrelation appears to exist for vitamin E and carotenoids (*70*). Beta-carotene itself is transported in plasma within the LDL particle (*71*) and may therefore play a protective role in limiting LDL oxidation. Esterbauer et al. (*62*) have found that there is a strong correlation between the content of LDL vitamin E and beta-carotene and the resistance of the LDL particle to oxidation. In addition, serum concentrations of beta carotene are greater in women than in men (*71*). There is a paucity of direct evidence from clinical trials to address the protection from CAD afforded by increasing dietary consumption of beta-carotene, although a number of investigators believe that such trials merit strong consideration.

6. Conclusions. CAD remains the leading cause of death in the United States and Western Europe. Demographic differences in the incidence of CAD have been interpreted as implicating dietary fat and cholesterol, although classical risk factors account for only about 65% of the variance in the occurrence of CAD. Recent experimental evidence has provided a fascinating hypothesis to explain the mechanism by which LDL cholesterol becomes atherogenic. The evidence for a role of oxidative processes in atherogenesis is now gaining acceptance, although a number of perplexing questions still exist. For example, why do nonhuman primates develop atherosclerosis on a diet containing saturated fat but are resistant to atherosclerosis when fed a polyunsaturated diet? Presumably, increased tissue content of polyunsaturated fatty acids would promote oxidative processes leading to modification of LDL and cholesteryl ester deposition in foam cells. Perhaps the answer may lie with the antioxidant status of the animals. This paradox calls critical attention to the evidence supporting the contention that LDL modification occurs *in vivo*. Some investigators question whether the information regarding oxidative modification of LDL *in vitro* is meaningful without more supporting evidence from *in vivo* observations. Based on the experimental evidence available thus far, most investigators would agree that a high priority in atherosclerosis research is to determine more conclusively whether oxidative modification of LDL is a key event in atherogenesis. From an epidemiological perspective there is some evidence supporting the antioxidant hypothesis although many investigators consider this evidence marginal. At this point, population-wide recommendations advocating nutritional intervention to increase intake of antioxidants seems from a scientific standpoint to be premature. Nevertheless, research on oxidative processes leading to atherosclerosis and the preventative role of antioxidants appears to hold great promise in further reducing incidence of CAD.

Dietary Oxidation Products and CAD

Research on dietary LOPS, atherosclerosis, and CAD has fully matured into a serious, if somewhat low-profile, research area which has in many ways paralleled research on MLDL. Dietary LOPS have been shown to be cytotoxic, to induce arterial injury, and enhance the atherosclerotic process and antioxidants are clearly important agents for the protection of food lipids (*72-78*). In addition, many antioxidants not consumed by free radical reactions by the time the food is consumed will be absorbed and transported by LDL, in effect increasing the resistance of the LDL particle to oxidation to the mLDL.

The primary deleterious affect of dietary LOPS, including oxysterols, appears to be arterial injury, based on the cytotoxic properties of these oxidation products. Arterial injury is believed to be one of the earliest steps in atherosclerosis (*see* References 77 and 78 for review). The possibility that dietary LOPS might influence the rate of formation of mLDL or the possibility that LDL can be modified by consumption of dietary LOPS is not known at this time but constitutes a research area of pivotal importance because it has been shown that dietary LOPS are absorbed from the intestine in both experimental animals and man (*see* Reference 78 for review).

It is well established that the typical diet in developed nations expose consumers to impressively high levels of LOPS. Addis and Warner (*78*) suggested that deep-fat-fried foods and foods containing powdered eggs represent the largest sources of LOPS in the human diet. Other potentially significant sources of LOPS, including oxysterols, include precooked uncured meats, freeze-dried meats, and sour cream-, butter-, and cheese-powders (*77*). Research has demonstrated the same oxysterols of foods in plasma HDL, LDL, and VLDL of fasted humans (*79*). There is obviously a great need for much more research into the types and quantities of LOPS in foods.

In Figure 7, representatives of a few LOPS which have been reported in foods and heated frying oils are listed. It is interesting and perhaps very significant that the compounds in Figure 7 are some of the same seen in mLDL. This fact plus the emerging recognition that mLDL is an important atherogenic factor demands that close attention be given to the possibility that the detrimental effects of dietary LOPS extend beyond arterial injury to the acceleration of plaque accumulation, and that antioxidants in foods, including synthetics like BHA and BHT, confer a dual benefit by reducing exposure of humans to LOPS and suppressing in vivo oxidation of LDL to mLDL.

Cancer

Ames (*78*), in a comprehensive review, noted the tendancy for lipid oxidation products and "active oxygen" to the possess procarcinogenic properties and antioxidants to possess anticarcinogenic activity. Wattenberg (*79*) listed the following compounds as "inhibitors of carcinogen-induced neoplasia:" ascorbic acid, α- and γ-tocopherol, gallic acid, propyl gallate, BHA, BHT, retinoids, carotenoids, phenols, caffeic acid and ferulic acid. The fact that all the foregoing compounds are antioxidants is somewhat in contrast to the "carcinogenic" properties of BHA reported by Ito et al. (*20*) but Wattenberg (*79*) cautioned that antioxidant vitamins are not necessarily effective as broad-spectrum anticarcinogens. The literature on the linkage between antioxidants and anticarcinogenesis is vast, difficult to interpret but interesting and promising enough to warrant further study and useful literature is available (*5, 76-83*).

Aging, Immune Responses, Toxin Interactions

One of the earliest properties of antioxidants discovered was an "anti-aging" effect. Free radicals are associated with aging and antioxidants stabilize free radicals. The relationship between active oxygen species and LOPS on one hand and cancer and

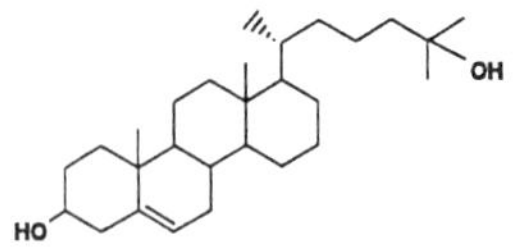

Cholest-5-ene-3 β , 25-diol
(25-hydroxycholesterol)

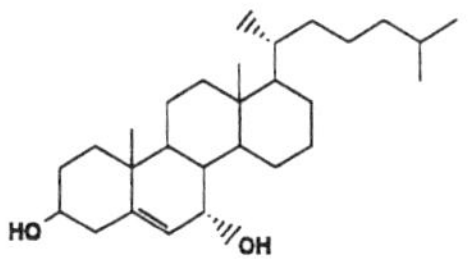

Cholest-5-ene-3β , 7α-diol
(7-hydroxycholesterol)

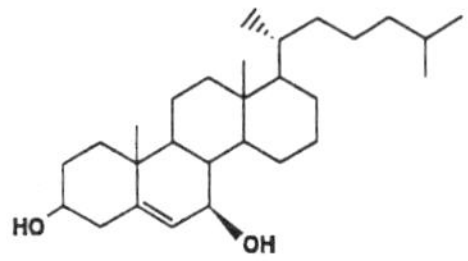

Cholest-5-ene-3 β, 7 β-diol
(7β -hydroxycholesterol)

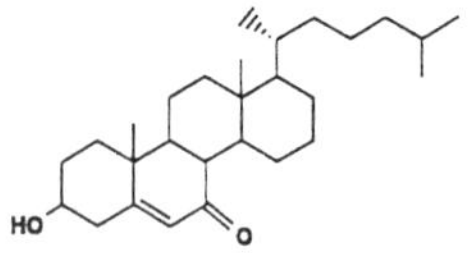

3 β-hydroxycholest-5-en-7-one
(7-ketocholesterol)

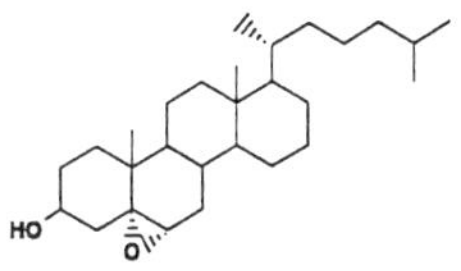

5,6α-epoxy-5α-cholestan-3 β-ol
(cholesterol-α-epoxide)

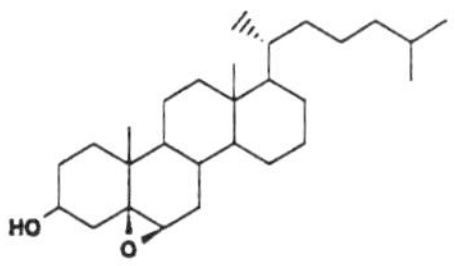

5,6β-epoxy-5β-cholestan-3β-ol
(cholesterol-β-epoxide)

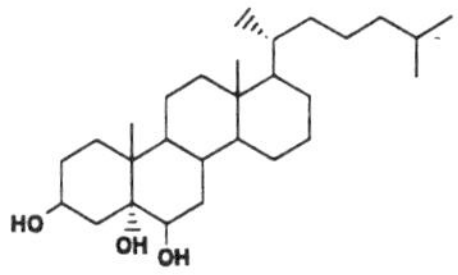

5α-cholestane-3β,5,6β-triol
(cholestanetriol)

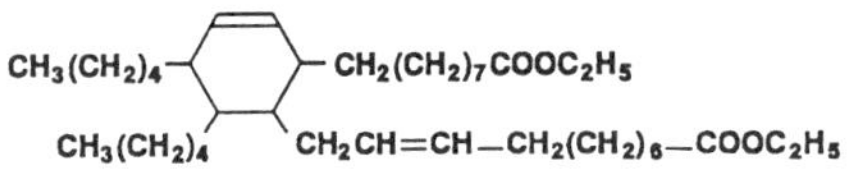

Dimer

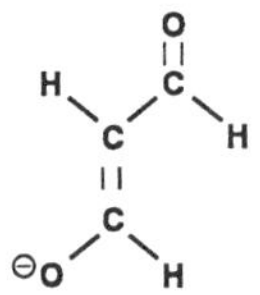

Malonaldehyde

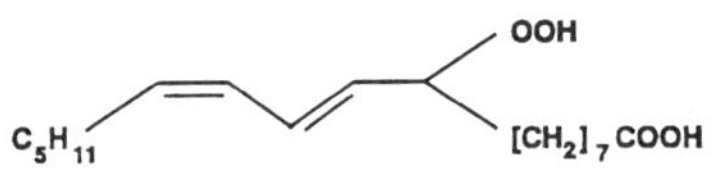

9-hydroperoxy-10-trans, 12-cis-octadecadienoic acid
(linoleic acid-9-hydroperoxide)

Figure 7. Some common lipid oxidation products found in foods that may be significant in terms of atherosclerosis and other diseases. Malonaldehyde exists in five resonance forms; only the ionic form is shown.

CAD on the other, has already been discussed but it should be emphasized that cancer and CAD are diseases of aging. Vitamin E has been recognized as the key defense against free radical damage. Selenium, a component of the glutathione peroxidase system, is also considered an antioxidant nutrient and deficiences of both selenium and vitamin E result in accelerated accumulation of lipofuscin, or age, pigment (*84*).

The interrelationships among vitamin E, aging and CAD are clear from a study by Hennig et al. (*85*). Endothelial cell culture age, susceptibility to oxidative injury and the protective effects of vitamin E were studied. Exposure of endothelial cell monolayers for 24 hours to 30 μM linoleic acid hydroperoxide caused increase albumin transfer, an indication of reduced barrier function. Pre-enrichment with 25 μM vitamin E consistently protected endothelial cells against oxidative damage, independent of cell age. The authors (*85*) suggested that vitamin E may play an important role in CAD-prevention because the maintenance of a robust barrier function by the arterial endothelial cells may restrict the influx of "cholesterol-rich lipoprotein remnants."

Antioxidant vitamins also appear to amplify the immune response of organisms (*86*). Extensive studies have been conducted on vitamins C and E and provitamin A (β-carotene). Deficiencies in the antioxidant vitamins are clearly associated with a compromised immune response but the amplification of immune response by high dosages of vitamins, although promising, is not clear in every case. Much more research is necessary.

Evidence exists that antioxidant vitamins and selenium help the animal organism to resist various hazardous elements in the environment. Vitamin E and selenium, even at modest levels, are quite effective at modifying the toxicity of mercury and silver (*87*). Dietary ascorbic acid supplementation will improve iron (II) absorption, thereby markedly reducing toxic effects of 5-200 ppm dietary cadmium (*88*). Vitamins C and E are very effective blockers of the nitrosation reaction in vivo and in foods (*89, 90*).

Antioxidants and Food Processing Issues

There are several, important emerging issues occurring in food processing which pertain to antioxidant usage. Foods are more highly processed and are subjected to longer storage times than ever before. As a result, oxidation may have greater opportunity to degrade food than in previous times. Recent reviews (*76, 78*) have outlined areas in food processing and preparation which may be linked to the production of high levels of LOPS. Of special concern are heated fats and the resultant deep-fried foods and powdered eggs. Of secondary concern are foods such as freeze-dried meats, powdered cheeses and pre-cooked, stored meat products. Addis and Warner (*78*) strongly suggest that a great deal more work is needed in the area but that governmental regulation of food LOPS is a possibility. In Europe, it is very common for frying fats in restaurants to be closely monitored for quality by government agencies.

Obviously, antioxidants are an important possible solution or partial solution to most of the foregoing problems. Effective use of antioxidants could provide a measure of protection against accumulation of LOPS in many foods, although deep-fried foods will require more than increased antioxidant use to lower LOPS because of the high temperatures involved. F.D.A. and U.S.D.A. will need to consider risks associated with increased antioxidant usage but should balance the risks with clear benefits to health that we have reviewed.

Risk/Benefit of Antioxidants. Even with the copious amounts of data available, it probably is not advisable to attempt a sophisticated, mathematical calculation of the risk/benefit ratio at this time and we shall not do so. However, we will propound a strong qualitative assessment of the available data.

Exposure of humans to megadoses of antioxidant vitamins and extremely high levels of synthetic antioxidants would be undesirable although, in our opinion, exposure to dietary LOPS is even more so. Attempting to find the correct balance is exceedingly complicated and may not be necessary if alternatives to antioxidants are employed to the fullest possible advantage. Examples include vacuum packaging, use of oxygen interceptors, omitting light and, in the case of heated oils, simply using the best available maintenance methods such as proper filtration and quality control. Nevertheless, some antioxidants will be required in combination with the foregoing technological advances to achieve truly low, inconsequential, levels of LOPS.

In terms of toxicity and carcinogenicity of antioxidants the data are, in our opinion, basically unimpressive. Ito (*20*) used levels 10,000-fold higher BHA than human intake (*21*) to produce deleterious effects in the rat and, at that, the forestomach was the organ affected. Humans do not possess a forestomach. Clearly, BHA is not a genotoxic carcinogen (*22*). Other synthetic antioxidants also appear unimpressive in terms of being detrimental biological agents. Vitamins C and E require astronomical levels to induce toxicity.

In contrast, the health benefits of antioxidants as well as their role in maintaining a high level of esthetic and functional properties in foodstuffs make them exceedingly critical to modern food processing. Antioxidants retard potentially health-significant deterioration of food by inhibiting lipid oxidation and also are responsible for numerous important health benefits to the consumer by retarding similar lipid oxidation reactions in vivo. Most notable among these effects is the retardation of the oxidation of LDL to mLDL, a reaction which appears to be a key factor in atherosclerosis.

Our conclusion is that, although much more research should be encouraged, it is premature and potentially detrimental to human health to ban the use of any currently-used U.S. antioxidants or to, as has been done recently in California, make it difficult for companies to use an antioxidant by placing it on an environmental "hit list". Uncertainty is a part of our existence: the absolute safety of synthetic and natural antioxidants may never be proven beyond any doubt. However, what is certain is that the risks of not using antioxidants far exceed the risks of using them.

Future Research Needs

One of the most exciting areas of research in antioxidants has to do with the possible benefits of antioxidants with regard to CAD and we strongly suggest that vigorous research activity on this relationship should continue. Both aspects of the antioxidant-CAD connection should be exhaustively explored. Antioxidants reduce LOPS levels in food and retard in vivo oxidation reactions. Are both of the foregoing effects truly significant in terms of human CAD and cancer? The possibilities are provocative but elucidating the facts will require much devoted research.

Another important, and closely related, research area has to do with deep-fried foods, clearly large sources of dietary LOPS. Are such levels high enough to have significant damaging impact on consumers and, if so, should regulations concerning oil quality and wholesomeness be promulgated?

As a corollary to the foregoing question, which methodology is most suitable for measuring heated oil deterioration? Also, what methods can be developed to improve the stability of heated oils?

It is hoped that the foregoing list of research ideas will receive prompt attention. However, the list is by no means complete and the authors hope that this review has stimulated many other research ideas as well.

Acknowledgments

Published as Paper No. 19,202 of the contribution series of the Minnesota Agricultural Experiment Station based on research conducted under Project No. 18-23, supported by Hatch funds.

Literature Cited

1. Finley, J.W. and Given, P. *Food Chem. Toxic.* **1986**, *24*, 999-1006.
2. Kanner, J., German, J.B. and Kinsella, J.E. *CRC Critical Rev. Food Sci. Nutr.* **1986**, *25*, 317-364.
3. Kahl, R. and Hildebrandt, A.G. *Food Chem. Toxic.* **1986**, *24*, 1007-1014.
4. Graf, E. and Eaton, J.W. *Free Radical Biol. Med.* **1990**, *8*, 61-69.
5. Aruoma, O.I., Evans, P.J., Kaur, H., Sutcliffe, L. and Halliwell, B. *Free Rad. Res. Comm.* **1990**, *10*, 143-157.
6. Miller, D.M., Buettern, G.R. and Aust, S.D. *Free Radical Biol. Med.* **1990**, *8*, 95-108.
7. Yuting, C., Rongliang, Z., Zhongjian, J. and Yong, Ju. *Free Radical Biol. Med.* **1990**, *9*, 19-21.
8. St. Angelo, A.J. and Ory, R.L. *Lipids* **1984**, *19*, 34-37.
9. Niki, E. In *Third conference on vitamin C*; Burns, J.J., Rivers, J.M., and Machlin, L.L., Eds.; *Annals New York Acad. Sci.* **1987**, *498*, 186-198.
10. Abdo, K.M., Rao, G., Montgomery, C.A., Dinowitz, M. and Kanagaling, K. *Food Chem. Toxic.* **1986**, *24*, 1043-1050.
11. Tomassi, G. and Silano, V. *Food Chem. Toxic.* **1986**, *24*, 1051-1061.
12. Parker, R.S. *Adv. Food Nutr. Res.* **1989**, *33*, 157-232.
13. Rivers, J.M. In *Third conference on vitamin C*; Burns, J.J., Rivers, J.M., and Machlin, L.L., Eds.; *Annals New York Acad. Sci.* **1987**, *498*, 186-198.
14. Champagne, E.T., Hinojosa, O. and Clemetson, C.A.B. *J. Food Sci.* **1990**, *55*, 1133-1136.

15. Van Esch, G.J. *Food Chem. Toxic.* **1986**, *24*, 1063-1065.
16. Van der Heijden, C.A., Janssen, P.J.C.M. and Strik, J.J. T.W.A. *Food Chem. Toxic.* **1986**, *24*, 1067-1070.
17. Wurtzen, G. and Olsen, P. *Food Chem. Toxic.* **1986**, *24*, 1121-1125.
18. Tobe, M., Furuya, T., Kawasaki, Y., Naito, K., Sekita, K., Matsumoto, K., Ochisi, T., Usui, A., Kokubo, T., Kanno, J. and Hayashi, Y. *Food Chem. Toxic.* **1986**, *24*, 1223-1228.
19. Wurtzen, G. and Olsen, P. *Food Chem. Toxic.* **1986**, *24*, 1229-1233.
20. Ito, N., Hirose, M., Fukushima, S., Truda, H., Shirai, T. and Tatematsu, M. *Food Chem. Toxic.* **1986**, *24*, 1071-1082.
21. Kirkpatrick, D.C. and Lauer, B.H. *Food Chem. Toxic.* **1986**, *24*, 1035-1037.
22. Nera, E.A., Iverson, F., Lok, E., Armstrong, C.L., Karpinski, K. and Clayson, D.B. *Toxicology* **1988**, *53*, 251-268.
23. *The Surgeon Generals Report on Nutrition and Health.* DHHS. Publication No. 88-50210. U.S. Government Printing Office. 1988.
24. Steinberg, D. *Circulation* **1989**, *80*, 1070-1078.
25. Goldstein, J. L., Basu, S.K., and Browm, M. S. In *Methods in Enzymology*; Fleisher, S. and Fleisher, B., Eds.; Academic Press: Orlando, Fl., 1983; Vol. 98; pp. 241-260, .
26. Multiple Risk Factor Intervention Trial Research Group: Multiple risk factor intervention trial. Risk factor changes and mortality results. *JAMA* **1982**, *248*, 1465-1477.
27. Steinbrecher, V.P., Zhang, H. and Lougheed, M. *Free Rad. Biol. Med.* **1990**, *9*, 155-168.
28. Mahley, R.W. *Circulation* **1985**, *72*, 943-948.
29. Sata, T., Havel, R.J. and Jones, A.L. *J. Lipid Res.* **1972**, *13*, 757-768.
30. Goldstein, J.L. and Brown, M.S. *Clin. Res.*, **1982**, *30*, 417-426.
31. Eisenberg, S. *J. Lipid Res.* **1984**, *25*, 1017-1058.
32. Glomset, J.A. *J. Lipid Res.* **1968**, *9*, 155-167.
33. Miller, G.J. and Miller, N.E. *Lancet* **1975**, *1*, 16-19.
34. Steinberg, D., Parthasarathy, S., Carew, T.E., Khoo, J.C. and Witzum, J.L. *N. Eng. J. Med.* **1989**, *320*, 915-924.
35. Aqel, N.M., Ball, R.Y. Waldman, H. and Mitchinson, M.J. *Atherosclerosis* **1984**, *53*, 265-271.
36. Rosenfeld, M.E., Tsukada, T., Gown, A.M. and Ross, R. *Arteriosclerosis* **1987**, *7*, 9-23.
37. Goldstein, J.L., Ho, Y.K., Basu, S.K. and Brown, M.S. *Proc. Natl. Acad. Sci. USA* **1979**, *76*, 333-337.
38. Haberland, M.E., Fogelman, A.M. and Edwards, P.A. *Proc. Natl. Acad. Sci. USA* **1982**, *79*, 1712-1716.
39. Henriksen, T., Mahoney, E.M. and Steinberg, D. *Arteriosclerosis* **1981**, *3*, 149-159, 1981.
40. Morel, D.W., DiCorleto, P.E. and Chisolm, G.M. *Arteriosclerosis* **1984**, *4*, 357-364.
41. Steinbrecher, V.P., Parthasarathy, S., Leake, D.S., Witzum, J.L. and Steinberg, D. *Proc. Natl. Acad. Sci. USA* **1984**, *83*, 3883-3887.
42. Morel, D.W., DiCorleto, P.E. and Chisolm, G.M. *Arteriosclerosis* **1984**, *4*, 357-364.

43. Morel, D.W., Hessler, J.R. and Chisolm, G.M. *J. Lipid Res.* **1983**, *24*, 1070-1076.
44. Zhang, H., Basra, H.J.K., and Steinbrecher, V.P. *J. Lipid Res.* **1990**, *31*, 1361-1370.
45. Fong, L.G., Parthasarathy, S., Witzum, J.L. and Steinberg, D. *J. Lipid Res.* **1987**, *28*, 1466-1477.
46. Steinbrecher, V.P. *J. Biol. Chem.* **1987**, *262*, 3603-3608.
47. Esterbauer, H., Jurgens, G., Quehenberger, O. and Koller, E. *J. Lipid Res.* **1987**, *28*, 495-509.
48. Steinbrecher, U.P., Witzum,J.L., Parthasarathy, S. and Steinberg, D. *Arteriosclerosis* **1987**, *7*, 135-143.
49. Steinbrecher, U.P., Lougheed, M., Kwan, W-C. and Dirks, M. *J. Biol. Chem.* **1989**, *264*, 15216-15223.
50. Jialal, I. and Chait, A. *J. Lipid Res.* **1989**, *30*, 1561-1568, 1989.
51. Warner, G.J., Addis, P.B., Emanuel, H.A., Wolfbauer, G. and Chait, A. *Cholesterol oxidation products in oxidatively low density lipoproteins*. Presented at 1990 meeting of Federation Amer. Soc. Expl. Biol., Washington, D.C.
52. Steinberg, D. *Atherosclerosis Rev.* **1988**, *18*, 1-23.
53. Ross, R. and Glomset, J.A. *N. Eng. J. Med.* **1976**, *295*, 369-377.
54. Jurgens, G., Hoff, H.F., Chisolm, G.M. and Esterbauer, H. *Chem. Phys. Lipids* **1987**, *45*, 315-336.
55. Quinn, M.T., Parthasarathy, S., Fong, L.G. and Steinberg, D. *Proc. Nat. Acad. Sci. USA* **1987**, *84*, 2995-2998.
56. Quinn, M.T., Parthasarathy, S. and Steinberg, D. *Proc. Nat. Acad. Sci. USA* **1988**, *85*, 2805-2809.
57. Carew, T.E. *Am. J. Cardiol.* **1989**, *64*, 18G-22G.
58. Carew, T.E., Schwenke, D.C. and Steinberg, D. *Proc. Natl. Acad. Sci. USA* **1987**, *84*, 7725-7729.
59. Marshall, F.N. *Artery* **1982**, *10*, 7-21.
60. Palinski, W., Rosenfeld, M.E., Yla-Heritvala, S., Gurtner, G.C., Socher, S.S., Butler, S., Parthasarathy, S., Carew, T.E., Steinberg, D. and Witzum, J.L. *Proc. Natl. Acad. Sci. USA* **1989**, *86*, 1372-1376.
61. Haberland, M.E., Fond, D. and Cheng, L. *Science* **1988**, *241*, 215-218.
62. Esterbauer, H., Striegl, G., Puhl, H., Oberreither, S., Rotheneder, M., El-Saadani, M. and Jurgens, G. *Ann. N.Y. Acad. Sci.* **1989**, *570*, 254-267.
63. Gey, K.F. and Puska, P. *Ann. N.Y. Acad. Sci.* **1989**, *570*, 268-282.
64. Riemersma, R.A., Wood, D.A., Macintyre, C.C.A., Elton, R., Gey, K.F. and Oliver, M.F. *Ann. N.Y. Acad. Sci.* **1989**, *570*, 291-295.
65. Kleijnen, J., Knipschild, P. and ter Reit, G. *Evr. J. Clin. Pharmacol.* **1989**, *37*, 541-544.
66. Rinsler, S.H., Bakst, H., Benjamin, Z.H., Bobb, A.L. and Travell, J. *Circulation* **1950**, *1*, 288-293.
67. Anderson, T.W. and Reid, W. *Am. J. Clin. Nutr.* **1974**, *27*, 1174-1178.
68. Duthie, G.G., Wahle, K.W.J., and James, W.P.T. *Nutr. Res. Rev.* **1989**, *2*, 51-62.
69. Dubik, M.A., Hunter, G.C., Casey, S.M. and Keen, C.L. *Proc. Soc. Exp. Biol. Med.* **1987**, *184*, 138-143.
70. Burton, G.W. *J. Nutr.* **1989**, *119*, 109-111.

71. Parker, R.S. *J. Nutr.* **1989**, *119*, 101-104.
72. Smith, L.L.; *Cholesterol Autoxidation*; Plenum Press, New York, 1981.
73. Addis, P.B. *Food Chem. Toxic.* **1984**, *24*, 1021-1030.
74. Yagi, K. *Chem. Phys. Lipids* **1987**, *43*, 337-351.
75. Addis, P.B., Csallany, A.S. and Kindom, S.E.; In *Xenobiotics in Foods and Feeds*; Finley, J.W. and Schwass, D.E., Eds; ACS Symposium Series No. 234; American Chemical Society: Washington, D.C., 1983.; pp. 85-98
76. Addis, P.B. and Park, S.W.; In *Food Toxicology: A Perspective on the Relative Risks*; Taylor, S.L. and Scanlan, R.A., Eds.; Marcel Dekker: New York. , 1989; pp. 297-330.
77. Smith, L.L. and Johnson, B.H. *Free Radical Biol. Med.* **1989**, *7*, 285-332.
78. Addis, P.B. and Warner G.J. In *Free Radicals and Food Additives.*; I.O. Auroma and B. Halliwell, Eds.; Taylor and Francis: London, 1991; pp. 77-119.
79. Addis, P.B., Emanuel, H.A., Bergman, S.D. and Zavoral, J.H. *Free Rad. Biol. Med.* **1989**, *7*, 179-182.
80. Ames, B.N. *Science* **1983**, *221*, 1256-1264.
81. Wattenberg, L.W. *Cancer Res.* **1985**, *45*, 1-8.
82. Temple, N.J. and Basu, T.K. *Nutr. Res.* **1988**, *8*, 685-701.
83. Ziegler, R.G. *J. Nutr.* **1989**, *119*, 116-122.
84. Csallany, A.S., Su, L.-C. and Menken, B.Z. *J. Nutr.* **1984**, *114*, 1582-1587.
85. Hennig, B., Boissoneault, and Wang, Y. *Internat. J. Vit. Nutr. Res.* **1989**, *59*, 273-279.
86. Bendich, A. *Nutrition and Immunology*; Alan R. Liss, Inc.: 1988; pp. 125-147.
87. Ganther, H.E. *Annals. New York Acad. Sci.* **1980**, *335*, 212-225.
88. Fox, M.R.S., Jacobs, R.M., Jones, A.O.L., Fry, Jr., B.E. and Stone, C.L. *Annals. New York Acad. Sci.* **1980**, *335*, 249-260.
89. Tannenbaum, S.R. and Mergens, W. *Annals. New York Acad. Sci.* **1980**, *335*, 267-275.
90. Hotchkiss, J.H. In *Food Toxicology: A perspective on the relative risks*; Taylor, S.L. and Scanlon, R.A., Eds; Marcel Dekker, Inc.: New York, 1989; pp. 57-100.

RECEIVED August 15, 1991

Chapter 31

Safety and Regulatory Status of Food, Drug, and Cosmetic Color Additives

Joseph F. Borzelleca[1] and John B. Hallagan[2]

[1]Department of Pharmacology, Medical College of Virginia, Richmond, VA 23298
[2]Daniel R. Thompson, P.C., 1620 I Street, NW, Suite 925, Washington, DC 20006

Color additives have long been used as a means of enhancing the esthetic value of foods, beverages and cosmetics, and for identifying drugs and other products. Archaeological evidence dates the use of color additives in cosmetics to 5000 B.C. The use of color additives in drugs is documented in ancient Egyptian writings and historic evidence of the use of color additives in foods is dated back to at least 1500 B.C.; natural substances such as turmeric, paprika and saffron, and inorganic mineral pigments were used. In the middle of the nineteenth century, synthetic organic dyes were developed creating a more economical and extensive array of colorants. Unfortunately, the use of color additives to adulterate products also has a long history. This misuse of color additives led to the imposition of controls by regulatory bodies and user associations.

In 1906, with the passage of the Federal Pure Food and Drug Act, certain color additives including FD&C Red No. 3 and FD&C Blue No. 2 were approved for food use. FD&C Yellow No. 5 was approved in 1916; FD&C Green No. 3 in 1927; and FD&C Yellow No. 6 and FD&C Blue No. 1 in 1929.

In 1908, the United States Department of Agriculture initiated a voluntary certification program for synthetic food color additives. Under the certification system, batch samples of color additives are submitted to the agency for analysis and confirmation that they comply with established specifications. The Federal Food, Drug and Cosmetic Act of 1938 (FFDCA) instituted mandatory certification and extended governmental control to colorings for drugs and cosmetics as well as foods. New scientific investigations of the safety of the color additives was promoted by the enactment of the 1960 Color Additive Amendments to the FFDCA and color additives have been the subject of extensive investigations since.

The seven currently approved FD&C color additives are identified in Table 1 and Figure 1. The total pounds certified by the U.S. Food and Drug Administration in recent years are presented in Figure 2. It is anticipated that the use of FD&C color additives will continue to increase with the introduction of new food, drug and cosmetic products.

A separate regulatory class of color additives are exempt from certification; this

0097–6156/92/0484–0377$06.00/0

Table 1. Regulatory Status of the FD&C Color Additives in the United States

Color Additive	*Dye*	*Lake*
FD&C Red No. 40	Permanently listed for all uses at GMP. 21 CFR §§74.340 74.1340, 74.2340 39 Fed Reg 44198 (23 December 1974)	Permanently listed for all uses at GMP. 21 CFR §§74.340, 74.1340, 74.2340 39 Fed Reg 44198 (23 December 1974)
FD&C Blue No. 1	Permanently listed for all uses at GMP. 21 CFR §§74.101, 74.1101, 74.2101 47 Fed Reg 42563 (28 September 1982)	Provisionally listed or all uses at GMP. 21 CFR §82.101
FD&C Blue No. 2	Permanently listed for all uses at GMP. 21 CFR §§74.102, 74.1102 48 Fed Reg 5252 (4 February 1983)	Provisionally listed for all uses at GMP. 21 CFR §82.102
FD&C Green No. 3	Permanently listed for all uses at GMP. 21 CFR §§74.203, 74.1203, 74.2203 47 Fed Reg 52140 (19 November 1982)	Provisionally listed for all uses at GMP. 21 CFR §82.203
FD&C Yellow No. 5	Permanently listed for all uses at GMP. 21 CFR §§74.705, 74.1705, 74.2705 50 Fed Reg 35774 (4 September 1985)	Provisionally listed for all uses at GMP. 21 CFR §82.705
FD&C Yellow No. 6	Permanently listed for all uses at GMP. 21 CFR §§74.2706, 74.1706, 74.2706 51 Fed Reg 41765 (19 November 1986)	Provisionally listed for all uses at GMP. 21 CFR §82.706
FD&C Red No. 3	Permanently listed for foods and ingested drugs at GMP. 21 CFR §§74.303, 74.1303 34 Fed Reg 7446 (8 May 1969) Delisted for cosmetic uses 55 Fed Reg 3516 (1 February 1990)	Delisted for all uses. 55 Fed Reg 3516 (1 February 1990)

FD&C Blue No. 1
Brilliant Blue FCF
Class: Triphenylmethane
CAS No.: 2650-18-2

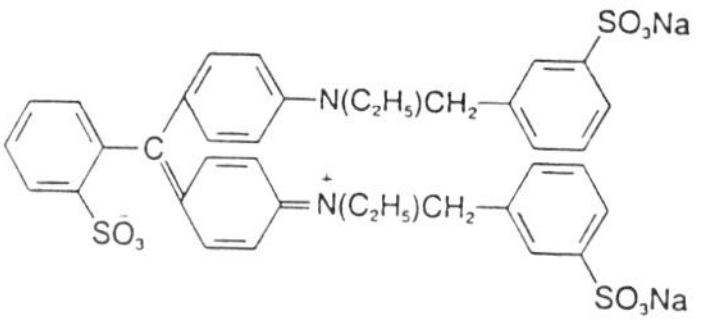

FD&C Blue No. 2
Indigo Carmine, Indigotine
Class: Indigoid
CAS No.: 860-22-0

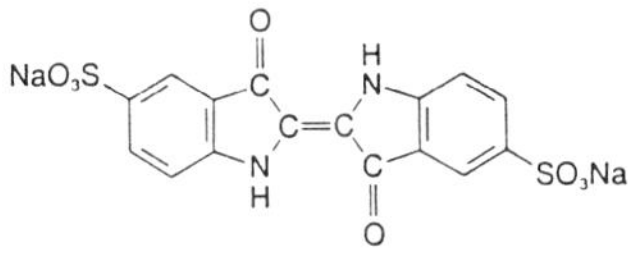

FD&C Green No. 3
Fast Green FCF
Class: Triphenylmethane
CAS No.: 2353-45-9

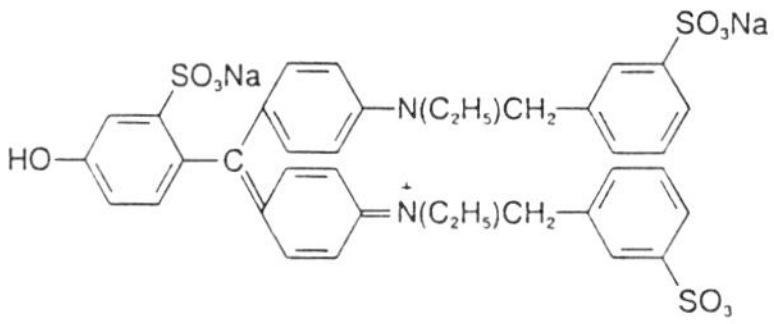

FD&C Red No. 3
Erythrosine
Class: Xanthene
CAS No.: 16423-68-0

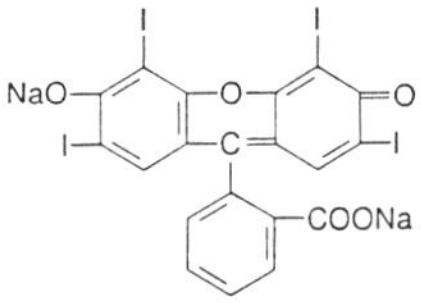

FD&C Red No. 40
Allura Red
Class: Monoazo
CAS No.: 25956-17-6

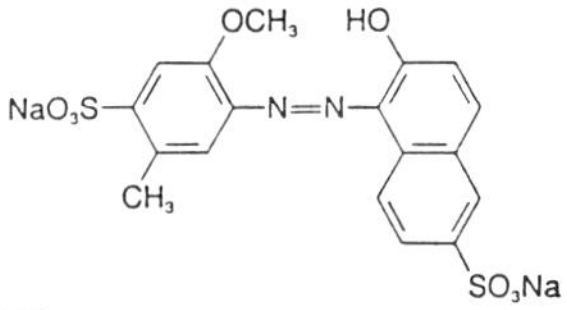

FD&C Yellow No. 5
Tartrazine
Class: Monoazo
CAS No.: 1934-21-0

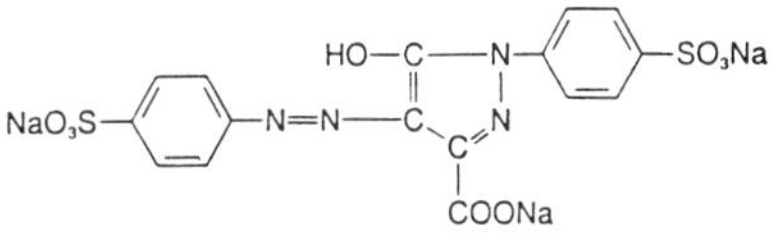

FD&C Yellow No. 6
Sunset Yellow FCF
Class: Monoazo
CAS No.: 2783-94-0

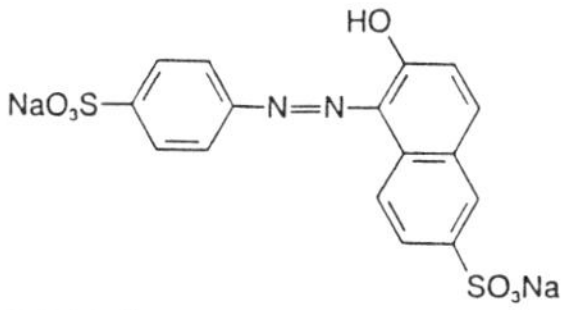

FIGURE 1. FD&C Color Additives

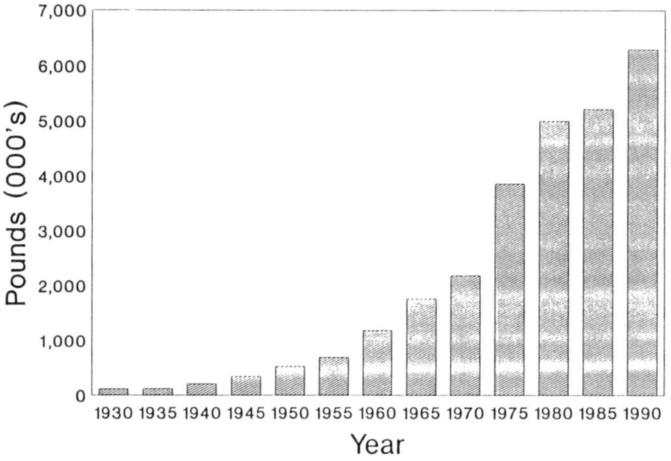

FIGURE 2. Total Pounds of FD&C Dyes Certified by Year

class includes many of the "natural" color additives including beta-carotene, annatto and carmine. The color additives exempt from certification have not been extensively tested but a long history of use in food suggests safety.

The history and status of the FD&C color additives were previously reviewed by Borzelleca et al., (*1*), Marmion (*2*) and Newsome (*3*). This brief review summarizes the large body of scientific information on the FD&C color additives by concentrating on the results of lifetime toxicity/carcinogenicity studies, reproduction and teratology studies, metabolism studies and genotoxicity assays. The regulatory status of the color additives is also summarized.

The Safety of the FD&C Color Additives

In general, the safety of the FD&C color additives has been evaluated in multiple species, acutely and chronically. Additional studies include genetic toxicity, reproduction and development, and absorption, distribution, biotransformation, excretion and kinetics in animals and humans. Special studies, such as mechanistic studies, have been conducted on several color additives in animals and humans. The weight of the available evidence indicates that the FD&C color additives are safe for their intended uses.

FD&C Red No. 40. FD&C Red No. 40 (Allura Red) is a monoazo color additive which has many applications including confectionery and candy products, alcoholic and nonalcoholic beverages, dairy products, meat and poultry products, bakery products and drugs and cosmetics. The amount certified in 1990 was 2,595,720 pounds.

It was negative in genotoxicity tests (*4-15*).

The acute oral toxicity is low; for example, the acute oral LD_{50} in rats is > 10,000 mg/kg (*16*).

Lifetime dietary administration studies of FD&C Red No. 40 in rats (*17*) and mice (*18*) demonstrated no evidence of carcinogenicity in either species. There were no consistent compound-related adverse effects in either species except a reduction in body weight in female rats that received the highest dose (a dietary concentration of 5.19%).

The no observable adverse effect levels reported were: rats -- males, 2829 mg/kg/day and females, 901 mg/kg/day; mice -- males, 7200 mg/kg//day and females, 8300 mg/kg/day.

There were no adverse effects on reproductive performance in a multigeneration study in rats (*19*). There were no adverse developmental effects in rats or rabbits (*20-23*).

FD&C Red No. 40 appears to be poorly absorbed from the gastrointestinal tract where it undergoes azo reduction (*24, 25*).

The acceptable daily intake established by the Joint Expert Committee on Food Additives of the World Health Organization (JECFA) in 1989 is 0-7.0 mg/kg/day. The maximum anticipated daily intake is calculated to be 0.19 mg/kg/day.

FD&C Yellow No. 5. FD&C Yellow No. 5 (Tartrazine) is a monoazo color additive used in candies, beverages, desserts, preserves, canned and frozen vegetables, drugs and cosmetics. The amount certified in 1990 was 1,642,914 pounds.

A weight of evidence analysis demonstrates that FD&C Yellow No. 5 is not genotoxic (*6-10, 13, 14, 26-41*).

The acute oral toxicity is low; for example, the acute oral LD_{50} in mice is 12,750 mg/kg (*42*).

Chronic toxicity/carcinogenicity studies (dietary administration) of FD&C Yellow No. 5 in rats and mice were reported by Borzelleca and Hallagan (*43, 44*); there was no evidence of carcinogenicity in either species nor evidence of consistent compound-related adverse effects.

The no observable adverse effect levels were: rats -- males, 2641 mg/kg/day and females, 3348 mg/kg/day; mice -- males, 8103 mg/kg/day and females, 9753 mg/kg/day. No carcinogenic or significant toxic effects were noted in rats that received FD&C Yellow No. 5 in drinking water at up to 2.0% for two years (*45*).

There were no adverse reproductive effects reported in a multigeneration study in rats (*46*) and there were no adverse developmental effects in rats or rabbits (*47*). There were no significant effects on the development of the central nervous system in the offspring of female rats fed FD&C Yellow No. 5 (*48*).

FD&C Yellow No. 5 undergoes bacterial azo reduction in the gastrointestinal tract of rats, rabbits, and humans (*49-54*). The major biotransformation product is sulfanilic acid (*55, 56*). Sulfanilic acid may have minor behavioral effects on young rats which the authors conclude cannot be extrapolated to humans (*57*).

The acceptable daily intake established by JECFA in 1964 is 0-7.5 mg/kg/day. The maximum anticipated daily intake is calculated to be 0.11 mg/kg/day.

FD&C Yellow No. 6. FD&C Yellow No. 6 (Sunset Yellow FCF)is a monoazo color additive used to color confectionery products, beverages, dessert powders, bakery products, dairy products and drugs and cosmetics. The amount certified in 1990 was 1,606,997 pounds.

It is not genotoxic by a weight of evidence analysis (*5-10, 13, 14, 30-33, 39-41, 58-68*).

The acute oral toxicity is low; for example, the acute oral LD_{50} in rats is > 10000 mg/kg (*69*).

There was no evidence of carcinogenicity in dietary administration lifetime studies in rats or mice (*70, 71*). An increased incidence of pelvic mineralization and chronic nephropathy was reported in female rats at the two highest levels fed (1.5% and 5.0%) in the CCMA study (*70*). The no observable adverse effect levels in the CCMA study were: rats -- males, 1635 mg/kg/day and females, 503 mg/kg/day; mice -- males, 2608 mg/kg/day and females, 11443 mg/kg/day. The no observable adverse effect level in the NTP studies (*71*) was a dietary concentration of 2.5% for rats and mice; compound consumption data are unavailable.

There were no adverse reproductive effects reported in a multigeneration study in rats and there were no adverse developmental effects in rats or rabbits (*46, 47*).

Like FD&C Yellow No. 5, FD&C Yellow No. 6 is poorly absorbed (<5%) and undergoes bacterial azo reduction in the gastrointestinal tract of rats with sulfanilic acid a major metabolite (*57, 72*).

The acceptable daily intake established by JECFA in 1982 is 0-2.5 mg/kg/day. The maximum anticipated daily intake is calculated to be 0.11 mg/kg/day.

FD&C Blue No. 2. FD&C Blue No. 2 (Indigo Carmine) is an indigoid color additive used in the production of candy and confectionery products, beverages, dessert powders, cereals, bakery products, snack foods, drugs and cosmetics. The amount certified in 1990 was 93,337 pounds.

FD&C Blue No. 2 is not genotoxic by a weight of evidence analysis (*5, 7, 10, 27, 32, 33, 36, 73-75*).

The acute oral toxicity is low; for example, the acute oral LD_{50} in rats is 2000 mg/kg (*76*).

Lifetime dietary administration toxicity/carcinogenicity studies in rats and mice were reported by Borzelleca, et al. (*77*) and Borzelleca and Hogan (*78*). There was no evidence of carcinogenicity in either species and there were no consistent compound-related adverse effects reported. In rats, food consumption showed a dose related increase.

A numerical increase in gliomas in male rats that received the highest level was determined to be not biologically significant. The no observed adverse effect levels were: rats -- males, 1282 mg/kg/day and females, 1592 mg/kg/day; mice -- males, 8259 mg/kg/day and females, 9456 mg/kg/day.

There were no adverse effects on reproduction reported in a multigeneration study in rats (*46*). There were no adverse developmental effects reported in rats or rabbits (*47, 79*).

FD&C Blue No. 2 is poorly absorbed (<5%) from the gastrointestinal tract of rats (*80*).

The acceptable daily intake established by JECFA in 1969 is 0-17.0 mg/kg/day. The maximum anticipated daily intake is calculated to be 0.009 mg/kg/day.

FD&C Blue No. 1. FD&C Blue No. 1 (Brilliant Blue FCF) is a triphenylmethane color additive used in the production of confectionery products, beverages and beverage powders, bakery products, dairy products, drugs, and cosmetics. The amount certified in 1990 was 233,418 pounds.

FD&C Blue No. 1 is not genotoxic by a weight of evidence analysis (*5, 7, 10, 27, 30, 31, 33-35, 38, 73, 81, 82*).

The acute oral toxicity is low; for example, the acute oral LD_{50} in the rat is 2000 mg/kg (*83*).

Lifetime toxicity/carcinogenicity studies (dietary administration) of FD&C Blue No. 1 in rats and mice were reported by Borzelleca et al. (*84*). There was no evidence of carcinogenicity in either species. There were no consistent compound related adverse effects in either species except a 15% reduction in body weight and decreased survival in female rats that received the highest dose (2.0%). The no observed adverse effects levels were: rats -- males, 1072 mg/kg/day and females, 631 mg/kg/day; mice -- males, 7354 mg/kg/day and females, 8966 mg/kg/day.

There were no adverse reproductive effects in a mutigeneration study in rats (*46*). There were no adverse developmental effects in rats or rabbits (*47*).

FD&C Blue No. 1 is poorly absorbed (<5%) from the gastrointestinal tract of rats (*85, 86*). It is not metabolized by rats,mice or guinea pigs (*87*).

The acceptable daily intake established by JECFA in 1969 is 0-12.5 mg/kg/day. The maximum anticipated daily intake is calculated to be 0.022 mg/kg/day.

FD&C Green No. 3. FD&C Green No. 3 (Fast Green FCF) is a triphenylmethane color additive used to color maraschino cherries, beverages, desserts, candies, bakery products, dairy products, drugs and cosmetics. The amount certified in 1990 was 7,493 pounds.

FD&C Green No. 3 is not genotoxic by a weight of evidence analysis (*7, 26, 27, 31-34, 38, 39, 74, 81, 82, 88-92*).

The acute oral toxicity is low; for example, the acute oral LD_{50} in rats is > 2000 mg/kg/day (*83*).

Lifetime toxicity/carcinogenicity studies (dietary administration) were conducted in rats and mice (*93*). There was no evidence of carcinogenicity in either species. There were no consistent compound-related adverse effects in either species except an increase in urinary bladder tumors in male rats that received the highest level; these were not considered compound-related. The no observed adverse effect levels reported were: rats -- males, 1486 mg/kg/day and females, 4021 mg/kg/day; mice -- males, 8806 mg/kg/day and females 11805 mg/kg/day.

There were no reported adverse reproductive effects in a multigeneration study in rats (*46*). There were no developmental effects in rats or rabbits (*47*).

FD&C Green No. 3 is poorly absorbed (<5%) from the gastrointestinal tract of rats (*86*).

The acceptable daily intake established by JECFA in 1986 is 0-25 mg/kg/day. The maximum anticipated daily intake is calculated to be 0.0003 mg/kg/day.

FD&C Red No. 3. FD&C Red No. 3 (Erythrosine) is a xanthene color additive used to color confectionery products, cherries, canned fruits and vegetables, fish products, baked goods and dairy products. The amount certified in 1990 was 182,596 pounds.

The potential genotoxic and clastogenic effects of FD&C Red No. 3 were critically reviewed by Lin and Brusick (*94*) and Brusick (*95*) who concluded that it is neither genotoxic nor clastogenic. The acute oral toxicity is low; for example, the acute oral LD_{50} in the rat is 7400 mg/kg (*96*).

Lifetime toxicity/carcinogenicity studies (dietary administration) of FD&C Red No. 3 in rats and mice were reported by Borzelleca et al. (*97*), and Borzelleca and Hallagan (*98*). There was no evidence of a direct carcinogenic effect in either species. In the study reported by Borzelleca et al. (*97*), male rats that received the highest dose (2464 mg/kg/day) had statistically significant increases in the incidence of thyroid follicular cell hypertrophy, hyperplasia and adenomas. A numerical increased incidence of thyroid follicular adenomas was reported in female rats in this study but the increase was not statistically significant. Food consumption increased in these studies in all treated groups in a dose-related manner. The no observed adverse effect levels were: rats -- males, 251 mg/kg/day and females, 641 mg/kg/day; mice -- males, 4759 mg/kg/day and females, 1834 mg/kg/day.

There were no adverse reproductive effects reported in a multigeneration study in rats (*46*). There were no adverse developmental effects in rats or rabbits (*47*).

FD&C Red No. 3 is poorly absorbed (<5%) from the gastrointestinal tract of the rat (*99, 100*). FD&C Red No. 3 is partially de-iodinated to lower iodinated fluoresceins (*101, 102*).

A number of mechanistic studies (biochemical and histological including histomorphometry) were conducted to evaluate the effects of this color additive on the rat thyroid and thyroid hormone economy (*103*). These studies demonstrate that high dietary concentrations of FD&C Red No. 3 inhibit the peripheral metabolism of thyroxine (T_4) to form triiodothyronine (T_3) resulting in biologically significant decreased levels of T_3 which in turn activates the pituitary to release increased amounts of thyrotropin (TSH); the role of TSH in rat thyroid oncogenesis is well established (*104*). The weight of the evidence indicates that increased serum TSH concentrations in rats consuming high dietary concentrations of FD&C Red No. 3 result in the development of rat thyroid follicular cell hypertrophy, hyperplasia and adenomas. No effect levels were established in short term mechanistic studies for each step in the progression of rat thyroid follicular cell changes (*103*).

The acceptable daily intake established by JECFA in 1990 is 0-0.1 mg/kg/day. The maximum anticipated daily intake is calculated to be 0.018 mg/kg/day.

Regulatory Status of the FD&C Color Additives

Color additives for use in foods, drugs and cosmetics are regulated by the Food and Drug Administration (FDA) under its authority derived from the 1960 Color Additive Amendments to the Federal Food, Drug, and Cosmetic Act (FFDCA). The Amendments shifted the burden to industry to prove that the FD&C color additives are safe. Sec. 706 of the Act established how a color additive can be "permanently listed". Permanent listing means that the FDA has completed its scientific review and concluded that the color additive is safe under Sec. 706(b)(4) of the Act. A variety of factors may be considered in FDA's safety evaluation including total anticipated exposure to the color additive, safety factors, and analytical methods (FFDCA Sec. 706 (b)(5)(A). The Delaney Clause prohibits the listing of color additives found by FDA to induce cancer in laboratory animals or man (FFDCA Sec. 706 (b)(5)(B)). The FDA defines safe as "convincing evidence that establishes with reasonable certainty that no harm will result from the intended use of the color additive" (21 CFR Sec. 70.3(i)(1989)).

The Color Additive Amendments of 1960 also included the "Transitional Provisions" which established the "provisional listing" of commercially established colors (Pub. L. 86-618, Sec. 203, Title II). Provisional listing permitted color additives approved when the Amendments were enacted to continue to be marketed while FDA determined whether they could be permanently listed. All approved color additives were placed on the provisional list in 1960 (25 Fed Reg 9759 (12 October 1960)) and were subsequently removed either because they were eventually permanently listed upon agency approval of a color additive petition, or they were delisted for various reasons.

There are two forms of the FD&C color additives, the dye and the lake. The dye is the color additive itself and is water soluble. The lake is the dye form attached to an aluminum or calcium substrate to make it insoluble.

All FD&C dyes are now permanently listed for use in foods, drugs, and cosmetics and the lakes of these color additives remain provisionally listed for all uses with two exceptions: (1) the dye and lake forms of FD&C Red No. 40 are permanently listed for all uses, and (2) the dye form of FD&C Red No. 3 is permanently listed for food and ingested drug uses only; the provisionally listed uses of the dye and all uses of the provisionally listed lake were terminated in 1990 because of FDA concerns that the color additive caused thyroid tumors in rats (55 Fed Reg 3516 (1 February 1990)). At some point the agency plans to propose termination of the permanently listed uses of FD&C Red No. 3 dye (cf. Table 1).

The lakes of the FD&C color additives, except for FD&C Red No. 40 and FD&C Red No. 3, remain provisionally listed because FDA has kept their approval separate from the dye form of the color additive for regulatory purposes. The agency first proposed to permanently list the FD&C lakes as a group in 1965 (30 Fed Reg 6490 (11 May 1965)). This proposal was subsequently withdrawn and a new proposal was published in 1979 also to permanently list the lakes as a group (44 Fed Reg 36411 (22 June 1979)).

There are few restrictions on the use of FD&C color additives. All are permitted for use at levels consistent with good manufacturing practice. Formerly, the FD&C color additives generally did not need to be specifically identified on food labels and could be declared by stating "colored", "color added" or another generic term (21 CFR Sec. 101.22(c)(1989)). The one exception was FD&C Yellow No. 5 which must be specifically declared in a product's ingredient statement because of concerns about allergenicity (21 CFR Sec. 74.705(d)(1989)). Whether FD&C Yellow No. 5 causes sensitivity reactions remains an open question but it appears that it does not cross-react with aspirin, precipitate asthma or induce hyperactivity. It appears that the color additive may cause urticaria in a very small percentage of the population (*105*).

In a recent development, the Nutrition Labeling and Education Act of 1990 amended Sec. 403(i) of the Food, Drug, and Cosmetic Act to require the specific declaration of FD&C color additives as individual ingredients on the ingredient label of foods. Color additives exempt from the certification can still be declared generically. The effective date for this change is 8 November 1991, one year after enactment, as specified in the amendments.

Most recently, new amendments enacted in August 1991 alter the effective date for the labeling change. The new amendments exempt from the 8 November 1991 effective date labels printed <u>before</u> 1 July 1991 <u>and</u> applied to food before 8 May 1993 (Pub.L. 102-108).

Literature Cited

1. Borzelleca, J.F.; Hallagan J.B.; & Reese, C.S.; In *Xenobiotics in Foods and Feeds*; J.W. Finley and D.E. Schwass, Eds.; ACS Symposium Series No. 234; American Chemical Society: Washington, DC., 1983; pp. 311.
2. Marmion, D.M.; *Handbook of U.S. Colorants for Foods. Drugs, and Cosmetics* 2nd Ed.; John Wiley & Sons: New York, New York, 1984.
3. Newsome, R.L. *Fd. Technol.* **1986**, 40 (7), 46.
4. Abrahamson, S. & Valencia, R. Evaluation of substances of interest for genetic damage using *Drosophila melanogaster*. Unpublished report, 1978.
5. Bonin, A.M. & Baker, R.S.U. *Fd. Tech. Australia* **1980**, 32, 608.
6. Brown, J.P. & Dietrich, P.S. *Mutat. Res.* **1983**, 116, 305.
7. Brown, J.P.; Roehm, G.W.; & Brown, R.J. *Mutat. Res.* **1978**, 56, 249.
8. Chung K.T. *Mutat. Res.* **1983**, 114, 269.
9. Chung, K.T.; Fulk, G.E.; & Andrews, A.W. *Environ. Microbio.* **1981**, 42, 641.
10. Haveland-Smith, R.B. & Combes, R.D. Fd. Cosmet. Toxicol. **1980**, *18*, 215.
11. Heberlein, G.T. Progress report on Contract 223-74-2107. Unpublished Report, 1977.
12. Litton Bionetics, Inc. Mutagenicity evaluation of NTR-Z-4576 (FD & C Red No. 40). Unpublished report, 1976.
13. Muzzall, J.M. & Cook, W.L. *Mut. Res.* 1979, 67, 1.
14. Prival, M.J.; Davis, V.M.; Peiperl, M.D.; & Bell, S.J. *Mutation Res.* **1988**, 206, 247.
15. U.S.F.D.A. Evaluation of final reports on heritable translocation test of ammonium saccharin, calcium saccharin and FD&C Red No. 40. Unpublished report, 1978.
16. Hazleton Laboratories, Inc. FD&C Red No. 40, acute oral toxicity in rats. Unpublished report, 1965.
17. Borzelleca, J.F.; Olson, J.W.; & Reno F.E. *Fd. Chem. Toxicol.* **1989**, 27, 701.
18. Borzelleca, J.F.; Olson, J.W.; & Reno, F.E. *Fd. Chem. Toxicol.* **1991**, *29*(5), 313-319.
19. Hazleton Laboratories, Inc. FD&C No. 40, two-generation reproduction study in rats. Unpublished report, 1969.
20. Collins, T.F.X.; Black, T.N.; Welsh, J.J.; & Brown, L.H. *Toxicol. Ind. Health* **1989**, 5, 937.
21. Collins, T.F.X.; Black, T.N.; Welsh, J.J.; & Brown, L.H. *Fd. Chem. Toxicol.* **1989**, 27, 707.
22. Collins, T.F.X. & Black, T.N. *Fd. Cosmet. Toxicol.* **1980**, 18, 561.
23. Hazleton Laboratories, Inc. FD&C No. 40, teratology studies in rabbits, Report No. 165-142. Unpublished report, 1974.
24. Hazleton Laboratories, Inc. FD&C No. 40, metabolic disposition of S-Allura Red AC in the dog and rat. Unpublished report, 1975.
25. Hazleton Laboratories, Inc. FD&C No. 40, determination of the metabolites of Allura Red AC in the rat and dog. Unpublished report, 1975.

26. Au, W. & Hsu, T.C. *Envir. Mutagen* **1979**, 1, 27.
27. Cameron, T.P.; Hughes, T.P.; Kirbey, P.E.; Fung, V.A.; & Dunkel, V.C. *Mutat. Res.* **1987**, 189, 223.
28. Combes, R.D. *Mutat. Res.* **1983**, *108*, 81-92.
29. Henschler, D. and Wild, D. *Archs. Toxicol.* **1986**, 59, 69.
30. Ishidate, M.; Sofuni, T.; & Yoshikawa, K. *Gann* **1981**, 27, 95.
31. Ishidate, M.; Sofuni, T.; Yoshikawa, K.; Hayashi, M.; Nohmi, T.; Sawada, M.; & Maisuoka, A. *Fd. Chem. Toxicol.* **1984**, *22*, 623.
32. Kada, T.; Tutikawa, K.; & Sadaie, Y. *Mutat. Res.* **1972**, 16, 165.
33. Kawachi, T.; Komatsu, T.; Kada, T.; Ishidate, M.; Sasaki, M.; Sugiyama, T.; & Tazima, Y. In *The Predictive Value of Short-term Screening Tests in Carcinogenicity Evaluation.;* G.M. Williams, Ed.; Elsevier: Amsterdam., 1980; pp. 253-267
34. Kornbrust , 0. & Barfknect, T. *Envir. Mutagen.* **1985**, 7, 101.
35. Longstaff, E.; McGregor, D.B.; Harris, W.I.; Robertson, J.A.; & Poole, A. *Dyes Pigments* **1984**, 5, 65.
36. Luck, H. & Rickerl, E. *Z. Lebensmittelunters u-Forsch.* **1960**, 112, 157.
37. Patterson, R.M. & Butler, J.S. *Fd. Chem. Toxicol.* **1982**, 20, 461.
38. Price, P.J.; Suk, W.A.; Freeman, A.E.; Lane, W.T.; Peters, R.L.; Vernon, M.L.; & Heuber, R.J. *Br. J. Cancer* **1978**, 21, 361.
39. Sankaranarayanan, N. & Murthy, M.S.S. *Mutat. Res.* **1979**, 67, 309.
40. Tarjan, V. & Kurti, M. *Mutat. Res.* **1982**, 97, 228.
41. Zhurkov, V.S. *Genetika* **1975**, 11, 146.
42. National Institute of Hygienic Sciences of Japan Unpublished data, 1964.
43. Borzelleca, J.F. & Hallagan, J.B. *Fd. Chem. Toxicol.* **1988**, 26, 189.
44. Borzelleca, J.F. & Hallagan, J.B. *Fd. Chem. Toxicol.* **1988**, 26, 179.
45. Maekawa, A.; Matsouka, C.; Onodera, H.; Tanigawa, H.; Furuta, K.; Kanno, J.; Jang, J.; & Hayashi, Y. *Fd. Chem. Toxicol.* **1987**, 25, 891.
46. Pierce, E.; Agersborg, H.; Borzelleca, J.; Burnett, C.; Eagle, E.; Ebert, A.; Kirschman, J.; & Scala, R. *Toxicol. appl. Pharmac.* **1974**, 29, 121.
47. Burnett, C.; Agersborg, H.; Borzelleca, J.; Eagle, E.; Ebert, A.; Pierce, E.; Kirschman, J.; & Scala, R. *Toxicol. appl. Pharmac.* **1974**, 29, 121.
48. Sobotka, T.J.; Brodie, R.E.; & Spaid, S.L. *J. Toxicol. Environ. Health.* **1977**, 2, 1211.
49. Allen, R. & Roxon, J. *Xenobiotica* **1974**, 4, 637.
50. Chung, K.T.; Fulk, G.; & Egan, M. *Appl. envir. Microbiol.* **1978**, 35, 558.
51. Dubin, P. & Wright, K. *Xenobiotica* **1975**, 5, 563.
52. Roxon, J.; Ryan, A.; & Wright, S. *Fd. Cosmet. Toxicol.* **1967**, 5, 367.
53. Roxon, J.; Ryan, A.; & Wright, S. *Fd. Cosmet. Toxicol.* **1967**, 5, 645.
54. Watabe, T.; Ozawa, N.; Kobayashi, F.; & Kurata, H. *Fd. Cosmet. Toxicol.* **1980**, 18, 349.
55. Jones, R.; Ryan, A.J.; & Wright, S.E. *Fd. Cosmet. Toxicol.* **1964**, 2, 447.
56. Ryan, A.J.; Welling, P.G.; & Wright, S.E. *Fd. Cosmet. Toxicol.* **1969**, 7, 287.
57. Goldenring, J.R.; Batter, D.K.; & Shaywitz, B.A. *Neurobehav. Toxicology Teratol.* **1982**, 4, 43.

58. Abe, S. & Sasaki, M. *J. Nat. Cancer Inst.* **1977**, 58, 1635.
59. Combes, R.D. (1983). *Mutat. Res.* **1983**, 103, 81.
60. Garner, R.C. & Nutman, C.A. *Mutat. Res.* **1977**, 44, 9.
61. Ishidate, M. & Odashima, S. *Mutat. Res.* **1977**, 48, 337.
62. Ivett, J.L.; Brown, B.M.; Rodgers, C.; Anderson, B.E.; Resnick, M.A.; & Zeiger, E. *Environ. Molecular Mutgen.* **1989**, *14*, 165.
63. Litton Bionetics, Inc. Mutagenicity of FD&C Yellow No. 6 in a mouse mutation assay. Unpublished report, 1985.
64. Litton Bionetics, Inc. Mutagenicity evaluation of FD&C Yellow No. 6 in the Ames *Salmonella*/microsome plate test. Unpublished report, 1985.
65. Litton Bionetics, Inc. Mutagenicity evaluation of FD&C Yellow No. 6 in an *in vitro* sister chromatid exchange assay in Chinese hamster ovary (CHO) cells. Unpublished report, 1985.
66. Litton Bionetics, Inc. Clastogenic evaluation of FD&C Yellow No. 6 in an *in vitro* cytogenetic assay measuring chromosomal aberration frequencies in Chinese hamster ovary (CHO) cells. Unpublished report, 1985.
67. Litton Bionetics, Inc. Evaluation of FD&C & C Yellow No. 6 in the rat primary hepatocyte unscheduled DNA synthesis assay. Unpublished report, 1985.
68. Zimmerman, F.K.; von Borstel, R.C.; von Halle, E.S.; Parry, J.M.; Siebert, D.; Zetterberg, G.; Barale, R.; & Loprieno, N. *Mutat. Res.* **1984**, 133, 199.
69. Gaunt, I.F.; Farmer, M.; Grasso, P.; & Gangolli, S.D. *Fd. Cosmet. Toxicol.* **1967**, 5, 747.
70. Certified Color Manufacturers Association. Lifetime toxicity/carcinogenicity studies of FD&C Yellow No. 6 in rats and mice. Unpublished reports, 1983.
71. NTP Technical Report - NCI/National Toxicology Program. Carcinogenesis Bioassay of FD&C Yellow No. 6. Tech. Rept. Series No. 208, 1981.
72. Radomski, J.L. & Mellinger, T.J. *J. Pharmacol. exp. Ther.* **1962**, 136, 259.
73. Auletta, A.E.; Kuzava, J. M.; & Parmar, A.S. *Mutat. Res.* **1977**, 56, 203.
74. Das, S.K. & Giri, A.K. *Cytobios* **1988**, 54, 25.
75. Fujita, H.; Mizuo, A.; & Hiraga, K. *Ann. Rep. Tokyo Metr. Res. Lab. Publ. Hlth.* **1976**, 27, 153.
76. U.S.F.D.A. Unpublished data, 1969.
77. Borzelleca, J.F.; Hogan, G.K.; & Koestner, A. *Fd. Chem. Toxicol.* **1985**, 23, 551.
78. Borzelleca, J.F. & Hogan, G.K. *Fd. Chem. Toxicol.* **1985**, 23, 719.
79. Borzelleca, J.F.; Goldenthal, E.I.; Wazeter, F.X.; & Schardein, J.L. *Fd. Chem. Toxicol.* **1987**, 25, 495.
80. Lethco, E.J. & Webb, J.M. *J. Pharmacol. exp. Ther.* **1966**, 154, 384.
81. Bonin, A.M.; Farquharson, J.B.; & Baker, R.S.U. *Mutat. Res.* **1981**, 89, 21.
82. Hayashi, M.; Kishi, M.; Sofuni, T.; & Ishidate, M., Jr. *Fd. Chem. Toxicol.* **1988**, 26, 487-500.
83. Lu, F.C. & Lavalle, A. *Can. pharm. J.* **1964**, 97, 30.
84. Borzelleca, J.F.; Depukat, K.; & Hallagan, J.B. *Fd. Chem. Toxicol.* **1990**, 28, 221-234.

85. Brown, J.P.; Dorsky, A.; Enderlin, F.E.; Hale, R.L.; Wright, V.A.; & Parkinson, T.M. *Fd. Cosmet. Toxicol.* **1980**, 18, 1.
86. Hess, S. & Fitzhugh, O. *J. Pharmacol. exp. Ther.* **1955**, 114, 38.
87. Phillips, J.C.; Mendis, D.; Eason, C.T.; & Gangolli, S.D. *Fd. Cosmet. Toxicol.* **1980**, 18, 1.
88. Litton Bionetics, Inc. Mutagenicity evaluation of FD&C Green No. 3 in the mouse lymphoma forward mutation assay. Unpublished report, 1982.
89. Litton Bionetics, Inc. Evaluation of FD&C Green No. 3 in primary rat hepatocyte unscheduled DNA synthesis assay. Unpublished report, 1982.
90. Litton Bionetics, Inc. Mutagenicity evaluation of FD&C Green No. 3 in the Ames *Salmonella* microsome plate test. Unpublished report, 1982.
91. Misra, R.N. & Misra, B. *Mutat. Res.* **1986**, 170, 75.
92. Rosenkranz, H.S. & Leifer, Z. Chem. Mutagens **1980**, 6, 109.
93. Borzelleca, J.F. Manuscript in preparation.
94. Lin, G. & Brusick, D. *Mutagenesis* **1986**, 1, 253.
95. Brusick, D. Addendum to review of the genotoxicity of FD&C Red No. 3. Unpublished report, 1989.
96. Butterworth, K.; Gaunt, I.; Grasso, P.; & Gangolli, S. *Fd. Cosmet. Toxicol.* **1976**, 14, 525.
97. Borzelleca, J.F.; Capen, C.C.; & Hallagan, J.B. *Fd. Chem. Toxicol.* **1987**, 25, 723.
98. Borzelleca, J.F. & Hallagan, J.B. *Fd. Chem. Toxicol.* **1987**, 25, 735.
99. Daniel, J. *Toxic. appl. Pharmac.* **1962**, 4, 572.
100. Webb, J.; Finck, M.; & Brouwer, E. *J. Pharmacol. exp. Ther.* **1962**, 137, 141.
101. Hazleton Laboratories, Inc. FD&C No. 3, metabolism in rats. Unpublished report, 1989.
102. Vought, R.; Brown, F.; & Wolff, J. *J. clin. Endcr. Metab.* **1972**, 34, 747.
103. Certified Color Manufacturers Association. The secondary mechanism of rat thyroid oncogenesis: the results of a 60-day study of the effects of FD&C Red No. 3 on thyroid hormone economy in male rats. Unpublished report, 1989.
104. Hill, R.N.; Erdreich, L.S.; Paynter, O.E.; Roberts, P.A.; Rosenthal, S.L.; & Wilkinson, C.F. *Fund. Appl. Toxicol.* **1989**, *12*, 629.
105. Robinson, G. *Fd. Chem. Toxicol.* **1988**, 26, 73.

RECEIVED December 2, 1991

Chapter 32

Safety Evaluation of Olestra

A Nonabsorbable Fat Replacement Derived from Fat

Carolyn M. Bergholz

The Procter & Gamble Company, Cincinnati, OH 45224

Olestra is the mixture of the hexa-, hepta-, and octa-esters of sucrose with long-chain fatty acids from any edible oil. Its physical properties are comparable to those of triglycerides, but it is not digested by lipolytic enzymes or absorbed and therefore is noncaloric. Technically it can replace fat in a wide variety of foods and can be used to make cooked, baked and fried foods lower in fat and calories. A Food Additive Petition is under review by the FDA which is comprised of results of extensive testing in animals and humans. The major areas of investigation are metabolism and absorption, chronic toxicity, mutagenicity, carcinogenicity, reproductive and developmental toxicity, safety for gastrointestinal tract, nutrition, and the potential for olestra to affect absorption of drugs. This testing involved studies in five different species of animals and over 30 clinical investigations. The results of this research support the safety of olestra for use in foods.

Olestra is the common and usual name proposed for the mixture of the hexa-, hepta-, and octa-esters of sucrose formed with long-chain fatty acids from any edible oil. Because its physical properties are like those of a triglyceride, it functions the same as fat in foods and performs the same as fat during cooking or frying. However, olestra is not digested by pancreatic enzymes and therefore is not absorbed from the gastrointestinal tract and contributes no calories. A Food Additive Petition is under review by the Food and Drug Administration (FDA) which requests approval to use olestra in place of fat for preparation of specific foods. This petition is comprised largely of reports of studies on the safety of olestra.

The safety evaluation of olestra, like that of any material, is determined by three fundamental considerations: the chemistry of the material, its biological properties

NOTE: Please address correspondence to K. D. Lawson, The Procter & Gamble Company, 6071 Center Hill Road, Cincinnati, OH 45224

0097–6156/92/0484–0391$06.00/0

and the expected exposure. This chapter reviews the olestra safety research program in the context of these considerations. The program is very extensive and is comprised of over 100 laboratory and clinical investigations. It is impossible, obviously, to present the results of the testing that support the safety of olestra in this review chapter. Rather, the intent is to provide an overview of the scope of the program, discuss the major areas of investigation, the most recent data and important conclusions, citing references for studies that have been published.

Chemistry

The raw materials used to make olestra are sucrose from sugar cane or beet sugar and fatty acids from edible oils. The structure of olestra is analogous to that of a fat, i.e., triglyceride molecule. Instead of a glycerol core with three fatty acid esters, olestra has a larger core, a sucrose molecule, with 6–8 fatty acid esters. The physical properties of olestra, like those of a triglyceride, are determined by the fatty acid composition, which varies depending upon the source oil used. A high proportion of long-chain and/or saturated fatty acids, for example, will increase the viscosity and raise the melting point of the olestra. The safety evaluation of olestra must provide assurance of safety for the full range of compositions specified for food use. Olestra is comprised largely of octaesters, as shown in the specifications listed in Table 1. There are also specifications for minor impurities arising from the starting materials or as by-products of the synthesis. These are the same as those for conventional fats and oils or other fatty acid derived food additives.

Table 1. Olestra Ester Distribution Specifications

Total Octa-, Hepta-, Hexaester	≥ 97%
Octaester	≥ 70%
Hexaester	≤ 1%
Penta- and Lower esters	≤ 0.5%

Because the anticipated uses of olestra include frying applications, there has been extensive investigation of the chemistry of heated olestra compared to heated fat. The results show that the olestra sucrose backbone is extremely heat stable and that reactions involving the fatty acid side chains are the same as those that occur during heating of triglyceride. This has been demonstrated by heating olestra for 7 days under deep fat frying conditions and using gas chromatography and mass spectroscopy to compare the fatty acid reaction products with those from the corresponding triglyceride heated under the same conditions. No unique products were detected at a level of sensitivity of 5 ppm (Henry, D. E.; Tallmadge, D. H.; Saunders, R. A.; and Gardner, D. R., Procter & Gamble, unpublished data).

Biological Properties

The most important biological property of olestra is its lack of absorption. This property, of course, is directly related to its chemical structure. Mattson and

Volpenhein in 1972 reported that a polyol backbone fully esterified with six or more fatty acids was completely resistant to digestion by pancreatic enzymes (*1*). One of the polyols tested was sucrose with eight fatty acid esters. Subsequent experiments in animals showed that a polyol with 6 - 8 fatty acid esters was not absorbed (*2*). These data and work of others established that fat must first be hydrolyzed to free fatty acids and 2-monoglycerides before it can form mixed micelles with bile salts and be absorbed from the intestinal lumen. The larger bulk of highly esterified sucrose apparently hinders enzymatic cleavage of the fatty acid ester bond in olestra.

Nonabsorption has a number of implications for design of a program to evaluate the safety of olestra for food use. First of all, it is important to determine the extent to which no absorption can be established. This is a challenge since it is impossible to prove zero. The kinds of approaches used will be discussed further below. The safety program must of course demonstrate the absence of any systemic toxicity after long-term ingestion. But importantly, nonabsorption implies that the gastrointestinal (GI) tract is the only organ system that is exposed to olestra. Therefore the safety evaluation must thoroughly assess the potential to affect the structure and function of the gastrointestinal system. A third implication of nonabsorption is the existence of a nonabsorbable lipid phase in the GI tract. This raises important questions about the potential to affect absorption of lipophilic constituents of the diet and lipophilic medications.

Absorption and Toxicology

Understanding the absorption, distribution, metabolism and excretion (ADME) of olestra, as with any new material, is fundamental to any safety evaluation. A number of different approaches have been utilized, all of which provide evidence that olestra is not digested and not absorbed from the gastrointestinal tract. Results of *in vitro* experiments showed that olestra was not hydrolyzed regardless of the fatty acid chain length or degree of unsaturation (*1*). Results of material balance studies (*2-4*) and early experiments with radiolabeled olestra were consistent with nonabsorption.

To learn where olestra would go if it were absorbed and if it would be metabolized, it was necessary to use intravenous (IV) dosing (*5–7*). Results showed that after IV administration of olestra to rats and monkeys, almost 70% of the olestra was rapidly taken up by the liver. Over time, unmetabolized olestra was slowly excreted via the bile into the stool. Therefore, if trace amounts of olestra were absorbed from the gastrointestinal tract, it would most likely accumulate in the liver. Based on this understanding, liver tissue from rats fed olestra at 9% (w/w) of the diet for two years was analyzed for olestra (*7*). The results showed that olestra did not accumulate in tissues. Based on the sensitivity of the analytical method and the kinetics of excretion, it was calculated that olestra would have been detected in the liver if more than 1×10^{-6} % of the total amount ingested were present. Similarly, no olestra was detected in any tissues from monkeys after 29 consecutive months of diet with 8% (w/w) olestra. The calculated limit of detection in this study was 4×10^{-5} % of the dose. Results of state-of-the-art ADME studies using high specific activity radiolabeled olestra continue to confirm that olestra does not accumulate in tissues above the limit of detection (Miller, K. W., Procter & Gamble, personal communications).

The potential for systemic toxicity was evaluated in subchronic and chronic feeding studies conducted in five different animal species with olestra at concentrations up to 15% (w/w) of the diet (Table 2). Results of all of these studies confirm that olestra is not toxic. Body weight gain, urine and blood chemistries, hematology and microscopic examination of all tissues showed no adverse effects related to ingestion of olestra (*8-11*). Results of two 2-year studies in rats also demonstrated that olestra is not carcinogenic (*8*). A long-term study in mice is in progress and is expected to again confirm that olestra does not cause tumors. Among these studies was a 91-day study of olestra heated under foodservice deep fat frying conditions (*11*).

Table 2. Feeding Studies Conducted in Animals to Evaluate the Safety of Olestra

Species	*Duration (days)*	*% (w/w) in Diet*
Subchronic:		
Rat	28	4, 8, 15
	91	4, 8, 15
	91	4, 8, 15
	91 (unheated)	3.5, 7.5
Hamster	28	5, 10
Dog	28	5, 10
	91	5, 10
Mouse	91	2.5, 5, 10
Long-Term:		
Dog	20	5, 10
Monkey	44	8
Rat	24	1, 5, 9
	24	9
Mouse	24 (in progress)	2.5, 5, 10

A series of short-term genotoxicity tests also demonstrated that olestra does not cause mutations, chromosomal aberrations, or affect DNA repair (*12*). A multi-generation feeding study in rats confirmed that olestra does not affect reproduction or embryonic development (*13*).

In summary, the lack of absorption and accumulation of olestra in tissues has been established at very low limits of detection. The results of a battery of basic toxicity evaluations demonstrate that olestra is not toxic, carcinogenic, or genotoxic and is not a reproductive toxin.

It is of interest to note that this toxicity evaluation is atypical in some respects. First, a toxic effect level was not identified. Actually this is not surprising since a nonabsorbed material is far less likely to cause toxicity. Secondly, it is impossible to feed olestra to animals at levels 100-times the intake expected from the use of olestra to replace fat in foods. For example, if expected intake over time is 6 grams per day, a rat would have to ingest 600 grams of olestra per day in addition to sufficient diet to meet energy and nutritional needs. For perspective, rats consume 150-200 grams of chow per week. In the feeding studies, olestra was tested at doses which exceeded the maximum high dose recommended for toxicity testing (5% by weight of the diet). The power of animal feeding studies for testing a macronutrient substitute rests on the wide number of parameters that can be measured and the ability to thoroughly examine tissues grossly and microscopically. The absence of adverse effects in any animal species tested in the basic toxicity evaluation opened the door to clinical evaluation of potential adverse health and nutritional effects in humans.

Gastrointestinal Safety

As discussed above, the GI tract is the only organ system exposed to measurable amounts of olestra. Therefore, the potential to affect the GI system was evaluated in both animal and human studies. In long-term feeding studies in animals, detailed microscopic examination of all segments of the GI tract showed that high dietary concentrations of olestra did not cause morphologic changes (*8-11*). Special stains and chemical analyses also showed that olestra was not present in associated lymphoid tissues.

Clinical studies have shown that the presence of olestra in the GI tract has no effect on gastric emptying time (*14*), GI motility or transit time through segments of the bowel (Aggarwal, A., Mayo Clinic, personal communications). A study in rats demonstrated that olestra does not stimulate secretion of pancreatic enzymes (*15*). Regulation of these GI functions is determined by the presence of free fatty acids from hydrolysis of fat, by calorie density and by water-soluble mediators. Olestra is not hydrolyzed, non-caloric and does not participate in the aqueous or mixed micellar phases in the intestinal lumen. It does not appear to provide signals that regulate digestion or to interfere with signals provided by other components of the diet.

Bile acid physiology is another important area of olestra safety research, since bile acids are important to digestion and absorption of fatty acids. Studies in rats, monkeys, and humans showed that excretion of fecal bile acids is normal and that olestra does not alter the composition of bile or reabsorption of bile acids from the intestinal lumen (*16*, *17*, St. Clair, R. W., Wake Forest University, unpublished data). Table 3 shows the concentration of primary and secondary bile acids in bile of monkeys fed olestra at 6% of the diet for one year compared to the profile of chow-fed controls. There were no significant differences, indicating that olestra did not affect synthesis or reabsorption of bile acids.

Extensive clinical research has shown that ingestion of foods made with olestra does not adversely affect bowel function, even with high intakes of 30-50 grams of olestra per day (*4,17-21*). Fecal water and electrolyte content is not affected in human or animals, confirming that olestra does not cause diarrhea. Stool consistency is sometimes softer due to the unabsorbed lipid-like material, but this is reported as a benefit (*4*). One reason bowel function and stool consistency are normal is that olestra has no osmotic effect in the colon and is not metabolized by colonic microflora. Microbial metabolism was investigated by *in vitro* anaerobic fecal culture systems with radiolabeled olestra. Results showed no degradation of fatty acids or change in ester distribution (*22*). Material balance studies also showed that olestra is excreted unchanged in feces, indicating that it is not metabolized by gut microflora (*4*).

Table 3. Biliary Bile Acid Composition in Monkeys Fed Olestra

Olestra in Diet (% w/w)	*n*	*Day of Study*	*Bile Acid (mol % ±SEM)* Lithocholic Acid	Chenodeoxycholic Acid	Deoxycholic Acid	Cholic Acid
6	5	316	6.6 ± 1.9	52.4 ± 7.5	15.6 ± 2.7	25.4 ± 7.7
	5	373	5.6 ± 1.2	55.1 ± 7.5	15.3 ± 3.2	24.0 ± 6.1
0	4	316	4.4 ± .84	58.6 ± 7.4	13.0 ± 4.4	24.0 ± 5.7
	3	373	3.9 ± .66	56.0 ± 4.7	16.5 ± 5.7	23.6 ± 2.2

Environmental safety studies have demonstrated that once olestra leaves the body it does not affect water treatment systems. It absorbs to sludge and can then be completely degraded aerobically by soil microorganisms following agricultural application of sludge (Greff, J. A., Procter & Gamble, personal communications).

Nutrition Research

The major function of the gastrointestinal system, of course, is digestion and absorption of nutrients. The potential for olestra to affect utilization of nutrients has been of primary importance in the evaluation of the safety of olestra. Utilization of water-soluble macro- and micro-nutrients, i.e., carbohydrates, amino acids, minerals and water-soluble vitamins would not be expected to be affected by olestra. Results of animal and clinical studies are consistent with this assessment. Data from animal studies show that growth and development are normal (*8,9,11,13*). Fasting blood sugar was not affected by olestra in animal studies (Jandacek, R. J.; Holcombe, B. N., Procter & Gamble, unpublished data) or in diabetic patients (*21*), confirming that olestra does not affect carbohydrate absorption. Other evidence that proteins and minerals are utilized normally in the presence of olestra comes from studies in rats of absorption of radiolabeled hydrophobic amino acids, (Gibson, W. B., Procter & Gamble, unpublished data), clinical studies showing no effect on fecal minerals or fecal nitrogen, and mineral levels in serum chemistries from animal and human studies (*8-11,18-21*).

Animal research showed very early that absorption of fat-soluble constituents in the diet, in particular cholesterol and fat-soluble vitamins, could be reduced by partitioning into olestra (*23,24*). The potential for this to occur and the extent to which absorption is decreased depends on a number of variables. Among them are the lipid solubility of the constituent, the solute concentration, the kinetics of water/oil partitioning, absorption kinetics, and importantly, the amount of olestra (*16,18-20,25*). Using radiolabeled cholesterol in olestra, it was demonstrated that 14g/day of olestra decreased absorption of dietary cholesterol about 9% (*26*). Olestra also decreases reabsorption of cholesterol in enterohepatic circulation (*16, 25*).

The fat-soluble vitamins vary in lipophilicity. To address the potential for the proposed uses of olestra to affect the nutritional status of fat-soluble vitamins, large base size, double blind, placebo-controlled clinical investigations have been conducted. The results of a 16-week study showed only a modest reduction in serum vitamin E levels after two weeks of ingestion of 18 grams of olestra per day (*27*). Continued ingestion for a total of 16 weeks gave no further reduction. Importantly, even with the modest reduction, serum alpha-tocopherol was at normal levels throughout the study.

In this same 16-week clinical investigation, functional prothrombin was measured by assay of the Simplastin:Ecarin (S/E) ratio. This assay has been shown to detect changes in vitamin K intake within days (*28*). The results of the 16-week investigation showed no change in the S/E ratio throughout the study, indicating 18 g/day of olestra does not affect vitamin K status as measured by this assay. These data are consistent with the results of another double-blind placebo-controlled study among 200 subjects which showed no difference from baseline in any measure of vitamin K status over six weeks of ingestion of olestra at 20 g/day (*29*).

During the 16-week study, there also was no difference from control in 25-hydroxy vitamin D status. This was predicted based on results of other clinical investigations (*18,30*). Vitamin A, as measured by plasma retinol levels, also showed no change, as expected (*18*). However, plasma retinol levels are not a sensitive measure of potential effects on vitamin A status, since most individuals have large stores of vitamin A in the liver which serve to maintain constant blood levels. Other approaches are being used to determine the dose-response effect on liver stores of vitamin A and potential for food uses to meaningfully affect status.

The results of animal and clinical test data must be considered in light of the proposed food uses and levels of fat replacement. These considerations determine the expected amount and pattern of chronic and single day intakes for the general population, various age groups and special subgroups, and are essential for assessment of the appropriateness of supplementation of a particular olestra food with a particular vitamin.

The goal is to ensure that foods made with olestra have full nutritional value, with the exception of less fat and fewer calories. Importantly, foods made with olestra will have the same taste and mouthfeel as those made with full calorie fats.

Drug Absorption

The safety evaluation of olestra has also included studies which established that olestra will not affect absorption and efficacy of lipophilic drugs. In one study (*31*), a single dose of propranolol, diazepam, norethindrone, or ethinyl estradiol was given with 18 grams of olestra or triglyceride. There were no differences in the absorption of any of the drugs in the two vehicles, as determined from the area under the serum concentration versus time curve. In another study (*32*), the oral contraceptives norgestrel and ethinyl estradiol were given daily over 28 days with a diet containing 18 g/day of olestra or triglyceride in a cross-over study involving 28 subjects. There were no significant differences between the two treatment periods in peak blood level, time to reach peak blood level, or area under the serum concentration-time curve for either drug. These drugs are among the most lipophilic oral medications, but are several orders of magnitude less fat-soluble than vitamin D, for example. Therefore, it is concluded that olestra will not affect absorption of even the most lipophilic drugs.

Conclusion

In conclusion, the safety evaluation of olestra is far broader than that normally conducted for a typical food additive. As in any safety evaluation, the animal and clinical studies conducted must be tailored to address the scientific questions posed by the chemical and biological properties of the material and the proposed uses. The current Food Additive approval process under the Food, Drug and Cosmetic Act allows for the flexibility needed to assure safety for the public and to allow innovations that will meet the needs of consumers.

Literature Cited

1. Mattson, F.H.; Volpenhein, R.A. *J. Lipid Res.* **1972**, *13*, 325-28.
2. Mattson, F.H.; Nolen, G.A. *J. Nutr.* **1972**, *102*, 1171-76.
3. Mattson, F.H.; Volpenhein, R.A. *J. Nutr.* **1972**, *102*, 1177-80.
4. Fallat, R.W.; Gluek, C.J.; Lutmer, R.; Mattson, F.H. *Am. J. Clin. Nutr.* **1976**, *29*, 1204-15.
5. Mattson, F.H.; Jandacek, R.J. *Lipids*, In Press.
6. Jandacek, R.J.; Holcombe, B.N. *Lipids*, In Press.
7. Wood, F.E.; DeMark, B.R.; Hollenbach, E.J.; Sargent, M.C.; Triebwasser, K.C. *Fd. Chem. Toxic.* **1991**, *29*, 231-36.
8. Wood, F.E.; Tierney, W.J.; Knezevich, A.L.; Bolte, H.F.; Maurer, J.K.; Bruce, R.D. *Fd. Chem. Toxic.* **1991**, *29*, 223-30.
9. Miller, K.W.; Wood, F.E.; Stuard, S.B.; Alden, C.L. *Fd. Chem.Toxic.* In Press.
10. Adams, M.R.; McMahan, M.R.; Mattson, F.H.; Clarkson, T.B. *Proc. Soc. Exp. Biol. Med.* **1981**, *167*, 346-53.
11. Miller, K.W; Long, P.H. *Fd. Chem. Toxic.* **1990**, *28*, 307-15.
12. Skare, K.L.; Skare, J.A.; Thompson, E.D. *Fd. Chem. Toxic.* **1990**, *28*, 69-73.
13. Nolen, G.A.; Wood, F.E.; Dierckman, T.A. *Fd. Chem. Toxic.* **1987**, *25*, 1-8.
14. Cortot, A.; Phillips, S.F.; Malagelada, J.R. *Gastroenterology* **1982**, *82*, 877-81.
15. Hager, M.H.; Schneeman, B.A. *J. Nutr.* **1986**, *116*, 2372-77.
16. Jandacek, R.J. *J. Drug Metab. Rev.* **1982**, *13*, 695-714.
17. Glueck, C.J.; Jandacek, R.J.; Subiah, M.T.R.; Gallon, L.; Yunker, R.; Allen, C.; Hogg, E.; Laskarzewski, P.M. *Am. J. Clin. Nutr.* **1980**, *33*, 2177-80.
18. Mellies, M.J.; Vitale, C.; Jandacek, R.J.; Lamkin, G.E.; Glueck, C.J. *Am. J. Clin. Nutr.* **1985**, *41*, 1-12.

19. Mellies, M.J.; Jandacek, R.J.; Taulbee, J.D.; Tewskbury, M.B.; et al. *Am. J. Clin. Nutr.* **1983**, *37*, 339-46.
20. Glueck, C.J.; Jandacek, R.J.; et al *Am. J. Clin. Nutr.* **1983**, *37*, 347-54.
21. Grundy, S.M.; Anastasia, J.V.; Kesaniemi, Y.A.; Abrams, J. *Am. J. Clin. Nutr.* **1986**, *44*, 620-29.
22. Nuck, B.A.; Federle, T.W. In *Abstracts of the Annual Meeting of the American Society for Microbiology*; American Society for Micribiology: Washington, 1990, p. 275.
23. Mattson, F.H.; Jandacek, R.J.; Webb, M.R. *J. Nutr.* **1976**, *106*, 747-52.
24. Mattson, F.H.; Hollenbach, E.J.; Kuehlthau, C.M. *J. Nutr.* **1979**, *109*, 1688-93.
25. Jandacek, R.J.; Mattson, F.H.; McNeely, S.; Gallon, L.; Yunker, R.; Glueck, C.J. *Am. J. Clin. Nutr.* **1980**, *33*, 251-59.
26. Jandacek, R.J.; Ramirez, M.M.; Crouse, J.R. *Metabolism* **1990**, *39*, 848-52.
27. Koonsvitsky, B.P.; Jones, D.Y.; Berry, D.A.; Jones, M.B. *Am. J. Clin. Nutr.* **1991**, *53*, 21.
28. Suttie, J.W.; Mummah-Schendel, B.S.; Shah, D.V.; Greger, J.L. *J. Am. Clin. Nutr.* **1989**, *47*, 299-304.
29. Jones, D.Y.; Miller, K.W.; Koonsvitsky, B.P.; Ebert, M.L.; Lin, P.Y.T.; Jones, M.B.; Will, B.H.; Suttie, J.W. *Am. J. Clin. Nutr.* **1991**, *53*, 943-6.

RECEIVED August 15, 1991

Chapter 33

Nitrate, Nitrite, and *N*-Nitroso Compounds

Food Safety and Biological Implications

Joseph H. Hotchkiss, Michael A. Helser, Chris M. Maragos, and Yih-Ming Weng

Institute of Food Science, Stocking Hall, Cornell University, Ithaca, NY 14853

Nitrogenous compounds such as amines, amides, guanidines, and ureas can react with oxides of nitrogen (NO_x) to yield N-nitroso compounds (NOC), of which over 270 are oncogenic in one or more of 41 species (*1*, *2*). Humans metabolize NOC similarly to animals and are, thus, unlikely to be resistant. For this reason human exposure to NOC has received considerable attention. NOC fall broadly into two categories based on chemical structure; the N-nitrosamines and N-nitrosamides (Figure 1).

Exposure to preformed NOC from occupational and lifestyle sources has been recently reviewed (*3-5*). Major exposures are from tobacco products, foods and beverages, cosmetics, occupational settings, and rubber products. Dietary sources such as nitrite cured meats and malt beverages are small compared to tobacco. There is evidence that exposure to NOC from the diet has declined in recent years (*5*).

In addition to exposure to environmental sources, NOC are formed *in vivo* through biological processes. The magnitude of endogenous nitrosation is unknown but is likely to be greater than exogenous exposure (*6*). Precursors to NOC from food, drugs, or tobacco are brought into contact in the oral cavity and stomach. Chemical nitrosation of most amines, and amides, is favored by acidic pHs characteristic of the stomach. The optimum nitrosation pH for most amines is between 2 and 4 whereas the nitrosation of amides, which proceeds through an alternate mechanism, increases with decreasing pH, with no optimum (*7*, *8*). Intragastric nitrosation may be a risk factor in gastric or other cancers (*3*, *9*, *10*). Endogenous nitrosation may also occur at sites other than the stomach. Intuitively, both physiological factors such as gastric pH, gastric nitrite concentration, and *in vivo* nitrate reductase activity, as well as dietary factors such as the amount of nitrate, ascorbic acid, and amine/amide precursor ingested should influence the amount NOC formed, and hence, cancer risk. This review will discuss the evidence that endogenous nitrosation occurs and the role that foods and diet play in the endogenous formation of NOC. These factors will then be contrasted against the occurrence of nitrate, nitrite, and NOC in foods.

0097–6156/92/0484–0400$06.00/0

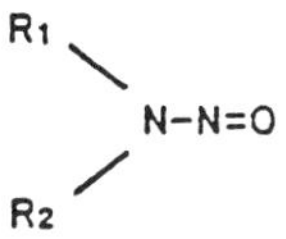

N-Nitrosamine

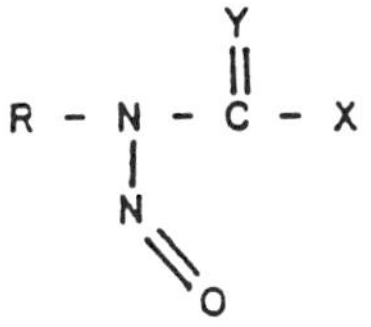

N-Nitrosamide

Y	X	
O	Alkyl, aryl	N-Nitrosamide
O	NH_2, NHR, NR_2	N-Nitrosourea
O	RO	N-Nitrosocarbamate
NH	NH_2, NHR, NR_2	N-Nitrosoguanidine

Figure 1. Generalized structure of N-nitroso compounds.

Biological Activity of N-Nitroso Compounds

Although NOC are acutely and subacutely toxic, embryotoxic, teratogenic, and mutagenic, it is their carcinogenicity that is of major interest. Preussmann and Stewart (*1*) have reviewed carcinogenicity and only a few points bear mentioning.

N-nitrosamides are direct mutagens (i.e., without activation) in most genetic indicators (*11*). N-nitrosamines are not mutagenic in bacteria without the presence of mammalian enzymes and are often only weak bacterial mutagens with metabolic activation. Usually the nitrosamine must be preincubated with bacteria and mammalian activating enzymes to show strong mutagenicity. For example, nitrosodimethylamine (NDMA) is not mutagenic on *S. typhimurium* TA100 unless a pre-incubation step is included (*12*).

Tumors can be induced by NOC in most organs including esophagus, lung, stomach, pancreas, kidney, urinary bladder, nasal cavities, trachea, brain, peripheral nerves, skin, hematopoietic tissues (*13*). Symmetrical aliphatic N-nitrosamines are hepatotoxic and hepatocarcinogenic in rodents including mice, rat, hamster and guinea pig while cyclic or unsymmetrical aliphatic are primarily esophageal carcinogens (*14*). The addition of a functional group (e.g., hydroxyl group) often changes the primary organ affected. Nitrosamides often initiate tumors at the site of application.

In general, the biological activity of NOC is thought to be related to alkylation of genetic macromolecules. N-nitrosamines are metabolically activated (oxidative dealkylation) or deactivated (reductive denitrosation) through hydroxylation at an α-carbon (*15, 16*). The resulting hydroxyalkyl moiety is eliminated as an aldehyde, and an unstable primary nitrosamine is formed. The latter tautomerizes to a diazonium hydroxide and ultimately to a carbonium ion (Figure 2). Nitrosamides spontaneously decompose to a carbonium ion at physiologic pH by a similar mechanism.

Exposure to Preformed NOC

N-nitrosamines are stable and formed facilely under several chemical conditions. For this reason, they have been found in a number of consumer products, occupational settings, and environments. N-nitrosamides are unstable under most biological and environmental conditions and have not been found. The largest individual exposure results from tobacco which contains carcinogenic nitrosamines in part-per-thousand amounts (*17*). Occupational settings can expose humans to high levels (*18*). Other exposures are orders of magnitude lower. Daily exposure from western foods is probably less than 1 μg/day/person (*5*).

Exposure to NOC through foods can be divided into two segments; preformed NOC in foods and NOC formed endogenously from food-derived precursors. Most western foods have been examined for NOC (*5*). The processes by which foods become contaminated with NOC have been discussed (*19*). These include:

1. The use of nitrate and/or nitrite as intentional food additives, both of which are added to fix the color of meats, inhibit oxidation, and prevent toxigenesis by *C. botulinum*.

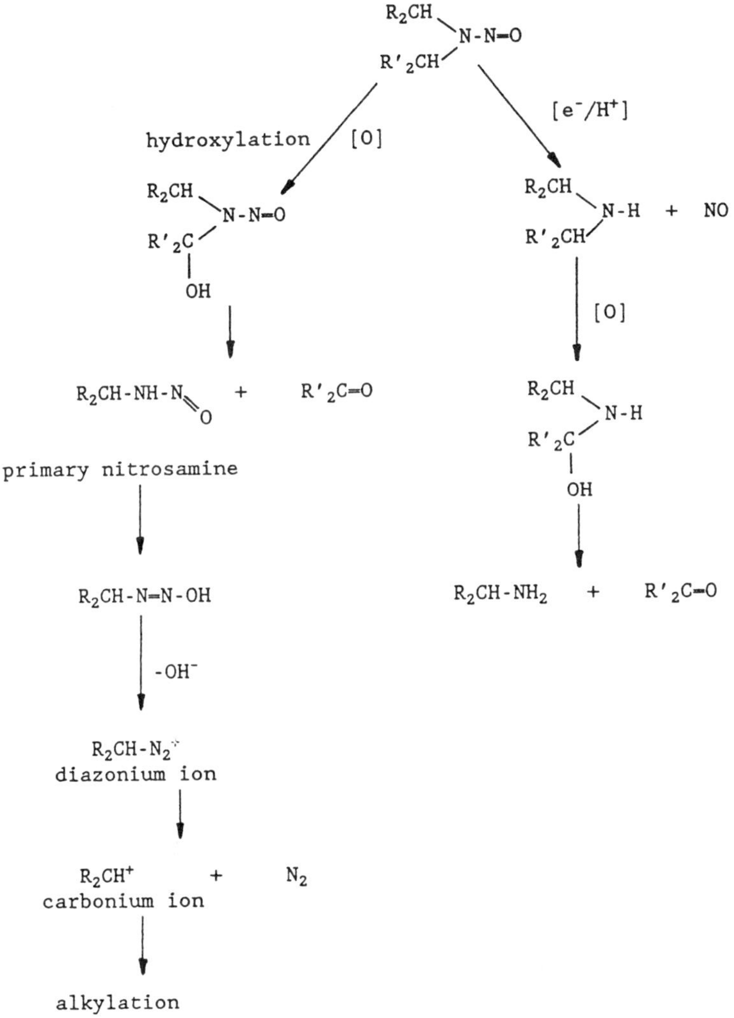

Figure 2. Metabolic activation (oxidative dealkylation and deactivation (reductive denitrosation) of N-nitrosamines (*15*, *16*).

2. Drying processes in which the drying air is heated by an open flame source. NO_x is generated in small amounts through the oxidation of N_2, which nitrosates amines in the foods. This is the mechanism for contamination of malted barley products [e.g., beer (*20*)].
3. NOC can migrate from food contact materials such as rubber bottle nipples (*21*).
4. NOC can also be added to foods if they are contaminants in a directly added substance such as spices (*22*)
5. Cooking over open flames (e.g., over natural gas flame) can result in NOC formation in foods by the same mechanism as drying.

Endogenous Formation

In addition to dietary sources, the endogenous formation of NOC has been conclusively demonstrated (*23*). There are several lines of evidence that indicate the NOC are formed within the body:

Epidemiology. It has been suggested that nitrate exposure is correlated with gastric cancer risk due to the endogenous formation of NOC (*24, 25*). However, not all epidemiological studies agree. Forman et al. (*26*) found a negative correlation between nitrate exposure and gastric cancer. One explanation is that many individuals exposed to high nitrate levels may also be consuming more vitamin C through vegetables rich both in nitrate and vitamin C (*27*). Vitamin C is an inhibitor of endogenous NOC formation (*28*). Consumption of nitrate from sources low in vitamin C (e.g., water) might be a higher risk.

***In vitro* Nitrosation Chemistry.** *In vitro* nitrosation carried out under gastric conditions of pH, concentration, temperature, and time have been reported. Typically, amines or amides are incubated with nitrosating agents in gastric aspirates or simulated gastric juice, and NOC formation measured. Gastric fluid from several species supports nitrosation (*29-33*). Sen et al. (*34*) incubated sodium nitrite (300 mg) and diethylamine (450 mg) in gastric juice from a variety of species. N-nitrosodiethylamine (NDEA) was formed in most cases, although greater concentrations were formed in gastric juice from humans (3000 μg NDEA/24 ml; pH 1.3) and rabbits (720 μg NDEA/100 ml; pH 2.0) than rats (400 μg NDEA/100 ml; pH 4.6). A minimum of 0.2 mg sodium nitrite and 50 mg diethylamine were required to form detectable amounts of NDEA (0.2 mg) in 5 ml. This work supported the hypothesis of gastric nitrosation occurrence and pointed out the importance of pH.

The formation of NOC *in vitro* has been reported using foods as sources of precursors. Siddiqi et al. (*35*) examined foods and teas from Kashmir (an Indian region of high risk for esophageal cancer) and found that 25 μM nitrite in artificial gastric juice gave significant amounts of NDMA, N-nitrosoproline (NPRO), N-nitrosothiazolidine-4-carboxylic acid (NTCA) and/or N-nitrosopipecolic acid (NPIPC), suggesting that consumption of these native foods may be a risk factor in the regional prevalence of gastric cancer.

Often these studies are carried out at precursor levels (particularly nitrite) which are much higher than possible in the human stomach, or they are carried out in simple solutions that have little relevance to *in vivo* conditions. For example, the normal human gastric concentration of nitrite is approximately 0.12 mg/L (*36*) which is orders of magnitude lower than most *in vitro* experiments.

Exposure to Precursors Resulting in Tumors. Several investigators have induced tumors in animals following administration of nitrite and amines or amides (*37-44*). The location and pathology of the resulting tumors are similar to those from administration of the corresponding preformed NOC. As many as 20 amines have been fed to rats in combination with nitrite. Of these, 13 induced significant numbers of tumors. Amides yielded a similar response. For example, co-feeding 36 or 72 mmol methylurea/kg diet and 14.5 mM nitrite in drinking water increased lung adenomas. Sodium ascorbate (58 mmol/kg diet) inhibited adenoma yield from the 36 mmol methylurea/kg regimen by 98%.

Excretion of NOC Following Exposure to Precursors. NOC or their adducts have been isolated from urine or tissues of animals and humans following administration of amines and nitrite (*45-48*). With a few exceptions (e.g., NPRO) most NOC are not excreted in urine due to rapid and extensive metabolism. Recently, we have recovered NDMA from ferret urine but only after animals were given 4-methyl-pyrazole to inhibit NDMA metabolism (*47*). The endogenous formation of N-nitrosoproline following administration of nitrate and L-proline has become a widely used procedure for measuring endogenous nitrosation in humans (*49,50,28*). Use of the 'Nitrosoproline test' as an index of endogenous nitrosation has recently been reviewed (*51*). Similar experiments have been conducted in ferrets (*52, 53*) and rats (*54, 48*).

DNA adducts have also been isolated from the urine of rats treated with precursors to NOC. Gombar et al. (*45*) administered various amounts of di(methyl-^{14}C)methylnitrosamine, and collected 24-h urine from rats. The amount of 7-(methyl-^{14}C)methylguanine recovered from urine was linearly related to the dose of di(methyl-^{14}C)methylnitrosamine administered. Using this relationship, the amount of NDMA formed following gavage with ^{14}C-aminopyrine and sodium nitrite was predicted (*45*). In rats, 7-methylguanine has also been isolated from stomach DNA following gavage with 30 μmol/kg (^{14}C)methylamine hydrochloride and 700 μmol/kg nitrite (*46*).

Direct Analyses of Gastric Contents. A direct approach to monitoring gastric nitrosation involves isolation of NOC from the stomach. Stomach contents have been collected either post-mortem (*55, 43, 56*) or by the placement of gastric fistulae (*57-59*). Most such studies have investigated the nitrosation of amines because of the lack of direct specific analytical methods for nitrosamides (*60- 62*). Mirvish and Chu (*55*) indirectly found nitrosomethylurea (NMU) or nitrosoethylurea (NEU) in the stomachs of rats given methylurea (291 μmol) or ethylurea (289 μmol) and nitrite (145 μmol). NMU formation could be prevented by instilling sodium bicarbonate (i.e., raising the pH to inhibit NMU formation) before administration of precursors. Similarly, Maekawa et al. (*43*) isolated 1-butyl-1-nitrosourea (BNU) 30 and 60 min after intubating pregnant rats with 861 µmol/kg butylurea and 725 µmol/kg sodium nitrite.

Lintas et al. (*58*) examined NDMA formation in dogs given pentagastrin. Precursors (870 µmol nitrite and 400 µmol dimethylamine (DMA)) were provided either in a semisynthetic diet or were instilled through the fistula. The first sample (at 3 min) contained the highest concentration of NDMA (6.1 µM) and a pH of 4.0. Gastric pH reached 2.0 in approximately 15 min. When the gastric pH was less than 5, nitrite disappeared rapidly from the stomach. The concentration of nitrite at 12 min was typically one-fourth to one-tenth that present at 3 min. NDMA concentration decreased more rapidly than the nitrite concentration suggesting NDMA may be absorbed through the gastric mucosa.

Another recent study also used fistulated dogs stimulated with pentagastrin (*59*). Formation of NPRO was monitored following instillation of 1.5 mM sodium nitrite, 9.0 mM proline and a nonabsorbable marker (PEG). Gastric nitrosoproline concentration maximized at 3 µM in 30 min then decreased to 2.5 µM at 60 min. Thiocyanate (1.0 mM) increased gastric NPRO concentration by 10 fold and 1.3 mM ascorbic acid reduced NPRO concentration to approximately 1.5 µM. No NPRO was detected when the ascorbic acid dose was increased to 3.7 mM (molar ratio ascorbate: nitrite of 2.5). When both 1.0 mM thiocyanate and 1.3 mM ascorbic acid were present, the NPRO concentration was lower than that produced with thiocyanate alone.

Licht et al. (*59*) developed a mathematical model to describe gastric nitrosation of proline in dogs. Kinetic data from *in vitro* studies were combined with mass balance equations to predict gastric nitrosation. The model adequately described the gastric NPRO concentration following addition of precursors. Inhibition by ascorbic acid and catalysis by thiocyanate were also demonstrated. Unfortunately the model has not been applied to NOC other than NPRO, which is non-carcinogenic. There is a need for an empirical data base against which to compare these predictions.

We recently completed a series of experiments which quantified the amounts and conditions under which N-nitrosotrimethylurea (NTMU) forms in a full-sized pig's stomach (*63*, *64*). Twenty-five to 125 µmol of nitrite increased NTMU formation linearly from 320 to 2560 nmol. Comparisons between pigs with disparate gastric pHs indicated a 4.5 fold greater formation at pH 1.9 than at 4.8. Ascorbic acid at 225 or 341 µmol inhibited TMU nitrosation by an average of 54 and 84% respectively. These data indicate that amide nitrosation occurs *in vivo* at gastric nitrite concentrations representative of humans.

Seven fruit and vegetable homogenates were tested for their effects on gastric TMU nitrosation. NTMU formation was reduced 93-94% by strawberries, 77% by brussel sprouts and kiwis, and less than 50% by orange juice and broccoli (*65*).

Nearly all *in vivo* experiments have also used precursor concentrations in excess of those encountered in human exposure. Previous experiments where amide nitrosation has been studied in rodents utilized nitrite in amounts from 15 µmol in guinea pigs (0.29 to 0.37 kg BW, *56*) to 145-725 µmol in rats (*55*, *43*). Assuming a rodent body weight of 300 g the corresponding nitrite amounts administered to a 50 kg human would be large: 2.5 to 121 mmol. Amounts used in the fistulated dog experiments ranged from 450 μmol to 14.5 mmol. In contrast the typical daily human exposure to nitrite is 90 to 250 μmol.

A number of groups have attempted to circumvent the analytical problems of nitrosamides in gastric juice by using 'total apparent NOC' methods (*66-68*). Such methods are group specific, but individual NOC are not quantified. The method may be subject to interferences from other components of biological samples such as C- and S-nitroso compounds and thionitrites.

Significance of Endogenous N-Nitrosation. Exposure to NOC from endogenous formation may be quantitatively greater than exogenous exposure. The extent of reaction depends on precursor concentration, the chemical properties of the amine or amide, the gastric pH, microbial flora, and the presence of modulators such as ascorbate (an inhibitor) or thiocyanate (a catalyst). Gastric nitrite concentrations are greatest in hypochlorhydric individuals (especially those consuming a diet high in vegetables containing nitrate) and normal individuals consuming foodstuffs high in nitrite, such as cured meats.

The extent to which endogenous nitrosation occurs under conditions relevant to human precursor exposure remains largely unknown. The high reactivity and rapid decomposition of nitrosamides, hinder their isolation and quantification. Analytical techniques for specific nitrosamides have only recently been developed (*69*). Consequently the majority of data on endogenous nitrosation are for stable nitrosamines (such as N-nitrosoproline) or 'total apparent NOC' (nonspecific determination). Amide nitrosation is unlikely to be modeled effectively by amino acid markers, due to the different mechanisms of nitrosation (*70*).

Exposure to Precursors

Nitrosatable Compounds in Food & Drugs. The demonstration that NOC are formed in the body raises interest in the amounts, types and sources of precursors to which humans are exposed. Nitrate and nitrite are intentional food additives and contaminants and NO_x compounds, which can nitrosate under a wide range of conditions, are present in the environment and are themselves formed endogenously (*71*). Several foods have the potential to form NOC *in vivo* (*9, 72, 73*). Dietary components such as choline may be metabolized to reactive precursors (dimethylamine) *in vivo* (*72*). Alkylureas and methylguanidine are present in fish products at levels from 20 to 180 mg/kg (*9*). Methylguanidine, which can yield methylnitrosourea upon nitrosation, may arise from creatine or creatinine. Shephard et al. (*73*) estimated the daily intake of nitrosatable precursors from the amounts of various food items consumed per person per year in Switzerland. Amides, in the form of peptides, and guanidine were the largest (approximately 100 g protein and 1 g creatine and creatinine per day) followed by primary amines and amino acids (100 mg/day), aryl amines, secondary amines, and ureas (1 to 10 mg/day).

The ease of nitrosation and the amount of a given precursor ingested will qualitatively and quantitatively influence *in vivo* nitrosation. Using estimates of daily precursor intake, nitrite exposure, and *in vitro* nitrosation rates, Shephard et al. (*73*) estimated *in vivo* nitrosation yields. Reaction was assumed to occur at the pH optimum (for amines) or at pH 2.0 (for amides), and gastric volume was assumed to be 1 L with constant reactant concentrations for 1 h. Two gastric nitrite concentrations were used, 1.7 and 72 μM. At 1.7 μM nitrite, the predicted NOC yields were greatest from protein and methylurea (800 and 400 pmol, respectively), followed by guanidines and arylamines (approximately 100 pmol). At the higher nitrite concentration (72 μM), yields were largest with N-methylaniline and aniline (arylamines), followed by amides (in protein) and creatinine. From this analysis, dietary protein, ureas (particularly methylurea), guanidines, and arylamines represent the greatest potential for nitrosation *in vivo*.

Amines are also prevalent. For example, monomethylamine (MMA) is a widespread component in fish and vegetables, and is readily nitrosated in the stomach. Moreover, MMA was detected in human gastric fluid at a level of 3.7 nmol/ml (*74*). Dimethylamine (DMA) is a degradation product of trimethylamine-oxide in marine fish. DMA is produced by the action of an endogenous enzyme found in gadoid fish (*75*). DMA is also formed during the toasting processing of oat flakes (*76*). Piperidine and pyrrolidine are nitrosamine precursors in black pepper (*77*). Trimethylamine (TMA) is a degradation product of trimethylamine-oxide in marine fish by the action of microbial enzymes (*75*). TMA occurs in human gastric fluid at 2.0 nmol/ml (*74*). Aminopyrine, an analgesic drug, is rapidly nitrosated to give NDMA (*78*) and is carcinogenic when administrated with nitrite (*79*).

Several amino acids containing a secondary amine function can be nitrosated. Proline, hydroxyproline, sarcosine, and D-fructose-L-amino acids (which are reaction products of non-enzymatic browning) are examples of amino acids which form NOC (*80*). Nitrosated product of D-fructose-L-tryptophan was mutagenic in *S. typhimurium*. Thermal decomposition of amino acids may result in the formation nitrosatable amines (*81*). N-nitrososarcosine, N-nitrosoproline and N-nitroso-4-hydroxyproline can decarboxylate to form NDMA, nitrosopyrrolidine (NPYR) and N-nitroso-3-hydroxypyrrolidine, respectively (*82*).

Other more complex nitrosatable amines can be derived from foods. Wakabayashi et al. (*83*) isolated β-carboline-3-carboxylic acid and its stereoisomer, both nitrosatable, from Japanese soy sauce. Several nitrosatable indoles and mutagen precursors have been isolated from Chinese cabbage (*84*, *85*).

Nitrate and nitrite in foods. Vegetables are the largest source of dietary nitrate. Knight at al. (*86*) estimate that vegetables contribute over 90% of nitrate exposure. Nitrate is present at significant levels (up to 64 mmol/kg) in a variety of vegetables, particularly celery, spinach, cabbage, and lettuce (*87*). Leaf and stem tissues accumulate the highest levels of nitrate, followed by roots (*88*). Turnip petioles were found to contain over 3% NO_3-N on a dry weight basis. Petioles of spinach, beet and kale accumulated about 2% nitrate. Nitrite was estimated to be present in concentrations corresponding to about 1% of nitrate in vegetables (*89*). Representative nitrate and nitrite levels of some fresh and pickled vegetables are listed in Tables 1 and 2.

The average western diet contains 1 to 2 mmol nitrate/person/day (*4*). High consumption of vegetables and/or high nitrate water can substantially increase this so

Table 1. Nitrate and Nitrite Contents in Vegetables

	Nitrate (mg/kg)	*Nitrite (mg/kg)*
Artichoke	12	0.4
Asparagus	44	0.6
Green beans	340	0.6
Lima beans	54	1.1
Beets	2400	4
Broccoli	740	1
Brussels sprouts	120	1
Cabbage	520	0.5
Carrots	200	0.8
Cauliflower	480	1.1
Celery	2300	0.5
Corn	45	2
Cucumber	110	0.5
Eggplant	270	0.5
Endive	1300	0.5
Kale/collard	800	1
Leek	510	NR
Lettuce	1700	0.4
Melon	360	NR
Mushroom	160	0.5
Onion	170	0.7
Parsley	1000	NR
Peas	28	0.6
Pepper, sweet	120	0.4
Potatoes		
White	110	0.6
Sweet	46	0.7
Pumpkin and squash	400	0.5
Radish	1900	0.2
Rhubarb	2100	NR
Spinach	1800	2.5
Tomatoes	58	NR
Turnip	390	NR
Turnip greens	6600	2.3

SOURCE: Reference 120.
NOTE: NR = not reported.

Table 2. Nitrate and Nitrite Contents in Japanese Pickled Vegetables

	Nitrite (mg/kg)	*Nitrate (mg/kg*
Eggplant, fermented in rice bran	0.6-3.4	2.1-94.3
Turnip, fermented in rice bran	1.2-29.7	130-510
Cucumber, fermented in rice bran	1.2-3.7	4.2-32.6
Cucumber seasoned with soy sauce	0.1-4.2	1.4-8.3
Chinese cabbage, salt-fermented	4.9	358
Chinese cabbage, kimchi	1.6	312
Nozawa-na, salted fermented turnip leaves	2.7-27.1	373-614
Taka-na, fermented broad leafed mustard	1.2	14.7
Karashi-na, fermented mustard leaves	4.3	561
Kizami-suguki, a kind of fermented turnip	2.1	176
Red turnip, salt-fermented	1.2-3.4	198-298
Oriental melon, kasuzuke	0.7-1.3	26.8-45.6
Radish root, fermented bettara-zuke	1	417
Radish root, takuan	1.8	283
Turnip, fermented, senmai-zake	2.8; 3.1	198; 212

SOURCE: Reference 96.

the interindividual variation in nitrate intake is large. The nitrate content of drinking water is generally low in the U.S., averaging 21 μM. However, levels averaging as high as 1600 μM (100 mg/L) have been reported (*6*). Nitrate and/or nitrite salts (potassium and sodium) are added to meats, poultry and fish in the low hundreds of mg per kg range as functional ingredients. Nitrate, when reduced by microorganisms, acts as the reservoir of nitrite during storage.

Nitrate and nitrite are also found in some foods to which they are not directly added, particularly fermented foods. For example, nitrate and nitrite have been found in oyster sauce, dried shrimp, shrimp sauce, shrimp paste, fish sauce, Chinese sausage, and dried squid (*90*). The nitrate and nitrite levels of some meat products were listed in Table 3. Dietary nitrite and nitrate intakes from several countries have been estimated (Table 4).

Diet represents the principal sources of exogenous exposure to nitrite (*91,92*). Cured meat products are the largest dietary source of preformed nitrite (*4*). For example, bacon sold in the United States is cured with 120 mg sodium nitrite per kg (*93*). The National Research Council (*6*) estimated the average residual nitrite content of cured meat products as 10 mg/kg. Higher levels (49 mg/kg) can be present in freshly cured meat (*94*). Bacon and ham may contain 15 to 70 mg nitrite/kg (*87*). Consumption of 100 g of meat with 50 mg/kg residual nitrite would expose an individual to 109 μmol nitrite. Nitrite is estimated to be 20 and 40 μmol/person/day for average and high cured meat diets, respectively (*6*). White (*91, 92*) using consumption data, estimated dietary nitrite exposure at 50 μmol/day.

Other nitrosating agents, such as, nitrogen oxides (NO, NO_2, N_2O_3 and N_2O_4) were found in the gases produced by combustion of fuels such as kerosene or natural gas (*95*). Nitrogen oxides are responsible for NOC in malt products (e.g., beer; *20*) and fish cooked on open flames (*96*). Products of N_2O_3 and methyl oleate were capable of nitrosating 2,6-dimethylmorpholine (*97*). When the products were added to ground, uncured pork and fried, NDMA and NPYR were detected. Ethylnitrite produced by combustion of kerosene or city gas was found to be an organic nitrosating species (*95*).

Nitrite in Gastric Fluid

The average nitrite content of fasting gastric fluid ranges from 1 to 7 μM for individuals with normal gastric function (*98-102*). The low concentrations have been attributed to low dietary intakes of nitrite, rapid absorption and oxidation to nitrate (*103*), and reaction with other gastric components (*104*). Gastric nitrite can, however, achieve higher concentrations following a meal. Forty minutes after eating a meal consisting of a fried egg (40 g), bread (32 g), butter (16 g), cheese (22 g), biscuits (17 g), milk (200 mL) and luncheon meat (80 g) the gastric nitrite concentration reached 300 μM (*104*).

The major source of nitrite exposure is the reduction of dietary nitrate. Ingested nitrate is rapidly absorbed from the stomach and approximately 25% is secreted in saliva. Of this, approximately 20% is reduced by oral microflora to nitrite (*106*). Therefore, the molar conversion of dietary nitrate to nitrite has been estimated to be 5 to 6% (*6, 107*). The daily exposure to nitrite resulting from reduction of ingested

Table 3. Nitrate and Nitrite Contents in Meat Products

	Nitrate (mg/kg)	*Nitrite (mg/kg)*
Unsmoked side bacon	134	12
Unsmoked back bacon	160	8
Peameal bacon	16	21
Smoked bacon	52	7
Corned beef	141	19
Cured corned beef	852	9
Corned beef brisket	90	3
Pickled beef	70	23
Canned corned beef	77	24
Ham	105	17
Smoked ham	138	50
Cured ham	767	35
Cooked ham	109	17
Canned ham	44	5
Cottage roll	553	28
Semicured ham	73	23
Unsmoked sausage	21	7
Smoked sausage	129	12
Wiener	97	7
Beef wiener	109	7
Luncheon meat	42	5
Pickle and Pimento loaf	51	4
Meat, macaroni, and cheese loaf	75	22
Mock chicken loaf	107	11
Salami	86	12
Beef salami	71	27
Bologna	77	19
Belitalia (garlic)	247	5
Pepperoni (beef)	149	23
Summer sausage	135	7
Ukranian sausage (Polish)	77	15
German sausage	71	17

SOURCE: Reference 120.

Table 4. Dietary nitrate and nitrite levels

*Nitrate**	*Nitrite**	*Country*	*Reference*
4.78	0.68	Japan	96
1.21	0.02	USA	6
1.53	0.03	Britain	86
0.48-8.06	0.01-0.11	Sweden	121
2.89	0.09	Netherlands	107

*mmoles/person/day.

nitrate, has been estimated as 264 μmol (*108*), 217 μmol (*73*), 173 μmol (*107*), and 91 μmol (*109*).

The oral microflora which catalyze nitrate reduction are also capable of colonizing the stomach provided the pH is sufficiently high (*110, 111, 101*). Individuals with an achlorhydric gastric environment may experience greater than 6% conversion of nitrate to nitrite. Furthermore nitrite is more stable at alkaline pH. As a result, fasting gastric nitrite concentrations are higher (22 to 613 μM) in hypochlorhydric individuals than in lower gastric pH individuals (*36, 112-115*). Daily nitrite exposure for individuals with a bacterially-colonized hypoacidic stomach has been projected to be as much as 1280 μmol (*109*).

Individuals under treatment for gastric ulcer with H_2 blockers or omeprazole, which raise gastric pH, have higher nitrate reductase bacteria numbers (*102*), higher gastric nitrite concentrations (*111, 116, 102, 117*) and higher levels of 'total' NOC (*111*).

Although the estimates of nitrite exposure described above include nitrite derived from exogenous nitrate, they do not include nitrite derived from endogenously synthesized nitrate. Nitrate balance studies have conclusively shown that 1-2 mmol nitrate is endogenously formed per day in healthy adults (*118, 28*). This amount is similar to dietary exposure. The contribution of endogenously formed nitrite and nitrate to gastric nitrite or endogenous NOC formation is unknown.

Conclusions

NOC have not been directly linked to human cancer but there is considerable indirect evidence that some portion of human cancer risk is related to NOC exposure. Humans are exposed to NOC through several vectors with tobacco being the largest. Foods also contain NOC albeit in much lower amounts. Several studies suggest that the formation of NOC within the body is the largest and most widespread route of exposure. Foods and diet influence endogenous formation by being the major source of precursors as well as a source of nitrosation inhibitors such as vitamin C. A combination of physiological factors such as gastric pH, residence time, and oral nitrate reductase activity may determine the amount of NOC formed within an individual. The formation of nitrosating agents via the arginine-nitric oxide pathway (*119*) may prove to be an additional and important source of NOC exposure.

Acknowledgments

Portions of the material contained in this article were contained in the theses submitted by the authors (MAH, CMM, Y-MW) to Cornell University.

Literature Cited

1. Preussmann, R.; Stewart, B.W. In *Chemical Carcinogens, Second Edition*; Searle, C.E., Ed.; ACS Monograph Series No. 182; American Chemical Society: Washington, DC, 1984; Vol. 2, pp 643-828.
2. Lijinsky, W. *Cancer Met. Rev.* **1987**, *6*, 301-356.
3. Tricker, A.R.; Preussmann, R. In *Nitrosamines Toxicology and Microbiology*; Hill, M.J., Ed.; VCH Publishers: New York, NY, 1988; pp 88-116.
4. Hotchkiss, J.H. In *Food Toxicology A Perspective on the Relative Risks*; Taylor S.L., Scanlan, R.A., Eds.; Marcel Dekker, Inc.: New York, 1988; pp 57-100.
5. Hotchkiss, J.H. *Cancer Surveys* **1989**, *8*, 295-321.
6. National Research Council. *The Health Effects of Nitrate, Nitrite, and N-Nitroso Compounds;* National Academy of Sciences Press: Washington, DC, 1981.
7. Mirvish, S.S. *Toxicol. Appl. Pharmacol.* **1975**, *31*, 325-351.
8. Williams, D.L.H. *Adv. Phys. Org. Chem.* **1983**, *19*, 381-428.
9. Mirvish, S.S. *J. Natl. Cancer Inst.* **1983**, *71*, 631-647.
10. Bartsch, H.; Ohshima, H.; Pignatelli B. *Mut. Res.* **1988**, *202*, 307-324.
11. Archer, M. C. *Cancer Surveys* **1989**, *8*, 241-250.
12. Yahagi, T.; Nagao, M.; Seino, Y.; Matsushima, T.; Sugimura, T.; Okada, M. *Mut. Res.* **1977**, *48*, 121-130.
13. Bogovski, P. In *Advances in Tumor Prevention, Detection and Characterization, Human Cancer, Its Characterization and Treatment*; Maltoni, C., ed.; Excerpta Medica: Amsterdam-Oxford-Princeton, 1979; Vol. 5, pp 105-113.
14. Nesnow, S.; Langenbach, R.; Mass, M.J. *Environ. Hlth. Perspect.* **1985**, *61*, 345-349.
15. Appel, K.E.; Ruhl, C.S.; Hildebrandt, A.G. *Chem. -Biol. Interactions* **1985**, *53*, 69-76.
16. Douglass, M.L.; Kabacoff, B.L.; Anderson, G.A.; Cheng, M.C. *J. Soc. Cosmet. Chem.* **1978**, *29*, 581-606.
17. Hecht, S.S.; Hoffmann, D. *Cancer Surveys* **1989**, *8*, 273-295.
18. Tricker, A.R.; Spiegelhalder, B.; Preussmann, R. *Cancer Surveys* **1989**, *8*, 251-272.
19. Hotchkiss, J.H. In *Nutritional and Toxicological Aspects of Food Safety*; Friedman, M., Ed.; Adv. Exp. Med. Biol. 1984; Vol. 177, pp 287-298.
20. Mangino, M.M.; Scanlan, R.A.; O'Brien, T.J. In *N-Nitroso Compounds*; Scanlan, R.A. Tannenbaum, S.R., Eds.; American Chemical Society: Washington, DC, 1981; p 229.
21. Havery, D.C.; Fazio, T. *Fd. Chem. Toxicol.* **1982**, *20*, 939-944.
22. Sen, N.P.; Donaldson, B.; Charbonneau, C.; Miles, W.F. *J. Agric. Fd. Chem.* **1974**, *22*, 1125.
23. Leaf, C.D.; Wishnok, J.S.; Tannenbaum, S.R. *Cancer Surveys* **1989**, *8(2)*, 323-334.
24. Hartman, P.E. *Environ. Mut.* **1983**, *5*, 111-121.

25. Correa, P.; Fontham, E.; Pickle, L.; Chen, V.; Lin, Y.; Haenszl, W. *J. Nat. Canc.* **1975**, *74(4)*, 645-652.
26. Forman, D.; Al-Dabbagh, S.; Doll, R. *Nature* **1985**, *313*, 620-625.
27. Forman, D. *British Med. J.* **1987**, *294*, 528-529.
28. Leaf, C.D.; Vecchio, A.J.; Roe, D.A.; Hotchkiss, J.H. *Carcinogenesis* **1987**, *8(6)*, 791-795.
29. Alam, B.S.; Saporoschetz, I.B.; Epstein S.S. *Nature* **1971**, *232*, 116-118.
30. Lane, R.P.; Bailey M.E. *Fd. Cosmet. Toxicol.* **1973**, *11*, 851-854.
31. Ziebarth, D. In *N-Nitroso Compounds in the Environment*; Bogovski, P.; Walker E.A., Davis, W., Eds.; IARC Scientific Publications No. 9, International Agency for Research on Cancer: Lyon, 1974; pp 137-141.
32. Tannenbaum, S.R.; Moran, D.; Falchuk, K.R.; Correa, P.; Cuello, C. *Cancer Lett.* **1981**, *14*, 131-136.
33. Schlemmer, K.H.; Eisenbrand G. *Arzneim. Forsch.* **1988**, *38*, 1365-1368.
34. Sen, N.P.; Smith, D.C.; Schwinghamer L. *Fd. Cosmet. Toxicol.* **1969**, *7*, 301.
35. Siddiqi, M.; Tricker, A.R.; Preussmann, R. *Cancer Lett.* **1988**, *39(1)*, 37-43.
36. Ruddell, W.S.J.; Blendis, L.M.; Walters C.L. *Gut* **1977**, *18*, 73-77.
37. Greenblatt, M.; Mirvish, S.S.; So B.T. *J. Natl. Cancer Inst.* **1971**, *46*, 1029-1034.
38. Greenblatt, M.; Kommineni, V.; Conrad, E.; Wallcave, L.; Lijinsky, W.; *Nature New Biol.* **1972**, *236*, 25-26.
39. Lijinsky, W. *Oncol.* **1980**, *37*, 223-226.
40. Yamamoto, K.; Nakajima, A.; Eimoto, H.; Tsutsumi, M.; Maruyama, H.; Dendra, A.; Nii, H.; Mori, Y.; Konishi Y. *Carcinogenesis* **1989**, *10*, 1607-1611.
41. Greenblatt, M.; Mirvish S.S. *J. Natl. Cancer Inst.* **1972**, *49*:119-124.
42. Rustia, M. *Cancer Res.* **1974**, *34*, 3232-3244.
43. Maekawa, A.; Ishiwata, H.; Odashima S. *Gann* **1977**, *68*, 81-87.
44. Ivankovic, S. and Preussmann R. *Naturwissen* **1970**, *57*, 460.
45. Gombar, C.T.; Zubroff, J.; Strahan, G.D.; Magee P.N. *Cancer Res.* **1983**, *43*, 5077-5080.
46. Huber, K.W.; Lutz W.K. *Carcinogenesis* **1984**, *5*, 1729-1732.
47. Perciballi, M.; Hotchkiss J.H. *Carcinogenesis* **1989**, *10*, 2303-2309.
48. Ohshima, H.; Mahon, G.A.T.; Wahrendorf, J.; Bartsch H. *Cancer Res.* **1983**, *43*, 5072-5076.
49. Ohshima, H. and Bartsch H. *Cancer Res.* **1981**, *41*, 3658-3662.
50. Wagner, D.A.; Shuker, D.E.G.; Bilmazes, C.; Obiedzinski, M.; Baker, I.; Young, V.R.; Tannenbaum, S.R. *Cancer Res.* **1985**, *45*, 6519-6522.
51. Bartsch, H.; Ohshima, H.; Pignatelli, B.; Calmels, S. *Cancer Surveys* **1989**, *8*, 335-362.
52. Perciballi, M.; Conboy, J.J.; Hotchkiss J.H. *Fd. Chem. Toxicol.* **1989**, *27*, 111-116.
53. Dull, B.J.; Hotchkiss, J.H.; Vecchio A.J. *Fd. Chem. Toxicol.* **1986**, *24*, 843-845.
54. Ohshima, H.; Bereziat, J.C.; Bartsch H. *Carcinogenesis* **1982**, *3*, 115-120.
55. Mirvish, S.S.; Chu C. *J. Natl. Cancer Inst.*, **1973**, *50*, 745-750.
56. Yamamoto, M.; Ishiwata, H.; Yamada, T.; Yoshihira, K.; Tanimura, A.; Tomita, I. *Fd. Chem. Toxicol.* **1987**, *25*, 663-668.

57. Mysliwy, T.S.; Wick, E.L.; Archer, M.C.; Shank, R.C.; Newberne, P.M. *Br. J. Cancer* **1974**, *30*, 279-283.
58. Lintas, C.; Clark, A.; Fox, J.; Tannenbaum, S.R.; Newberne P.M. *Carcinogenesis* **1982**, *3*, 161-165.
59. Licht, W.R.; Tannenbaum, S.R.; Deen, W.M. *Carcinogenesis* **1988**, *9*, 365-372.
60. Hotchkiss, J.H. *J. Assoc. Off. Anal. Chem.* **1981**, *64*, 1037-1054.
61. Scanlan, R.A.; Reyes, F.G. *Fd. Technol.* **1985**, *39*, 95-99.
62. Massey, R.C. In *Nitrosamines Toxicology and Microbiology*; Hill, M.J., Ed.; VCH Publishers: New York, 1988; pp 16-47.
63. Maragos, C.M.; Hotchkiss, J.H.; Fubini, S.L. *Carcinogenesis* **1990**, *11*(9), 1587-1591.
64. Maragos, C.M.; Klausner, K.A.; Shapiro, K.B.; Hotchkiss, J.H. *Carcinogenesis* **1990**, *12*, 141-143.
65. Maragos, C.M. Ph.D. Thesis, Cornell University, 1990.
66. Walters, C.L.; Downes, M.J.; Edwards, M.W.; Smith, P.L.R. *Analyst* **1978**, *103*, 1127-1133.
67. Bavin, P.M.G.; Darkin, D.W.; Viney, N.J. In *N-Nitroso Compounds: Occurrence and Biological Effects*; Bartsch, H; O'Neill, I.K.; Castignaro, M.; Okada, M., Eds.; IARC Scientific Publication No. 41; International Agency for Research on Cancer: Lyon, 1982; pp 337-344.
68. Pignatelli, B.; Richard, I.; Bourgade, M.C.; Bartsch, H. *Analyst* **1987**, *112*, 945-949.
69. Conboy, J.J.; Hotchkiss J.H. *Analyst* **1989**, *114*, 155-159.
70. Adam, B.; Schlag, P.; Friedl, P.; Preussmann, R.; Eisenbrand, G. *Gut* **1989**, *30(8)*, 1068-1075.
71. Marletta, M.A. *Trends Bioc.* **1989**, *14(12)*, 488-492.
72. Zeisel, S.H.; DaCosta, K.A.; Fox J.G. *Biochem. J.* **1985**, *232*, 403-408.
73. Shephard, S.E.; Schlatter, C.; Lutz W.K. *Fd. Chem. Toxicol.* **1987**, *25*, 91-108.
74. Zeisel, S.H.; DaCosta, K.A.; Lamont J.T. *Carcinogenesis* **1988**, *9*, 179-181.
75. Lundstrom, R.C. and Racicot, L.D. *J. Assoc. Off. Anal. Chem.* **1983**, *66(5)*, 1158-1163.
76. Hrdlicka, J.; Janicek, G. *Nature* **1964**, *204*, 1201.
77. Sen, N.P.; Miles, W.F.; Donaldson, B.A.; Panalaks, T.; Iyengar, J.R. *Nature* **1973**, *245*, 104.
78. Barale, R.; Zucconi D.; Loprieno, N. *Mut. Res.* **1981**, *85*, 57-70.
79. Lijinsky, W. *Fd. Chem. Toxicol.* **1984**, *22*, 715-720.
80. Pool, B.L.; Roper, H.; Roper, S.; Romruen, K. *Fd. Chem. Toxic.* **1984**, *22(10)*, 797-801.
81. Maga, J.A. *CRC Critical Review in Food Science and Nutrition* **1978**, *10*, 373-403.
82. Janzowski, C.; Eisenbrand, G.; Preussmann, R.; *Fd. Cosmet. Toxicol.* **1978**, *16*, 343-348.
83. Wakabayashi, K.; Ochiai, M.; Saito, H.; Tsuda, M.; Suwa, Y.; Nagao, M.; Sugimura, T. *Proc. Natl. Acad. Sci. USA* **1983**, *80*, 2912-2916.
84. Wakabayashi, K.; Nagao, M.; Ochiai, M.; Tahira, T.; Yamaizumi, Z.; Sugimura, T. *Mut. Res.* **1985**, *143*, 17-21.
85. Wakabayashi, K.; Nagao, M.; Tahira, T.; Yamaizumi, Z.; Katayama, M.; Marumo, S.; Sugimura, T. *Mutagenesis* **1986**, *1(6)*, 423-426.

86. Knight, T.M.; Forman, D.; Al-Dabbagh, S.A.; Doll, R. *Fd. Chem. Toxic.* **1987**, *25(4)*, 277-285.
87. Hill, M.J. In *Nitrosamines -Toxicology and Microbiology*; Hill, M.J., Ed.; VCH Publishers: New York, **1988**; pp 142-162.
88. Lorenz, O.A. In *Nitrogen in the environment*; Nielsen, D.R.; MacDonald, J.G., Eds.; Academic Press: New York, 1978; Vol. 2, Soil-plant-nitrogen relationships, pp. 201-219.
89. Bostrom C.-E.; Tammelin C.-E. *J. Environ. Pathol. Toxicol. Oncol.* **1987**, *7(4)*, 109-112.
90. Fong, Y.Y.; Chan, W.C. *Fd. Cosmet. Toxicol.* **1977**, *15*, 143-145.
91. White, J.W. Jr. *J. Agric. Fd. Chem.* **1975**, *23*, 886-891.
92. White, J.W. Jr. *Erratum. J. Agric. Fd. Chem.* **1976**, *24*, 202.
93. USDA. *U.S. Dept. of Agriculture, Fed. Reg.* **1978**, *43*, 20992.
94. Lee, K.L.; Greger, J.L.; Consaul, J.R.; Graham, K.L.; Chinn, B.L. *Am. J. Clin. Nutr.* **1986**, *44*, 188-194.
95. Matsui, M.; Ishibashi, T.; Kawabata, T. *Bulletin of the Japanese Society of Scientific Fisheries* **1984**, *50(1)*, 155-159.
96. Kawabata, T.; Ohshima, H.; Uibu, J.; Nakamura, M.; Matsui, M. and Hamano, M. 1979. In *Naturally occurring carcinogens-mutagens and modulators of carcinogenesis*; Miller, E.C. et al., Eds.; Japan Sci. Soc. Press: Tokyo/Univ.; Park Press: Baltimore, 1979; pp 195-209.
97. Ross, H.D.; Henion, J.; Babish, J.G.; Hotchkiss, J.H. *Fd. Chem.* **1987**, *23*, 207-222.
98. Walters, C.L.; Dyke, C.S.; Saxby, M.J.; Walker, R. In *Environmental N-nitroso compounds analysis and formation;* Walker, E.A.; Griciute, L., Bogovski, P., Eds.; IARC Scientific Publications No. 14, International Agency for Research on Cancer: Lyon, 1976; pp 181-193.
99. Ruddell, W.S.J.; Bone, E.S.; Hill, M.J.; Walters, C.L.; Blendis, L, M. *Lancet* **1978**, *1*, 521-523.
100. Tannenbaum, S.R.; Moran, D.; Rand, W.; Cuello, C.; Correa, P. *J. Natl. Cancer Inst.* **1979**, *62*, 9-12.
101. Mueller, R.L.; Hagel, H.J.; Wild, H.; Ruppin, H.; Domschke, W. *Oncol.* **1986**, *43*, 50-53.
102. Sharma, B.K.; Santana, I.A.; Wood, E.C.; Walt, R.P.; Pereira, M.; Noone, P.; Smith, P.L.R.; Walters, C.L.; Pounder C.E. *Br. Med. J.* **1984**, *289*, 717-719.
103. Leach, S. In *Nitrosamines Toxicology and Microbiology*; M.J. Hill, Ed.; VCH Publishers: New York, **1988**; pp 69-87.
104. Licht, W.R.; Schultz, D.S.; Fox, J.G.; Tannenbaum, S.R.; Deen, W.M. 1986. *Carcinogenesis* **1986**, *7*, 1681-1687.
105. Walters, C.L.; Carr, F.P.A.; Dyke, C.S.; Saxby, M.J.; Smith, P.L.R. *Fd. Chem. Toxicol.* **1979**, *17*, 473-479.
106. Spiegelhalder, B.; Eisenbrand, G.; Preussmann R. *Fd. Cosmet. Toxicol.* **1976**, *14*, 545-548.
107. Stephany, R.W.; Schuller P.L. *Oncol.* **1980**, *37*, 203-210.
108. Walters, C.L. *Oncol.* **1980**, *37*, 289-296.

109. Hartman, P.E. In *Banbury Report No. 12*; Magee, P.N., Ed.; Cold Spring Harbor Laboratory: New York, **1982**; pp 415-431.
110. Bartholomew, B.A.; Hill, M.J.; Hudson, M.J.; Ruddell, W.S.J; Walters, C.L. In *N-Nitroso Compounds: Analysis, Formation and Occurrence*; Walker, E.A.; Castegnaro, M.; Griciute, L.; Borzsonyi, M., Eds.; IARC Scientific Publications No. 31; International Agency for Research on Cancer: Lyon, 1980; p 595.
111. Reed, P.I.; Smith, P.L.R.; Haines, K.; House, F.R.; Walters, C.L. *Lancet* **1981**, *2*, 550-552.
112. Schlag, P.; Böckler, R.; Ulrich, H.; Peter, M.; Merkle, P.; Herfarth, C. *Lancet* **1980**, *1*, 727-729.
113. Dolby, J.M.; Webster, A.D.B.; Borriello, S.P.; Barclay, F.E.; Bartholomew, B.A.; Hill M.J. *Scand. J. Gastroenterol.* **1984**, *19*, 105-110.
114. Eisenbrand, G.; Adam, B.; Peter, M.; Malfertheiner, P.; Schlag P. In *N-Nitroso Compounds: Occurrence, Biological Effects and Relevance to Human Cancer*; O'Neill, I.K.; von Borstel, R.C.; Long, J.E.; Miller, C.T.; Bartsch, H., Eds.; IARC Scientific Publications No. 57. International Agency for Research on Cancer: Lyon, 1984; pp 963-968.
115. Hall, C.N.; Darkin, D.; Viney, N.; Cook, A.; Kirkham, J.S.; Northfield, T.C. In *The Relevance of N-Nitroso Compounds to Human Cancer*; Bartsch, H.; O'Neill, I.K.; Schulte-Herman, R., Eds.; IARC Scientific Publications No. 84, International Agency for Research on Cancer: Lyon, 1987; pp 527-530.
116. Milton-Thompson, G.L.; Lightfoot, N.F.; Ahmet, Z.; Hunt, R.H.; Barnard, J.; Bavin, P.M.G.; Brimblecombe, R.W.; Darkin, D.W.; Moore, P.J.; Viney N. *Lancet* **1982**, *2*, 1091-1095.
117. Elder, J.B.; Burdett, K.; Smith, P.L.R.; Walters, C.L; Reed, P.I. In *N-Nitroso Compounds: Occurrence, Biological Effects and Relevance to Human Cancer*; O'Neill, I.K.; von Borstel, R.C.; Long, J.E.; Miller, C.T.; Bartsch, H., Eds.; IARC Scientific Publications No. 57. International Agency for Research on Cancer: Lyon, 1984; pp 969-974.
118. Wagner, D.A.; Schultz, D.S.; Deen, W.M.; Young, V.R.; Tannenbaum, S.R. *Cancer Res.* **1983**, *43*, 1921-1925.
119. Hibbs, J.B.; Taintor, R.R.; Vavrin, Z. *Science* **1987**, *235*, 473-476.
120. Choi, B.C.K. *Am. J. Epidemiol.* **1985**, *121(5)*, 737-743.
121. Slorach, S.A. *J. Environ. Pathol. Toxicol. Oncol.* **1987**, *7*, 137-150.

RECEIVED August 15, 1991

Chapter 34

Ethyl Carbamate in Alcoholic Beverages and Fermented Foods

Gregory W. Diachenko, Benjamin J. Canas, Frank L. Joe, and Michael DiNovi

Division of Food Chemistry and Technology, U.S. Food and Drug Administration, 200 C Street, SW, Washington, DC 20204

Significant research and regulatory activity have been focused on ethyl carbamate (EC) as a result of Canada's establishment of regulatory limits for EC in alcoholic beverages. Industry, academic, and government laboratories have developed analytical methods for EC and confirmed its presence in a wide variety of alcoholic beverages. Although EC is an animal carcinogen, insufficient toxicological data are available for a meaningful human risk assessment. The Food and Drug Administration has requested additional toxicological studies and established voluntary EC reduction programs with the United States wine and distilled spirits industries. These programs aim at identifying factors contributing to EC formation and reducing EC to the lowest levels that are technologically feasible. This chapter presents recent industry and government data on EC levels in alcoholic beverages and fermented foods and an initial assessment of EC intake from these sources. Results of studies conducted by industry as part of their voluntary EC reduction programs are also presented.

Since December 1985, when Canada announced regulatory limits for ethyl carbamate (EC) in alcoholic beverages, considerable research and regulatory activity have been focused on this compound. The reported presence of EC in alcoholic beverages was of immediate food safety concern because it is a well-known animal carcinogen. Following the Canadian findings and regulatory action, the U.S. Food and Drug Administration (FDA) initiated a wide range of activities. These activities included a limited survey of alcoholic beverages, evaluation of EC toxicity data, assessment of other sources of EC exposure, and most important, working with industry to reduce EC levels.

The current status of the toxicological evaluation can be addressed by noting the conclusion of the Cancer Assessment Committee of FDA's Center for Food Safety and Applied Nutrition, which stated, "... that the data were not sufficient to perform a meaningful assessment of the risk posed by urethane in alcoholic beverages ..." (*1*). At FDA's request, the National Toxicology Program (NTP) has initiated toxicological research to provide the data needed for a quantitative estimate of the risk due to EC in alcoholic beverages. While awaiting the results of NTP's toxicological studies, FDA has addressed the exposure side of the risk assessment equation. The remainder of this paper will focus on providing an overview of the efforts of FDA and industry to determine EC concentrations, total EC exposure, and ways to reduce EC in alcoholic beverages to the lowest level that is technologically feasible.

FDA and the Bureau of Alcohol, Tobacco, and Firearms (BATF) initially surveyed domestic and imported alcoholic beverages for EC to confirm the Canadian reports and assess the frequency and levels of EC's occurrence. Table 1 summarizes the findings of EC in a wide range of beverages collected from January 1986 through August 1987 (*2*). The average levels for various product classes varied dramatically, with fruit brandy, bourbon whiskey, sherry, and sake having the highest values. These higher-level product classes also included the greatest percentages of samples that exceeded the following Canadian EC regulatory limits: 150 ppb in distilled spirits; 100 ppb in dessert wines (>14% alcohol, e.g., port and sherry); 30 ppb in table wines (<14% alcohol); 400 ppb in fruit brandy and liqueur; and 200 ppb in sake. The U.S. data were consistent with the Canadian findings (*3*) and suggested the need to reduce EC levels and generate the toxicological and exposure data necessary for FDA to evaluate the risk posed by EC.

To enable a more complete assessment of the total human exposure to EC, FDA (*4, 5*) developed analytical methods and analyzed a wide variety of fermented foods. The selected foods included many that had previously been reported by Ough (*6*) to contain low concentrations of EC. Canas et al. (*4, 5*) analyzed a greater number and variety of fermented foods, and their results are summarized in Tables 2 and 3. Average levels of EC were generally <5 ppb, with the exception of soy sauce. Imported beer and ale, soy sauce, bread, instant miso soup and yeast spread (small-volume ethnic or health foods), and wine vinegar contained the highest EC concentrations. Heating some fermented foods may also increase EC levels. Canas et al. (*4*) have reported that toasting increased the average EC content of bread by 2.6 times. Foods produced at least partially by yeast fermentation (Table 2) seemed to be much more likely to contain EC than those fermented by lactic acid bacteria, acetic acid bacteria, or molds (Table 3). The non-yeast fermented products ranged from cheese to fermented meats, with EC being undetected in almost all samples.

Average EC levels found in the various categories of fermented foods (Tables 2 and 3) and alcoholic beverages (Table 1) were used to estimate per capita exposure to EC. The EC exposure results listed in Table 4 were obtained by combining the average EC concentrations with food intake data derived from various sources (*7-9*). Distilled spirits and wines were found to contribute most of the average per capita exposure to EC and therefore merit special attention. When converted to percentages (Table 5), these two categories provide 92% of the total exposure to EC from foods and beverages. Beer, ale, and other malt beverages contributed 4%, with bakery products and other fermented foods yielding the remaining 4%. It should be noted that the alcoholic beverage levels were taken from the previously presented 1986-87 FDA/BATF data. These exposure percentages are expected to change in the future as industry's voluntary programs result in significant EC reductions in retail products.

Table 1. Summary of FDA and BATF Data on Ethyl Carbamate Levels (ppb) in Alcoholic Beverages[a]

Product Class	*N*[b]	*Average*	*Median*	*% Exceeding Canadian Limit*[c]
Brandy (Grape)	45	41.9	25.0	0
Brandy (Fruit)	89	1197.0	704.0	58
Bourbon (Stock)	150	173.9	116.0	53
Bourbon (Retail)	114	155.6	126.5	38
Rum	29	21.2	16.0	0
Liqueur	162	104.4	19.5	5.5
Scotch	48	48.9	46.5	0
Sherry	30	128.8	89.5	47
Port	23	59.0	29.0	17
Grape Wine	362	12.8	8.0	8.6
Sake	16	286.2	265.5	63

[a]As of 8/87; Diachenko et al., 1988 (*2*).
[b]Number of samples.
[c]Canadian regulatory limits: distilled spirits — 150 ppb; port and sherry — 100 ppb; grape wine — 30 ppb; fruit brandy — 400 ppb; sake — 200 ppb.

Table 2. Summary of FDA Data on Ethyl Carbamate in Fermented Foods and Beverages (Yeast Fermentation)

		Ethyl Carbamate (ppb)		
Product Class	*N*[a]	*Range*	*Average*[b]	*Median*
Malt Beverage (USA)	32	ND[c]–1	0.3	ND
Malt Beverage (Import)	38	ND–13	2.7	2
Bread	34	ND–8	1.7	1
Doughnuts	6	ND	ND	ND
Soy Sauce	20	ND–84	16.5	9
Soy Bean Paste (Miso)	1	4	-	-
Instant Soup (Dry Miso)	4	ND–6	3.0	3
Yeast Spread	1	51	-	-
Apple Cider	8	ND–3	0.4	ND
Wine Vinegar	6	4–26	8.8	7

SOURCE: Data obtained from published and unpublished reports of Canas et al. (*4,5*).
[a]N = number of samples.
[b]Average has been calculated assuming "0" for "ND" values.
[c]ND = below 1–1.5 ppb (ng/g) limit of detection.

Table 3. Summary of FDA Data on Ethyl Carbamate in Fermented Foods and Beverages (Non-Yeast Fermentation)

		Ethyl Carbamate (ppb)		
Product Class	*N*[a]	*Range*	*Average*[b]	*Median*
Misc. Sauces	6	ND[c]	ND	ND
Olives	6	ND–2	0.3	ND
Pickles & Relish	6	ND	ND	ND
Apple Cider Vinegar	2	ND	ND	ND
Powder Vinegar	1	ND	-	-
Cheese	17	ND	ND	ND
Yogurt	14	ND–3	0.4	ND
Cultured Buttermilk	3	ND	ND	ND
Sour Cream	4	ND	ND	ND
Orange Juice & Drink	10	ND	ND	ND
Salami & Pepperoni	6	ND	ND	ND
Tea	6	ND	ND	ND

SOURCE: Data obtained from published and unpublished reports of Canas et al. (*4*,*5*).
[a]N = number of samples.
[b]Average has been calculated assuming "0" for "ND" values.
[c]ND = below 1-1.5 ppb (ng/g) limit of detection.

Table 4. Estimated Per Capita Exposures to Ethyl Carbamate (EC) From Food

Food	*EC Level*[a] *(ppb)*	*Food Intake (g/person/day)*	*EC Exposure (μg/person/day)*
Beer, Ale, Malt Bev.[b]			0.14
Domestic	0.3	336	
Imported	2.7	14	
Wines[b]			0.78
Domestic	23.4	25.8	
Imported	21.5	8.2	
Dist. Spirits[b]			2.8
Domestic	122	18.5	
Imported	84	6.5	
Bakery Products[c]			0.13
White	3.0	39.9	
Dark (wheat)	1.0	10.7	
Other	1.0	3.4	
Yogurt[c]	0.4	4.3	0.002
Soy Sauce[c]	16.5	1.65	0.03
Olives[c]	0.3	0.3	0.00008
Inst. Soup (Miso)[c]	3.0	4.2	0.01
Wine Vinegar[d]	8.8	1.0	0.009
Apple Cider (alc.)[e]	0.4	25.8	0.01
Total			3.88

[a]Average EC levels are taken from Tables 1-3, or from subcategories of raw data used to prepare these tables.

[b]The food intake figures are derived by using the domestic and imported proportions of total beer, wine, or distilled spirits intake from the Statistical Abstracts of the United States (*7*), 1987 (1985 intake, per capita disappearance).

[c]Intake figures are derived from the Market Research Corporation of America (MRCA) 14-day average survey (5-year menu census, 1982-87) (*8*) and U.S. Department of Agriculture (USDA) 1977-78 survey portion sizes (*9*).

[d]Intake figure is taken from USDA 1977-78 survey (*9*). Wine vinegar has been assumed to be an equivalent substitute for italian salad dressing.

[e]Intake figure is that for domestic wine; *see* footnote b. Hard cider was deemed to be an equivalent substitute for wine.

Table 5. Relative Exposure to Ethyl Carbamate (EC) From Food and Alcoholic Beverage Sources

Food Category	*EC Exposure*[a] *(μg/person/day)*	*% Daily Exposure*
Beer, Ale, Malt Beverages	0.14	4
Distilled Spirits & Wines	3.58	92
Bakery Products & Other Fermented Foods	0.16	4
Total	3.88	

[a]EC exposure data are taken from Table 4 and grouped into larger food categories.

Soon after learning of the Canadian actions, FDA worked with the domestic industry to develop voluntary industry programs aimed at reducing EC in their alcoholic products to the maximum extent that is technologically feasible. The basic goals of these programs are (1) to reduce EC below target levels for various products; (2) to monitor EC levels to measure progress; (3) to investigate the sources and mechanism of EC formation; and (4) to investigate practical process changes to reduce EC levels.

On December 24, 1987, FDA formally accepted a proposal by the Distilled Spirits Council of the United States (DISCUS) to reduce EC levels to 125 ppb in all new whiskey produced as of January 1, 1989. DISCUS also made a commitment to conduct research on EC formation and processing modifications that would make it possible to reduce EC levels to this goal (*1*). Recently, research by MacKenzie et al. (*10*) identified a series of cyanide-related precursors involved in EC formation in Scotch whiskey. They indicated that the precursors included cyanide and copper cyanide complex anions, lactonitrile and isobutyraldehyde cyanohydrin, and cyanate and thiocyanate anions. DISCUS members have also developed several changes in the distillation process that have reduced EC levels in the final product of some distillers (*11*), including the following: (1) using copper packing in the upper part of the distillation still to improve the separation efficiency of the distillation process; (2) eliminating entrainment (carryover of liquid feedstock (beer) with distilled vapors) to prevent carryover of EC or its precursors into the high wines; (3) improving cleanup procedures and increasing the frequency of still cleanout or rinsing to minimize EC precursor buildup; and (4) tightening controls on the operating parameters of the still such as adjustment of the rates of beer fed to the still.

As part of their voluntary agreement with FDA to determine their progress in lowering EC levels, DISCUS has provided quarterly summaries of their members' daily analyses of new distillate (*12*). Their monitoring data indicate considerable variability in weekly averages both within and between distillers. However, the use of numerous variations of the recommended processing changes has enabled all 16 active bourbon whiskey distillers to achieve average EC levels that are below their target of 125 ppb in almost all new production at the point of distillation. Many distillers have achieved weekly averages below 30-40 ppb. Unfortunately, a complicating factor has recently been discovered, as some companies have found that EC levels in a few batches may increase over time (*1*). As a result, FDA has requested that industry revise its sampling and analysis protocols to ensure that future data reflect EC levels found in aged products as purchased by consumers. DISCUS members are also investigating ways to eliminate unidentified volatile precursors that appear to cause these post-distillation EC increases.

On January 25, 1988, FDA formally accepted a voluntary program proposal submitted by the Wine Institute and American Association of Vintners, representing the U.S. wine industry. In addition to providing a commitment for research and monitoring, this program set target goals for EC levels in table and dessert wines. The agreement provided that (a) starting with the 1988 harvest, the volume-weighted average level of EC in table wines (<14% alcohol) would not be greater than 15 ppb;

(b) starting with the 1989 harvest, the volume-weighted average level of EC in dessert wines (>14% alcohol) would not exceed 60 ppb; and (c) starting with wines produced from the 1995 harvest, no more than 1% of the volume of table wine would contain >25 ppb EC and no more than 1% of the volume of dessert wine would contain >90 ppb (*1*). Although these targets are below the Canadian regulatory limits, the commitment was also made to conduct an annual survey, and if possible, accelerate the target date or modify the goals to reduce EC levels to the greatest extent possible.

To move toward attaining these goals, the wine industry supported extensive research on the mechanism of EC formation and also investigated practical measures to reduce EC levels. Ough (*13*) demonstrated that urea, citrulline, and carbamyl phosphate could all react with ethanol to produce EC. Several other precursors for EC formation were also investigated, and after thorough testing Ough concluded that urea, a natural by-product of yeast metabolism, is the main precursor of EC in wines (*14*). He also demonstrated that arginine was the major amino acid metabolized by yeast to produce urea. Any urea that is not utilized by the yeast during fermentation can then react with ethanol over time to form EC. If the juice or musk originally contained high levels of nitrogenous compounds that are metabolized by yeast before urea, more urea would remain after fermentation to form higher EC concentrations. Ough (*13*) also demonstrated that citrulline, another amino acid found in various grapes, is a precursor of EC in wines, although not to as great an extent as urea.

Based on these research findings, the wine industry made the following recommendations to all U.S. wineries concerning ways to minimize development of EC in wines (*15*; Wine Institute, personal communication, 1989): (1) encourage grape growers to minimize fertilization of vineyards, as research indicates that heavily fertilized vineyards tend to contain relatively high levels of arginine and other nitrogenous compounds; (2) analyze grapes from heavily fertilized vineyards for total α-amino acids, as research suggests that grapes exceeding 1500 ppm may develop significant EC potentials during fermentation; (3) if possible, use prise de mousse yeast to produce wine with lower levels of urea and EC, especially if the juice has high levels of total α-amino acids; and (4) fortify dessert wines at the point in fermentation when urea levels are lowest, as EC potentials are significantly affected by the urea levels at the time the fermentation is halted by fortification. A more recent EC-reduction process has also been developed that recommends the addition of urease, an enzyme that removes the unmetabolized urea, to wines which contain higher post-fermentation urea levels (Wine Institute, personal communication, 1989). As these recommendations for reducing EC are instituted, it is expected that average EC levels and the percentage of commercial wines with higher EC concentrations will be lowered.

In 1989 the wine industry conducted its initial EC survey to establish a baseline for measuring future progress. The results of this survey are summarized in Table 6 (*16*). The 5 ppb mean and volume-weighted average values for the 193 samples of bottled table wines (<14% alcohol) from 43 different wineries were well below the target goal of 15 ppb (on a volume-weighted average basis). This initial survey of table wines from 1989 or earlier harvests indicated that none exceeded the 25 ppb target goal for the 1995 harvest. Dessert wines containing >14% alcohol were also below their 1989

Table 6. Summary of 1989 Wine Industry Ethyl Carbamate Survey

		Ethyl Carbamate (ppb)			
Wine Class	*N*[a]	*Range*	*Mean*	*Weighted Average*[b]	*% Exceeding Goal*[c]
<14% alcohol	193	1-24	5	5	0
>14% alcohol	37	1-862	80	33	24

SOURCE: Unpublished data provided by the Wine Institute (*16*).

[a]N = number of samples.

[b]Weighted Average represents the average level computed by the Wine Institute using a weighting factor that takes into account the market share or production volume represented by each winery.

[c]Percentage of survey samples that exceeded the wine industry target goals for the 1995 harvest, which are that no more than 1% of the volume of table wine (<14% alcohol) would be above 25 ppb and no more than 1% of the volume of dessert wine (>14% alcohol) would be above 90 ppb (*1*).

target of 60 ppb on a volume-weighted average basis. However, because of a number of high-level samples, ranging up to 862 ppb, the mean level was more than twice the volume-weighted average that takes into account the market share represented by each winery. In this case, 24% of the 37 samples of dessert wine exceeded the 90 ppb goal for the 1995 harvest. It is important to note that these wines probably represent products that were produced before any of the previously mentioned recommendations for reducing EC levels in wine were implemented. The wine industry has been requested to reconsider their target levels based on these encouraging results.

FDA has also alerted all countries that export alcoholic beverages to the U.S. of the need to develop EC programs with target goals similar to those adopted by the U.S. industry. Research papers published by workers in Britain, France, Germany, and Switzerland suggest that some progress has been made in understanding and controlling EC formation within these countries. Tanner (*17*), of the Swiss Federal Research Institute, has made recommendations for reducing EC in stone fruit brandies, which had the highest EC levels of any alcoholic beverage category. His first recommendation, similar to that of the U.S. industry, was to conduct more controlled distillations. This process change was probably intended to improve the efficiency of EC separation. His other recommendations were to distill the mash within 2 months and to stop crushing the stones or pits of the stone fruits. These recommendations apparently focus on reducing the release of potential EC precursors from the pits, such as cyanate and cyanic acid, which can react with vicinal dicarbonyl compounds such as diacetyl in the presence of ethanol to form EC. At this time we have no data on how effective these process recommendations have been in reducing EC levels in fruit brandy.

In summary, it appears that both the U.S. distilled spirits and U.S. wine industries have made significant progress toward understanding and controlling EC formation and achieving their target goals. FDA will be involved in several future EC-related activities that will include (1) surveys of alcoholic beverages (by BATF) to monitor EC levels in commercial products; (2) monitoring results of the industries' sampling programs and progress in lowering EC levels; (3) evaluating the results of the National Toxicology Program and any other new toxicological studies, when available; and finally, (4) combining new toxicological information with updated exposure data to perform a quantitative assessment of the risk posed by EC in alcoholic beverages. FDA will also continue to work closely with the distilled spirits and wine industries in their efforts to reduce EC to the lowest levels technologically feasible.

Literature Cited

1. Urethane in Alcoholic Beverages; Research and Survey Reports; Availability, *Fed. Regist.* March 23, **1990**, *55*, 10816-10817.
2. Diachenko, G. W.; Canas, B. J.; Joe, F. L., Jr.; Havery, D. C. Presented at the International Office of Vine and Wine Meeting on Ethyl Carbamate, Paris, May 31, 1988.
3. Lau, B. P.-Y.; Weber, D.; Page, D. B. *J. Chromatogr.* **1987**, *402*, 233-241.
4. Canas, B. J.; Havery, D. C.; Robinson, L. R.; Sullivan, M. P.; Joe, F. L., Jr.; Diachenko, G. W. *J. Assoc. Off. Anal. Chem.* **1989**, *72*, 873-876.

5. Canas, B. J.; Joe, F. L., Jr.; Diachenko, G. W. Presented at the 104th Annual Meeting of the Association of Official Analytical Chemists, New Orleans, LA, September 1990; paper 126.
6. Ough, C. S. *J. Agric. Food Chem.* **1976**, *24*, 323-328.
7. *Statistical Abstracts of the United States*, U. S. Government Printing Office: Washington, DC, 1987.
8. *Frequency Distributions of the 14-Day Average Quantity of Food Consumed by Age of Eater,* Market Research Corporation of America report under FDA contract No. 223-87-2088, 1988.
9. *Foods Commonly Eaten by Individuals*, U.S. Department of Agriculture Home Economics Research Report 44: Hyattsville, MD, 1982.
10. MacKenzie, W. M.; Clyne, A. H.; MacDonald, L. S. *J. Inst. Brew.* **1990**, *96*, 223-232.
11. Comments of the Distilled Spirits Council of the U.S. (Docket No. 86P-0482/CP, EC File II, June 8, 1987), available from the FDA Hearing Clerk, Rockville, MD.
12. Quarterly Reports from the Distilled Spirits Council of the U.S. (Docket No. 89N-0483, letters from DISCUS to S. Delgado, July 14 and October 3, 1989 and January 30, 1990), available from Industry Activities Section (HFF-326), FDA, Washington, DC.
13. Ough, C. S. *Proceedings of the Institute of Pathobiology Toxicology Forum*, Aspen, CO, 1986, pp. 387-392.
14. Ough, C. S.; Crowell, E. A.; Mooney, L. A. *Am. J. Enol. Vitic.* **1988**, *39*(3), 243-249.
15. Letter from the Wine Institute to FDA's C. Coker, dated March 10, 1989 (Docket No. 89N-0483), available from Industry Activities Section (HFF-326), FDA, Washington, DC.
16. Letter from the Wine Institute to FDA's C. Coker, dated November 8, 1989 (Docket No. 89N-0483), available from the Industry Activities Section (HFF-326), FDA, Washington, DC.
17. Tanner, H. *Schweiz. Z. Obst. Weinbau* **1986**, *122*(9), 260-262.

RECEIVED September 4, 1991

Chapter 35

Composition and Safety Evaluation of Potato Berries, Potato and Tomato Seeds, Potatoes, and Potato Alkaloids

Mendel Friedman

Food Safety Research Unit, Western Regional Research Center, Agricultural Research Service, U.S. Department of Agriculture, 800 Buchanan Street, Albany, CA 94710

Potatoes and potato berries frequently contain antinutritional and toxic compounds including inhibitors of trypsin, chymotrypsin, and carboxypeptidase; hemagglutinins; and glycoalkaloids. The glycoalkaloid content of the berries is about 10 times that of a commercial potato variety. Male mice were fed freeze-dried potato berries at 1, 5, 10, 20, and 40% of the diet and solasodine up to 1600 mg per kg diet. Gross clinical observations, body weight, and feed intake were recorded weekly. At 14 days, all surviving animals were autopsied, organs weighed, and blood chemistry analyzed. Selected tissues from mice fed the control, solasodine, and 20% potato berry diets were examined histologically. All mice fed the 40% potato berry diet died. Mice fed 20% or less potato berries and those fed the solasodine diet gained less weight than controls. Additional effects were noted mainly in the solasodine-fed animals. These included elevated serum alkaline phosphatase, glutamic-pyruvic transaminase (GPT), and glutamic-oxaloacetic transaminase (GOT); elevated liver weight as percent of body weight; decreased body-weight gain; and increased incidence of liver cholangiohepatitis and gastric gland dilation/degeneration. In a separate study, potato and tomato seeds were fed at 1, 2, and 4% and potatoes at up to 40% of the diet. The only significant effects noted were a decrease in serum GPT in mice fed all levels of the seeds and an increase in pancreatic weights in mice fed the high levels of potatoes. The pancreatic effect may result from the trypsin inhibitors in the potatoes. Solanaceae alkaloids are also treported to inhibit cholinesterase, disrupt cell membranes, and induce teratogenicity. Possible research strategies in food safety, nutrition, and plant physiology to minimize adverse effects of alkaloids are discussed.

The *Solanaceae* or "nightshade" family of plants contains many plants important to man. These include agricultural crops such as capsicum, eggplant, potato, and tomato;

the toxic weeds (black nightshade and Jimsonweed); and tobacco. These plants produce antinutritional and toxic compounds, both during growth and after harvest. These compounds include glycoalkaloids (Figures 1 and 2, Table 1), trypsin and carboxypeptidase inhibitors, and hemagglutinins (lectins) (*1-7*).

Relatively high concentrations of glycoalkaloids have been found in *Solanaceae* plants consumed by man, such as potatoes and tomatoes (*8*). Levels are especially high in green and damaged potatoes and immature green tomatoes (*3-7*). Glycoalkaloids are far more toxic to man than to other animals studied (*9-14*). Levels of 3 to 6 mg/kg are reported lethal, a toxic potency comparable to strychnine, although there appears to be considerable individual variation in the susceptibility of animals and humans (*13*). The toxicity may be due to adverse effects such as anticholinesterase activity of the glycoalkaloids on the central nervous system (*14-15*) and to disruption of cell membranes adversely affecting the digestive system and general body metabolism (*16*). Inducement of teratogenicity by these alkaloids (*10*) makes pregnant women particularly susceptible (*17-19*). There also are reports that extracts of potatoes are more toxic than one would expect on the basis of their alkaloidal content (*13*), suggesting the presence of additional toxic compounds of unknown structure and function (*20-21*). Worldwide, up to 25% of the potato crop has to be discarded because glycoalkaloid levels are above the maximum deemed to be safe (*13*).

The possible human toxicity of the *Solanum* glycoalkaloids, including about 30 reported deaths and over 2000 cases of poisoning, has been documented (*12,13*). These reports and the apparent relationship between at-birth prevalence of human malformation and the severity of blighted potatoes in the diet of the mothers (*19*), has led to the establishment of guidelines limiting the glycoalkaloid content of new potato cultivars. According to Morris and Lee (*13*), these guidelines may be too high.

As part of a program designed to improve food safety through identification and elimination of naturally occuring *Solanaceae* toxicants, we evaluated the safety of potato berries, potatoes, and potato and tomato seeds added to a 10% protein casein diet fed to mice for 14 days. This study is part of a broader objective designed to (a) identify and characterize *Solanaceae* alkaloids, glycoalkaloids, hydrolysis products, and biosynthetic intermediates; (b) to develop a relative toxicity (potency) scale for these compounds; (c) to identify and characterize key enzymes in the biosynthetic pathways; and (d) to suppress the synthesis of those enzymes which catalyze the formation of the more toxic compounds using antisense RNA and related molecular biology techniques. This overview discusses some of these concepts.

Materials and Methods

Alkaloids, trypsin, chymotrypsin, carboxypeptidase, N-α-tosyl-arginine methyl ester (TAME), N-benzoyl-L-tyrosine ethyl ester (BTEE), [Tris (hydroxymethyl)-aminomethane](Tris), and other reagents were obtained from Sigma Chemical Co. (St. Lous, MO). Potato berries and experimental potato varieties were kindly provided by D. Corsini, Cereal and Vegetable Crop Production Laboratory, ARS, USDA, Aberdeen, Idaho; potato seeds by J. Pavek, USDA, University of Idaho, Aberdeen, Idaho; and tomato seeds by R. Mitchell, NK Company, Gilroy, California.

Table 1. Steroidal Glycoalkaloids

Steroidal glycosides	*Sugar moiety*	*Glycoside structure*
Solanidine glycoalkaloids		Rham
α-Solanine	Solatriose	A: R-Gal
		Glu
β-Solanine	Solabiose	B: R-Gal-Glu
α-Solanine	Galactose	C: R-Gal
		Rham-a
α-Chaconine	Chacotriose	D: R-Glu
		Rham-b
β_1-Chaconine	Chacobiose	E: R-Glu-Rham-a
β_2-Chaconine	Chacobiose	F: R-Glu-Rham-b
Δ-Chaconine	Glucose	G: R-Glu
		Glu
Dehydrocommersonine	Commertetraose	H: R-Gal-Glu
		Glu
Demissidine glycosides		Glu
Demissine	Lycotetraose	I: R-Gal-Glu
		Xyl
Commersonine	Commertetraose	As H
Leptinidine glycosides		
Leptinine I	Chacotriose	As D
Leptinine II	Solatriose	As A
Acetylleptinidine glycosides		
Leptine I	Chacotriose	As D
Leptine II	Solatriose	As A
Tomatidenol glycosides		
α-Solamarine	Solatriose	As A
β-Solamarine	Chacotriose	As D
Solasodine glycosides		
Solasonine	Solatriose	As A
Solamarine	Chacotriose	As D
Tomatidine glycosides		
α-Tomatine	Lycotetraose	As I
Sisunine (neotomatine)	Commertetraose	As H
		3-1 2-1 2-1
Solardixine		-Gal-Glu-Glu-Rham
Solasurine		-Glu-Rham
Solashabanine		Gal, 3 Glu, Rham
Solaradinine		Gal, 4 Glu, Rham
Solapersine		Gal, Glu, 2 Xyl
Solatifoline		Glu, Gal, Rham
Sisunine		Gal, 3 Glu
Solaverbascine		
N-methyl-solasodine		
12-β-Hydroxysolasodine		
Solanocapsine		

SOURCE: Adapted from References *6, 8, 61, 62*

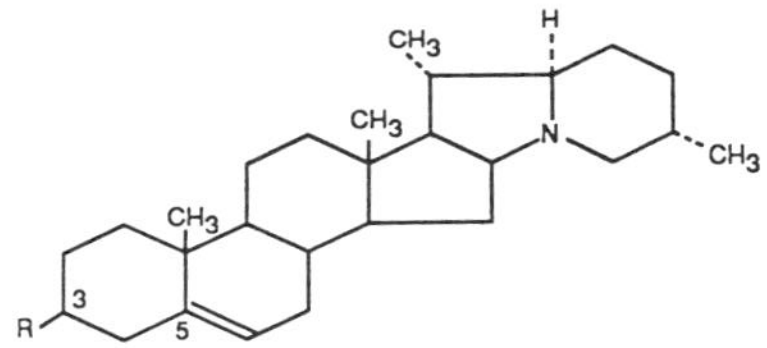

22αH, 25βH-SOLANID-5-EN-3β-R

Solanidine	R = OH
Solanthrene	R = —, Δ3
Solanine	R = galactose — glucose │ rhamnose
Chaconine	R = glucose — rhamnose │ rhamnose

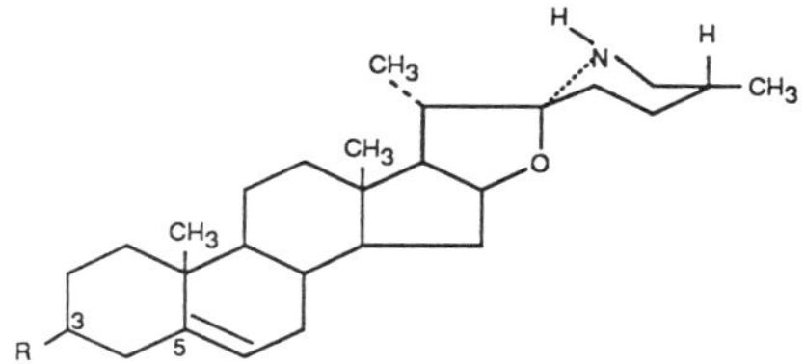

(25R)-22αN-SPIROSAL-5-EN-3β-R

Solasodine	R = OH
Solasodiene	R = —, Δ3
Soladulcidine	R = OH, 5α
Solasonine	R = galactose — glucose │ rhamnose

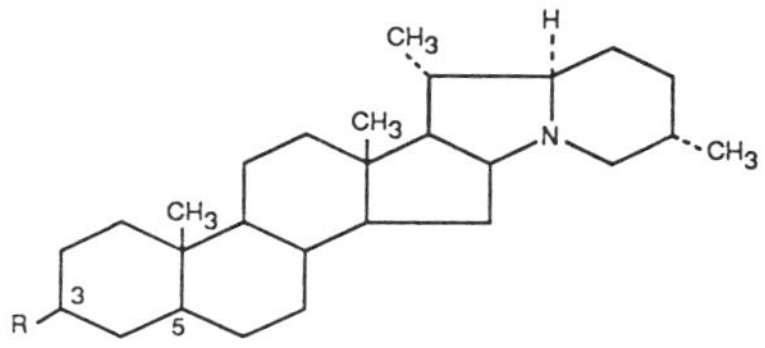

5α, 22αH, 25βH-SOLANIDAN-3β-R

Demissidine	R = OH
Demissine	R = galactose — glucose — xylose │ glucose

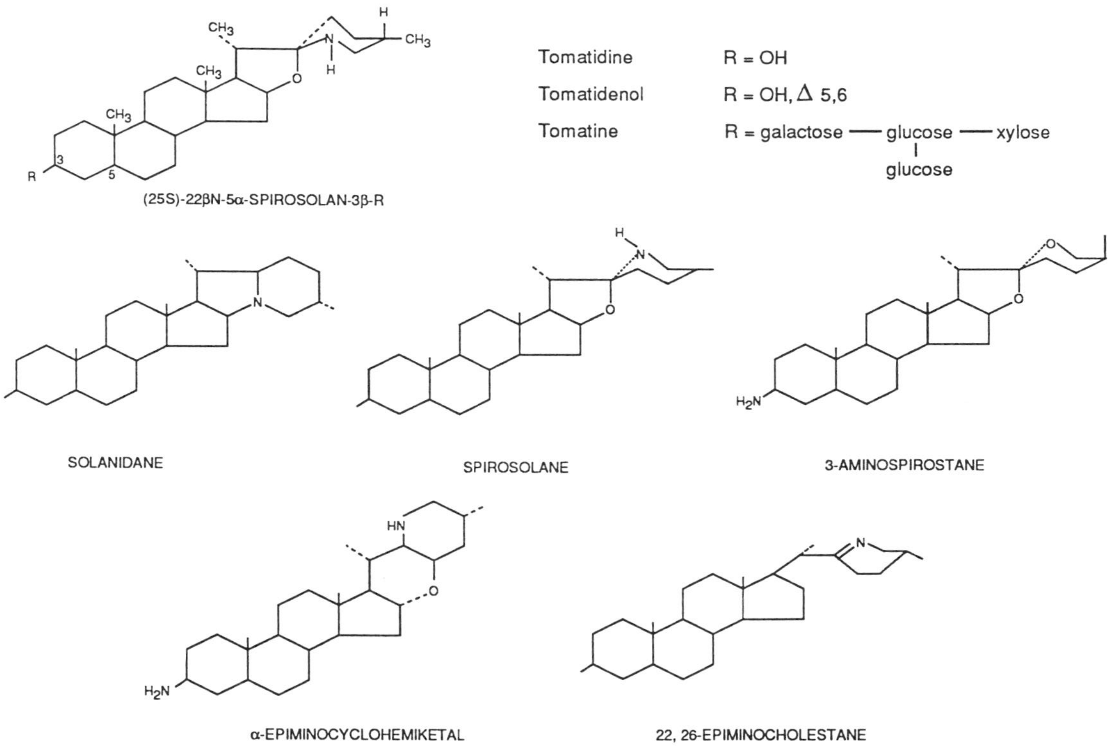

Figure 1. Structures of Solanum alkaloids. Top, chemical structures of Solanum glycoalkaloids; bottom, structures of steroid part of major classes of Solanum alkaloids (aglycones) (*6, 8, 19*). The more than 60 steroidal alkaloids of toxicological interest all have the C-27 skeleton of cholestane

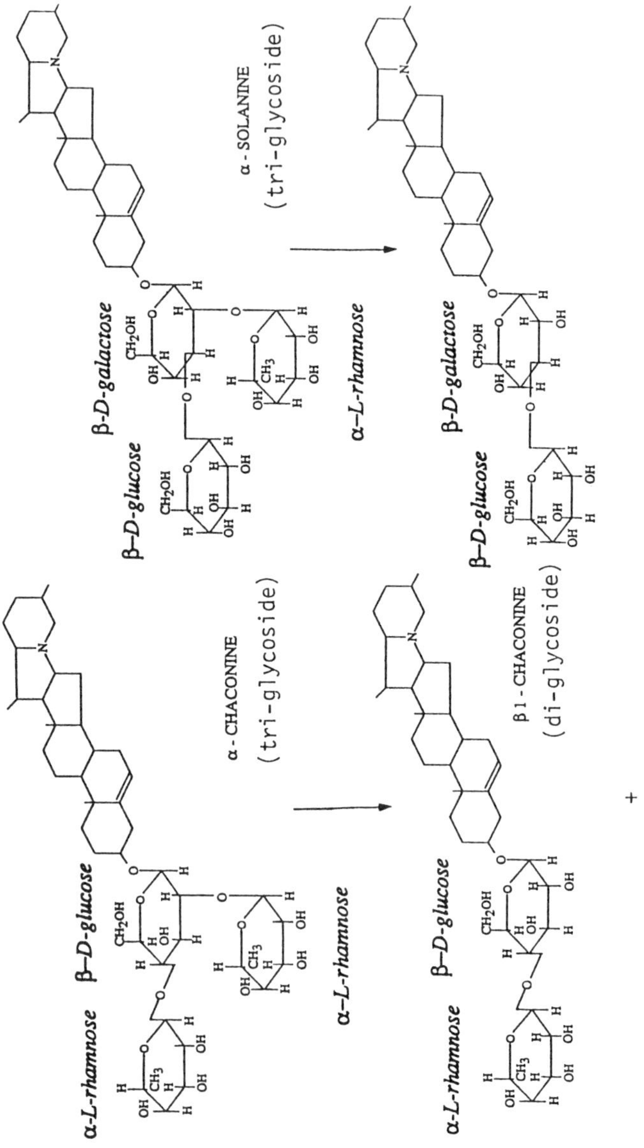
α-SOLANINE
(tri-glycoside)
β-D-galactose
β-D-glucose
α-L-rhamnose
β-D-galactose
β-D-glucose
α-CHACONINE
(tri-glycoside)
α-L-rhamnose
β-D-glucose
α-L-rhamnose
β1-CHACONINE
(di-glycoside)
β-D-glucose
α-L-rhamnose
+

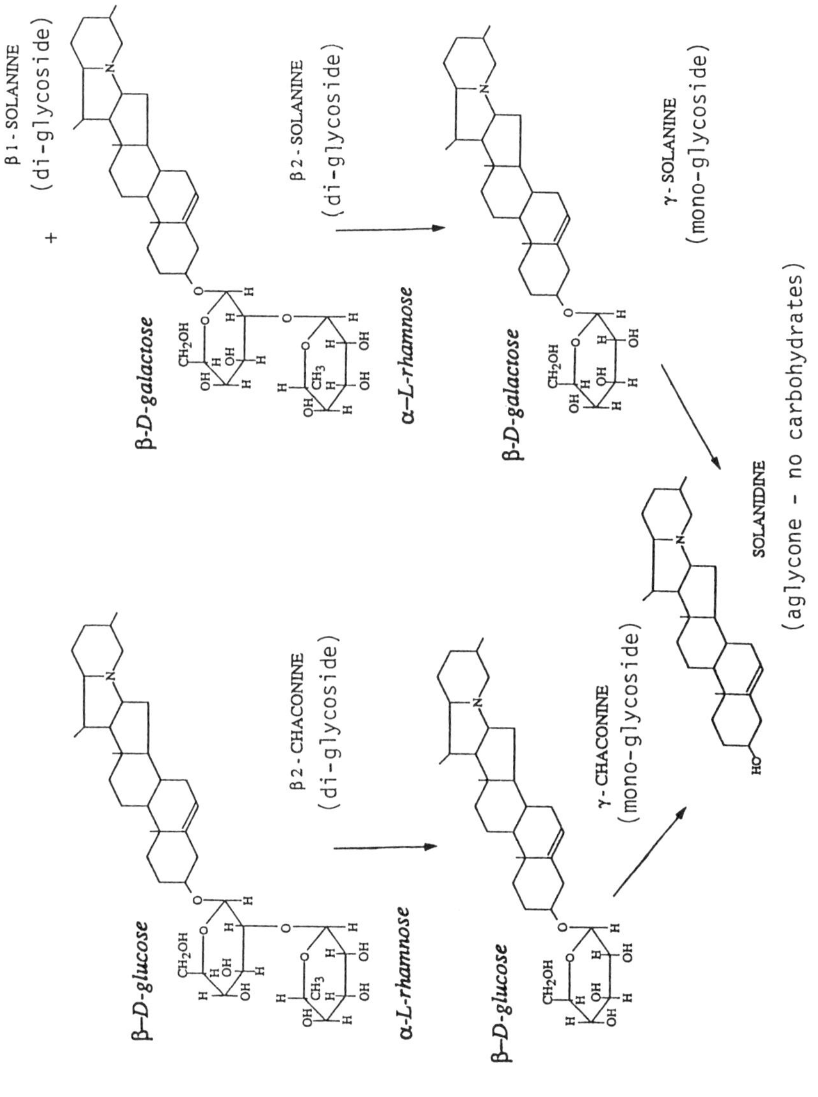

Figure 2. Structures of α-chaconine and α-solanine and hydrolysis products.

All compositional and feeding studies were carried out with powdered samples prepared as follows: frozen potato berries were crushed with a metal block and potatoes were cut into small cubes and then freeze-dried by lyophilization. Potato and tomato seeds were ground in an Omni mill which was kept cold with dry ice.

Compositional Analyses. Procedures for amino acid, enzyme inhibitor, and lectin contents of potato berries, potato seeds, and potatoes were adapted from the literature (*2, 2-26*)

Amino Acid Composition. The following three analyses with flour containing about 5 mg of protein (N X 6.25) were used to establish the amino acid composition (*23*): (a) standard hydrolysis with 6 N HCl for 24 h in evacuated sealed tubes; (b) hydrolysis with 6 N HCl after performic acid oxidation to measure half-cystine and methionine content as cysteic acid and methionine sulfone, respectively; (c) basic hydrolysis by barium hydroxide to measure tryptophan content (*22*). The reproducibility of these analyses is estimated to be ± 3%, based on past experience (*23-24*).

Trypsin Inhibitor. Potato powder (100 mg) was suspended in 10 mL Tris-HCl (pH 8.1) buffer with stirring for 1 hr at room temperature. The suspension was centrifuged for 10 min and the supernatant was diluted 1:2 with Tris buffer. Twenty µL of this solution were used for the inhibition assay. The following conditions were used: temperature 25°C; buffer, 0.046 M Tris-HCl containing 0.0115 M $CaCl_2$ pH 8.1; substrate, 10 mM TAME (37.9 mg/10 mL H_20); enzyme, 1 mg/mL, 1 mM HCl. The enzyme solution was diluted to 10-20 µg/mL. In the absence of inhibitor, 2.6 mL buffer and 0.3 mL TAME were added to a 3 mL cuvette followed by 0.1 mL diluted enzyme solution. The absorbance was recorded at 247 nm (A_{247}) for 3 min on a Perkin-Elmer Lambda 6 spectrophotometer. The increase in absorbance was then determined from the initial linear portion of the curve. In the presence of inhibitors, 2.6 mL buffer, 0.1 mL enzyme solution, and 20 µL of inhibitor solution (prepared to give 50% inhibition) were pre-incubated for 6 min. The reaction was started by adding 0.3 mL TAME; A_{247} was then recorded for 3 min. Values were based on sample dilutions yielding 40 to 60% inhibition (*25*). Enzyme activity is defined by the following equation: Units/mg = (ΔA_{247}/min X 1000 X 3)/ (540 X mg enzyme used).
A trypsin unit (TU) is defined as the amount of trypsin that catalyzes the hydrolysis of 1µmol of substrate/min. A trypsin inhibitor unit (TIU) is the reduction in activity of trypsin by 1 TU.

Chymotrypsin Inhibitor Assay. Potato powder (200 mg) was suspended in 10 mL Tris-HCl (pH 8.1) buffer with stirring for 1 hr at room temperature. The suspension was centrifuged for 10 min. Twenty µL of the supernatant was used for the inhibition assay. The following conditions were used: buffer, 0.08 M Tris-HCl containing 0.1 M $CaCl_2$, pH 7.8; substrate, 1.07 mM BTEE (8.4 mg/25 mL 50% methanol); enzyme, 1 mg/mL, 1 mM HCl. The enzyme solution was diluted to a concentration of 10-20 µg/mL. In the absence of inhibitor, 1.5 mL buffer, 1.4 mL BTEE, and 0.1 mL enzyme solution were added and the increase in absorbance at 256 nm (A_{256}) was recorded for

3 min. The ΔA_{256}/min was then calculated from the initial linear portion of the curve. In the presence of inhibitor, 1.5 mL buffer, 0.1 mL enzyme solution, and 20 µL inhibitor were incubated for 6 min before adding 1.4 mL BTEE and recording as above. Buffer plus substrate served as controls for all measurements. Values were based on sample dilutions yielding 40 to 60% inhibition (*25*). Enzyme activity is defined by the following equation: Units/mg = (ΔA_{256}/min X 1000 X 3)/(964 X mg enzyme used).

One chymotrypsin unit (CU) is defined as the amount of chymotrypsin that catalyzes the hydrolysis of 1 µmol of substrate/min. A chymotrypsin inhibitor unit (CIU) is the reduction in activity of chymotrypsin by 1 CU.

Carboxypeptidase Inhibitor Assay. Potato powder (300 mg) was suspended in 10 mL of Tris-HCl (pH 7.5) buffer with stirring for 1 hr at room temperature. The suspension was centrifuged for 10 min. One hundred µL of the supernatant was used for the inhibition assay. The method was adapted from Worthington (*25*). Reagents were 0.025 M Tris-HCl buffer containing 0.5 NaCl, pH 7.5; 1 mM hippuryl-L-phenylalanine in 0.025 M Tris-HCl buffer, and 10% LiCl.

The enzyme was dissolved in 10% LiCl to a concentration of 1-3 units/mL. The rate of hydrolysis of hippuryl-L-phenylalanine by carboxypeptidase in the absence and presence of inhibitors was determined by measuring the increase in absorbance at 254 nm. Specifically, into each cuvette was pipetted 2.9 mL of substrate solution, which was then incubated at 25°C for 3-4 min to reach temperature equilibrium. The diluted enzyme solution (0.1 mL) was then added and the absorbance at 254 nm was recorded for 3-5 min. The change in absorbance (ΔA_{254}) was calculated from the initial portion of the curve.

One carboxypeptidase unit is defined as the amount of carboxypeptidase that catalyzes the hydrolysis of one micromole of hippuryl-L-phenylalanine per min at pH 7.5 and 25°C.

Pure soybean Kunitz trypsin inhibitor (KTI) and pure soy Bowman-Birk chymotrypsin inhibitor (BBI) were used as standards with each trypsin and chymotrypsin inhibition assay. The calculated values, which are averages from 2 to 3 separate determinations, are based on the individual inhibitor control values. For carboxypeptidase, the extent of inhibition by the potato flours was estimated from the observed decrease in enzyme activity without the use of a pure inhibitor standard to correct for any day-to-day variability in the assay.

Lectin (Hemagglutinin) Assay. The sample (100 mg) was mixed with 5.0 mL of the phosphate buffered saline (PBS), pH 7.2 buffer. Lectin was extracted by stirring for 1 h at room temperature. After extraction, the resulting slurry was immediately chilled and centrifuged at 9000 g for 5 min in a Beckman Microfuge (Beckman Instruments, Palo Alto, CA). When necessary, the extracts were diluted with isotonic phosphate buffer (PBS, 0.05 M NaH_2P0_4 and 0.15 M NaCl, pH 7.2) before plating so that incipient activity would fall midrange in the plated series.

Samples (50 µL) of glutaraldehyde-stabilized rabbit red blood cells (Sigma R-1629) diluted with a buffer to 1% hematocrit were added to equal volumes of serially diluted extracts and a buffer blank. Agglutination was observed visually after 1 h (*26*).

Lectin activity is calculated as the reciprocal of the minimum amount (mg/mL) of soy flour required to cause agglutination of blood cells under these test conditions. This value is derived from the minimum experimental value (µg/50mL) which produces hemagglutination. For the experimental value, the lower the number the more potent the activity, whereas the opposite is true for the calculated values. The results of four separate assays conducted on each sample were averaged.

Feeding Studies, Histology, and Clinical Chemistry. Mice feeding studies were carried out for 14 days as previously described (*27-30*). Clinical chemistry was provided by W.M. Spangler, D.V.M. (California Pathology Consultants, West Sacramento, CA) and histology by Spangler and M.A. Stedham, D.V.M. (Pathology Associates, Frederick, MD), as previously described (*31*).

Specifically, mice (Swiss Webster strain, Simonsen Laboratories, Inc., Gilroy, CA) were housed singly or two per cage. The cages were polycarbonate with stainless steel wire tops and pine shavings for litter. Feed and water were provided *ad libitum*. The temperature of the animal room was 22 ± 10%. The light cycle was 6AM - 6PM light and 6PM - 6AM dark, as regulated by an automatic timer. Animals were assigned so that all treatment groups had nearly the same initial body weight.

The effect on weight gain and other parameters was examined by feeding mice 10% protein (N casein X 6.25 = 11.9% actual casein) diets fortified with various levels of freeze-dried potato berries and potatoes, milled potato and tomato seeds, and the alkaloid solasodine, as listed in the tables.

Animals surviving to the end of the study were anesthesized using ether and killed by exsanguination via an axillary space incision and severance of the brachial artery. A blood sample was obtained from the axillary space incision. Serum samples were prepared, frozen on dry ice and stored at -80°C. Animals were subjected to autopsy in which selected tissues were examined grossly and weighed. For animals from three diets of the potato berry study (control, solasodine, and 20% potato berry) a portion of each liver was homogenized in three volumes (w:v) cold 0.01 M KCl buffer. The homogenate was centrifuged at 20,000 rpm for 20 minutes and the supernatant frozen on dry ice and stored at -80°C for later analysis of N-demethylase activity. The caudate lobes of these livers were retained for histological evaluation. Tissues for histological evaluation were fixed in 10% buffered neutral formalin, embedded in paraffin, sectioned, and stained with hematoxylin and eosin. For surviving animals, the liver, stomach, and testes of mice fed potato berries and the liver, stomach, and kidneys of mice fed potato seeds were histologically examined. Evalutions were made "blind" without reference to treatment. In addition, the livers of three mice fed the 40% potato berry diet and removed from the study when found moribund were also evaluated.

Serum cholinesterase (ChE) was determined according to the method of Ellman et al. (*32*). A Boehringer Mannheim Diagnostic Model 8600 automated clinical chemistry analyzer was used for all other serum clinical chemistries. Liver N-demethylase was determined using the method of Dalton and DiSalvo for the Technicon AutoAnalyzer II (*33*).

Embryotoxicity . The Frog Embryo Teratogenesis Assay (*Xenopus*) was used to evaluate the embryotoxicity of *Solanaceae* alkaloids and glycoalkaloids (*34-35*).

Statistical Methods. Lesion incidence and mortality data were analyzed by Fisher's Exact Test, one-tailed (*36*). Feed and water consumption and body weight gain were screened for outliers and subjected to analysis of variance and comparison of means using Duncan's Multiple-Range Test (*37*). Other data were analyzed for effects of treatment using the General Linear Model Procedure of SAS (*38*). Analysis of co-variance for continuous effects of dose was used to determine if overall dose-related responses had occurred.

Results

Amino Acids. Tables 2 and 3 show the complete amino acid composition of potato berries, potato seeds, and potatoes, with tomato seeds included for comparison (*39*). The values of the amino acid scoring pattern of the essential amino acids for an ideal protein, as defined by the Agricultural Organization of the United Nations (*40*), are also shown for comparison in Table 3. Noteworthy is that: (a) the protein of all of these products is high quality by the FAO requirements; (b) all potato products have a similar amino acid composition, with about one half of the total consisting of aspartic acid (plus asparagine); (c) potato berries and potatoes, but not potato seeds contain significant amounts of Δ-amino-butyric acid (GABA); and (d) nitrogen (protein) content of the tomato seeds is more than twice that of potato seeds and about twice that of the other potato products (*see* footnotes to tables).

Enzyme Inhibitors. Table 4 shows that potatoes, but not potato berries, contain significant amounts of inhibitors of digestive enzymes such as carboxypeptidase A, trypsin, and chymotrypsin.

Lectins. Table 5 shows that the lectin content of potato berries is about 2 to 3 times that of potatoes but only about one fourth that of raw soy flour. Note that the lower the number of the activity, X, the higher the potency, whereas for the parameter 1/X (*see* last column in Table 5), activity is directly related to the size of this reciprocal.

Glycoalkaloids. Table 6 lists, for potato berries and two varieties of potatoes, the a-chaconine and α-solanine content determined by an HPLC procedure and the total glycoalkaloid content determined by colorimetric (bromphenol) method. The data show that (a) the α-chaconine level of the berries is 7 to 15 times greater than that of the potatoes; and (c) the total glycoalkaloid content of the berries is 7 to 16 times greater than in the two potato varieties. The cited values are based on improved assays developed in the course of this work, details of which will be published separately (*41*).

Table 2. Amino Acid Content of Potato Berries, Potato Seeds, Potatoes, and Tomato Seeds (A = g/100g; B = g/16g N)

Amino Acid	*Potato berries* (A)	(B)	*Potato seeds* (A)	(B)	*Commercial Potatoes* (A)	(B)	*Tomato seeds* (A)	(B)
Asp	5.78	40.6	2.27	9.35	6.02	51.3	3.25	9.9
Thr	0.30	2.07	0.82	3.38	0.29	2.43	1.00	3.0
Ser	0.43	3.03	1.15	4.74	0.34	2.85	1.54	4.7
Glu	2.52	17.70	5.33	21.9	2.09	17.8	6.01	18.4
Pro	n.d[g]	n.d.	n.d	n.d	0.46	3.94	n.d	n.d
Gly	0.42	2.96	1.22	5.01	0.23	1.98	1.44	4.4
Ala	0.46	3.21	1.09	4.48	0.25	2.13	1.28	3.9
Val	0.39	2.73	1.19	4.90	0.57	4.87	1.32	4.0
Cys[a]	0.22	1.56	0.42	1.72	0.11	0.93	0.50	1.5
Met[b]	0.17	1.20	0.52	2.16	0.15	1.23	0.62	1.9
Ile	0.38	2.64	0.89	3.65	0.34	2.92	1.05	3.2
Leu	0.55	3.86	1.51	6.22	0.44	3.75	1.77	5.0
Tyr	0.39	2.75	0.91	3.76	0.36	3.05	1.44	4.4
Phe	0.45	3.15	1.15	4.73	0.48	4.11	1.26	3.2
His	0.26	1.84	0.56	2.31	0.20	1.73	0.70	2.1
Lys	0.63	4.45	1.29	5.32	0.59	5.02	1.97	6.0
Arg	0.52	3.66	2.07	8.51	0.62	5.31	2.86	8.7
Try[c]	0.014	0.102	0.13	0.53	0.063	0.536	0.15	0.4
γ-ABA[d]	0.18	1.27	none	none	0.25	2.15	none	none
X[e]	—	—	—	—	0.02	0.17	—	—
X[f]	—	—	—	—	0.02	0.16	—	—
Total	4.1	98.8	22.52	92.7	13.6	118.7	28.10	86

NOTE: Nitrogen (N) content, % potato berries, 2.28; potato seeds, 3.89; commercial potatoes (flesh), 1.78; tomato seeds, 5.22.

[a]Determined as cysteic acid after performic acid oxidation.

[b]Determined as methionine sulfone after performic acid oxidation.

[c]Determined after hydrolysis by barium hydroxide.

[d]γ-Amino-butyric acid.

[e]Unknown peak eluting before Asp.

[f]Unknown peak eluting before Lys.

[g]Not detected by 590 nm photometer.

Table 3. Amino Acid Content (g/16 g N) of Different Potato Varieties

Amino Acid	Commercial		Experimental variety (Lenape)		Experimental variety (No. 2461)		FAO[a]
	Flesh	Peel	Flesh	Peel	Flesh	Peel	
Asp	51.3	55.2	42.6	47.0	45.4	43.5	
Thr	2.43	2.64	3.18	3.26	3.01	3.27	4.0
Ser	2.85	3.01	3.34	3.63	3.21	3.48	
Glu	17.8	11.2	16.6	13.5	17.3	13.1	
Pro	3.94	3.66	4.49	4.94	3.50	3.20	
Gly	1.98	2.43	2.92	3.15	2.73	2.93	
Ala	2.13	2.48	2.45	2.78	2.63	3.06	
Val	4.87	4.48	4.24	4.39	4.36	4.30	5.0
Cys	0.93	0.95	1.31	1.60	1.11	1.11	3.5[b]
Met	1.23	0.99	1.04	1.30	1.33	1.09	
Ile	2.92	3.09	2.95	3.37	3.07	3.33	4.0
Leu	3.75	4.56	5.37	5.81	5.20	5.56	7.0
Tyr	3.05	2.51	2.80	2.63	2.68	2.70	6.0[c]
Phe	4.11	4.25	4.22	4.17	3.77	3.96	
His	1.73	1.57	1.52	1.51	1.53	1.52	
Lys	5.02	4.18	1.89	4.80	4.69	4.61	5.5
Arg	5.31	3.57	5.80	3.66	6.37	3.45	
Try	0.54	0.11	0.470	0.26	0.13	0.48	
γ-ABA	2.15	1.46	2.52	1.89	2.44	1.74	
X	0.17	0.21	—	—	—	—	
X	0.16	—	—	—	0.14	—	
Total	118.4	112.6	112.7	113.6	114.6	106.4	

NOTE: Nitrogen (N) content, %: commercial flesh, 1.78; commercial peel, 2.59; Lenape flesh, 1.55; Lenape peel, 1.78; No 2461 flesh, 1.86; No 2461 peel, 2.14.

[a]Provisional amino acid scoring pattern for an ideal protein (*2,40*).

[b]Cys + Met. [c]Tyr + Phe.

Table 4. Carboxypeptidase A, Chymotrypsin, and Trypsin Inhibition by Freeze-dried Potatoes and Potato Berries (units inhibited/g sample)

Sample	*Carboxypeptidase A*	*Chymotrypsin*	*Trypsin*
Russett potatoes[a]	144 ± 8.7	260 ±5.5	768 ± 32
Potatoes (No 3194)[b]	94 ± 9.9	409 ± 17	1280 ± 67
Potato berries	0	0	0

NOTE: All values are averages from three separate determinations ± standard deviation.

[a] Obtained from a local store; [b] Experimental variety.

Table 5. Lectin (Hemagglutinin) Activity of Potato Berries, Potatoes, and Raw Soy Flour Against Rabbit Red Blood Cells

Sample	*Activity (X)*[a]	*1/X (10^3)*
Potato berries	13 ± 2.5 (5)[d]	77 ± 15
Potatoes - No 3194[b]	38 ± 4.7 (5)	42 ±7.4
Potatoes - Russet[c]	24 ± 4.2 (5)	26 ±3.2
Raw soy flour	3.5 ± 0.7 (3)	290 ± 5.8

[a]Minimum amount of freeze-dried flour (in μg/50μL) causing hemagglutination after 1 hr;
[b] Experimental variety;
[c] Obtained from a local store;
[d] Average ± standard deviation. Values in parentheses are number of separate determinations.

Table 6. Glycoalkaloid Content of Potato Berries and Potatoes

	α-Chaconine	*α-Solanine*	*Total glycoalkaloid content*
	(mg/100 g fresh weight)[a]		*(mg/100 g fresh weight)*[b]
Potato berries	22.1± 1.43[c]	15.9 ± 0.80	44.6 ± 2.71
Potatoes[d] (No 3194)	3.68 ± 0.43	1.96 ± 0.05	6.37 ± 0.71
Potatoes[e]	1.35 ± 0.08	0.65 ± 0.02	2.86 ± 0.54

[a]Determined by an HPLC procedure (*41*);
[b]Determined by a bromphenol blue colorimetric procedure;
[c]All values are averages from three separate determinations ± standard deviation;
[d]Experimental variety;
[e]Commercial variety obtained from a local store.

Mortality and Clinical Signs. Table 7 lists the ingredients of the control diet used in all feeding studies. All animals fed 40% potato berries died or were removed from the study when they became moribund. This mortality (5/5) significantly exceeded the zero mortality of all other groups ($P \leq 0.05$) (Table 8). Clinical conditions observed in animals fed this high level of berries included rough fur, lethargy, and hunched body posture. All other animals appeared healthy and exhibited no gross clinical signs related to the diet.

Body Weight. Controls gained significantly more weight than animals fed 20% or more potato berries (Tables 8-9) and those fed the solasodine diet ($P \leq 0.01$, Fig. 3). Animals fed potato seed diets gained at a rate equal to that of those fed the control diet (data not shown).
Mice fed commercial-variety potatoes at 5% in the diet gained significantly less weight than the control animals ($P < 0.05$) (Table 9). (At $P < 0.01$, however, potatoes up to 40% in the diet did not affect weight gain). Relative pancreas weights were significantly elevated ($P < 0.05$) by 20% or more experimental potatoes and by 10% or more commercial potatoes in the diet. This is probably a trypsin inhibitor effect (*30,43,44*). Serum GPT was significantly elevated ($P < 0.05$) in mice fed the commercial potato diet at 40%.

Feed Consumption. For animals fed 20 or 40% potato berries, feed consumption could not be determined accurately due to feed spillage. Overall 14-day feed consumption (g/day) was less than that of the control in mice fed solasodine or 10% berries ($P \leq 0.01$), but not in those fed 1 or 5% berries (Table 9). However, on a body weight basis (g/kg body weight/day) there were no differences from the control. Feed consumption of animals fed the potato and tomato seed diets was unaffected (data not shown).

Organ Weights. Relative weights (Tables 10, 11) of the liver, kidneys, testes, and pancreas did not differ significantly from control mice fed potato or tomato seed at up to 4% in the diet or potato berries at up to 20% in the diet. For animals fed solasodine, the relative liver weight was significantly elevated ($P \leq 0.01$). Pancreatic weights of mice fed potato diets were greater than those of controls.

Clinical Chemistries. Solasodine in the diet resulted in markedly increased levels of serum alkaline phosphatase, glutamic-pyruvic transaminase (GPT), and glutamic-oxaloacetic transaminase (GOT) (Tables 12, 13). The increase in bilirubin, while less dramatic, was significant. Statistical analysis of dietary effects on GOT, GPT, and alkaline phosphatase in mice fed the potato preparations did not include the solasodine positive control; inclusion of this group created problems with the heterogeneity of variances, which violated the assumptions of analysis of variance. In mice fed the potato berries at 20%, serum concentrations of GOT, GPT, and liver N-demethylase were significantly elevated ($P \leq 0.01$). Cholinesterase was unaffected. For animals fed potato seeds at 1, 2, or 4% in the diet, the GPT levels were significantly less than in the controls ($P \leq 0.05$). Other values were unaffected.

Table 7. Control Diet Formulation

Ingredient	*Percent in Diet*
Casein	11.9
Corn oil	8.0
Water	5.0
Fiber (Alphacel)	3.0
Vitamin mixture[a]	2.0
Choline chloride	0.2
Mineral mixture, AIN-76[b]	5.0
Cornstarch: Dextrose[c]	64.9
Total	**100.0**

[a] 1 kg of the vitamin mixture contains: 900,000 i.u. A acetate; 100,000 i.u. D_3; 5,500 i.u. DL-α-tocopherol acetate; 1 g menadione; 4.5 g nicotinic acid; 1 g riboflavin; 1 g pyridoxine HCl; 1 g thiamine HCl; 3 g calcium pantothenate; 20 mg D-biotin; 200 mg folic acid; 1.35 mg B_{12}; 5 g inositol; 45 g ascorbic acid;

[b] AIN (1977) (*42*);

[c] Ratio was 2:1 for potato berry study and 1:2 for potato seed study.Adjustment produced neglible effects on food consumption.

Table 8. Feed Consumption and Body Weight Gains of Mice Fed Casein Diets Containing Potato Berry, Potato (No 3194) or Solasodine

	Body weight gain		*Feed consumption*				
			g/mouse/day		*g/Kg weight*		
Diets	*0-7 days*	*0-14 days*	*0-7 days*	*0-14 days*	*0-7days*	*0-14 days*	*Mortality*
Casein (10% protein)	7.4^{abA}	11.8^{bcAB}	3.8^{bcABC}	4.6^{abA}	205^{cBC}	220^{bcB}	0/5
Casein + potato berries (%)							
1	7.4abA	13.2^{abAB}	4.1^{abA}	4.9^{aA}	219^{cBC}	228^{bAB}	0/5
5	$6.8b^{AB}$	12.8^{abcAB}	3.7^{bcdABC}	4.8^{abA}	203^{cC}	227^{bAB}	0/5
10	$5.6c^{BC}$	11.4^{cB}	3.2^{dC}	3.9^{cdB}	189^{cC}	197^{cB}	0/5
20	$3.4d^{B}$	8.2^{dC}	3.6^{bcdABC}	4.3^{bcdAB}	250^{bB}	254^{aA}	0/5
40	$-.5e^{E}$	—	4.4^{aA}	—	397^{aA}	—	5/5
Casein + potatoes (%)							
1	$7.0bA^{B}$	13.0^{abcAB}	3.9^{abAB}	4.8^{abA}	213^{cBC}	224^{bcAB}	0/5
5	$7.8a^{bA}$	14.0^{aA}	4.1^{abA}	4.9^{aA}	213^{cBC}	218^{bcB}	0/5
10	$7.4a^{bA}$	13.2^{abAB}	3.9^{abAB}	4.7^{abA}	211^{cBC}	219^{bcB}	0/5
20	$7.0b^{AB}$	12.4^{abcAB}	3.7^{bcABC}	4.4^{abA}	203^{cC}	209^{bcB}	0/5
40	$8.4a^{A}$	13.0^{abcAB}	4.0^{abA}	4.7^{abA}	206^{cBC}	213^{bcB}	0/5
Solasodine (80 mg/100 g)	$4.2d^{cD}$	9.2^{dC}	3.3^{cdBC}	3.9^{dB}	212^{cBC}	214^{bcB}	0/5

NOTE: Means with a letter in common do not differ significantly [$P \leq 0.01$, upper case; Duncan's Multiple Range Test (SAS Institute, Inc., 1987)]. Values are for groups of 5 mice except the 1% potato berry group where statistically outlying data for one mouse was excluded from GOT, alkaline phosphatase, and total bilirubin means (values = 30, 83, and 1.1 respectively), and the GTP mean for the 5% berry group is for 4 mice (insufficient sample). See *also* Tables 9, 10, 12 and 13.

Table 9. Body Weight Gain and Feed Consumption of Mice Fed Casein and Casein Plus Solasodine Diets

	Casein[a]	*Diet 2*[b]	*Diet 3*[c]	*Diet 4*[d]	*Diet 5*[e]	*Diet 6*[f]	*Diet 7*[g]
Body weight gain (g)							
Days 0-7	7.0[abA]	6.8[abA]	7.0[abA]	7.6[abA]	8.0[aA]	6.2[bA]	4.2[cB]
Days 0-14	13.4[aA]	12.4[aA]	12.6[aA]	13.6[aA]	14.2[aA]	12.4[aA]	8.2[bB]
Feed consumption (g/mouse/day):							
Days 0-7	3.2[abAB]	3.1[bcAB]	3.2[abAB]	3.5[aA]	3.5[aA]	3.3[abAB]	2.8[cb]
Days 0-14	4.7[a]	4.2[a]	4.4[a]	4.7[a]	4.4[a]	4.2[a]	4.2[a]
Feed consumption (g/kg body weight):							
Days 0-7	214[a]	205[a]	216[a]	228[a]	223[a]	225[a]	216[a]
Days 0-14	251[ab]	230[b]	242[ab]	248[ab]	240[ab]	241[ab]	278[a]

[a] 10% protein based on N content;
[b] 5 mg solasodine/100g diet;
[c] 10mg solasodine/100g diet;
[d] 20mg solasodine/100g diet;
[e] 40mg solasodine/100g diet;
[f] 80mg solasodine/100g diet;
[g] 160mg solasodine/100g diet.

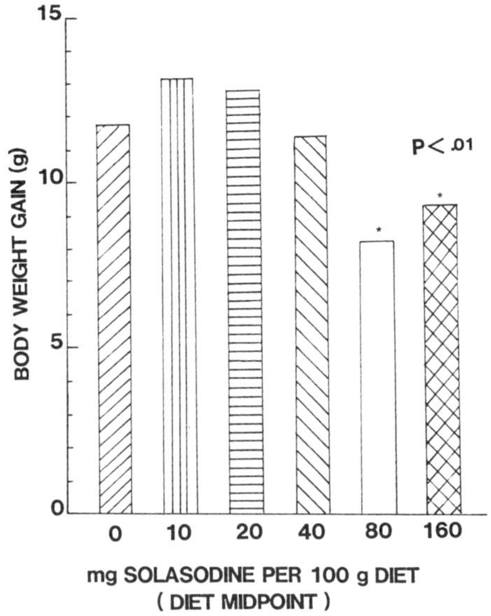

Figure 3. Body weight gains in mice fed casein and casein-plus-solasodine diets.

Table 10. Organ Weights of Mice Fed Casein Diets Containing Potato Berries, Potato (No 3194), or Solasodine

	Absolute Organ Weights (g)				*Relative Organ to Body Weights (%)*			
Diet	*Liver*	*Kidneys*	*Testes*	*Pancreas*	*Liver*	*Kidneys*	*Testes*	*Pancreas*
Casein (10% protein)	1.3^{bAB}	$.34^{abcAB}$	$.17^{abA}$	$.20^{dC}$	5.6^{bcdB}	1.5^{a}	$.73^{a}$	0.89^{cdBC}
Casein + potato berries (%)								
1	1.3^{bAB}	$.38^{abA}$	$.15^{abA}$	$.29^{abABC}$	5.1^{cdB}	1.6^{a}	$.61^{a}$	1.21^{abcABC}
5	1.5^{abAB}	$.36^{abAB}$	$.18^{aA}$	$.20^{dC}$	6.0^{bcB}	1.5^{a}	$.73^{a}$	0.83^{dC}
10	1.3^{bAB}	$.33^{bcAB}$	$.15^{abA}$	$.26^{bcdABC}$	5.6^{bcdB}	1.4^{a}	$.66^{a}$	$1.14^{abcdABC}$
20	1.2^{bAB}	$.28^{cB}$	$.14^{bA}$	$.22^{cdBC}$	6.3^{bB}	1.4^{a}	$.73^{a}$	$1.11^{abcdABC}$
40	—	—	—	—	—	—	—	—
Casein + potatoes (%)								
1	1.3^{bAB}	$.39^{abA}$	$.17^{abA}$	$.29^{abABC}$	5.3^{cdB}	1.6^{a}	$.68^{a}$	1.18^{abcABC}
5	1.4^{abAB}	$.38^{abA}$	$.16^{abA}$	$.27^{bcABC}$	5.5^{bcdB}	1.5^{a}	$.63^{a}$	1.09^{bcdABC}
10	1.4^{abAB}	$.40^{aA}$	$.15^{abA}$	$.30^{abABC}$	5.6^{bcdB}	1.7^{a}	$.62^{a}$	1.21^{abcABC}
20	1.3^{bAB}	$.34^{abcAB}$	$.15^{abA}$	$.31^{abAB}$	5.4^{bcdB}	1.4^{a}	$.62^{a}$	1.31^{abAB}
40	1.2^{Bb}	$.34^{ABCab}$	$.18^{Aa}$	$.35^{Aa}$	5.0^{dB}	1.4^{a}	$.74^{a}$	1.44^{aA}
Solasodine (80 mg/100 g)	1.6^{aA}	$.34^{abcAB}$	$.16^{abA}$	$.21^{cdBC}$	7.6^{aA}	1.6^{a}	$.76^{a}$	1.04^{bcdABC}

See footnote to Table 8.

Table 11. Organ Weights of Mice Fed Casein and Casein (10% Protein) Plus Solasodine (160 mg per 100 g) Diets

	Casein	*Casein + Solasodine*
Absolute organ weight (g)		
Brain	0.469[a]	0.446[a]
Liver	1.217[bA]	1.540[aA]
Kidneys	0.325[a]	0.275[a]
Relative organ to body weight (%)		
Brain	1.982[a]	2.484[a]
Liver	5.121[bB]	8.421[aA]
Kidneys	1.366[a]	1.479[a]

See footnote to Table 8.

Table 12. Clinical Chemistries in Mice Fed Casein Diets Containing Potato Berries, Solasodine, or Potato Seed

Diets	*GOT (IU/L)*	*Alkaline GPT (IU/L)*	*Liver phosphatase (IU/L)*	*N-demethylase (IU/L)*	*Cholinesterase (moles/min)*	*Bilirubin (mg/100mL)*	*Urea nitrogen (mg/100mL)*
Experiment 1							
Casein (10% protein)	101^{b}	26^{cB}	197^{a}	362^{bB}	3.1^{a}	0.12^{b}	-
Potato berries (%)							
1	106ab	26^{cB}	229^{a}	-	3.1^{a}	0.15^{ab}	-
5	$136a^{b}$	39^{bcB}	228^{a}	-	3.1^{a}	0.12^{b}-	
10	$114a^{b}$	48^{bAB}	235^{a}	-	2.9^{a}	0.14^{ab}	-
20	152a	69^{aA}	252^{a}	496^{aA}	2.7^{a}	0.20^{ab}	-
Solasodine (80 mg/100 g)	374	329	617	$441a^{bAB}$	3.1^{a}	0.26^{a}	-
Experiment 2							
Casein	123^{a}	44^{a}	241^{a}	-	-	0.02^{a}	16.4^{a}
Potato seed (%)							
1	119a	32^{b}	286^{a}	-	-	0.04^{a}	18.2^{a}
2	112^{a}	30^{b}	268^{a}	-	-	0.02^{a}	17.8^{a}
4	134^{a}	32^{b}	253^{a}	-	-	0.04^{a}	15.6^{a}

NOTE: GPT = glutamic-pyruvic transaminase; GOT = glutamic-oxaloacetic transaminase.
See footnote to Table 8.

Table 13. Clinical Chemistries in Mice Fed Casein Diets Containing Potato Berries, Potatoes (No 3194), or Solasodine

Diets	*GOT (IU/L)*	*GPT (IU/L)*	*Alkaline phosphatase (IU/L)*	*Bilirubin (mg/100 mL)*
Casein (10% protein)	101[bB]	26[bB]	197[bB]	.12[bcB]
Potato berries (%)				
1	106[bB]	26[bB]	229[bB]	.15[bcAB]
5	136[bB]	39[bB]	228[bB]	.12[bcB]
10	114[bB]	48[bB]	235[bB]	.14[bcAB]
20	152[bB]	69[bB]	252[bB]	.20[abAB]
40	—	—	—	—
Potatoes (%)				
1	89[bB]	25[bB]	215[bB]	.12[bcB]
5	102[bB]	25[bB]	233[bB]	.10[cB]
10	114[bB]	34[bB]	246[bB]	.10[cB]
20	112[bB]	31[bB]	265[bB]	.10[cB]
40	117[bB]	38[bB]	245[bB]	.12[bcB]
Solasodine (80 mg/100 g)	374[aA]	329[aA]	617[aB]	.26[aA]

See footnote to Table 8.

Histology. The liver, stomach, and testes of mice controls, mice fed solasodine, and mice fed the potato berry diet were histologically examined. The incidence of liver cholangiohepatitis (4/5) and of gastric gland dilation/degeneration (3/5) in mice fed solasodine (Table 14) was significantly ($P \leq 0.01$) above the zero incidence of the controls and the animals fed potato berries. There were no other significant findings. For the potato seed study, the liver, stomach, and kidneys were examined. There were no lesions resulting from ingestion of the potato seeds. Tomato seeds fed up to 4% in the diet were also nontoxic.

Discussion

Several reported pharmacological and toxicological effects of *Solanum* alkaloids will be briefly summarized to help place our recent findings into proper perspective.

Toxicology of Alkaloids There is a large species-variation in the susceptibility to glycoalkaloids. The glycoalkaloids appear to be much more toxic to man than to other animals studied. The human burden of glycoalkaloids is somewhat cumulative since their residence time in certain organs, such as the liver, is quite long. In fact, glycoalkaloids from the diet which are stored in the body might be mobilized at times of increased metabolic stress (pregnancy, starvation, debilitating illness) with deleterious effects (*45*). Adverse effects of glycoalkaloids include anticholinesterase activity on the central nervous system; disruption of cell membranes, adversely affecting the digestive system and general body metabolism; interference with calcium transport, embryotoxicity, and teratogenicity.

a. Inhibition of Cholinesterase. Organophosphorus insecticides and nerve gases exert their primary toxicologic effects by preventing the hydrolysis of acetylcholine, which is involved in the transmission of nerve impulses in the peripheral nervous system, spinal cord, and brain. The resulting accumulation of acetylcholine induces neuromuscular block. The solanaceous alkaloids also are moderate inhibitors of cholinesterase (*14-15*). This inhibition may provide a biochemical basis for some of the observed toxicological manifestations of the alkaloids, especially of embryotoxicity and teratogenicity. However, additional studies are needed to demonstrate this possibility.

b. Disruption of Cell Membranes. An unanswered question is whether the toxicity/repellant action of glycoalkaloids is related to their ability to disrupt cell membranes. Roddick and colleagues (*16*) studied the disruption of liposome membranes by potato and tomato glycoalkaloids. They report that a mixture of chaconine and solanine acting synergistically, is more effective than the individual alkaloids in lysing rabbit erythrocytes, red beet cells, and *Penicillum notatum* protoplasts. They also report that the nature of the carbohydrate side chain in different glycoalkaloids appears to influence the cell disruption process.

This biochemical parameter needs to be extended further and defined in order to establish whether it can serve as an index of mammalian toxicity. One approach is to develop a relative scale of cell disruption in rabbit and human erythrocytes and to relate the observed findings to relative toxicities of the glycoalkaloids.

Table 14. Histology Data of Mice Fed Casein and Casein Plus Solasodine Diets

	Number of animals			
Stomach:				
Gastric gland degeneration (diffuse)	0/5[a]	0/5[b]	2/5[c]	3/5[d]
Liver:				
Cholangiohepatitis (diffuse chronic)	0/5	0/5	1/5	3/5
Cholangiohepatitis (diffuse subacute)	0/5	0/5	2/5	1/5
Cholangiohepatitis (pooled)	0/5	0/5	3/5	4/5
Hepatic necrosis (multifocal acute)	0/5	0/5	0/5	1/5

NOTE: No significant changes: duodenum, ileum, jejunum, cecum, colon, kidney, and pancreas

[a] 10% protein from casein control;

[b] 10% protein + 40 mg solasodine/100 g diet;

[c] 10% protein + 80 mg solasodine/100 g diet;

[d] 10% protein + 160 mg solasodine/100 g diet.

c. Dietary Effects. Dietary constituents in the food and the process of digestion and metabolism can be expected to modify the adverse manifestations of the glycoalkaloids. For example, Renwick and colleagues (*18*) reported that vitamins lower the incidence of glycoalkaloid-induced spina bifida. A need, therefore, exists to define the role of nutrients and non-nutritive fiber in the toxicities of glycoalkaloids.

Also needing clarification are the relative hydrolytic stabilities during the digestion of oligosaccharide side chains attached to the glycoalkaloids. In a relevant study, Nilsson et al. (*46*) showed that oligosaccharides undergo partial to complete hydrolysis, depending on their structure, when exposed *in vitro* to human gastric juice or to the intestinal mucosa of the rat. Experiments are needed to define the digestion products of glycoalkaloids in order to relate the process to observed toxic manifestations. If carbohydrates of solanine and chaconine are cleaved at different rates during digestion in the rat intestine, then the real target of our toxicity studies should be the resulting hydrolysis products.

d. Spina Bifida, Embryotoxicity, Teratogenicity. According to the Merck Manual (*47*), spina bifida (the defective closure of the vertebral column) is one of the most serious neural tube defects compatible with prolonged life. The malformation is fairly common in the North American white population, with an incidence of about 1.5 per 1000 live births. The incidence seems to be partly environmentally related, and is much higher in some parts of the country than in others (*48*).

The literature (*9-10, 17, 19*) shows that glycoalkaloids have the ability to induce spina bifida, anencephaly (absence of part of the brain and skull), embryotoxicity, and teratogenicity. According to Morris and Lee (*13*), the most likely candidate(s) for the teratogenic agent is the potato glycoalkaloid or a metabolic product, possibly transformation products induced by cytochrome P-450 (*49,50*).

Although there appears to be a large variation in susceptibility to glycoalkaloid-induced teratogenicity among various animal species, possibly due to differences of the mother-to-fetus transport among the species (*51*), the evidence indicates that glycoalkaloids suppress fertility, enhance infant mortality, and induce neurological defects (*52*).

The biochemical mechanisms of the neurological impairment is largely unknown. Structural, stereochemical, and electronic configurations of the glycoalkaloid molecule seem to be paramount in influencing the teratogenic response (*10*). These authors suggest that those compounds with a basic nitrogen atom in the F-ring, shared or unshared with ring E, and with bonding capabilities a to the steroid plane may be suspect as teratogens. This plausible hypothesis merits further study to demonstrate its predictability, since it is generally recognized that stereochemical and structural features govern chemical and physiological properties of steroidal compounds (*53*). I propose the following approaches to attempt to clarify the mechanism of teratogenicity and suggest ways to prevent it:

(1) Establish whether the true teratogens are native glycoalkaloids or their digestive or metabolic products. Develop a chemical structure-developmental toxicity correlation.

(2) Establish relative susceptibilities of the glycoalkaloids to activation or modification by the cytochrome P-450 oxidase system in the liver. Since

some of the glycoalkaloids have been shown to be hepatotoxic, this information will be useful for designing strategies to lessen both hepatotoxicity and teratogenicity.

(3) Define the effect of dietary modification on the teratogenic process. The key question is whether nutrient-glycoalkaloid interactions can reduce embryotoxicity and teratogenicity.

(4) Since the nucleophilic strengths of the unshared electrons on the steroid ring may influence binding affinities of the glycoalkaloids to cell receptor sites, ascertain if the basicities (pK values) and metal ion affinities of the ring nitrogens are useful indicators of toxicity, including teratogenicity (*54*).

(5) Carry out computer modeling of glycoalkaloid-receptor site interactions and computer graphic studies (*55*) to define electronic and stereochemical features which govern toxicity and teratogenicity of the glycoalkaloid molecules. This information will be useful for predicting the safety of known (and unknown) alkaloids and for designing measures to counteract toxic manifestations by competitive inhibition of cell receptor binding sites. (Such modeling studies are widely used and have led to the discoveries of the drugs AZT for AIDS and mercaptopurine for cancer.)

Recent Findings

Our results show that solasodine, the aglycone of solasonine, a steroidal glycoalkaloid present in various *Solanum* species (*56*), adversely affected body-weight gain as well as the livers of male mice fed the compound at relatively high levels in the diet for two weeks, as evidenced by gross liver weight, histopathological evaluation, and increased levels of GPT, GOT, alkaline phosphatase, and total bilirubin in the serum. The absence of serum cholinesterase inhibition is consistent with the results of Roddick (*16*), who reported that solasodine did not inhibit human or bovine cholinesterase activity *in vitro*. Potato berries contain high levels of glycoalkaloids (*57-58*; Table 5). At 20% in the diet, potato berries resulted in reduced weight gain comparable to that resulting from 1600 mg solasodine/kg diet. However, other symptoms of glycoalkaloid exposure — liver pathology and increase in relative liver weight — did not occur. Elevations in serum chemistries were slight. When potato berries were fed at 40% in the diet, the animals did not survive. Histopathological examination of the livers of the three mice removed when moribund revealed no significant changes and the cause of death in these animals was not determined.

Potato seeds, produced by the potato berry on ripening, produced no ill effects when fed at 4% in the diet. (The slightly significant reduction in serum GPT levels in mice fed the seed diets is probably spurious.)

Since we do not know the exact weight ratio of seed to flesh in the berries, we cannot unequivocally state whether seeds rather than flesh contain toxic compounds responsible for the adverse effects of the berries.

A related study on the safety of seeds from another *Solanaceae* plant, (*Datura stramonium*) or Jimsonweed, shows that these seeds are toxic at the 4% dietary level (*31*).

Since high levels of the potato alkaloid solasodine, when fed to mice as part of a nutritionally adequate casein diet, produced liver damage and reduced body weight gain but no mortality, our results imply that the adverse effects of feeding high levels of potato berries added to the same casein diet may be due to the alkaloid content of the berries or to other toxic compounds.

We have used growth assays in mice for a number of years to assess the antinutritional and toxic potential of new amino acids formed during food processing (*27-18, 59-60*). We have detected and monitored antinutritional changes involving losses in nutritive value through destruction of proteins, amino acids, carbohydrates, and other nutrients and the formation of antinutrients or toxic materials. The purpose is to obtain an objective measurement of nutritional value or to detect the presence of toxic material in food. Reduced weight gain in mice fed a nutritionally adequate casein diet supplemented with potentially toxic materials, such as the potato alkaloid solasodine or potato berries, is a major criterion of toxicity. Other criteria include feed consumption, clinical chemistry, and histopathology.

Our results show that the mouse provides a useful animal model to study toxic manifestation of *Solanaceae* alkaloids and potato products. Mouse bioassays have a major advantage. They require about one-fifth of the test material needed in rats.

The FETAX (Frog Embryo Teratogenesis Assay: *Xenopus*) was used to evaluate the embryotoxicity and potential teratogenicity of potato alkaloids and related compounds. According to Dawson et al. (34), this i*n vitro* assay is both time and cost effective and correlates with more extensive teratogenicity evaluations in live animals.

Friedman et al. (*35*) showed that *Solanaceae* alkaloids and glycoalkaloids are embryotoxic in the FETAX assay (Figure 4). The data suggest that α-chaconine is teratogenic and more embryotoxic than α-solanine, in terms of the median lethal concentration after 96 h exposure (96-hr LC_{50}) and the concentration inducing gross terata in 50% of the surviving animals (96-hr EC_{50}-malformation), and the minimum concentration needed to inhibit the growth of the embryos. Since these two compounds differ only in the nature of the carbohydrate side chain attached to the 3-0H group of solanidine, the side chain appears to be an important factor governing teratogenicity. The aglycones demissidine, solanidine, and solasodine were less toxic than the glycosides α-chaconine and α-solanine. The *in vitro* teratogenesis assay should be useful for predicting the teratogenic potential of *Solanaceae* alkaloids, glycoalkaloids, and related natural and processing induced food toxicants.

Future Studies

Most of the pharmacological and toxicological studies on glycoalkaloids have been done with either α-chaconine or α-solanine. Very little biological data are available on the hydrolysis products (metabolites) of these glycoalkaloids (Figures 2 and 5) or on some twenty other glycoalkaloids present in different varieties of potatoes and tomatoes (Table 1). Biological studies should focus on two primary toxicity endpoints: (a) embryotoxicity-teratogenicity; and (b) hepatic dysfunction including the recently discovered induction of hepatic ornithine decarboxylase (ODC) by potato alkaloids

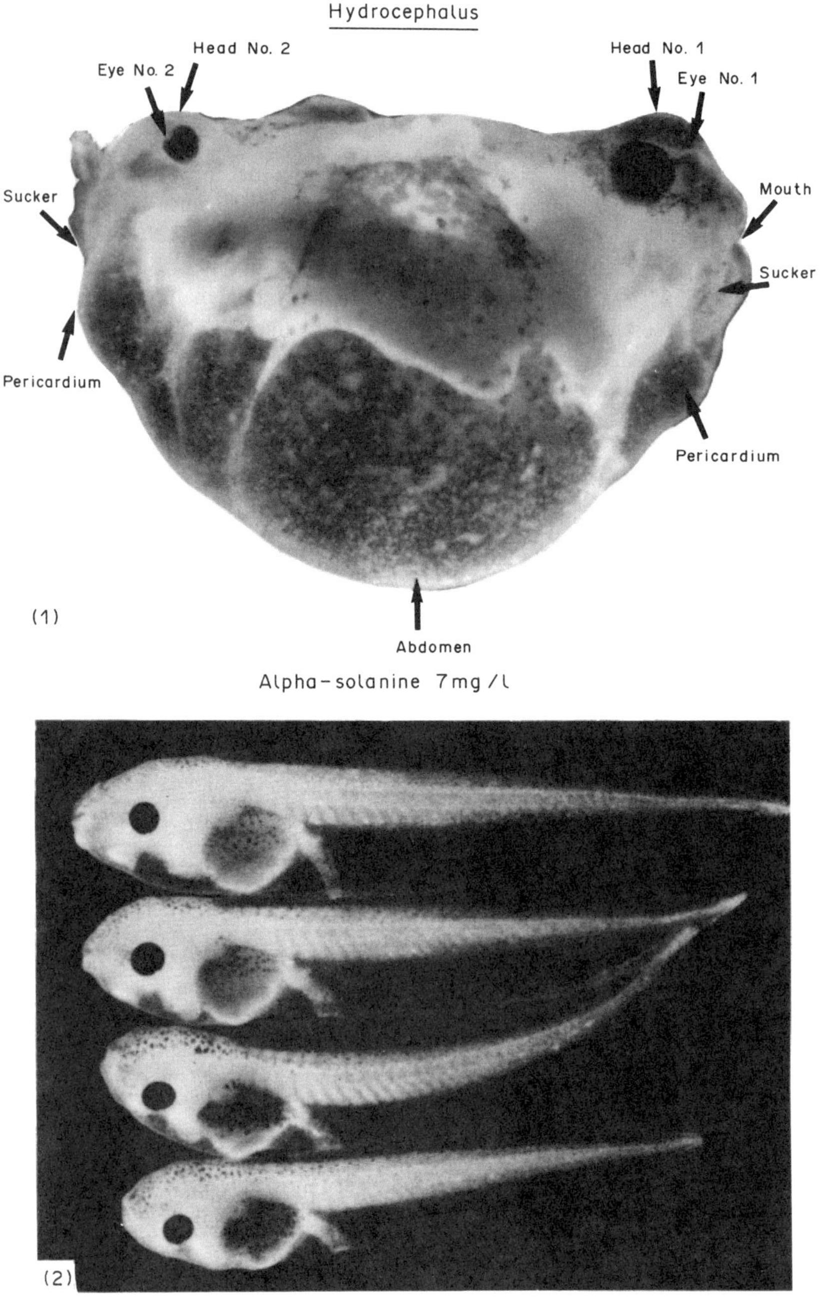

Figure 4. Upper: Two-headed frog embryo caused by exposure to α-solanine for 96 h. Lower: Typical malformations induced by increasing concentrations of α-solanine (*35*). *See* ref. 35 for a discussion of embryonic potencies of potato alkaloids.

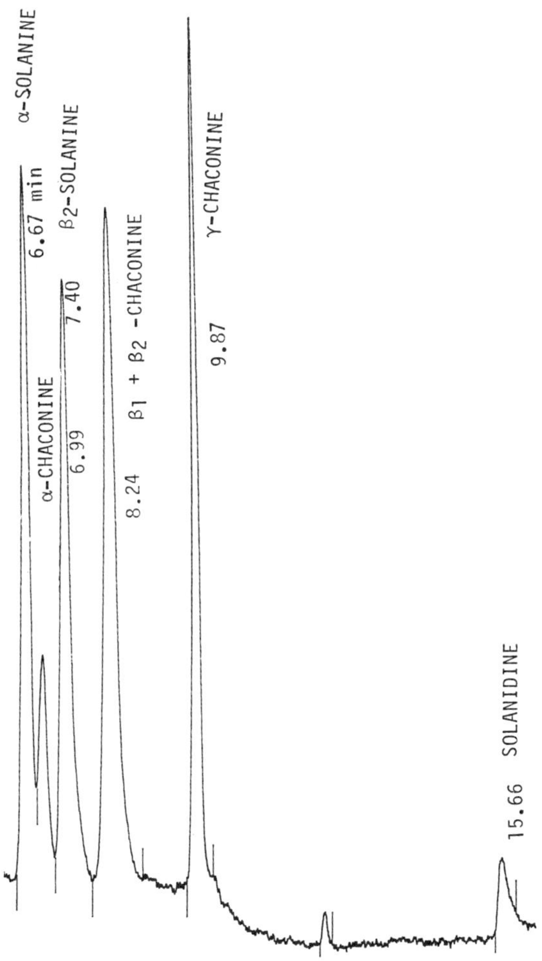

Figure 5. HPLC of acid hydrolysis products from a mixture of α-chaconine and α-solanine (41).

(*61*). If the *in vitro* FETAX assay of structurally different *Solanum* alkaloids could be related to *in vivo* teratogenicities in higher animal species, then the relatively simple and inexpensive frog assay could be used to screen plant foods for potential teratogenicity. Only compounds which are teratogenic in the *in vitro* test would require confirmation with pregnant animals.

Because of their acute toxicity, the identification and characterization of glycoalkaloid toxicants and their biosynthesis in potatoes (*62-67*; Figure 6) has long been the subject of research, with considerable progress made. However, a critical analysis of the literature reveals that many problems remain to be solved. For example, the need exists to develop a better understanding of why structurally related glycoalkaloids vary greatly in their pharmacological and toxicological activities.

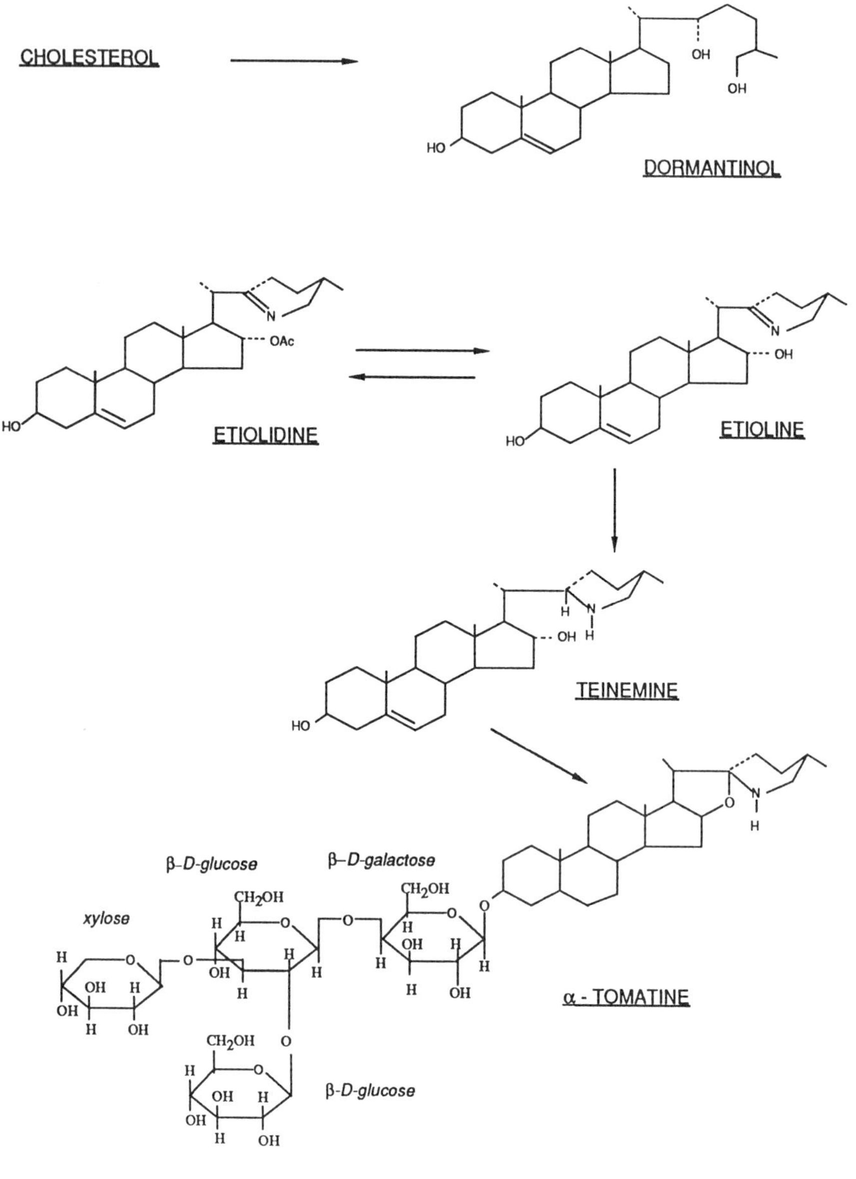

Figure 6. Biosynthesis of potato and tomato alkaloids. Although tentative assignments have been made to the structures of the intermediates, little informaiton is available on the enzymes which catalyze the indicated transformations. (*See* References 6 and 62-67).

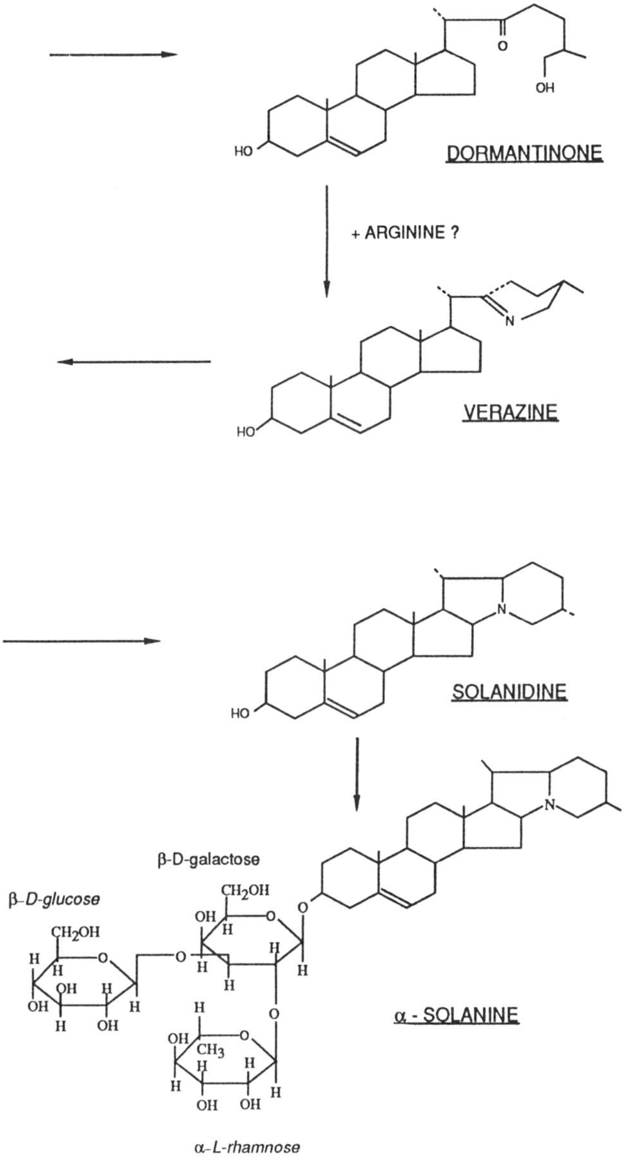
OH
O
HO
DORMANTINONE
+ ARGININE ?
N
HO
VERAZINE
N
HO
SOLANIDINE
N
β-D-galactose
β-D-glucose
CH2OH
CH2OH
OH
O
H
O
H
H
O
O
H
H
H
OH
H
OH
H
H
H
H
OH
O
α - SOLANINE
OH
O
CH3
H
H
H
H
OH
OH
α-L-rhamnose

Lessening of the toxicity of potato cultivars by direct genetic modification will require a significant effort in the chemistry and toxicology of the variety of glycoalkaloids present in the potato tuber. We have identified an enzyme, glucosyl transferase, the inactivation of which should result in greatly decreased levels of glycoalkaloid toxicants (*62*).

The implementation of the gene inactivation aspect of our multi-disciplinary studies would be greatly enhanced by a thorough characterization of the biosynthesis, chemistry, biochemistry, and safety of these toxic principles. The more information we have on the specific glycoalkaloids involved in animal and human toxicity, the more effective we will be at preventing their accumulation in potatoes and other plants.

The described findings with potato alkaloids complement related studies on the chemisty and toxicology of anthraquinones, atropine, ergot alkaloids, and scopolamine in toxic weed seeds (*2,31,50,68-71*).

Acknowledgments

I thank MacDonald C. Calhoun and Glenda M. Dugan for the feeding studies and for summarizing the results; Andrew Stapleton for computer graphing of structural formulae; Bruce Mackey for the statistical analyses; Prof. John A. Bantle, and my colleagues whose names appear in the cited references, for excellent scientific collaboration.

Literature Cited

1. Brown, W.E.; Graham, J.S.; Lee, J.S.; Ryan, C.A. In *Nutritional and Toxicological Significance of Enzyme Inhibitors in Foods*; Friedman, M., Ed.; Plenum Press, New York, 1986, pp 281-290.
2. Friedman, M.; Levin, C.E. *J. Agric. Food Chem.* **1989**, *37*, 998-1005.
3. Jadhav, S.J.; Sharma, R.P.; Salunkhe, D.K. *CRC Crit. Rev. Toxicol.* **1981**, *9*, 21-104.
4. Maga, J.A. *CRC Crit. Rev. Food Sci. Nutr.* **1980**, *12*, 371-405.
5. McCay, C.M.; McCay, J.B.; Smith, O. In *Potato Processing*; Talburt, W.F.; Smith, O., Eds., AVI: Westport, CT, 1987; pp 287-331.
6. Van Gelder, W.M.J. Doctoral Thesis and references cited therein, Agricultural University, Wageningen, Holland, 1989.
7. Woolfe, J.A. *The Potato in the Human Diet*; Cambridge University Press: Cambridge, England, 1987.

7a. Swallow, A. J. In *Nutritional and Toxicological Consequences of Food Processing*; Friedman, M., Ed.; Plenum Press: New York, 1991; pp 11-32.

8. Osman, S.F. *Food Chem.* **1983**, *11*, 235-247.
9. Baker, D.C.; Keeler, R.F.; Gaffield, W. *Tox. Pathol.* **1988**, *16*, 333-339.
10. Gaffield, W.; Keeler, R.F. In *Nutritional and Toxicological Aspects of Food Safety;* Friedman, M., Ed.; Plenum Press: New York, 1984; pp 241-251.
11. Harvey, M.H.; McMillan, M.; Morgan, M.R.A.; Chan, W.S. *Human Toxicol.* **1985**, *4*, 187-194.
12. McMillan, M.; Thompson, J.C. *Quart. J. Med.* **1979**, *48*, 227-243.

13. Morris, S.C.; Lee, T.H. *Food Technol. Australia* **1984**, *36*, 118-124.
14. Roddick, J.G. *Phytochemistry* **1989**, *28*, 2631-2634.
15. Bushway, R.J.M.; Savage, S.A.; Ferguson, B.S. *Am. Potato J.* **1987**, *64*, 409-413.
16. Roddick, J.G.; Rijenberg, A.; Osman, F. *J. Chem. Ecol.* **1988**, *14*, 889-902.
17. Nevin, N.C.; Merrett, J.D. *Br. J. Prev. Soc. Med.* **1976**, *29*, 11-15.
18. Renwick, J.H.; Claringbold, D.B.; Earthy, M.E.; Few, J.D.; McLean, A.C.S. *Teratology* **1984**, *30*, 371-381.
19. Renwick, J.H. In *Solanaceae: Biology and Systematics;* D'Arcy, G., Ed.; Columbia University Press, New York, 1986; pp 569-576.
20. Dick, R.; Baumann, U.; Zimmerli, B. *Mitt. Gebiete Lebensm. Hyg.* **1987**, *78*, 200-207.
21. Molnar-Perl, I.; Friedman, J. *J. Agric. Food Chem.* **1990**, *38*, 1652-1656.
22. Friedman, M.; Cuq, J.L. *J. Agric. Food Chem.* **1988**, *36*, 1079-1093.
23. Friedman, M.; Noma, A.T.; Wagner, J.R. *Anal. Biochem.* **1979**, *98*, 388-397.
24. Menefee, E.; Friedman, J. *J. Protein Chem.* **1985**, *4*, 333-341.
25. Worthington Biochemical Products. Worthington Diagnostic System, Freehold, New Jersey. 1982.
26. Wallace, J.; Friedman, M. *Nutr. Reports Int.* **1985**, *32*, 743-748.
27. Friedman, M.; Gumbmann, M.R. *J. Nutrition* **1981**, *111*, 1362-1369.
28. Friedman, M.; Gumbmann, M.R. *J. Nutrition* **1988**, *118*, 388-397.
29. Friedman, M.; Gumbmann, M.R.; Ziderman, I.I. *J. Nutrition* **1987**, *117*, 508-518.
30. Gumbmann, M.R.; Friedman, M. *J. Nutrition* **1987**, *117*, 108-1023.
31. Dugan, G.M.; Gumbmann, M.R.; Friedman, M. *Food Chem Toxicol.* **1989**, *27*, 501-510.
32. Ellman, G.L.; Courtney, D.; Andres, Jr. V.; Featherstone, R.M. *Biochem. Pharmacol.* **1961**, *7*, 88-95.
33. Dalton, C.; DiSalvo, D. *Technicon Quarterly* **1972**, *4*, 20-24.
34. Dawson, D.A.; Fort, D.J.; Newell, D.L.; Bantle, J.A. *Drug and Chem. Toxicol.* **1989**, *12*, 67-75.
35. Friedman, M.; Rayburn, J.R.; Bantle, J.A. *Food Chem. Toxicol.* **1991**, *29*, 537-547.
36. Bliss, C.I. In *Statistics in Biology;* McGraw-Hill, New York. Vol. 1, 1967, 53-91.
37. Duncan, D.B. *Biometrics* 1955, *11*, 1-42.
38. SAS Institute. In *SAS/STAT Guide for Personal Computers*; SSA Institute, Cary, NC, Version 6 Ed. 1987, 549-640.
39. Cantarelli, P.R.; Plama, E.R.; Caruso, J.G.B. *Acta Alimentaria* **1989**, *18*, 13-18.
40. Mercer, L.P.; Dodds, S.J.; Smith, D.L. In *Absorption and Utilization of Amino Acids*; Friedman, M., Ed.; CRC: Boca Raton, FL, 1989; Volume 1, Chapter 1.
41. Friedman, M.; Dao, L. *J. Agric. Food Chem.* **1991**, submitted.
42. AIN Report of the American Institute of Nutrition Ad Hoc Committee on Standards for Nutritional Studies. *J. Nutr* **1977**, *107*, 1340-1348.
43. Friedman, M.; Gumbmann, M.R.; Brandon, D.L.; Bates, A.H. In *Food Proteins;* Kinsella, J.E.; Soucie, W.G., Eds.; American Oil Chemists Society: Champaign, Illinois, 1989; pp 296-328.

44. Brandon, D.L.; Bates, A.H.; Friedman, M. In *Nutritional and Toxicological Consequences of Food Processing*; Friedman, M., Ed.; Plenum: New York, 1991; pp 321-337.
45. Claringbold, W.D.B.; Few, J.D.; Renwick, J.H. *Xenobiotica* **1982**, *12*, 293-302.
46. Nilsson, U.; Oste, R.; Jagerstad, M.; Birkhed, D. *J. Nutrition* **1988**, *118*, 1325-1330.
47. *The Merck Manual of Diagnosis and Therapy*; Merck & Co.: Rahway, New Jersey, 15th Ed.; 1987, p 1951.
48. Milunsky, A.; Hershel, J.; Jick, S.S.; Bruel, C.L.; MacLaughlin, D.D.; Rothman, K.J.; Willet, W. *J. Amer. Med. Assoc* **1989**, *262*, 2847-2852.
49. Dalvi, R.R.; Jones, R.D.H. *Current Science* **1986**, *55*, 558-561.
50. Crawford, L.; Friedman, M. *Toxicol. Letters,* **1990**, *54* (2-3), 175-181.
51. Watanabe, T.; Endo, A. *J. Nutrition* **1989**, *119*, 255-261.
52 Allen, J.R.; Marlar, R.J.; Chesney, C.F.; Helgeson, J.P.; Kelman, A.; Weckel, K.G.; Traisman, E.; White, Jr., J.W. *Teratology* **1976**, *15*, 17-24.
53. Kupchan, S.M.; Eriksen, S.P.; Friedman, M. *J. Amer. Chem. Soc.* **1962**, *84*, 4159-4160; *ibid.* **1966**, *88*, 343-346.
54. Pearce, K.N.; Friedman, M. *J. Agric. Food Chem.* **1988**, *37*, 707-717.
55. Martin, Y.C.; Kutter, E.; Austel, V. *Modern Drug Research - Paths to Better and Safer Drugs,* Mercel Dekker, New York, 1989.
56. *The Merck Index*; Merck & Co.; Rahway, New Jersey, 11th Ed. **1989**; p 1372.
57. Coxon, D.T. *J. Sci. Food Agric.* **1981**, *32*, 412-414.
58. Jones, P.G.; Fenwick, R.G. *J. Sci. Food Agric.* **1981**, *32*, 419-421.
59. Friedman, M.; Finot, P.A. *J. Agric. Food Chem.* **1990**, *38*, 2011-2020.
60. Oste, R.E.; Friedman, M. *J. Agric. Food Chem.* **1990**, *38*, 1687-1690.
61. Caldwell, K.A.; Grosjean, O.K.; Henika, R.E.; Friedman, M. *Food Chem. Toxicol.* **1991**, *29,* 531-535.
62. Stapleton, A.; Allen, P.V.; Friedman, M.; Belknap, B.L. *J. Agric. Food Chem.* **1991**, *39*, 1187-1193.
62a. Stapleton, A.; Allen, P.V.; Tao, H.R.; Belknap, W.R.; Friedman, M. *Protein Expression and Purification* **1991**, submitted.
63. Aubert, S.; Daunay, M.C.; Pochard, E. *Agronomie* **1989**, *9*, 641-651.
64. Dopke, W.; Duday, S.; Matos, N. *Zeitschrift Chem.* **1987**, *27*, 65.
65. Kaneko, K.; Tanaka, M.W.; Mitsuhashi, H. *Phytochemistry* **1976**, *15*, 1391-1393.
66. Cordell, C.A. *Introduction to Alkaloids - A Biogenetic Approach*; Wiley-Interscience, New York, 1981; p 920.
67. Heftmann, E. *Phytochemistry* **1983**, *22*, 1843-1860.
68. Crawford, L.; McDonald, G.M.; Friedman, M. *J. Agric. Food Chem.* **1990**, *38*, 2169-2175.
69. Friedman, M.; Dao, L. *J. Agric. Food Chem.* **1990**, *38*, 805-808.
70. Friedman, M.; Dao, L.; Gumbmann, M.R. *J. Agric. Food Chem.* **1989**, *37*, 708-712.
71. Friedman, M.; Henika, P.R. *J. Agric. Food Chem.* **1991**, *39*, 494-501.

RECEIVED December 3, 1991

INDEXES

Author Index

Affiliation Index

Subject Index

N

O

P

Production: Paula M. Bérard
Indexing: Deborah H. Steiner
Acquisition: Barbara C. Tansill
Cover design: Amy Meyer Phifer

Printed and bound by Maple Press, York, PA